Injury Prevention: Meeting the Challenge

INJURY PREVENTION:
Meeting the Challenge
A report of the National Committee for Injury Prevention and Control
Prepared by Education Development Center, Inc. (EDC)

Chairperson
Bernard Guyer, MD, MPH

Members
Kathleen H. Acree, MD, MPH, JD
Phyllis F. Agran, MD
Joan Ascheim, MSN, CPNP
Richard Cales, MD
Henry C. Cleveland, MD
Susan S. Gallagher, MPH
Bruce M. Gans, MD
Kristine Gebbie, MSN
Susan Goodwin Gerberich, PhD
Joseph Greensher, MD
William Hollinshead, MD, MPH
Greg Istre, MD
Paul B. Jones
Albert I. King, PhD
Susan McLoughlin, MSN, CPNP
Mark Moore, PhD
Woodrow A. Myers, MD
Deborah Prothrow-Stith, MD
Gordon R. Reeve, PhD
Frederick Rivara, MD, MPH
Mark L. Rosenberg, MD, MPP
David A. Sleet, PhD
Rudolph L. Sutton, MPH
Stephen Teret, JD, MPH
Patricia Waller, PhD
Harold (Hank) B. Weiss, MS, MPH
M. Patricia West, MSW

Ex officio members
Arthur Funke, PhD
Jerry Hershovitz, BA
Joan White Quinlan, MA

Education Development Center, Inc. (EDC) Project Staff
Cheryl J. Vince, EdM, Vice President, EDC
Patricia J. Molloy, MSW, Project Director
Stu Cohen, BA, Senior Research Associate
Christine Blaber, EdM, Associate Project Director
Marc Posner, PhD, Associate Project Director
Daphne Northrop, BA, Associate Project Director
Betty Mutney, BS, Administrative Assistant
Ruth Rappaport, MS, Conference Coordinator

Injury Prevention: Meeting the Challenge

The National Committee for
Injury Prevention and Control

American Journal of Preventive Medicine

Library of Congress Cataloging-in-Publication Data

National Committee for Injury Prevention and Control (U.S.)
  Injury prevention: meeting the challenge/The National Committee for
Injury Prevention and Control.
  p. cm.
  "Published by Oxford University Press as a supplement to the
American Journal of Preventive Medicine, Volume 5, Number 3, 1989"—
Copr. p.
  Includes index.
  ISBN 0-19-506248-5 (alk. paper)
  1. Wounds and injuries—United States—Prevention. 2. Accidents—
United States—Prevention. 3. Emergency medical services—
Government policy—United States. 4. Wounds and injuries—United
States—Statistics. I. Title.
RD93.8.N35 1989
617.1—dc20

Published by Oxford University Press
200 Madison Ave., New York, NY 10016, as a supplement to the
American Journal of Preventive Medicine, Volume 5, Number 3, 1989.

OXFORD UNIVERSITY PRESS Oxford, New York, Toronto, Delhi,
Bombay, Calcutta, Madras, Karachi, Petaling Jaya, Singapore, Hong
Kong, Tokyo, Nairobi, Dar es Salaam, Cape Town, Melbourne,
Auckland, and associated companies in Beirut, Berlin, Ibadan,
Nicosia

Oxford is a registered trademark of Oxford University Press.

Printed in the United States of America on acid-free paper.

9 8 7 6 5 4 3 2 1

DEPARTMENT OF HEALTH & HUMAN SERVICES

Public Health Service

The Surgeon General of the
Public Health Service
Washington DC 20201

May, 1989

Greetings:

This is a book for people who are concerned about injuries and who want to act to do something about them. If you are not already concerned or motivated to act, I predict that you will be before you have finished <u>Injury Prevention: Meeting the Challenge</u>. The National Committee that produced this book and the federal agencies that supported it hope that your concern will lead to action --community leadership, participation of your agency or organization, membership in an advocacy group, or informed citizen activities.

I urge each State and locality to examine current injury prevention efforts and to make every effort to strengthen the resources and the will of state government to reduce this most costly problem (in both human and dollar terms). But increased determination by government at every level is not enough. There must be significant changes in basic public attitudes and behaviors concerning injuries, as there have been in regard to tobacco, alcohol, exercise, and sexual behaviors, for example. We must accept that the injuries associated with motor vehicles are not "accidents" and that much can be done to reduce them. We must realize that violence in the forms of abuse, assault, or suicide is not only within the purview of the police and the criminal justice system but also of the health system. An informed and aroused public can change the behavior of each of us, but more importantly, it must lead to community outrage and action in regard to unsafe playgrounds, automobiles, highways, work places, toys, homes, and use of handguns.

In 1985, a conference of injury prevention practitioners agreed on the need for a book that would serve as a resource to practitioners and would-be practitioners. It was thought that this book would be a companion piece to <u>Injury in America</u>, published earlier that year by the National Academy of Sciences. <u>Injury in America</u> focuses on the problem of injuries in terms of the gaps in our basic knowledge, and the need for research to fill those gaps. <u>Injury Prevention: Meeting the Challenge</u> focuses on the knowledge and other resources that exist now, and how they can be mobilized and applied in a well designed, scientifically sound manner to ensure the effectiveness of community injury prevention and control efforts. It also focuses on the generation of new knowledge through such efforts, because it recognizes that research science is not the only source of useful knowledge. Careful monitoring by surveillance systems, for instance, can contribute valuable knowledge for internal use by the program itself and for the field, in general.

The product represents the successful collaboration of three federal agencies: the Office of Maternal and Child Health and Resources Development of the Health Resources and Services Administration, the Division of Injury Epidemiology and Control of the Centers for Disease Control, and the National Highway Traffic Safety Administration of the U.S. Department of Transportation. These three must be the core of a much broader coalition of federal agencies if the prevention and control of injuries is to be successfully addressed at that level.

The content for the book was provided by the ad hoc 31 member National Committee for Injury Prevention and Control. Members came from public and private organizations and national, state, and local levels of injury practice, research and teaching, with recognized knowledge in all aspects of the problem. The task of converting their work into a useful and readable book was accomplished by the staff of Education Development Center, Inc.

The book is divided into three parts. The first part is a primer on the process of identifying, strengthening and using existing resources. These chapters stress the importance of citizen involvement; of coalitions and other collaboration among key agencies and organizations; of data and other knowledge of the injury problems in the particular state and locality; and of local problems and conditions. The challenging task is to adapt and combine those research findings in light of local data and other resources, in order to design, monitor, evaluate, and modify a community program, and keep it on course toward the goal of reduced incidence of injuries and reduced mortality, morbidity, and disability.

Part two of this book is a report on the state of the art and knowledge in regard to major determinants of injuries. This information provides an excellent starting point for a planning group in a particular community. It will help such a group avoid many pitfalls, but it will not give them a blueprint for their program. Each program must be tailored to fit the unique problem characteristics, conditions, and populations of the particular setting and then "tried on" (implemented, monitored, and evaluated) repeatedly. This will provide management with the information necessary to determine how the program must be modified, even as it is being implemented, in order to realize better or sooner the goal of prevention and control. Plans should include provision for such periodic "adjustments" throughout the life of the program.

Part three describes the community response to persons who are severely injured. While it is obviously most desirable that we minimize the number and severity of injuries, we must also minimize the negative consequences of those injuries. We can reduce the number of deaths that result from injuries. We can reduce the severity and duration of disability, the number of days of hospitalization, the period of absence from work or school or normal daily living. And we must remember the enormous negative consequences that injuries have for persons close to the victims. Emergency medical service systems and trauma care systems, in particular, are the community's response to the severely injured individual and his or her family. The strengthening of these systems must be a high priority.

I commend the National Committee that volunteered its time and expertise to the preparation of this book. I believe this publication will take its place among the short list of major milestones toward the goal of injury prevention and control.

C. Everett Koop, M.D., Sc.D.
Surgeon General, U.S Public Health Service

Contents

Illustrations and Tables

Members of the National Committee for Injury Prevention and Control

Chairperson
Bernard Guyer, MD, MPH
Director
New England Injury Prevention Research Center
Department of Maternal and Child Health
Harvard School of Public Health
Boston, Massachusetts

Members
Kathleen H. Acree, MD, MPH, JD
Assistant Chief
California Department of Health Services
Division of Preventive Medical Services
Sacramento, California

Phyllis F. Agran, MD
Director, Pediatric Gastroenterology
Department of Pediatrics
University of California, Irvine
Irvine, California

Joan Ascheim, MSN, CPNP
Child Health Program Chief
New Hampshire Division of Public Health Services
Health and Human Services Building
Concord, New Hampshire

Richard Cales, MD
Chief, Emergency Services
San Francisco General Hospital
San Francisco, California

Henry C. Cleveland, MD
Chief of Trauma Service
St. Anthony Hospital Systems
Denver, Colorado

Susan S. Gallagher, MPH
Director
Childhood Injury Prevention Resource Center
Maternal and Child Health
Harvard School of Public Health
Boston, Massachusetts

Bruce M. Gans, MD
President and CEO
Rehabilitation Institute
Detroit Medical Center
Detroit, Michigan

Kristine Gebbie, MSN
Assistant Director for Health
Administrator of Oregon Health Division
Oregon State Health Division
Portland, Oregon

Susan Goodwin Gerberich, PhD
Head of Injury Prevention and Control Program

School of Public Health
University of Minnesota
Minneapolis, Minnesota

Joseph Greensher, MD
Chairman, Department of Pediatrics
Winthrop University Hospital
Mineola, New York

William Hollinshead, MD, MPH
Medical Director
Divisions of Family Health
Rhode Island Department of Health
Providence, Rhode Island

Greg Istre, MD
State Epidemiologist
Medical Chief of Epidemiology Service
Oklahoma State Department of Health
Oklahoma City, Oklahoma

Paul B. Jones
Governors' Highway Safety Representative
Raleigh, North Carolina

Albert I. King, PhD
Director
Bioengineering Center
Detroit Neurotrauma Research Center
Wayne State University
Detroit, Michigan

Susan McLoughlin, MSN, CPNP
Senior Staff Specialist
American Nurses' Association
Practice Program and Council Services
Kansas City, Missouri

Mark Moore, PhD
Guggenheim Professor of Criminal Justice Policy and
  Management
John F. Kennedy School of Government
Harvard University
Boston, Massachusetts

Woodrow A. Myers, MD
Commissioner
Indiana State Board of Health
Indianapolis, Indiana

Deborah Prothrow-Stith, MD
Commissioner
Massachusetts Department of Public Health
Boston, Massachusetts

Gordon R. Reeve, PhD
Chief, Bureau of Disease Intervention
Indiana State Board of Health
Indianapolis, Indiana

Frederick Rivara, MD, MPH
Director
Harborview Injury Prevention Research Center
Harborview Medical Center
Seattle, Washington

Mark L. Rosenberg, MD, MPP
Special Assistant to the Deputy Director (AIDS)
Centers for Disease Control
Atlanta, Georgia

David A. Sleet, PhD
Professor of Health Science/Public Health
San Diego State University
School of Public Health
Division of Health Promotion
San Diego, California

Rudolph L. Sutton, MPH
Program Administrator
Philadelphia Injury Prevention Program
Vector Control Division
Philadelphia Department of Public Health
Philadelphia, Pennsylvania

Stephen Teret, JD, MPH
Director
Johns Hopkins Injury Prevention Research Center
Johns Hopkins University School of Hygiene and Public Health
Baltimore, Maryland

Patricia Waller, PhD
Director
University of Michigan Transportation Research Institute
Ann Arbor, Michigan

Harold (Hank) B. Weiss, MS, MPH
Director, Injury Control Unit
Wisconsin Department of Health & Social Science
Division of Health
Madison, Wisconsin

M. Patricia West, MSW
Director, Injury Prevention and Control Program
Colorado Department of Health
Denver, Colorado

Ex officio members

Arthur Funke, PhD
Chief Psychologist
Bureau of Maternal and Child Health and Resources
  Development
United States Public Health Service
Rockville, Maryland

Jerry Hershovitz, BA
Chief, Program Services Section
Division of Injury Epidemiology and Control
Center of Environmental Health
Centers for Disease Control
Atlanta, Georgia

Joan White Quinlan, MA
Prevention Program Coordinator
National Highway Traffic Safety Administration
Washington, D.C.

COMMITTEE WORKING GROUPS

Problem Identification and Community Planning (Chapter 1)
Susan Gallagher, Co-Chair
Hank Weiss, Co-Chair
Arthur Funke
Kristine Gebbie
Jerry Hershovitz
William Hollinshead
Paul Jones
Woodrow Meyers
Joan Quinlan
Rudolph Sutton
Patricia West
Stu Cohen, Daphne Northrop, Staff

Data Collection and Analysis (Chapters 2 and 3)
Greg Istre, Chair
Gordon Reeve, Vice-Chair
Phyllis Agran
Richard Cales
Bruce Gans
Susan Gerberich
Albert King
Mark Rosenberg
Patricia Waller
Patricia Molloy, Marc Posner, Staff

Program Implementation and Evaluation (Chapters 4 and 5)
Fred Rivara, Chair
David Sleet, Vice-Chair
Kathleen Acree
Joan Ascheim
Henry Cleveland
Joseph Greensher
Susan McLoughlin
Mark Moore
Deborah Prothrow-Stith
Stephen Teret
Christine Blaber, Cheryl Vince, Staff

Traffic Injuries (Chapter 6)
Pat Waller, Chair
Paul Jones, Vice-Chair
Phyllis Agran
Richard Cales
Henry Cleveland
Bruce Gans
Kristine Gebbie
Albert King
Joan Quinlan
David Sleet
Stephen Teret
Patricia West
Stu Cohen, Christine Blaber, Staff

Residential, Recreational, and Occupational Injuries (Chapters 7–9)
Joan Ascheim, Chair
Joseph Greensher, Vice-Chair

Kathleen Acree
Susan Gallagher
Susan Gerberich
Jerry Hershovitz
William Hollinshead
Susan McLoughlin
Gordon Reeve
Fred Rivara
Rudolph Sutton
Hank Weiss
Patricia Molloy, Marc Posner, Staff

Interpersonal Violence and Suicide (Chapters 10–17)
Mark Rosenberg, Chair
Deborah Prothrow-Stith, Vice-Chair

Arthur Funke
Greg Istre
Mark Moore
Woodrow Myers
Christine Blaber, Stu Cohen, Daphne Northrop, Marc Posner,
    Cheryl Vince, Staff

Trauma Care Systems (Chapter 18)
Richard Cales, Chair
Phyllis Agran
Henry Cleveland
Bruce Gans
Kristine Gebbie
Hank Weiss
Stu Cohen, Staff

Acknowledgments

The preparation of this report took 2 years and involved the active and intense collaboration of a great many individuals. The National Committee for Injury Prevention and Control is indebted to each of them.

This book came into existence because leadership at the federal level was responsive to the needs of local injury prevention practitioners for applied information. Susan Gallagher, MPH, and Hank Weiss, MS, MPH, first identified the need for such a document. As practitioners then working in health departments, they described how many colleagues understood the magnitude of the injury problem, but not how to respond.

Three major federal agencies took up the challenge. We acknowledge the steadfast guidance and support provided by the Bureau of Maternal and Child Health and Resources Development (BMCH) of the Health Resources and Services Administration, the Division of Injury Epidemiology and Control (DIEC) of the Centers for Disease Control, and the National Highway Traffic Safety Administration (NHTSA). In particular, David Heppel, MD, Director of the Division of Maternal, Child, and Infant Health, Stuart T. Brown, MD, Director of the DIEC, and Marilena Amoni, MPA, Policy Advisor (NHTSA), have been enthusiastic about the committee's work and ever responsive to the members' needs.

The federal agencies were represented on the committee by three ex officio members: Arthur Funke, PhD, of BMCH (who also served as project officer), Jerry Hershovitz, BA, of DIEC, and Joan White Quinlan, MA, of NHTSA. Not only did they make the resources of their own agencies available to us and organize the review of drafts by experts in many disciplinary areas, they also tirelessly reviewed material and faithfully participated in our meetings. In a project focused on the need for multidisciplinary and interagency collaboration, they provided a superb example of how effective such collaboration can be.

We thank the many federal agency reviewers who took the time to read and comment on individual chapters and to provide additional material for our consideration: Loran Archer, Catherine Bell, Lew Buchanan, Philip Graitcer, Jim Hedlund, Patricia Honchar, Virginia Litres, Polly A. Marchbanks, Philip McClain, Daniel McGee, James A. Mercy, Herb Miller, Jackie Moore, E. Chukwudi Onwuachi-Saunders, Daniel Pollock, Philip Rhodes, Nancy Rubenstein, Jeffrey Sacks, L. Rachid Salmi, Linda E. Saltzman, Richard Sattin, Thomas Scheib, Arthur Schletty, Richard Smith, Suzanne Smith, Daniel Sosin, Stephen Thacker, and Richard Waxweiler.

We thank Bernard Guyer, MD, for his willingness to undertake, and his skill in performing, the role of committee chair. That we were able to accomplish our many complex tasks, achieve consensus, and do so on time is a tribute to his leadership and skill in facilitation.

To Cheryl J. Vince, Vice President of Education Development Center, Inc., who served as the committee's chief of staff, Project Director Patricia J. Molloy, and their colleagues at EDC we owe a special debt. They organized the committee's work, reviewed and digested a mountain of background material, staffed the committee's working groups, and drafted and revised chapters of the book to reflect the developing consensus within the committee. And to Janet Whitla, EDC President, and the Board of Trustees, we appreciate the working environment for staff that made such a quality product possible and for their organization's long-standing commitment to injury prevention.

The "Introduction: A History of Injury Prevention" was written by Stu Cohen, as were chapters 6 (with assistance from Christine Blaber), 10, 12, 16, 18, and the "Guide to Chapters 6–17." As the project's principal writer, Mr. Cohen helped shape the book's style and edited the individual chapters. Chapters 1 and 14 were written by Daphne Northrop. The data chapters, 2 and 3, were written by Marc Posner, as were chapters 8, 11, and 17. Christine Blaber wrote chapters 4 and 5, and Patricia Molloy wrote chapter 7. We would also like to thank consultants Jennifer Helmick, Ruth McCambridge, and Lisa Lightman, who worked on chapters 9, 13, and 15, respectively. Kimberly Hamrick assisted in the production of the appendixes.

Cynthia Lang of EDC provided editorial assistance at several stages of the project. The complete manuscript was edited by Barbara Herzstein. To Dorothy Meckel and Roberta Winston fell the task of bringing order to the references.

We thank Jeffrey House, Vice President and Executive Editor, Susan Keiser, Journals Manager, and Stephen Chasteen, Production Editor, at Oxford University Press for their enthusiastic support and assistance at every stage of the process.

Betty Mutney, the project's administrative assistant, managed with great skill and patience to keep the project on track and to prevent the chaos that so many collaborators and so much paper might otherwise have created. Ruth Rappaport, EDC's Conference Coordinator, not only arranged meeting sites that were enjoyable and conducive to our work but also, with the assistance of her fellow musicians in the Poodles and the Hurtin' Four, entertained the committee and reminded us how deeply ingrained is the subject of injury in American folklore and music.

Finally, a book about the state of the art in so multidisciplinary a field as injury prevention could succeed only with cooperation from an enormous number of persons. Indeed, this project has been marked by the large number of practitioners and researchers willing to share their experiences, thoughts, documents, and time with us. In addition to the federal agency staff members already acknowledged, the alphabetical list below acknowledges the contributions of many of these persons. Inevitably, in a large project of two years' duration, the list will not be complete. To anyone whose name is omitted, we apologize.

To all of those named or inadvertently omitted, we say "thank you": Nicholas Ashford, Nancy Baer, David Baker, Sandy Belk, Abe Bergman, Jetta Bernier, Larraine Bernstein, Marie Bond, Peter Brigham, Angela Browne, Diane Butkus, Nancy Carrey-Beaver, Kent J. Chabotar, Larry Cohen, Beverly Coleman-Miller, Lew Colwell, Forrest Council, Maureen DeJong, Edward DeVos, Jeffrey Diver, M. Teresa Dowling, Howard Dubowitz, Jim Enriquez, Jeff Feck, William Field, Leslie Fisher, William Foege, Janet Fuchs, Terry Fulmer, Ann Garland, John Graham, Vivian Guilfoy, Gary Gurian, Cecile Heimann, Janet Holden, Beth Hume, Joseph Hunt, Melanie Hwalek, Trudy Karlson, Murray Katcher, Judy Ladd, Garry Lapidus, Frank Lardo, Fran Lindsay, Linda Lloyd, Andrew McGuire, Carmen Maiocco, Susan Makintubee, William Marine, Hildy Meyers, Marcia Meyers, Sylvia Micik, Amy Miller, Janice Mirabassi, Frederick Morrissey, Dennis Murphy, Kathleen Myrick, Eli Newberger, David Orrick, Jane Quinlan, Shirley Reed-Randolph, Sally Repka, Richard Retting, Donna Richardson, Gladys Riddell, Nadina Riggsbee, George Rodman, Cynthia Rodgers, George Rogers, Lisa Rogers, Al Ros, Robert Sanders, Susan Schecter, Philip Shattuck, Jeff Simon, Gordon Smith, Robert Spengler, Charlotte Spiegel, Ron Spiro, Howard Spivak, Susan Standfast, Susan Steinmetz, Marty Steyer, Denis Lee Tucker, Robert Verhalen, Candace Waldron, Richard Wasserman, Mary Weissman, Debra Whitcomb, Garen Wintemute, Amy Wishner, Rosalie Wolf, Alan Woolf, Suzanne Yoffe, Janice Yuwiler, and the reference department staffs at the National Center for Health Statistics, the SCIPP Library (Boston), the Boston Public Library, the Francis A. Countway Library of Medicine at Harvard University, the Insurance Institute for Highway Safety, and the National Institute for Occupational Safety and Health (NIOSH).

General Recommendations

Injury is probably the most underrecognized major public health problem facing the nation today, and the study of injury represents unparalleled opportunities for reducing morbidity and mortality and for realizing significant savings in both financial and human terms—all in return for a relatively modest investment.

—National Academy of Sciences, *Injury Control* (1988)

In 1985 a landmark report of the Committee on Trauma Research, *Injury in America,* documented the magnitude of the injury problem, set forth a research agenda, and called upon Congress to establish a center for injury control within the federal government. In reviewing the progress made since 1985, the National Academy of Sciences' Committee to Review the Status and Progress of the Injury Control Program at the Centers for Disease Control (1988) pointed both to successes and to how much remains to be done. The message is clear: injury remains a major public health problem. Prevention and control efforts must expand to reflect the magnitude of the challenge.

Injuries kill more than 142,000 Americans each year and cause more than 62 million persons to require medical attention. They are the single greatest killer of individuals from ages 1–44 and cost the nation approximately $133.2 billion each year. Injuries are an enormous public health problem, but they are both understandable and preventable.

Injuries affect all segments of the population, but the burden is borne disproportionately by the poor and minorities. Underlying social, environmental, and economic conditions exacerbate these disparities. Programs to prevent injuries must recognize this and work to improve the conditions that lead to this burden. Community-based action is critical to the success of these programs.

Alcohol and other drugs play a role in nearly half of all injury deaths. Preventing substance abuse could dramatically reduce the number and severity of motor vehicle and fire injuries as well as suicides and homicides. Social attitudes and media portrayal of drugs and alcohol are a major component of this problem; changing these attitudes is a challenge for all sectors of society.

Injury can be studied just as diseases are. By charting their occurrence, identifying those at risk, and intervening to prevent injury we can reduce death and disability. We know from research and practice in injury control who is at risk for particular kinds of injuries and what works to prevent them. A comprehensive view of the state of the art in injury prevention demonstrates that, while questions remain, we already know enough to act. Indeed, if the interventions recommended in this book were put into general practice, the result would be a dramatic saving in lives, health, and resources. Moreover, it is only by acting and evaluating the impact of such actions that we will learn more. For both reasons, then—to save lives and to learn more about saving lives—injury prevention programs must be undertaken now.

Our understanding of the injury problem and our capability to act are the results of continuing efforts by researchers, educators, and practitioners from public health, behavioral science, criminal justice, medicine and nursing, engineering, traffic and public safety, biomechanics, epidemiology, social service, mental health, and other disciplines. It is only through further multidisciplinary collaboration that we can continue to build the science and practice of injury prevention. This is especially critical in the area of interpersonal violence and suicide.

We look to public health because its strong multidisciplinary base can help provide leadership. Public health officials must declare injury prevention a priority. Using a systems approach, they can collaborate with others to define common goals and a common language with which to understand and control injuries.

The National Committee for Injury Prevention and Control puts forth in this book a process by which injury prevention efforts are most likely to be effective. This process includes using data to define the local injury problem, collaborating with others to develop a program based on these findings, selecting a mix of interventions that reflects the state of the art in injury control, and evaluating the program's achievement of process and outcome objectives.

The committee, following its mandate to review

the state of the art in injury prevention, recommends that the following measures be adopted.

Recommendations:

• Decision makers in the private and public sectors must recognize the magnitude of the injury problem and declare injury control a priority. They must strengthen interventions that work, increase enforcement of safety legislation and regulation, and further the use of engineering approaches to reduce hazards in the environment.

• Funding for injury prevention and control research and practice programs should be commensurate with the importance of injury as the largest cause of death and disability of children and young adults in the United States. Responsibility for providing these resources must be shared by public and private sources, including federal and state governments, foundations, and corporations. To achieve this, federal agencies must include injury control within their mandate and strengthen their regulatory and programmatic responsibilities. As a guideline, we suggest that the minimum federal share for injury prevention research and practice approach $125 million by 1992.

• Government, private foundations, industry, and community organizations should vigorously support training for injury prevention and control. Training is needed for health care workers, including physicians and nurses; public safety officials, including police and firefighters; teachers and early childhood educators; engineers, architects, and city planners; and state and local practitioners. In addition, journalists and other media professionals need to be educated about the opportunities for promoting prevention when reporting injury events.

• National leadership is needed to forge partnerships across the many disciplines involved in injury control and among state and local public health leaders. Often these disciplines are separated by theory, training, and vocabulary. To achieve lasting results, funders should require representatives of these disciplines to work together in a concerted and coordinated fashion. Injury prevention should be integrated into routine agency activities and the staff duties of many disciplines.

• Injury surveillance activities that build upon and improve existing data collection systems should be established at the national, state, and local levels. Improving the availability and quality of morbidity data should be given priority. Recognizing the enormity of this task, we suggest initial action be taken through two avenues by requiring inclusion of E codes as a separate data element in hospital discharge data and by requiring reporting

of selected injuries, starting with spinal cord injuries, as part of the National Notifiable Disease System.

• Each state and large metropolitan area should designate a lead agency devoted to injury prevention and control and an organizational unit within the agency with full-time staff to address injuries. To ensure a permanent base and adequate funding, the lead agency should include injury prevention and control programming as a line item in its budget. The lead agency should foster coordination and collaboration with other agencies and organizations leading to a comprehensive injury prevention and control program. A task force may facilitate the development and implementation of a plan to address the injury problem in a comprehensive and coordinated manner.

• National leadership is essential to affect social attitudes about and media portrayal of alcohol and other drugs. Public and private organizations should collaborate to effect legislation, regulatory change, and public education to prevent substance abuse and coordinate programs to combat it.

• Researchers and program developers should develop and test new interventions to address interpersonal violence and suicide. Decision makers and practitioners must recognize that these injuries are more than a criminal justice or mental health problem. They constitute major public health problems that can be understood and prevented through the same strategies and techniques as other injuries.

• Decision makers and funders must recognize that the evaluation of injury programs is an integral part of the management of community injury problems and must require such components in programs. When funding is limited, programs can select interventions that are known to be effective with a similar population and then limit their evaluation. Programs that implement untried interventions, or ones that have yielded conflicting evaluation findings, however, should include extensive evaluation measures.

• Program developers should design interventions that apply the best information available regarding engineering, biomechanics, behavior change, and enforcement strategies. At present there are few models and much uncertainty about the effectiveness of many available countermeasures. Therefore, the greatest need is to design interventions with specific, measurable objectives, evaluate the interventions, and disseminate the results widely.

• Research on injury prevention must address the identification and modification of behavioral factors that contribute to injuries. Interventions based on the same theories that have been successfully applied to the prevention or cessation of smoking and alcohol and drug use and to encourage exercise and medication compliance should be applied to injury control and the results evaluated.

• Firearms are involved in 30,000 deaths and 900,000 nonfatal injuries each year. Because of this tremendous toll, the following measures related to firearms should be considered: restrictive licensing, reinforcement of existing restrictions on gun purchases with waiting periods to permit effective background checks, strict enforcement of existing laws, and changes in the design of firearms to make them safer. With respect to handguns, we also support federal, state, and local initiatives to restrict the manufacture, sale, possession, and carrying of handguns.

• Support for the development of comprehensive trauma care systems by all levels of government and by the health care and public health professions must be a priority. Trauma care systems have been proven effective in reducing injury-related mortality and morbidity. They are an essential component of a systematic approach to injury prevention from primary prevention through rehabilitation.

Introduction: A History of Injury Prevention

In the United States on an average day, more than 170,000 men, women, and children are injured seriously enough to need medical care; nearly 400 die as a result of their injuries.[1] (We define injury as any unintentional or intentional damage to the body resulting from acute exposure to thermal, mechanical, electrical, or chemical energy or from the absence of such essentials as heat or oxygen. The terms "injury" and "trauma" are used interchangeably in this book.) The injury toll each day is greater than the populations of Charleston, South Carolina; Elmira, New York; and Galveston, Texas, combined. The annual cost of injury in 1987 has been estimated at $133.2 billion.[2] But injuries are not inevitable. They are preventable.

Injury is the single greatest killer of Americans between the ages of 1 and 44.[3] Every year, nonfatal injuries cause one in three of us to seek medical attention or render us unable to perform normal activities. That flood can be abated, if not completely halted.[3] Motor vehicle crashes, house fires, drownings, assaults, and all the other ways in which injuries occur are not, as we used to think, "accidents"—random, uncontrollable acts of fate. They are understandable, predictable, and preventable.

That is the central theme of *Injury Prevention: Meeting the Challenge*. During the last decades, an increasingly sophisticated science of injury prevention and control has developed in this country. It relies on the same methods and tools that public health and medicine once used to understand and eliminate smallpox. Through detailed studies of patterns of injury, we are learning how injuries occur and who in the population is most at risk. Building on this new knowledge, we are developing and testing specific, targeted interventions and implementing them in communities around the country. Were we to apply the lessons of the science of injury prevention in a truly comprehensive way, we would see an enormous reduction in death, disability, and cost to individuals, government, and the private sector.

We still have much to learn, however. The study of injury prevention is incomplete. We need to appreciate the toll and allocate commensurate resources. We must unite all the disciplines and institutions that have a part to play in injury prevention —public and private, state and local.

This book is about where we are in the struggle to control and prevent injuries and how our knowledge can be used on the state and local levels to reduce the injury burdens of communities. It explores the process of preventing injury and specific countermeasures to reduce injury.

How do injuries occur? What is their toll? What role can public health play in collaborating with traffic safety, criminal justice, and other fields, including members of the private sector? What is the association between alcohol consumption and injury? What are the primary strategies for intervening against injuries? How can these strategies be combined into programs that are effective in reducing the injury toll? What is the appropriate balance between the community's health needs and the individual's choice? How do social and individual attitudes affect our injury burden? In this introduction, we answer these questions and provide a context for what follows. And, having considered where we are today and how we arrived there, we look toward the future and suggest what we hope can be achieved over the next several years through the collective efforts of the disciplines involved in injury prevention.

THE MAGNITUDE OF THE PROBLEM

In 1985, the latest year for which official figures are available, more than 142,000 Americans died as a result of injuries,[1] and nonfatal injuries serious enough to require medical attention affected an estimated 62.5 million persons.[4] Though enormous, the numbers are neither well publicized nor widely appreciated. "Injury," concluded a recent National Academy of Sciences study, "is probably the most underrecognized major public health problem facing the nation today."[5]

Unlike cancer, cardiovascular disease, and other chronic diseases, injuries disproportionately strike the young. Table 1 summarizes the percentage of injuries as the cause of death in each age group during 1985.

Because the burden of injury falls disproportionately on the young, comparing the total number of

Table 1. Injury deaths by age group, 1985

Age group (years)	Percentage of population	Percentage of deaths caused by injury
1–4	5.9	44
5–14	14.2	52
15–24	16.5	63
25–44	30.9	40
45–64	18.8	6
65+	11.9	2

Adapted from reference 4.

injury deaths with deaths from other causes can be misleading. It is also important to consider how the deaths of so many young people affect the future. What, in an aggregate sense, might they have contributed? The effect of this premature mortality is reflected in the measurement of years of potential life lost (YPLL) by each death occurring before age 65.[6] YPLL measures the potential life lost for persons between the ages of 1 and 65 at the time of death, derived from the number of deaths in each age category as reported by the NCHS.

Table 2 shows that injuries are responsible for more years of potential life lost than cancer and cardiovascular disease combined. In fact, injuries accounted for nearly one-third of the 11.8 million years of potential life lost from all causes during the year.[7]

Severe but nonfatal injuries exact a different price. Like ripples spreading out from a stone thrown in a quiet pond, their effects can be far-reaching.

Near Seattle, on an early January morning in 1985, John (not his real name) tried to jump his motorcycle over a hole in the road. He missed, fell, and struck his head on the pavement. He was not wearing a helmet. The 27-year-old was incoherent and combative when the paramedics found him and took him to the hospital. He was operated on for a hemorrhage in the brain and spent 23 days in the intensive care unit. After another 10 days, he was transferred to the rehabilitation unit. His attention was impaired, his memory and cognitive skills seriously affected, and he was now blind.

John's medical care cost $51,000; immediate rehabilitative services added another $14,000. The bills

Table 2. Causes of death and years of potential life lost (YPLL), 1985

Cause	YPLL
Injury	3,476,752
Cancer	1,813,245
Heart disease	1,600,265

Adapted from reference 7.

were paid by Medicaid. John was transferred to a rehabilitation facility in Michigan, near his parents' home. It is unlikely that he will ever again perform meaningful labor, and he will require continuing, costly rehabilitative care in the foreseeable future.

John's was one of 105 motorcycle injury cases included in a recent Seattle-based study. "Total direct costs for these 105 patients," concluded the researchers, "were more than $2.7 million, with an average of $25,764 per patient. Only 60% of the direct costs were accounted for by the initial hospital care; 23% of costs were for rehabilitation care or readmission for treatment of acute problems. The majority (63.4%) of care was paid for by *public funds*, with Medicaid accounting for more than half of all charges" (emphasis added).[8]

There were approximately 62.5 million injuries serious enough to require medical attention in 1985. The cost of injuries for that year has been estimated at more than $107 billion.[9] (Note that there was an increase in costs of nearly 25% between 1985 and 1987, as indicated by the figure previously cited for 1987.) A comparative analysis of the total 1975 economic costs of major health impairments found that motor vehicle injuries were more costly than heart disease, twice as costly as stroke, and exceeded only by the costs of cancer.[10] However, motor vehicle injuries account for approximately half of all injury deaths. Had the analysis included all injuries, rather than motor vehicle injuries alone, this study probably would have revealed injury to be the single most costly health problem.

We may not know exactly how much injuries cost. Indeed, the costs of a family's or an individual's pain and suffering are incalculable. But we do know that the dollar cost is great and that it is spread throughout our society to individuals and families, businesses of every size, and governments at every level. And this is true of all of the major types of injury discussed in this volume: traffic injuries, residential, recreational, and occupational injuries; and injuries resulting from interpersonal violence and suicide.

Special note must be taken of the effect of the use and abuse of alcohol on the magnitude of the injury problem. The influence of alcohol can be measured across virtually all types of injuries. "Almost half of fatally injured drivers and substantial proportions of adult passengers and pedestrians killed in motor-vehicle crashes—as well as in falls, drownings, fires, assaults, and suicides—have blood alcohol concentrations of 0.10% or higher," (i.e., are legally intoxicated).[11] Alcohol abuse has been associated with railroad and aviation crashes as well.[12] And a 1987 report by the Transportation

Research Board cited 750 fatal crashes annually in which a commercial vehicle driver had been drinking.[13] Table 3 shows the estimated number of injury deaths attributable to alcohol in 1980.

Although it is widely recognized that alcohol use is associated with an increased probability of injury, it has long been believed that once an injury-producing event occurs, alcohol might protect an individual against serious trauma. Recent research demonstrates, however, that alcohol use can exacerbate the effects of injury. With regard to traffic crashes, researchers have demonstrated that alcohol increases an individual's vulnerability to injury in any given type of collision.[14]

INJURY AS A PUBLIC HEALTH PROBLEM

In order to calculate the costs of injury, we assume that there is a group of events that share essential characteristics, as do the many different diseases called "cancer." To prevent injury, it is necessary to understand how the members of that group fit together. What is it that unites such a broad range of events as traffic collisions, falls on the stairs, children choking on toys, gunshot wounds, and fingers and limbs lost to unguarded machinery?

The realization that injury is a significant problem for public health, that it can be understood with the same tools we have directed against disease, and that the elements underlying all injuries make possible broad prevention and control strategies is recent. It is important to understand the context in which it was developed.

For much of this century, injury prevention efforts focused on the assumed shortcomings of the victims. As shown in Chapters 6 and 7, the traffic safety movement in the 1920s and the home safety movement in the 1950s directed much of their energy to such educational measures as the production and distribution of pamphlets and posters. Although this emphasis changed dramatically, especially in traffic safety, it was still possible to find vestiges of the emphasis on individual shortcomings as recently as the early 1960s, as in this example, which focused entirely on self-blame:

> Once a sense of personal responsibility for accident causation can be created in the minds of people, great progress will have been made. Then the sequel to an accident will no longer be an orgy of self-pity for having been the unhappy victim of an uncontrollable event. Instead the sequel can be a character-building period of self-evaluation during which acts of personal stupidity, carelessness, and indifference may be identified. Hopefully, the accident-causing sequence of events will not be permitted to recur.[16]

The modern view of injury does not eliminate personal responsibility, but it does assign greater weight to other factors. The first of many landmarks came from Hugh De Haven, a World War I pilot (and crash survivor) turned physiology researcher. Searching for ways to reduce the toll of death in airplane and auto crashes, De Haven studied cases in which individuals had plunged, in free fall, distances of 50 to 150 feet without sustaining serious injury. It was not the force, per se, that produced injury, De Haven found, but the structural environment that controlled deceleration of the force and its distribution over the body. If the fall could not be prevented, then "Structural provisions to reduce impact and distribute pressures can enhance survival and modify injury within wide limits in aircraft and automobile accidents," he concluded.[17]

Hugh De Haven was the first researcher to understand the importance of injury thresholds in biomechanical exchanges. His work helped provide the theoretical underpinnings for the idea that

Table 3. Estimated number of injury deaths attributed to alcohol, 1980

Cause of death	Number of deaths	Alcohol-related	%
Railway crashes	632	63	10
Motor vehicle crashes	51,930	25,965	50
Other traffic deaths	232	46	20
Water transport deaths	1,429	286	20
Aviation deaths	1,494	149	10
Falls	13,294	3,324	25
Fires	5,822	1,455	25
Natural/environmental factors	3,194	799	25
Submersion, suffocation	10,216	3,576	35
Other unintentional	8,744	2,186	25
Suicide	26,896	8,061	30
Homicide	23,967	11,984	50

Adapted from reference 15.

human beings in automobiles could be "crash-packaged" through the use of safety belts, air bags, and structural changes to the vehicle. By measuring the body's ability to withstand increases in mechanical energy, De Haven helped to establish the central importance of biomechanics in injury prevention research. This area of investigation continues today, drawing its expertise from engineering, physiology, medicine, biology, and anatomy.

De Haven's seminal article was published in 1942. The next landmark in injury prevention came in 1949, when Dr. John E. Gordon suggested that injuries behaved like classic infectious diseases and were characterized by epidemic episodes, seasonal variation, long-term trends, and demographic distribution. Therefore, they could be studied through the same techniques. Most important, each injury, like each disease outbreak, was the product not of one cause but "of forces from at least three sources, which are the host . . . the agent itself, and the environment in which host and agent find themselves."[18]

Gordon's description of the agent of injury (his examples included glass-paneled doors, faulty ladders, and playful pups) was unsatisfying, however. If there were virtually as many distinct agents as there were injuries themselves, prevention would be an impossible task. Little more than a decade after Gordon's article appeared, the puzzle of how to define the agents of injury was solved by an experimental psychologist at Cornell University.

James J. Gibson was not an injury specialist. His general concern was with human and animal behavior relative to the environment. Man, he wrote,

> responds to the flux of energies which surround him —gravitational and mechanical, radiant, thermal, and chemical. . . . Injuries to a living organism can be produced only by some energy interchange. Consequently, a most effective way of classifying sources of energy is according to the forms of physical energy involved. The analysis can thus be exhaustive and conceptually clear. Physical energy is either mechanical, thermal, radiant, chemical, or electrical.[19]

Having arrived at the same conclusion as Gibson, Dr. William Haddon, Jr., of the New York State Health Department modified the energy-transfer analysis to include "negative agents" for injuries produced by the absence of such necessary elements as oxygen or heat.[20] Thus, frostbite results from the absence of necessary energy.

Haddon's landmark contributions extended Gordon's analysis to the development of preventive approaches. We will encounter Haddon's "phase-factor matrix" a number of times in this book. The matrix is actually a series of matrices developed for different purposes and makes explicit the preventive value of the epidemiologic view of injury. In the matrix, the host, agent (or vector), and environment are seen as factors that interact over time to cause injury. (To avoid confusing the actual agent of injury—energy—and the mechanism by which that energy is transferred, the matrix illustrated specifies the vector. The automobile is a vector for physical energy, just as a faulty appliance cord can be a vector for injurious electrical energy.) A Haddon matrix designed around traffic injuries appears in Figure 1.

The precrash phase includes everything that determines whether a crash will take place (the driver's lack of ability or alcohol-related impairment, the car's malfunctioning brakes, the poorly lit curve in the road, or society's attitudes about alcohol consumption). The crash phase includes everything that determines whether an injury results from the crash: Are the occupants wearing safety belts? Is the car large or small? Is the pole they hit designed to break away? Are laws about child safety seats enforced? The postcrash phase determines whether the severity of the injury's consequences can be reduced: Can the bleeding be stopped? How quickly do the paramedics arrive? How effective is emergency room care? Does society support the development of trauma care systems? Haddon's analysis suggested that preventing an injury may require modifying only one of the elements, and there is no essential priority determining which one must be modified (the weakest link in the chain will do). For example, as will be seen in Chapter 6, one potentially effective measure against impaired driving is an in-car breath testing device that prevents the car from starting if the driver is drunk—a modification of the vector, rather than the host. In another major contribution, Haddon later developed a series of 10 classes of injury countermeasures, to which we return later when discussing the main types of injury prevention measures.

The contributions of De Haven, Gordon, Gibson, Haddon, and others helped to shift injury prevention away from an early, naive preoccupation with distributing educational pamphlets and posters and toward modifying the environments in which injuries occur. By developing new laws and enforcement mechanisms and through new technologies and engineering changes in products, injury experts from a broad range of disciplines sought to protect people from coming into contact with injurious amounts of energy.

This emphasis on modifying vectors and environments addressed circumstances in which it was dif-

Phase	Host (human)	Vector (vehicle)	Physical environment	Socioeconomic environment
Precrash	Driver vision Alcohol intoxication Experience and judgment Amount of travel	Brakes, tires Center of gravity Jackknife tendency Speed of travel Ease of control Load characteristics	Visibility of hazards Road curvature and gradient Surface coefficient of friction Divided highways, one-way streets Intersections, access control Signalization	Attitudes about alcohol Laws related to impaired driving Speed limits Support for injury prevention efforts
Crash	Safety belt use Osteoporosis	Speed capability Vehicle size Automatic restraints Placement, hardness, and sharpness of contact surfaces Load containment	Recovery areas Guard rails Characteristics of fixed objects Median barriers Roadside embankments Speed limits	Attitudes about safety belt use Laws about safety belt use Enforcement of child safety seat laws Motorcycle helmet use laws
Postcrash	Age Physical condition	Fuel system integrity	Emergency communication systems Distance to and quality of emergency medical services Rehabilitation programs	Support for trauma care systems Training of EMS personnel

Figure 1. The Haddon Matrix. Adapted from references 71 and 72.

ficult or impossible for an individual to prevent injury through his or her own behavior (i.e., even the most careful driver cannot reduce the potential for injury in a collison if the vehicle lacks safety belts, adequate padding, a collapsible steering column, and other protective devices). But human behavior and personal responsibility remained "undeniably important in injury causation."[21]

That view has been reinforced in recent years as psychologists and others have begun to contribute to an understanding of the behavioral and social causes of injuries.

> Within psychological domains, there are multiple origins of . . . injuries, and therefore a rich variety of prevention or intervention strategies. Researchers are beginning to examine behavioral approaches for altering children's and care givers' unsafe behavior to become safer. . . . The behavioral approach appears effective for motivating change of individuals' unsafe

behaviors that may not be affected by other approaches.[22]

At this time, behavioral psychology's contribution to the prevention of injury is in its infancy, but there is much we can learn from its successful application to other health problems. Before discussing the several approaches to injury prevention in greater detail, however, it is necessary to understand that while injury is a public health problem, injury prevention cannot be solely a public health responsibility.

The Role of Public Health and the Need for Collaboration

Injury is a public health problem because of its magnitude and because of its consequences for the health of Americans. Traffic injuries alone have produced more fatalities than all the wars in which the United States has fought, combined. No health problem responsible for so much death and disability could be defined as anything other than a public health problem. And, as we have seen, injury is a public health problem because public health methods, practitioners, and agencies can contribute to its understanding and prevention. The work of De Haven, Gordon, Gibson, Haddon, and other pioneers demonstrated that injuries could be understood with the same techniques of epidemiology that had been applied, with increasing success, to infectious diseases since the mid-19th century.[23] And, as Chapter 10 indicates, these are the very reasons why interpersonal violence and suicide are increasingly being understood as major causes of injury that must be addressed through the same methods.

Collection and analysis of data about health problems are one of the primary functions of public health agencies. By collecting and analyzing data about injuries, as is done for infectious diseases—where, when, and how they occur, and to whom—it is possible to understand patterns of occurrence, to identify risk groups for specific injuries, and to use the information as the basis for designing preventive measures. This is the foundation of the data-based approach to the design, implementation, and evaluation of prevention programs around which this book is based. Collection and analysis of data are discussed in Chapters 2 and 3; program design, implementation, and evaluation are the subjects of Chapters 4 and 5.

In addition to their data collection and analysis capabilities, public health agencies can offer practical experience in the successful management of communitywide health problems through the design, implementation, and evaluation of community-based prevention programs. And, in its recognition that health problems have multiple causes and are therefore multidisciplinary by nature, public health understands the need to coordinate and participate in fashioning multidisciplinary solutions.

Public health is only one of a number of participants—and sometimes one of the most recent arrivals—when it comes to injury prevention. If the preventability of injury is one of this book's central themes, so too is the critical need for collaboration among the many individuals and institutions whose expertise is a prerequisite for success.

Public health agencies, in particular state health departments, "have been involved sporadically over the past 50 years in childhood injury prevention and control activities."[24] However, the effort to prevent the greatest source of injury-related deaths —traffic injuries—has long been led by engineering, criminal justice, and traffic safety agencies. Preventing injuries caused by interpersonal violence and suicide was the concern of criminal justice and, more recently, of social service and mental health specialists long before public health recognized that violence could be understood through the same techniques as other sources of injury.

The point is not who got there first, but how to draw upon the expertise and the contributions each participant can make. How to foster and manage collaborative efforts are among the topics discussed in Chapters 1 and 5. It is an assumption of this book that state and local health departments can play a central role in developing or implementing injury prevention programs. It is also assumed that state and local health practitioners will participate in injury prevention efforts that begin and are housed in other departments and agencies. Where programs begin is a function of leadership, and leadership in injury prevention arises because individuals care enough to lead. A Tennessee pediatrician, an epidemiologist in the Vermont health department, two parents in California, and an Indian Health Service worker are only a few of the injury prevention leaders whose stories are told in these pages.

STRATEGIES FOR INJURY PREVENTION

Just as the occurrence of an injury requires the interaction of several factors, preventing one may require a mixture of countermeasures or interventions (the terms are used synonymously). One of the earliest attempts to systematize the process of considering injury prevention measures was Haddon's list of 10 countermeasures, mentioned earlier. Beginning in 1962, Haddon developed and refined a list of 10 general strategies designed to interfere with the energy transfer/injury process:

1. Prevent the creation of the hazard (stop producing poisons).
2. Reduce the amount of the hazard (package toxic drugs in smaller, safe amounts).
3. Prevent the release of a hazard that already exists (make bathtubs less slippery).
4. Modify the rate or spatial distribution of the hazard (require automobile air bags).
5. Separate, in time or space, the hazard from that which is to be protected (use sidewalks to separate pedestrians from automobiles).
6. Separate the hazard from that which is to be protected by a material barrier (insulate electrical cords).
7. Modify relevant basic qualities of the hazard

(make crib slat spacings too narrow to strangle a child).

8. Make what is to be protected more resistant to damage from the hazard (improve the host's physical condition through appropriate nutrition and exercise programs).

9. Begin to counter the damage already done by the hazard (provide emergency medical care).

10. Stabilize, repair, and rehabilitate the object of the damage (provide acute care and rehabilitation facilities).[20]

These 10 strategies were not intended as a formula for *choosing* countermeasures so much as an aid to thinking about them logically and systematically.

Other aids, such as PRECEDE (a diagnostic health promotion model focused on determinants of behavior change), have been developed in recent years. The PRECEDE model suggests that three types of variables should be addressed to influence health behaviors. *Predisposing* variables are antecedent to behavior and include relevant knowledge, beliefs, and values. *Enabling* variables include the availability and accessibility of personal and community resources required to perform the behavior. *Reinforcing* variables are factors subsequent to behavior that provide rewards, incentives, or punishments for continuation of the behavior. Any injury behavior may be seen as a function of the collective influence of these three factors.[25] The use of the Haddon countermeasures and the PRECEDE model in helping to design programs and select specific interventions is detailed in Chapter 4.

Interventions can be characterized as either "passive" or "active" in nature. Passive (or automatic) countermeasures require little individual action on the part of those being protected. The automobile air bag is a classic example. In a crash, the air bag automatically inflates to cushion the driver and prevent injury. Nonautomatic safety belts, on the other hand, are active interventions that require the wearer to buckle up each time to be protected. Passive measures requiring no action are often described as being "better" or "more effective" than active countermeasures.[26] Passive measures such as air bags are often difficult to implement because they require either legislative or regulatory changes directed at specific product modifications. New legislation may require educating both the public and their legislators as well as other decision makers. Look, for example, at the more than 2-decade-long battle for air bags, which are only now becoming available.[27]

Although the ideal passive measure would protect all members of the population without any action on their part, *truly* passive interventions are rare. The issue, therefore, is not allegiance to one type of intervention but the need for flexibility in combining strategies to arrive at the most effective mix.[21,28]

Intervening successfully against injuries may involve the passage and enforcement of new laws or the increased enforcement of existing laws, the education of the population at large or of targeted groups, efforts to alter specific injury-related behaviors, or changes in the design of products or of the physical environment. In this book these approaches are categorized as legislation/enforcement, education/behavior change, and engineering/technology. These are not mutually exclusive categories. The approaches can often be combined effectively. Child safety seats provide an excellent example.

Since 1978, every state has passed a law requiring that children (generally under the age of 4) riding in motor vehicles be restrained in federally approved child safety seats. These laws are interventions of the legislation/enforcement type. The seats are an engineering/technology countermeasure known to be extremely effective when used properly.[29] But the seats frequently are used incorrectly.[30] Education was an important factor in the passage of these laws and in encouraging parents to obtain and use the seats correctly. Clearly, education/behavior change interventions are critical in maintaining compliance with and thus maximizing protection from child safety seat laws.

Providing effective protection for automobile occupants, in fact, requires a mix of strategies. Developing and implementing this programmatic mix has required the combined efforts of pediatricians, public health practitioners, legislators, traffic safety specialists, educators, psychologists, public safety officials, researchers, manufacturers, parents, and other health care professionals. The best-known successes of these efforts have come through the use of legislative/enforcement and engineering/technology interventions.

Among the legislative/enforcement interventions discussed in Chapters 6–17 are increased taxes on alcoholic beverages, automobile safety belt laws, laws designed to reduce impaired driving, the Poison Prevention Packaging Act of 1970, the Children's Sleepwear Statute of 1971, smoke detector laws, and handgun control measures. Roadway countermeasures to protect pedestrians, safety belts and child safety seats, motorcycle and bicycle helmets, smoke detectors, automatic sprinkler systems, and better street lighting are among the engineering/technology countermeasures discussed.

Education/behavior change interventions are less well known within the field of injury prevention and thus require a somewhat more detailed discussion. These are interventions that respond to the fact that injuries result from both environmental and behavioral causes. Although "passive" approaches usually are preferred to "active" ones, it is difficult to envision an environmental change that is without any behavioral component. Although education and behavior change are treated as one approach, it is important to understand the differences between the two activities.

Education about injury risk and the importance of risk-reduction behaviors is a gradual process. The effects of educational campaigns may not be seen for a year or more. Further studies are needed to evaluate the longer-term benefit of educational programs and how these approaches can be combined with others to accelerate behavior change. Education interventions intended to change knowledge and attitudes that influence decisions about potentially injurious behavior can be targeted to take into account the characteristics and demographics of and appropriate message content for specific groups.

Behavioral science has much to offer the field of injury control in understanding the determinants of injury behavior and in developing effective strategies for behavior change.[31-34] Behavioral research suggests several promising techniques on which interventions can be based: participative education,[35-37] incentives,[38-40] behavioral feedback,[41,42] and modeling.[31,43] Because the use of these techniques is new, further research is necessary to determine their long-term effects.

The behavioral approach views changes in behavior as significant steps toward the reduction of injury rates. Although psychological research and applications have been slow to develop in the injury field, intervention has increasingly been based on the recognition that injury prevention involves individual actions related to protection, caution, and risk. The role that decision-making and lifestyle choices play in injury control should not be underestimated. Only then will it be possible to ensure that interventions such as laws on the use of safety belts and child safety seats provide protection for the largest number of persons.

The limited success of education/behavior change interventions in modifying injury prevention behavior to date may be a result in part of the failure to understand the behavioral causes of injuries and properly to apply what is known to develop effective interventions. The application of current behavioral theory to injury control has the potential for improving our understanding of the behavioral determinants of injury and the development of effective strategies for prevention. While "there may be no one theory or model that can be universally applied," the synthesis of components from many different models may result in new and useful approaches to the problem.[44]

Legislation/enforcement and engineering/technology interventions do have the potential to reach and protect the greatest number of persons, and they should be employed whenever feasible. But, as is demonstrated in Chapter 5, education—of both policymakers and the general population—is often an antecedent of action. It is also necessary for public acceptance of new legislation and as a means of increasing compliance with the law. And it is a primary way of redressing user rejection or misuse of injury prevention technologies (including safety belts and child safety seats).

PUBLIC HEALTH AND INDIVIDUAL CHOICE

Some interventions seek to help people alter their behaviors or adopt new, safer ones. Others rely on more "coercive" methods of legislation and enforcement. This has been a source of controversy and heated political battles over such public health measures as alcohol excise taxes, fluoridation, safety belt laws, and the requirement that motorcycle riders wear helmets. Where is the balance between public health and individual choice?

When John crashed his motorcycle in Seattle, the seriousness of his injury and his projected lifelong disability were directly related to his not having worn a helmet. Before 1977, he would have been legally obligated to wear one, but in that year the state of Washington repealed its helmet law.

This was part of a national movement. From 1966 to 1975, under the pressure of congressional legislation and federal regulation, 49 states adopted mandatory helmet laws. (California was the lone exception, although Utah's law was not in strict compliance with federal requirements and the Illinois statute was overturned by the state supreme court in 1969.) The effectiveness of helmets in reducing fatalities and serious injuries was never in question and was demonstrated repeatedly during the years when the laws were enforced.[45]

In 1975 the federal government moved to withhold highway safety funds (and some highway construction funds, as well) from California, Utah, and Illinois for their failure to mandate helmet use. The action was halted when Congress prohibited the

Secretary of Transportation from requiring helmet use laws. Within three years, 27 states completely rescinded or substantially weakened their helmet laws. Predictably, helmet use declined, and motorcycle fatalities and serious injuries increased.[45]

For the motorcyle riders and their supporters who had long opposed helmet laws, and for those in Congress who agreed, the issue was clear. Helmet laws were "an outrageous example of an overbearing, overprotective bureaucracy gratuitously trying to restrict the freedom of choice of the individual and using the clout of federal money to try to impose its view on a state government."[46] Certainly most persons would agree that compelling a motorcycle rider to wear a helmet is an infringement of his or her personal choice. The question is whether it is an appropriate one.

The right of government to protect the citizenry from harm by adopting, in the name of public health, measures that restrict individual liberty is well established. This "police power" (a term coined by Chief Justice John Marshall in 1824)[47] has been defined as "the power vested in the legislature by the constitution, to make, ordain, and establish all manner of wholesome and reasonable laws . . . for the good and welfare of the commonwealth, and of the subjects of the same."[48] For example, a landmark 20th-century case, *Jacobson v. Massachusetts*, ruled that an individual could not avoid immunization against disease because it infringed upon his freedom. The potential harm to the community was ruled to be greater than the infringement.[49]

Opponents of helmet laws (and other similar measures such as minimum age laws for the purchase and public possession of alcohol or mandatory safety belt laws) have not, by and large, challenged the government's authority to protect the public from harm through the exercise of the police powers doctrine. They have, instead, asked where the public harm is if an individual's behavior affects only him- or herself?

One answer to this fundamental question has come from the courts. Helmet laws have been challenged in many states and upheld 30 times in the states' highest courts; the U.S. Supreme Court has refused to overturn them.[50] In one court's celebrated comment:

> From the moment of the injury, society picks the person up off the highway; delivers him to a municipal hospital and municipal doctors; provides him with unemployment compensation, if after recovery, he cannot replace his lost job and, if the injury causes permanent disability, may assume the responsibility for his and his family's continued subsistence. We do

not understand the state of mind that permits the plaintiff to think that only he himself is concerned.[51]

Certainly John's case and those of the other 104 injured motorcyclists in the Seattle study bear out the court's logic. The total cost of all injuries—an estimated $107 billion for 1985, the year under study—is borne by the entire society. This is an essentially economic rationale, a resource allocation argument, for injury prevention. Because individual liberty is a fundamental element of our social contract, however, the economic argument is not entirely satisfying. For one thing, it neglects the possibility that infringing on individual liberty may entail economic costs and that prevention might occasionally use more resources and be more expensive than the problem it seeks to ameliorate.[52]

In yet another view, public health is one of the public, collective goods to which people aspire when they organize in societies with shared loyalties and obligations. In this vision, "public health is part of the basic glue that cements the democratic community together, a form of group solidarity in the face of man's most ancient foe."[53] For the message, "the life you save may be your own," the proponents of the public-health-as-common-good view would substitute "the lives we save together might include your own."[54]

ATTITUDES AND INJURY

The experience of the helmet use law also points the way to another critical element in understanding injuries in America: the extent to which their occurrence is affected by people's attitudes. Perhaps the most pervasive of these—what Haddon, Suchman, and Klein called "the last folklore subscribed to by rational men"—is the belief that injuries are "accidents." "People who protect themselves against disease by inoculation and the observance of appropriate nutritional and personal habits nevertheless tend to look upon [injuries] as either a punishment for misdeeds or as unwarranted blows delivered by a capricious fate."[55]

The problem is exacerbated by the way the media cover injuries. Injury is everywhere, but nowhere at the same time. Car crashes and house fires may be staples for newspapers and broadcast news reports. But each incident is reported as a separate, unique event—an "accident." Little attempt is made to look beyond that day's headline to patterns of injury or to prevention. The importance of working with the media to address this problem and strategies for doing so are highlighted in Chapter 5.

Almost two decades ago, two researchers investigated the differences in death rates from tornadoes in the South, where they were high, and in the North, where they were low. After adjusting for population density, tornado activity, housing type, time of day the tornadoes occurred, and other variables, the differences persisted. A survey of people's beliefs in each area helped to explain the variation. Southerners were much more inclined to believe that their lives were controlled by God or luck than were their northern counterparts. They also tended to put much less faith in the weather service's ability to forecast tornadoes and were less likely to heed warnings.[56]

In addition, there is a general acceptance in American society of the view that risk taking is essentially a good thing.[57] Indeed, risk taking is glorified in the media. Further, the kinds of risks involved are generally physical rather than emotional or intellectual. As many injuries occur as a result of involvement in activities that are pleasurable, many Americans will resist any attempts at control if they perceive the risks as acceptable.

Although attempts have been made to identify individuals who are more likely (through genetics, physiology, socialization, or other factors) to take risks, "investigators have had relatively little success in detecting consistent differences in risk-taking propensity."[58] Indeed, researchers have learned that, in general, each person sees him- or herself as being less at risk than others. Thus, most people view themselves as safer-than-average drivers.[59]

Nor is the answer a simple one: reducing risk taking reduces injury. Many risks are beneficial to society, and not all risks result in injury.[60] How much of the social value of risk taking is worth preserving and at what cost? As Chapter 7 makes clear, if we are to tolerate no risks at all, few sports, whether professional or amateur, would be permitted. Clearly, there are competing values at play.[61]

Automobiles, alcohol, and handguns are the three commodities connected with an overwhelming majority of fatal and serious injuries. Confronting and altering attitudes about each of them and the ways in which they are to be used are important concerns for health professionals. Education and behavior change interventions can be important in accomplishing these goals.

Other attitudes condition the violent behavior that leads to so many injuries. Violence is portrayed as an acceptable and often successful instrument of conflict resolution. Murder and suicide are endemic on motion picture and television screens. The ideologies of racism and male dominance are deeply in-grained. Many social norms and public models encourage rather than inhibit interpersonal or self-directed violence. These issues are discussed further in Chapters 6–17.

Growing Support for Injury Prevention

Clearly there is much work to be done in injury prevention and control. Fortunately, for a number of important reasons, this is a most favorable time in which to proceed.

• A national agency infrastructure exists to stimulate and support activities.

• Funding for injury research and prevention programs (although still not commensurate with the problem) has increased dramatically.

• Many effective or promising countermeasures, through this support and encouragement, have been or are being tried.

• These countermeasures and the information about their effectiveness constitute a scientific basis for action.

Finally, and most important, many state and local health departments, traffic and public safety agencies, and communities themselves are poised to take action against injuries. The National Head Injury Foundation, Remove Intoxicated Drivers (RID), Mothers Against Drunk Driving (MADD), and the California-based Drowning Prevention Foundation are only a few examples of injury coalitions. (Others, as well as a discussion of coalition building, can be found in Chapter 5.)

A comprehensive history of injury prevention has yet to be written, but we have pointed to some of the intellectual landmarks along the way in the growing understanding of injury. Important events in the practice of prevention and control, such as the community-based demonstration projects sponsored by the W. K. Kellogg Foundation in 1953 and the early work of the U.S. Public Health Service, have recently been summarized by Fisher.[24] The history of preventive efforts (including federal legislation) against traffic injuries; residential, recreational, and occupational injuries; and injuries of interpersonal violence and suicide are discussed in Chapters 6–17.

The most recent burst of energy in injury prevention and control began with 1983 congressional action that authorized the Secretary of Transportation to ask the prestigious National Academy of Sciences to report on the problem of injury. What was known about injury? What research should be done to learn more? And what federal institutional arrangements would improve the process?

The academy's National Research Council, in collaboration with the Institute of Medicine and the Committee on Trauma Research, convened a panel of experts and in 1985 released *Injury in America: A Continuing Public Health Problem*.[3] The report proved to be enormously influential. It laid out an agenda for research in epidemiology, prevention, biomechanics, treatment, and rehabilitation. Most important, it called not only for increased spending but also for the coordination of separate efforts and the creation of an institutional base: "The committee recommends the establishment of a center for injury control within the federal government. The Centers for Disease Control of the Department of Health and Human Services is recommended as the location for that center."[3]

Almost immediately, Congress acted to implement the academy's recommendation. A 3-year pilot program was established at the Centers for Disease Control (CDC) and is now housed in the Division of Injury Epidemiology and Control of the CDC's Center for Environmental Health and Injury Control. It included a program of extramural research grants and funding for the establishment of five Injury Prevention Research Centers around the country. The program received $32 million in funding during the three years.[62]

The new initiative at CDC complemented longstanding injury prevention activities in the Department of Transportation's National Highway Traffic Safety Administration (NHTSA) and work in the Division of Maternal and Child Health (DMCH) of the Department of Health and Human Services. The National Highway Traffic Safety Administration (originally called the National Highway Safety Bureau) was established by Congress in 1966 and empowered to set safety standards for new cars, beginning with 1968 models. Under its first administrator, Dr. William Haddon, Jr., NHTSA issued "standards that have made cars safer ever since: laminated windshields, collapsible steering assemblies, dashboard padding, improved door locks, dual braking systems, and many other automatic safety features."[63] With a mandate to reduce the enormous traffic injury burden, NHTSA has issued federal motor vehicle safety standards, conducted vehicle defect investigations, provided essential data through the Fatal Accident Reporting System (FARS) and National Accident Sampling System (NASS), conducted research, and sponsored demonstration projects throughout the country in occupant protection, in reducing alcohol-related traffic injuries, and in motorcycle, pedestrian, and bicycle safety. In addition to pioneering in the development of emergency medical services and trauma care systems and in providing services and training to law enforcement agencies, NHTSA has provided both funding and technical assistance for projects carried out through state offices of highway safety.

THE ORIGINS OF THIS BOOK

In 1979 DMCH funded three pilot injury prevention projects in Virginia, California, and Massachusetts. By 1988 the reorganized Bureau of Maternal and Child Health and Resources Development was sponsoring 20 demonstration projects affecting injury prevention in 15 states plus 12 projects based in pediatric emergency medical services through its Maternal and Child Health Improvement Projects (MCHIP) and incentive grant programs. The New England Network to Prevent Childhood Injuries, a DMCH project, was the first to attempt to provide technical assistance in injury prevention on a regional basis. Other federal agencies, including the Indian Health Service, also became more active in injury prevention during this period.

In February 1986 DMCH brought together all of its injury prevention grantees to talk about their progress in establishing injury prevention programs and their needs for the future. What was needed, many said, was a guide to practice—a reference to the *how* and *what* of injury prevention programs.

Dr. Vince L. Hutchins, DMCH's director, agreed.

> The research and demonstration efforts of today will provide a solid future base for programming. But, until all of that information becomes available, we need a resource that will encourage best practices today . . . to develop an approach to addressing community injury prevention problems based on the current state of the art, using sound data-based planning and evaluation. We do not yet have all of the answers, but we do have some.[64]

A similar feeling was percolating at NHTSA and CDC. "We felt it was time for such an effort," said Dr. Stuart T. Brown, Director of CDC's Division of Injury Epidemiology and Control. "First, the research community had been invigorated and projects begun. Now, we had to ask 'who will use the results of this research and how? How is it going to make a difference to the health of the American public?' "[65]

Injury Prevention: Meeting the Challenge

Injury in America: A Continuing Public Health Problem focused on issues of injury research, as described above. What Hutchins, Brown, and many others envisioned was a new publication addressed to injury practitioners and those with an interest in the practical aspects of prevention. In late 1986 DMCH, with assistance from CDC and NHTSA, established the National Committee for Injury Prevention and Control (NCIPC). The committee's charge was to investigate and report on the state of the art in injury prevention and to make recommendations for continued progress. The project was managed and staffed by Education Development Center, Inc. (EDC), of Newton, Massachusetts. *Injury Prevention: Meeting the Challenge* by the NCIPC and EDC is the result of that collaboration.

The committee was made up of 31 experts from many areas of injury prevention (state health department practitioners, injury prevention researchers, engineers, physicians, nurses, highway safety officials, and representatives of the sponsoring federal agencies, among others). The two major areas of its investigation are reflected in the book's structure. First, it examines the process of design, implementation, and evaluation of injury prevention programs; then it reviews and evaluates specific injury countermeasures.

The first five chapters, "Getting Started," "Learning from Data," "Working with Data," "Program Design and Evaluation," and "Program Implementation," explain the process of injury prevention from the moment leaders emerge to address an injury problem to the time of disseminating a project's final results. These chapters explore how to identify injury problems using available national and local data, how to design a program and determine how and what to evaluate, how to select the most practical and appropriate interventions, and how to manage the program. They also discuss raising funds for injury prevention efforts, building coalitions, and working with the media. This section not only provides much valuable practical information but also suggests how practitioners can go about getting expert advice in such technical areas as data analysis and evaluation design.

Following and complementing these chapters, the intervention section reviews traffic injuries; residential, recreational, and occupational injuries; and interpersonal violence and suicide. Each of the 12 chapters details the state of the art in preventive countermeasures to answer the question: What works in injury prevention? NCIPC members and EDC staff, organized into working groups (the groups are listed on p. 000), reviewed hundreds of interventions. Each is discussed and categorized in terms of its efficacy ("proven effective," "promising," "ineffective," or "unknown") and the committee's recommendation for its future application is provided.

In a final section, the committee considers some important aspects of trauma care (including medical rehabilitation). The role of trauma care within the systems approach to injury prevention is explored, as are the ways in which injury prevention practitioners and emergency medical service specialists can collaborate effectively.

Injury Prevention: Meeting the Challenge is a response to the need for a reference to help individuals carry out injury prevention and control work. But it is a resource, not a prescription. It presents options and alternatives in reducing injuries. It tells users what they need to know to make choices as well as surveys what some of the relevant choices are.

The state of the art in injury prevention and control is promising, if uneven. Too few interventions have been thoroughly evaluated, but many show great promise. Their continued use is important, but so too is further evaluation to determine what works. One conclusion of this review is inescapable: We know enough to act, and we know that doing so can dramatically reduce the burden of injury. One estimate for childhood injury, for example, is that the implementation of only 12 currently available interventions (air bags, child safety seats, motorcycle and bicycle helmets, smoke detectors, and the elimination of handguns, among others) could reduce deaths by 29%.[66]

Looking Toward the Future

Two years after *Injury in America* had borne fruit, Congress again approached the National Academy of Sciences, this time for a report on the progress made under its mandate to CDC. The academy's document, *Injury Control: A Review of the Status and Progress of the Injury Control Program at the Centers for Disease Control,* was published recently.[67] It concludes that substantial progress has been made, despite severely limited resources. It calls for increased support for the initiative and its establishment as a permanent, expanded program, perhaps on the model of CDC's National Institute of Occupational Safety and Health. Subsequently, there was an increase in the program's budget. For fiscal 1989, Congress has appropriated $20 million to the Department of Health and Human Services for CDC's injury program.

Injury Prevention: Meeting the Challenge is about injury prevention today, how far we have come, and how far we have yet to go. We can point to some indications of where the nation might be in the next several years, as for example the 1990 Health Objectives.[68] These are the benchmarks of a national agenda to promote health and reduce disease. The most recent assessment of the injury-related objectives predicts that the following objectives (over the baseline year of 1978) have been or will be met by 1990:

• A motor vehicle fatality objective of 18.0 per 100,000 population (down from 23.6)

• A motor vehicle fatality rate for children under the age of 15 of 5.5 per 100,000 (down from 9.0)

• A home injury fatality rate for children under the age of 15 of 5.0 per 100,000 (down from 6.0)

• Residential fire deaths reduced to 4,500 per year (from 5,401)

In addition, it is projected that several will not be met:

• Reducing the death rate from falls to 2.0 per 100,000 (from 6.2)

• Reducing the death rate from drowning to 1.5 per 100,000 (from 3.2)

• Reducing the rate of suicide among people 15–24 years of age to 11.0 per 100,000 (from 12.1)

• Reducing the death rate from homicide among black males 15–24 years of age to below 60.0 per 100,000 (from 70.7)[69]

This level of progress, insufficient but progress nonetheless, is within our grasp. The question is whether we can do better. Can agencies and communities use the processes and interventions outlined in this book to improve the lives of Americans? One indication that this is possible is contained in a survey of state health departments published by the Childhood Injury Prevention Resource Center at the Harvard School of Public Health.[70]

The authors conclude that in the period from 1981 to 1987, injury prevention programming at the state level tripled, injury activities were integrated into a wider variety of state health departmental divisions, injuries of interpersonal and self-directed violence became part of the domain of some state health departments, the development of systems to report injuries increased, and staffing levels appear to have expanded. However, despite this progress, which the authors characterize as piecemeal and lacking the coordination of a systems approach, "few state health departments had established comprehensive injury prevention programs with long- and short-term action plans, and staff need additional training to ensure the development of effective programs."[70] It is the committee's hope that *Injury Prevention: Meeting the Challenge* will help fill the need to train persons working in health departments and elsewhere in how to plan, implement, and evaluate programs that will prove to be effective in reducing injuries.

In short, much progress remains to be made, but there is a substantial foundation on which to make it. Injury is among the oldest health problems faced by humanity. Never before, however, has there been as much interest in preventing injury, whether among university researchers, government officials, health professionals, business, labor, or concerned community members. How leaders in injury prevention emerge, and how they and their agencies and communities harness the energy and growing expertise and resources that exist, will determine whether "preventing injuries" remains an expression of hope or becomes a reality.

REFERENCES

1. National Center for Health Statistics. Advance report of final mortality statistics, 1985. Washington, DC: U.S. Government Printing Office, 1987.

2. National Safety Council. Accident facts, 1988. Chicago: National Safety Council, 1988.

3. National Research Council. Injury in America: A Continuing Public Health Problem. Washington, DC: National Academy Press, 1985.

4. National Center for Health Statistics. Current estimates from the National Health Interview Survey, United States, 1985. Washington, DC: U.S. Government Printing Office, 1986.

5. National Academy of Sciences. Injury control. Washington, DC: National Academy Press, 1988:7.

6. Centers for Disease Control. Premature mortality in the United States: public health issues in the use of years of potential life lost. Morbid Mortal Weekly Rep 1986;35 (suppl):2S.

7. Centers for Disease Control. Estimated years of potential life lost before age 65 and cause-specific mortality, by cause of death: U.S., 1985. Morbid Mortal Weekly Rep 1987;36:447 (Table 5).

8. Rivara FP, Dicker BG, Bergman AB, Dacey R, Herman C. The public cost of motorcycle trauma. JAMA 1988;260: 221–3.

9. National Safety Council. Accident facts, 1986. Chicago: National Safety Council, 1986.

10. Hartunian NS, Smart CN, Thompson MS. The incidence and economic costs of major health impairments. Lexington, Massachusetts: Lexington Books, 1981.

11. Waller J. Alcohol and unintentional injury. In: Kissin B, Begleiter H, eds. The biology of alcoholism, vol. 4. New York: Plenum, 307–49.

12. Secretary of Health and Human Services. Sixth special report to the U.S. Congress on Alcohol and Health, January 1987. Washington, DC: Department of Health and Human Services, 1987:10.

13. Transportation Research Board. Zero alcohol and other options: limits for truck and bus drivers. Washington, DC: National Research Council, 1987.

14. Waller PF, Stewart RJ, Hansen AR, et al. The potentiating effects of alcohol on driver injury. JAMA 1986;256:1461–6.

15. Ravenholt RT. Addiction mortality in the United States, 1980: tobacco, alcohol, and other substances. Pop Dev Rev 1984;10:697–724.

16. Chapman AL. Accident prevention. In: Halsey MN, ed. New York: McGraw-Hill; 1961:vi.

17. De Haven H. Mechanical analysis of survival in falls from heights of fifty to one hundred and fifty feet. War Med 1942:586–96.

18. Gordon JE. The epidemiology of accidents. Am J Public Health 1949;39:504–15.

19. Gibson JJ. The contribution of experimental psychology to the formulation of the problem of safety: a brief for basic research. Reprinted from: Behavioral approaches to accident research. New York: New York Association for the Aid of Crippled Children; 1961:77–89.

20. Haddon W. Advances in the epidemiology of injuries as a basis for public policy. Public Health Rep 1980;95:411–21.

21. Haddon W, Baker SP. Injury control. In: Clark D, MacMahon B, eds. Preventive and community medicine. Boston: Little, Brown; 1981:10.

22. Roberts MC, Brooks PH. Children's injuries: issues in prevention and public policy. J Soc Issues 1987;43(2):1–12.

23. MacMahon B, Pugh TF. Epidemiology: principles and methods. Boston: Little, Brown, 1970.

24. Fisher L. Childhood injuries: causes, preventive theories and case studies. J Environ Health 1988;50:355–60.

25. Green LW, Kreuter MW, Deeds SG, Partridge KB. Health education planning: a diagnostic approach. Palo Alto, California: Mayfield Publishing, 1980.

26. Robertson LS. Injuries: causes, control strategies, and public policy. Lexington, Massachusetts: Lexington Books, 1984.

27. Roberts MC. Public health and health psychology: two cats of Kilkenny? Prof Psychol Res Pract 1987;18:145–49.

28. Sleet DA. Occupant protection and health promotion. Health Educ Q 1984;11:109–111.

29. Agran PF, Dunkle DE, Winn DG. Effects of legislation on motor vehicle injuries to children. Am J Dis Child 1987;141:959–64.

30. National Transportation Safety Board. Safety study: child passenger protection against death, disability, and disfigurement in motor vehicle accidents. Washington, DC: National Transportation Safety Board, 1983.

31. Geller ES, Elder JP, Hovell MF, Sleet DA. Behavioral approaches to drinking-driving interventions. In: Ward W, ed. Advances in health education and promotion, vol. 3. Greenwich, Connecticut: JAI Press (in press).

32. Sleet DA, Hollenbach K, Hovell M. Applying behavioral principles to motor vehicle occupant protection. Educ Treat Child 1986;9:320–33.

33. Geller ES. A behavioral science approach to transportation safety. Bull NY Acad Med 1988;64:632–61.

34. Roberts MC, Fanurik D, Layfield DA. Behavioral approaches to prevention of childhood injuries. J Soc Issues 1987;43:105–118.

35. Geller ES, Rudd JR, Kalsher MJ, Streff FM, Lehman GR. Employer-based programs to motivate safety belt use: a review of short- and long-term effects. J Safety Res 1987;18:1–17.

36. Jones RT, Ollendick TH, McLaughlin KJ, Williams CE. Elaborative and behavioral reversal in the acquisition of fire emergency skills and the reduction of fear of fire. Behav Ther (in press).

37. Weinstein ND, Grubb PD, Vautier JS. Increasing seat belt use: an intervention emphasizing risk susceptibility. J Appl Psychol 1986;71(2):285–90.

38. Campbell BJ, Hunter WW, Stutts JC. The use of economic incentives and education to modify safety belt use behavior of high school students. Health Educ 1984;15(5):30–3.

39. Geller ES. Motivating safety belt use with incentives: a critical review of the past and a look to the future. In: Advances in belt restraint systems: design, performance and usage. Warrendale, Pennsylvania: Society of Automotive Engineers, 1984:127–52.

40. Roberts M, Fanurik D, Wilson D. A community program to reward children's use of seat belts. Am J Community Psychol 1988;16:395–407.

41. Evans L. Factors controlling traffic crashes. J Appl Behav Sci 1987;23:201–18.

42. Van Houten R, Nau PA. Feedback interventions and driving speed: a parametric and comparative analysis. J Appl Behav Anal 1983;16:253–81.

43. Geller ES. Preventing injuries and death from vehicle crashes: encouraging belts and discouraging booze. In: Edwards J, Tindale RS, Heath L, Posavac E, eds. Social influence processes and prevention. New York: Plenum (in press).

44. Kolbe LJ. The application of health behavior research, health education, and health promotion. In: Gochman D, ed. Health behavior. New York: Plenum, 1988:387.

45. National Highway Traffic Safety Administration. A report to the Congress on the effect of motorcycle helmet use law repeal: a case for helmet use. Washington, DC: U.S. Department of Transportation; 1980.

46. Cranston A (D-Calif.). 121 Cong. Rec. 40261 (1975).

47. Gibbons v. Ogden, 22 U.S. (9 Wheaton) 1 (1824).

48. Commonwealth v. Alger, 61 Mass. (7 Cushing) 53 (1851).

49. Jacobson v. Massachusetts, 197 U.S. 11 (1905).

50. Teret SP, Gaare R. The law and the public's health. In: Gaare R, ed. BioLaw. Frederick, Maryland: University Publications of America; 1986:29–50.

51. Simon v. Sargent, 346F. Supp. 279 (Mass., 1972), affirmed in 409 U.S. 1020 (1972).

52. Childress JF. Priorities in biomedical ethics. Philadelphia: Westminster Press, 1981.

53. Beauchamp DE. The political theory of injury control. Presented at the Conference on Injury Prevention. Berkeley, California: Prevention Research Center, 1986.

54. Beauchamp DE. Community: the neglected tradition of public health. Briarcliff Manor, New York: Hastings Center, 1985:28–36.

55. Haddon W, Suchman EA, Klein D. Accident research: methods and approaches. New York: Harper and Row, 1964.

56. Sims JH, Baumann DD. The tornado threat: coping styles in the north and south. Science 1972:1386–91.

57. Waller JA, Klein D. Society, energy, and injury: inevitable triad? In: Research directions toward the reduction of injury. Washington, D.C.: U.S. Department of Health, Education, and Welfare, 1971.

58. Fischhoff B. Making decisions about AIDS. Presented at the Vermont Conference on Primary Prevention of Psychopathology. Burlington, Vermont: University of Vermont; 1988(Sep):6.

59. Weinstein ND. Unrealistic optimism about future events. J Pers Soc Psychol 1980;39:806–20.

60. Fischhoff B, Lichtenstein S, Slovic P, Derby S, Keeney R. Acceptable risk. London: Cambridge University Press, 1981.

61. MacLean D, ed. Values at risk. Totowa, New Jersey: Rowman and Allanheld, 1986.

62. National Academy of Sciences. Injury control. Washington, DC: National Academy Press, 1988:12.

63. Claybrook J et al. Retreat from safety. New York: Pantheon, 1984.

64. Hutchins VL. Remarks to the National Committee for Injury Prevention and Control. Washington, DC: National Committee for Injury Prevention and Control, 1987.

65. Brown ST. Remarks to the National Committee for Injury Prevention and Control. Washington, DC: National Committee for Injury Prevention and Control, 1987.

66. Rivara FP. Traumatic deaths of children in the United States: currently available prevention strategies. Pediatrics 1985;75:456–62.

67. Centers for Disease Control. Injury control: a review of the status and progress of the injury control program at the Centers for Disease Control. Washington, DC: National Academy Press, 1988.

68. U.S. Public Health Service. Promoting health, preventing disease: objectives for the nation. Washington, DC: U.S. Department of Health and Human Services, 1980.

69. Centers for Disease Control. Progress toward achieving the national 1990 objectives for injury prevention and control. Morbid Mortal Weekly Rep 1988;37:138–49.

70. Harrington C, Gallagher SS, Burgess LL, Guyer B. Injury prevention programs in state health departments: a national survey. Boston: Childhood Injury Prevention Resource Center, 1988.

71. Baker SP, O'Neill B, Karpf RS. The injury fact book. Lexington, Massachusetts: Lexington Books, 1984.

72. Haddon W. Options for the prevention of motor vehicle crash injury. Isr J Med 1980;16:45–68.

Part 1: The Process of Injury Prevention and Control

Chapter 1: Getting Started

The rescue of 18-month-old Jessica McClure from an abandoned well in the backyard of a Texas day-care center was one of the most gripping and poignant news stories of 1987. As hundreds of workers labored to release her, millions of American television viewers witnessed the drama, thanks to diligent coverage by the nation's three networks.

Michelle Snow's name is far less familiar to the public. She died at the age of 7 after being hit on the head with a foot-long lawn dart, a toy designed to be tossed at targets on the grass. Following her death, Michelle's father, David, undertook a crusade to have the product removed from the market. One year later, the Consumer Product Safety Commission voted to ban darts with metal tips by the end of 1988.

Less known still is Dr. Robert Spengler, assistant state epidemiologist in Vermont, whose scientific curiosity led him to study the costs and magnitude of injury in his state. Findings in hand, he is developing statewide activities to reduce the toll.

Periodically, an injury surges to a nation's or a community's attention, eliciting great concern and empathy. But only sporadically is the concern converted into a systematic, ongoing effort to reduce death and disability. These three stories illustrate more than the different ways injury emerges as a public concern; they illustrate the willingness and inspiration to respond that can make the difference between fleeting publicity and sustained impact on the problem. Jessica's town and nation were gripped by shared horror at her predicament, then relieved and appeased by the rescue. In contrast, the other two stories are about people who understood that injuries are not isolated or random events but predictable, preventable occurrences. Our nation is dotted with David Snows and Robert Spenglers, who see that individuals, communities, and governments can respond effectively to the injury problem. And they see themselves as the ones to make change happen.

This book is about crossing the bridge from recognizing that injuries can be prevented to creating programs that *work* to prevent them. There are many ways to respond to the injury problem. We focus on a programmatic response, a systematic approach to making injury prevention a part of community services. State and local health agencies are ideally placed to coordinate these efforts, and there is a variety of ways to accomplish this goal. This chapter introduces a process for developing injury programs that is neither a sequence nor a standard formula. It is a description of the ingredients of success that each community can mix to suit its own needs. These include assuming leadership, identifying the injury problem, devising a strategic plan to address it, translating the plan into action, and managing that action to achieve a reduction in injuries. We also discuss common obstacles to negotiating a way through this process successfully.

This chapter and the four that follow deal with the process of injury prevention and control. They are based on interviews with many experienced injury program managers, practitioners, and researchers. (A complete list of the interviews appears at the end of each chapter.)

Three points should be kept in mind:

- It is not necessary to reinvent the wheel. While the prevention of injury is not yet the nation's highest priority, programs across the country are thriving. There is a lot to be learned from the varied experiences of others.
- There are no quick and easy solutions. Becoming a leader in injury prevention and creating a public constituency can take several years, but running a program to prevent a *specific* injury is considerably more manageable.
- One person can make a difference. To date, advances in injury prevention can be credited more to individual leadership than to agency initiative. This chapter introduces individuals whose concern about injuries and deaths has led them to take action (S. Gallagher, personal communication).

INDIVIDUAL LEADERSHIP

What is the bridge between recognizing the problem and planning a program of action? Leadership. Injuries are a major public health problem that will not be resolved without concerted leadership and action. Creative and steady leadership, many believe, is even more valuable than money.

The seeds of what is now a nationwide enactment of child passenger safety seat laws were planted by one such leader. Tennessee pediatrician Robert S. Sanders described himself as "a guy in the right place at the right time to push and shove a bit."[1] Push and shove he did, for three years, until he managed to secure passage of the world's first child safety seat law in 1978. When he started out, he knew little about how to influence the legislative process. Today he is fondly called "Dr. Seatbelt" by legislators, lobbyists, and the news media. Dr. Sanders's strategy was a mix of varied approaches, persistence, and collaboration with many colleagues.

The same philosophy is at work in Vermont. When Robert Spengler saw that almost 80% of the deaths in the 5–24-year age group were from injuries, he began to plot his course of action.[2] An epidemiologist who had long specialized in cancer and birth defects, Spengler's interests shifted when he discovered the statistics and the immediacy of cause and effect in injury incidents. "I became intrigued when I saw the numbers. It was a dose of reality. Compared to other things I've dealt with, it was obvious that injury was a major public health problem" (personal communication). He presented his information at a health department seminar, generating enough interest to form Vermont Injury Prevention and Surveillance (VIPS), a group of more than a dozen health department directors who are interested in injury prevention. The state commissioner of health gave Spengler the go-ahead to involve others in the department and in the community, laying the groundwork for increased activity.

Rudolph Sutton of the Philadelphia Injury Prevention Program took a different approach. "We waited two years before we talked to our commissioner. We built a political constituency first—proving that there was a recognized problem—and then we met with the commissioner" (personal communication).

Keeping these injury prevention initiatives alive requires leaders who provide coordination, keep access to data and decision makers open, promote decisions based on those data, maintain enthusiasm, and remain sensitive to political realities. It is important to note that these leaders are self-selected; no one assigned them. In addition, the programs and activities spawned by these leaders were initiated before funding was secured.

When the injury prevention activist is a community member with no affiliation with an agency, the agencies nevertheless have an opportunity to be a partner and work cooperatively to create a systematic response, as is illustrated by David Snow's and Bob Sanders's efforts. An agency can arm activists with data, direct them toward worthwhile projects, and identify and help them overcome potential obstacles.

Individual leadership and activism can have many positive outcomes, as Michelle Snow's father demonstrated, but it alone cannot create a program. It must be bolstered by the support of an agency or community group, but one group cannot do it all. Injury prevention falls into the domains of many different professions. It is the concern of many in traffic and public safety, health, criminal justice, education, social service, engineering, and social science. To be successful, many groups must work together in a concerted and coordinated fashion over time. Piecemeal approaches are not effective.

AGENCY LEADERSHIP AND SUPPORT

Without a lead agency or organization, injury prevention efforts by assorted groups and individuals are often unfocused and duplicative. "Time, energy and money will be expended on isolated efforts, but the critical mass of resources needed to be effective will not be brought together."[3] A lead agency or organization can serve as a community or statewide focal point for injury prevention expertise, offering assistance and resources. It can also serve as a broker among the varied agencies and interest groups and can function as the source of energy and a manager of a coalition of resources.

In Colorado, the Injury Prevention Network (IPN), based at the department of health, is an example of a statewide group taking the lead. With members representing state agencies, universities, medical professionals, and businesses, the 145-member IPN works to educate citizens about injury prevention, finds data sources, and creates mechanisms to disseminate that data. Begun in 1982 with no financial support except in-kind contributions, the group is now funded (through federal prevention block grant dollars) as part of the state injury prevention and control program.

Various agencies can take the lead in injury prevention. In many states, the governor's representative for highway safety leads the effort to reduce traffic-related injuries and deaths. Joining forces with them can be extremely important and useful. In some states, the health department's division of maternal and child health leads childhood injury prevention efforts. In others, health promotion, epidemiology, environmental health, or emergency medical services divisions take the lead.[4] It is also possible for a community group or a nonprofit advocacy group to provide a home for a program.

"*Where* the injury control effort is placed is not a

critical issue in the start-up phase," said Hank Weiss, Director of Wisconsin's Injury Prevention Unit. "In the early stages, an injury control program should go where it is best sustained and supported."[5] State and local health departments may be the natural home for injury prevention and control, but they are, nationwide, facing a fiscal crisis.[6]

For six midwestern states, the momentum for injury prevention is coming from the regional office of the U.S. Public Health Service (USPHS). By sponsoring a regional workshop on strengthening injury prevention efforts, the USPHS hopes to prompt both in-state and regional commitment to the problem of injuries. A participant at that meeting, Shirley Reed-Randolph, Associate Director of the Office of Health Services in the Illinois Department of Public Health, has seized the opportunity. She now leads an injury control working group within the department that includes representatives from numerous divisions of the department. In addition, John Lumpkin, Associate Director of the Office of Health Regulation, has taken the lead to form a regionwide coalition on injury prevention. (For a discussion of coalitions, see Chapter 5.)

As these examples show, direction in injury prevention can come from the local, state, or regional level. Although the tasks of preventing injury may differ at each level, each can complement the others. For example, developing a statewide program to support injury prevention efforts is a job for a state agency, whereas carrying out projects at the local level can better be done by organizations that are closer to the community.[7]

Only with competent leadership can agencies perform their roles. Lead agency staff "must have the skills to organize political, community and professional leaders to use their influence and power to carry out injury prevention strategies."[3] Often, acquiring these skills comes only from experience.

IDENTIFYING THE PROBLEM

Problem identification is the process of deciding which injury you want to prevent in what population and with what resources. This requires gathering information that will be the raw material from which goals are set, objectives named, and interventions selected. The sequence of events in problem identification is not important; the accomplishment of the tasks is. Figure 1 illustrates the elements of problem identification. The tasks are to determine the nature of the injury problem, the characteristics of the population, the resources available to address it, the community's perception of the problem, and the political environment. Dis-

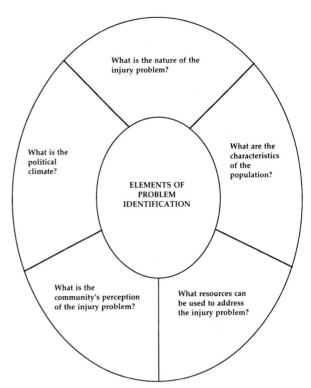

Figure 1. Problem identification.

played as a circle, this is a process that can be entered at any point.

What Is the Nature of the Injury Problem?

Any number of things can prompt a person or group to decide to take steps to prevent injury: compelling data, a well-publicized injury or death, the availability of federal funds, a directive from an informed supervisor, a call from a motivated citizen. Whatever prompts the initiative, data collection and analysis help determine which injuries need to be addressed. Data can also be used to convince community members and decision makers of the importance of the problem in their community.

Data can portray the overall injury problem or tell more about an injury problem aleady suspected, as happened in Dade County, Florida. There, the injury control program grew out of county rodent control inspectors' observations that poor housing conditions were jeopardizing the safety of residents. The health department collected data specifically on injuries caused by home hazards and confirmed the problem. In Connecticut, on the other hand, it was only through a broad investigation of death data that the Coalition to Prevent Childhood Injuries identified child homicide as a leading cause of death for certain age groups. It then was possible to target child homicide for further study and development of preventive efforts.[8]

An investigation of patterns of injury in a community can determine the following patterns:

• *Problematic injuries:* Which injuries occur most frequently and how severe are they? This provides a framework for deciding which injuries to target and suggests the organizations and interests that need to be involved.

• *Populations at risk:* Which groups experience a disproportionate number of target injuries? This is useful in selecting a target population.

• *Injury causes:* What circumstances contribute to the injuries? This provides a basis for selecting appropriate interventions.[9]

• *Injury costs:* What is the economic burden of injuries? Do the costs warrant spending money on prevention and control? Who pays the costs?

"In order to avoid wasting time and resources on an extensive data collection effort, be sure to search out what is already known about . . . injuries in your community."[3] This will provide indications of what additional local data might be needed to understand your community's injury burden. Chapter 2, "Learning from Data," and Chapter 3, "Working with Data," discuss local and national data sources and their uses.

Data from published studies and reports are valuable, but nothing is more compelling than current, accurate local data. The Missouri Division of Health's burn prevention program, created in 1969, is an example of matching a problem with a solution. The data collection, done through at-home interviews with community residents, uncovered who was being burned, why and how they were burned, the extent of their injuries, what kind of first aid was received, and what some possible solutions might be. "When these data were combined with knowledge of the accident processes encountered, it was hoped that light would be shed on the causes, effects, and solutions of the problems of fire and burning injuries. In retrospect the information obtained was invaluable in planning, development, and execution of a prevention and control program."[10]

What Is the Community's Perception of the Injury Problem?

Data can tell a lot, but they cannot determine which injuries most concern people or compel them to act. Only a thorough understanding of the community will provide that. "Too often, health programs are 'dropped' into the middle of a community with little attempt to understand the political, sociocultural, and economic environment into which they must fit if they are to survive and to effect their goals."[11]

Community outreach "involves the creation and maintenance of open communications pathways between the program and its target groups and between the program and surrounding organizations."[11]

What if data reveal that falls are the leading cause of injury in young children but the community is up in arms over the more publicized (though less frequent) problem of school bus crashes? "Lend an ear. Sometimes you have to work on issues that your data show are not the most serious ones if you want the community to accept the concept of injury prevention and have confidence in your agency," said Susan Gallagher, Director of the Harvard Childhood Injury Prevention Resource Center (personal communication). By using people's perceptions of problems creatively, it is possible to educate them about injury. The community's concern for one type of injury can be used as an entry to explaining the toll of all other injuries. Although Jessica McClure's injury is not a common one, for example, it could have been used to illustrate children's high risk for injury or day-care safety issues and to develop interest in prevention programs. "Setting priorities for local injury programs means more than counting numbers to determine frequency and severity. Concerns about any kind of injury can lay the groundwork for a focus on more critical injury problems in the community," said Gallagher (personal communication). In addition, a community needs assessment can benefit an agency by raising its visibility and generating understanding of and support for other needed programs.[12] Telephone surveys can be useful tools in conducting a needs assessment.

The Philadelphia Injury Prevention Program (PIPP), a comprehensive injury control program based in the Philadelphia Department of Public Health and funded by the city and the Centers for Disease Control, made community assessment a priority from the start. Amy Wishner, Coordinator of Surveillance and Prevention for PIPP, reported, "People in our community were interested in pedestrian injury out of proportion to what our data were showing. Nonetheless, we devoted a fair amount of attention to the problem. Because it is so visible, unlike falls, stabs, or assaults, it influenced the direction we took" (personal communication). To take best advantage of community interest in preventing a certain kind of injury, some background work should be done early. Certain information should be available, such as data about the community and ideas about what program(s) would work.

Along with collecting facts, it is important to discover what concerns engage and motivate the community. "We spoke with anyone who had done community work. We called people in other health department divisions to learn about their experiences in the community and obtain community contacts. Then we went out and met those people, explained what we wanted to do, got feedback, and asked for more names of people to talk to," reported Wishner. "Community support for injury prevention programs is not only advisable, it is essential. . . . Our ultimate goal was helping people to help themselves. People were responsive to this." Wishner said they emphasized local control over the program. "We said, 'We can't do it without your participation. If you don't want it, it's not going to be here.' It was important to be explicit that they had that control" (personal communication).

It is important to find out what kinds of injury prevention projects have been tried before in the community, whether they have worked, and whether they are still in place. How were the projects received? Are residents aware of effective injury prevention methods? It is also important to learn about the community through the local newspaper, perhaps by clipping stories that illustrate both the community's injury problem and the need for more information about it.

Ideally, injury prevention programs should be preceded by an analysis of what the target population knows about injuries and their prevention; which individuals and groups are at the highest risk for what injuries; what types of injuries occur; and when, where, and under what circumstances they occur. By comparing data over time, one can observe changing injury patterns, determine whether interventions have affected those patterns, and perhaps identify alternate interventions.

At the same time, subjecting a community's heartfelt concern about a certain kind of injury to time-consuming data analysis can extinguish the initiative to commit resources and time to the program. It is important to strike a balance, designing the program carefully but also putting it into effect as rapidly as possible.

What Resources Can Be Used to Address the Injury Problem?

In the early 1980s, traffic safety was a serious problem in San Antonio, Texas. City authorities decided that both the government and citizens must play a role in improving safety. "Target '90—Goals for San Antonio," a committee formed by Mayor Henry Cisneros, recommended that the lead be taken by an existing grass-roots organization, Community for Automobile Responsibility and Safety (CARS). With CARS coordinating a multifaceted program (covering drunken and drugged driving, safety belt usage, pedestrian safety, motorcycle safety, and courteous driving), an extremely broad range of city groups is active. Collaborators include the police department, schools, municipal offices, military bases, courts, the state department of transportation and public highways, numerous private businesses, and health and civic organizations.

Involving the community and getting to know its resources is a key step in getting an injury prevention program off the ground. The first tasks are to become acquainted with others who are interested or active in injury prevention in local, state, and federal government, in other community organizations, and in the private sector. Generate their support. Use the relationships to advance or mobilize an injury prevention program, and locate existing data sources and legislation to prevent injuries. Find out what other agencies have pursued or are pursuing. In some cases, agencies can co-apply for funding. Figure 2 lists some of the agencies that should be contacted.

It is essential to make contact with more than the obvious safety and government groups. Talking to individuals such as the community librarian or a service club president can be helpful. He or she can identify the "go-getters," not just the titular leaders. "Discover the informal community power structure; key leaders may be found in unlikely places."[9]

Physicians, especially pediatricians, can be influential in building community commitment to injury prevention. "As an advocate for child safety, the pediatrician can use professional credibility and influence to mobilize community efforts in injury prevention. The pediatrician may take a leadership role and initiate a program, or may provide support to established groups in reaching common goals."[13]

If influential leaders have not been included in the process from the beginning and are not sympathetic to the problem, they could sabotage the effort. For example, in one New England town a powerful local general practitioner, who had not been contacted personally at the initiation of a community injury program sponsored by the state, opposed the program at every turn, insisting that the community did not have an injury problem.

Another task is to find out where financial resources exist and whether there are resources that have never been tapped or ones that have been solicited but rejected. Similarly, it is wise to conduct an inventory of the community's media outlets, in-

Fire departments	State alcohol and drug abuse authority
Police departments	Children's services agencies
Elderly services agencies	Red Cross
Hospitals	Poison control centers
Emergency medical services	Local businesses, churches, and labor groups
Group medical practices	Parent/teacher associations
Voluntary agencies	American Academy of Pediatrics' chair for Accident and Poison Prevention Committee
School nurses and teachers	
Professional organizations for physicians, nurses, health care workers	Health maintenance organizations
Medical schools	Insurance companies
Schools of public health	Social and civic groups
Schools of nursing	Community officials (city council, mayors, legislators, judges)
Pharmacists	
Health and other educators	Legislators
Department or board of health	Architects
Department of motor vehicles	Engineers
Department of parks and recreation	Mothers Against Drunk Driving (MADD), Students Against Drunken Driving (SADD) chapters
Department of social services	
Department of transportation	Regional offices of federal agencies: U.S. Consumer Product Safety Commission, National Highway Traffic Safety Administration, U.S. Public Health Service
Department of mental health	
Department of education	

Figure 2. Possible collaborators in injury prevention.

cluding newspapers and radio, television, and cable television stations. It is important to build relationships with these organizations early on.

Has a community task force or coalition been assembled around a specific injury cause, such as drunken driving? If so, have they made recommendations? Find out if the community has accepted or rejected injury prevention efforts in the past, and why.

What Are the Characteristics of the Population?

The demographic, geographic, and economic makeup of a community guides many program decisions. Learn as much as possible about the population, its environment, the various neighborhoods, and where residents receive medical care. Questions to ask: What percentage of the population is elderly people and children? What services are available to them? Are there particular places or intersections where many injuries of a certain type

occur? Are there streets with a high incidence of motor vehicle crashes? What occupations are common in the community, and what are the hazards associated with them? Do specific recreational hazards exist? Does the condition of housing contribute to injuries? Does a common transportation mode or weather condition create a special risk?

What Is the Political Climate?

It is essential to understand how public policy is formulated in the community. As a public health school dean noted more than 35 years ago,

"We must have knowledge . . . of community organization, of the power structure of that community, of the political structure, of health laws and regulations, of attitudes that determine acceptance or rejection of change and development. We must have sophisticated knowledge of education and educational methods, of mores and morals that affect the growth and development of community consciousness and community action."[14]

There are several questions to pose: Who in the community can influence and authorize public action on or point media attention to the injury problem? Who can seek funds, who has the power to allocate funds, and how can those decisions be influenced? Which would be an effective lead agency? Who are the political, corporate, religious, and social/civic leaders in the community? Who can reach them to enlist their support? Is there relevant injury prevention legislation?

Although vital, progressive programs can thrive (albeit with harder work) without the active support of a commissioner, department head, or other supervisor, their support is an important political asset. One injury prevention worker in a state health department said that her commissioner's sentiment about her federally funded injury prevention initiative was, "If you want to do it, go ahead." He would sign key letters, proposals, or reports, but he would never name injury as a priority or fund it. The lack of support did not impede the program as long as it had outside funding, but after that support ended, his marginal interest could be damaging. However, by taking even this degree of support and running with it, a staff member can create a program useful and popular enough to get continued funding. Successful efforts often need support from more than one funding source. Leadership entails persistent, creative persuasion of colleagues, supervisors, the media, and legislators about the importance of injury prevention.

THE SYSTEMS APPROACH TO DEVELOPING A STRATEGIC PLAN

The information collected to identify the injury problem and understand the community can also be used to develop a strategic plan to control and prevent injuries. Because a successful program will require change on the part of individuals, agencies, and environments, the program must address all three. "Injury control demands a systems approach because of the very nature of the multiple, concurrent actions that must be taken together by both individuals and agencies."[3]

The systems approach reflects a recognition of the comprehensive nature of the injury problem and the need to develop equally comprehensive solutions. Neither new nor complex in theory, this approach has been used in campaigns to reduce infectious disease and alcohol-impaired driving.

A good model for injury prevention comes from the development of trauma care systems. Their development, beginning in the 1970s, grew from the recognition that effective, often life-saving, care could not be rendered if the hospital were the sole focus of attention. Specialized trauma care centers or units could not be an end in themselves because "in the absence of a system you end up with no control over who gets to the trauma center or how" (R. Cales, personal communication). Therefore, these well-staffed, well-equipped facilities "are of little use . . . unless they are integrated into a trauma care system that provides continuous treatment from the moment of injury through discharge or death."[15]

In providing effective trauma care, the systems approach dictates that a lead agency bring together and coordinate institutions and agencies involved in communications, personnel training, transportation, and hospital management. A statewide office, such as the health department, may often be the leader in some injury prevention efforts but at other times will be called on to collaborate with other lead agencies.

We will learn about trauma care systems in Chapter 18, but the lesson for injury prevention and control is clear. In developing a program, it is critical to draw on and coordinate all the institutions and individuals recognized in the problem identification phase. In this way, even a small, highly focused program can incorporate the comprehensive view that injury prevention requires for success.[16,17]

At both the state and local levels, collaborators in an injury prevention effort must perform eight essential functions:

• Draw on the cooperation of a multidisciplinary group of community members.
• Coordinate existing local and state resources for injury prevention.
• Initiate the examination of injury data and support prevention and control strategies (including support for legislation).
• Create a statewide plan to promote the development of injury control programs at the local level.
• Provide initial state funding of programs based in local communities when possible and help identify other sources of funding.
• Provide technical assistance and training to involved groups.
• Stimulate injury prevention research and training.
• Be aware of the incidence and distribution of injuries over time.

Public health departments increasingly are taking a lead role in injury prevention. Their strengths are as follows:

• *Conveners* bring together groups and experts interested in injury prevention and control.

• *Consensus builders* help these groups make decisions about approach and focus.

• *Communicators* get the results of research and data analysis out to those who can translate it to action.

In planning, be creative about ways to "piggyback" on an existing program. This can save money, assure access to a population, and get things moving quickly. "If you make injury prevention a separate function from what everyone else normally does, it will disappear when program funding dries up," noted Susan Gallagher (personal communication). As described earlier, the Dade County Health Department's Division of Environmental Health started one of the nation's first injury prevention programs by applying the knowledge and resources of its community-based rodent control program. Its door-to-door survey to discover the level of hazards was considerably easier to do because environmental health staff were known to the community and were welcomed into residents' homes. Survey findings narrowed the program's focus to preventing home fires and tap water scalds. Preventive measures, including a smoke detector giveaway and installation program and routinely lowering tap water temperatures, were incorporated into the staff's regular home visits.

An affiliation between the La Crosse (Wisconsin) County Health Department and the Hmong Mutual Assistance Association made possible a state-funded injury prevention program designed to benefit the large population of Hmong people from Laos who have settled in La Crosse. The mutual assistance association, which over the years has worked with the county health department on a health screening program, turned to the county when it realized that housing conditions—faulty wiring, poorly lit stairwells, and general disrepair —were putting the refugee population at risk for injury. At the association's request, the county health department secured funding from the state division of health to form the Environmental Health Education Program. Through home visits, refugees are educated about home management and safety practices. "We had participated in mutual programs with the county health department in the past and also had continued to discuss with them periodically the needs of the Hmong population," said Denis Lee Tucker, Executive Associate Director of the Hmong Mutual Assistance Association (personal communication).

A written plan helps guide a cooperative effort. It should reflect the information gathered about the community and its members and the specific inju-

ries, hazards, and target groups. A written plan holds the program accountable and ensures that steps are not taken randomly. It also provides a basis for evaluation and modification of program efforts. "The existence of a concrete plan with specific objectives and timelines helps to ensure systematic and ongoing attention to community outreach efforts—which, unfortunately, tend to get 'lost in the shuffle' of day-to-day program implementation."[11]

FUNDING AND STAFFING

Because decision makers, practitioners, and the public have only a limited awareness of the magnitude of the injury problem—and are often unaware of the science of injury prevention—injury prevention has not received funding, either private or public, commensurate with the severity of the problem. To date, few federal start-up funds have been replaced or supplemented by ongoing state or private support. While funding issues vary for each program, some general principles about soliciting funds from both private and public sources have emerged.

Injury prevention professionals can often find initial funding in the form of "in-kind" contributions. "You can start by making injury prevention a part-time piece of somebody's job. With that step, you've given the effort a name and a permanent address. That's the key to getting the whole thing started," said Kristine Gebbie, Administrator of the Oregon Health Division (personal communication). Contributions of services in kind, such as printing, use of mailing labels, use of meeting facilities, and lending of letterhead or stature in the community for visibility and presence, are also valuable. In Colorado, the injury prevention program did not have money for its own newsletter but was able to persuade another program to publish injury prevention information in its publication.

In evaluating the start-up of nine community injury prevention programs, the Massachusetts Statewide Comprehensive Injury Prevention Program (SCIPP) cited staffing issues as "the most critical and most troublesome."[18] Either existing staff were being stretched too far, or hiring new staff was excessively time-consuming. Thus, "developing a sense of 'ownership' or investment in injury prevention activities—particularly for those staff whose main responsibilities lie elsewhere—is a challenging yet essential task."[18] Toward this end, SCIPP recommended drawing on staff expertise, involving them in decisions, and asking for their feedback.

The number of paid staff needed by an injury

prevention program varies depending on the size of the community, the scope of the problem, and the resources available from other agencies or volunteer groups. Although there is no prescription for minimum staffing levels, expertise is certainly needed in program coordination and injury data surveillance.[16]

In injury prevention, a little bit of money can go a long way. The New England Network to Prevent Childhood Injuries funds special injury prevention projects based at the health departments of each of the six member states. While the average annual funding level of $12,500 per state is modest, it has given states the means and incentive to begin analyzing their data and implementing interventions based on their findings and can lead to broader activities. In Rhode Island, for example, a network-funded project was supplemented by a $35,000 grant from the state legislature to pursue three additional approaches to the study of suicide.

Because injury prevention is not broadly thought of as a public health issue, it is especially vulnerable to the vagaries of government and private funding. States do not allocate resources by data alone and consequently do not give injury prevention programming top priority. Often they must divide scarce federal and state funds among numerous competing programs that are promoted through the media or by concerned advocacy and citizen groups. In addition, changes in leadership can wipe out months or even years of cultivating the interest of a decision maker. And finally, it is important to note that this low ranking of injury and lack of funding coincide with a public health funding crisis that shows no sign of letting up and affects many concerns other than injury prevention.

One way to change the tendency of state and local health departments to rank injury prevention low is to question funding priorities. Ask that they be based on measures such as years of potential life lost, mortality rate, morbidity rate, costs of the conditions affected, effectiveness of the intervention, costs of the intervention (as opposed to costs of nonintervention), and availability of other general or federal funds.

Through a state's distribution of federal block grant money, a state health department may be able to offer incentives to include injury prevention to local agencies. The Wisconsin Health Department offered grants to local health departments with the stipulation that they address one of the department's priorities, injury being one of them. "The state health department can set priorities and bring injury prevention to the attention of local agencies, especially when dollars are attached. Without in-

jury prevention listed as a priority, I believe local health departments would have been much less likely to work in that area," said Hank Weiss, Director of the Injury Prevention Unit (personal communication).

Several other injury prevention programs across the country have funding success stories to share. On a large scale, the North Carolina legislature recently appropriated $1.2 million for a statewide Health Promotion and Disease Prevention Program for fiscal year 1988. Almost all of this money was awarded to local health departments and other agencies to establish prevention programs, with injury, cardiovascular disease, and cancer identified as areas to address. The first steps toward this request were taken three years ago. By grouping injuries with two other major causes of death, supporters of the bill were able to secure funding for all three areas.

This movement in North Carolina began with pediatrician and county health director Dr. Thad Wester, who had been active in injury prevention in his county for many years. In 1985 Wester convinced his state legislator to introduce a bill to form a legislative study committee on health promotion and disease prevention in North Carolina. Over several months, the committee heard testimony from dozens of public health experts. Supporters of the bill wrote letters and made telephone and personal contact with legislators on the study committee. In 1987 the legislature approved the funding measure. Communities across the state submitted competitive bids to use these funds to support health promotion/injury prevention programs.

On a smaller scale, many community-based programs have been just as successful in persuading decision makers to allocate funds for injury prevention. When state funding ran out for a multicity lead poisoning prevention program in southeastern Massachusetts (which also included other injuries in its prevention efforts), Carmen Maiocco, director, turned to the city office of community development with a funding request. That year, the New Bedford–based office funded the non-lead aspects of the injury prevention program at a level of $3,300; the next year, the office in neighboring Fall River awarded similar funding.

Identifying and pursuing private and public sources of continuing program support beyond initial funding presents a major challenge to most community-based and state-level programs. Limited federal and state funds are "forcing health departments to explore alternative sources of funding for injury control programs, including cooperative ventures with other public and private organizations and programs supported by grants."[19]

The key to securing ongoing funding lies in careful identification of potential funders followed by persistent hard work to convince them of the match between the program and their mission. A successful search for funds for health promotion projects depends on several key factors: a skilled proposal writer, a comprehensive plan of action, letters of support from key persons and organizations, familiarity with sources of funds, and an understanding of the various ways to apply for funds.[20]

While fund-raising may not be easy or pleasant, injury prevention professionals should try to tap into the dollars available from both government and private sources for health research and programming. Ideas should be prepared and cooperative agreements forged before funding availability is announced. As a rule, government funds are usually granted with strict application requirements and controls; private funds tend to allow greater flexibility.[20]

The target audience—the group the program is designed to reach—may trigger ideas about potential donors. If preventing medicinal poisonings among children under the age of 5 in a specific county is a goal, fund-raising might target local businesses that cater to children, pharmaceutical companies, civic groups, and county government agencies. Before applying for funds, document the following:

• Why is the program needed? Include data on the problem and a description of the target population. If possible, document the costs of injury, either as one individual's injury costs or as the annual state costs for a category of injury.

• Who in the community supports the program? Are there strong bipartisan relationships with key organizations and individuals?

• Is the program appropriate for the organization's size, staffing, and previous involvement?

• Be as specific as possible about what is needed. For example, "We need $3,000 to purchase smoke detectors that will be made available on a sliding scale fee basis to health clinic clients."

Other factors to consider include:

• What groups or businesses might supply funds? Learn as much as possible about potential funders. Do they have funds available? What are their priorities and eligibility requirements?

• What are the goals of the agency or foundation?

• Is it possible to link up with an existing program?[20]

• Define feasible and measurable objectives and describe how the program will be evaluated.[21,22]

Public Funding

To request program funding from most of the agencies listed in Figure 3, prepare a formal written proposal. If you do not have experience developing proposals, consult with someone in your agency or a collaborating agency. In addition, numerous publications provide information on seeking program funding.[21]

Public funding varies over time, and state and federal agencies should be contacted for current program initiatives. National conferences on injury prevention such as those sponsored by the Centers for Disease Control can provide useful, current information on funding opportunities. Publications such as the *Federal Register* and *Commerce Business Daily* contain announcements of future funding availability.

Private Funding

Private sources of funds can be sought from foundations, businesses, health associations and facilities, churches, educational institutions, and community groups. Before developing a proposal for a potential funder, inquire about the funder's re-

National Level
• **Department of Health and Human Services**
  • **National Institute on Alcohol Abuse and Alcoholism**
  • **Alcohol, Drug Abuse, and Mental Health Administration**
  • **Centers for Disease Control**
  • **Office of Disease Prevention and Health Promotion**
  • **Health Resources and Services Administration**
  • **Bureau of Maternal and Child Health and Resources Development**
  • **Office of the Assistant Secretary for Health**
  • **Office of Minority Health**
• **Department of Transportation: National Highway Traffic Safety Administration**
• **Consumer Product Safety Commission**
• **Department of Justice**
• **Department of Education**
State Level
• **Health department**
• **Office of highway safety**
• **Department of education**
• **Department of mental health**

Figure 3. Potential funders of injury prevention programs.

quirements and procedures. A foundation's annual report will describe its concerns and priorities.

One useful resource is the Foundation Center, a clearinghouse on all the nation's foundations. The nearest foundation library can be located by calling (800) 424-9836. The center publishes several directories, including *The Foundation Directory*, which can be used to identify foundations that are concerned with issues that match your program. This directory may be available through a local university or public library.

CHALLENGES FACING INJURY PREVENTION PROGRAMS

Launching an effective injury prevention program is a challenging task that demands perseverance, commitment by an individual or agency, investigation, creativity, and sometimes a fair amount of good luck. Not surprisingly, practitioners often stumble over some common obstacles, regardless of the community. Drawing from the knowledge and expertise of seasoned injury prevention professionals, the following section examines these challenges and their potential solutions.[23] Chapters 2 through 5 address the solution to these problems at the program level.

Challenge No. 1: Key decision makers and the general public may be unaware of the magnitude of the injury problem in comparison with other public health problems.

The lack of information—and abundance of misinformation—about the injury problem and its preventability were alluded to in the introduction to this book. Getting public and political support for lifesaving measures such as mandatory safety belt use laws requires broad education on the problems and the benefits to be gained by such measures.

Potential solutions: As Chapters 2 and 3 suggest, it is crucial to translate injury morbidity and mortality statistics into easily understandable formats and share the data with agency officials and the community at large. The data are compelling: as the major killer of all Americans between the ages of 1 and 44, injury is a public health problem of enormous proportions. But data cannot speak. Leadership involves becoming a spokesperson with command of the information and an ability to persuade others of the need to act immediately to prevent injuries.

In some agencies, injury prevention program directors have difficulty gaining access to those who make decisions about the allocation of agency funds. In this situation, citizen advocacy groups or individual persons can be encouraged to request a meeting with the agency head to discuss the community's injury problem. Advocacy group members can also speak openly about their concern that the agency devote adequate resources to injury prevention, for example at the annual public hearings on the allocation and priorities for state block grant funds. Prominent community members or others who can supply case studies, such as physicians, can be especially persuasive in meetings.

There are many formats for summarizing injury data. One useful technique (especially for childhood injuries) is to compare rates of injury morbidity and mortality to other causes of disease and death. Statistics of years of potential life lost as a result of injuries are also compelling. (See Chapters 2 and 4 for suggestions on appropriate formats for presenting injury data.)

Combining hard data on injury incidence and the costs of injuries with anecdotes about real families whose lives have been touched by injury is another technique for conveying the impact of injury on society and on the individual. The news media's role in this effort is discussed in Chapter 5.

Challenge No. 2: Many injury prevention programs suffer from a lack of timely, high-quality data on nonfatal injuries in their community.

Chapter 2 explores the problems associated with obtaining accurate community-based data on injury morbidity. Without a true picture of the community's fatal *and* nonfatal injury problem, program planners cannot design appropriate and effective prevention efforts.

Potential solutions: For most injury causes, the best source of morbidity information is the admission and emergency room data of local hospitals and police reports. However, if discharge data do not contain information on external cause of injury codes (E codes) and emergency room logs do not include the cause of the injury, their usefulness is extremely limited. The efforts of injury prevention practitioners to mandate such coding and to improve record keeping are discussed in Chapter 2. Emphasizing the lack of usable data may capture the interest of a decision maker.

In some instances, it is feasible for a community to extrapolate from state or national morbidity data to form a picture of the local injury problem. See Chapter 2 for more information on extrapolating from other data sources.

Challenge No. 3: Well-documented, effective interventions are lacking for some injury causes (e.g., suicide) and populations (e.g., adolescents).

Because the science of injury prevention and control is relatively young and the injury problem so complex, we do not have well-tested interventions for every cause and population. As a result, some programs face the task of designing and implementing untested countermeasures, and decision makers are reluctant to allocate resources for such efforts.

Potential solutions: Chapters 6–17 review the state of the art in injury prevention. Practitioners can also contact experienced injury prevention researchers and practitioners to find out what they know about intervention efforts that have not been summarized in the literature. In addition, researchers and practitioners from other fields who have dealt with behavior change for other health problems are a valuable resource.

Informal networking can help identify little-known programs that may have been evaluated and shown to be effective but whose results were never written up or were documented but not disseminated. (The large number of such examples emphasizes the necessity for *all* programs to document and disseminate the findings of their evaluation efforts.)

If both formal and informal research turns up no appropriate or comparable interventions, it is necessary to develop a list of potential interventions, drawing on the techniques laid out in Chapter 4. Brainstorming works best when the group includes people with slightly different perspectives on the issue. For example, to create a list of potential interventions for adolescent suicide, one could convene an ad hoc committee of individuals from the following fields: injury prevention, developmental psychology, adolescent medicine, mental health, child welfare, youth services, education, criminal justice, and advertising/media relations.[24] Chapter 4 also describes how to reduce the resulting list of interventions to the few that are most appropriate and most likely to succeed with the target population.

Challenge No. 4: Our society glorifies risk taking; safety has a poor public image.

In many segments of the population (especially among young males), it is considered heroic to put oneself at risk of injury. In many communities, other values supersede reduction of injury, and thus it is difficult to motivate community members to learn about injuries and change their behavior. In some settings, taking risks with one's own safety and that of others by carrying a loaded handgun, driving after drinking, or speeding is the norm. In the face of such attitudes, the effectiveness of many interventions (be they legislation/enforcement, education/behavior change, or engineering/technology) is greatly hampered.

Potential solutions: As stated in the introduction, legislation/enforcement and engineering/technology interventions work only indirectly to change attitudes. Education/behavior change strategies may be the most successful approach over the long term to challenge the notion that it is courageous to expose oneself to injury. A combination of approaches is needed to reduce injury and change attitudes.

Education about injury risk and the importance of risk-reduction behaviors is a gradual process. An analogy can be drawn to the anti-drunken-driving campaign that began in this country in the late 1960s. Even with the broad-based support expressed in the 1980s from scores of public and private institutions, it has taken this intense, large-scale educational effort (which drew heavily on grass-roots support and the media) two decades to convince many Americans that drinking and driving is dangerous, unglamorous behavior. Legislation and regulation in those years have also helped to shape Americans' attitudes.

Challenge No. 5: In most communities, few professionals are trained in injury epidemiology and injury prevention.

Especially outside major metropolitan areas, professionals who are knowledgeable about injuries and committed to prevention often feel as if they are working in a vacuum; their colleagues are simply uninterested in or unprepared to act on the injury problem.

Potential solutions: To increase the number of professionals capable of addressing injuries, the subject of injury epidemiology and injury prevention must be integrated into the curricula of a wide variety of disciplines. One recent report recommends that such education be incorporated into the training of health care workers (especially physicians and nurses), public safety officials (especially traffic safety and police officers and firefighters), teachers, early childhood educators, engineers, architects, and city planners.[16]

Few educational institutions now offer training in the study of injuries and their prevention. Many people now working in injury prevention and control report that they learned about injury epidemiology and prevention on the job. If formal education is to cover injury prevention, administrators and professors in undergraduate and graduate programs must become informed about injuries and have access to written and audiovisual teaching materials. Efforts are now under way to produce such materials and pilot-test them in a number of different educational settings.[25] There is also a need for continuing education courses on injuries for professionals who have already completed their training.

Creation of an injury prevention forum through an agency, association, or medical society can reduce isolation and increase mutual support. Newcomers to injury prevention can draw on the knowledge of experienced professionals in the field. One source is the five injury prevention research centers funded by the Centers for Disease Control (see Appendix F). Conferences, summer courses, and personal contacts are also useful.[26]

Challenge No. 6: The survival of many injury prevention programs is threatened by uncertain funding and/or the absence of support from agency heads or the community.

Ongoing financial support of and commitment to an injury prevention program are not easy to attain. Once achieved, they are maintained only through hard work.

Potential solutions: The process of making injury prevention an ongoing commitment varies widely. Some programs place great emphasis on building grass-roots community support, while others focus on factors within the lead agency that affect program funding. In the long run, both are important. Chapter 5 addresses the steps that can be taken to achieve long-term survival for injury prevention.

SUMMARY

This chapter describes the process of assessing the "big picture" of the injury problem in a community. Subsequent chapters in this section narrow that scope to building individual injury prevention programs. The main points to remember in starting a program are:

• Leadership, at both the individual and agency levels, is the bridge between recognizing that injury is a problem and planning a program of prevention.

• The tasks of leaders and lead agencies are to understand the nature of the community's injury problem, the characteristics of the population, the resources available to address the injury problem, the community's perception of the problem, and the political climate.

• A successful prevention program demands a systems approach involving individuals, agencies, and environments.

• Injury prevention professionals can often find initial funding in the form of in-kind contributions.

• Injury prevention professionals, facing inadequate funding at all levels, must call for a reassessment of funding priorities.

INTERVIEW SOURCES

Richard Cales, MD, Chief, Emergency Services, San Francisco General Hospital, California, September 2, 1988.

Susan Gallagher, MPH, Director, Harvard Childhood Injury Prevention Resource Center, Boston, Massachusetts, June 30, 1988.

Kristine Gebbie, MSN, Administrator, Oregon Health Division, Portland, July 7, 1988.

Robert Spengler, MD, Assistant State Epidemiologist, Montpelier, Vermont, August 10, 1988.

Rudolph L. Sutton, MPH, Philadelphia Injury Prevention Program, Pennsylvania, September 14, 1988.

Denis Lee Tucker, Executive Associate Director, Hmong Mutual Assistance Association, La Crosse, Wisconsin, July 1988.

Hank Weiss, MS, MPH, Director, Wisconsin Injury Prevention Unit, Madison, October 5, 1988.

Amy Wishner, RN, MSN, Coordinator of Surveillance and Prevention, Philadelphia Injury Prevention Project, Pennsylvania, July 1988.

REFERENCES

1. Huston P. He has done more to save lives . . . Contemp Pediatr 1988;5:84–94.

2. Vermont Department of Health. Injury related deaths and hospitalizations in Vermont. Burlington, Vermont: Vermont Department of Health, 1988.

3. Micik S, Yuwiler J, Walker C. Preventing childhood injuries: a guide for public health services. In: Childhood injury prevention project, North County Health Services. 2nd ed. San Marcos, California: North County Health Services, 1987:28.

4. Harrington C, Gallagher S, Burgess L, Guyer B. Injury prevention programs in state health departments: a national survey. In: Childhood Injury Prevention Resource Center, Harvard School of Public Health. Boston: Harvard School of Public Health, 1988:11.

5. Weiss H, Schmidt W. Developing a statewide injury prevention program: the Wisconsin experience—past, present and future. Paper presented at the 114th annual meeting of the American Public Health Association. Las Vegas, Nevada: American Public Health Association, 1986:4.

6. Miller CA, Moos MK. Local health departments: fifteen case studies. Chapel Hill, North Carolina: American Public Health Association, 1981:xi.

7. Institute of Medicine, Division of Health Care Services, Committee for the Study of the Future of Public Health. The future of public health. Washington, DC: National Academy Press, 1988:8–9.

8. Lapidus G. Child homicide in Connecticut. Hartford, Connecticut: Connecticut Coalition to Prevent Childhood Injuries, 1987.

9. Birch & Davis Associates, Inc. Developing childhood injury prevention programs: an administrative guide for state maternal and child health (Title V) programs. Silver Spring, Maryland: U.S. Department of Health and Human Services, Division of Maternal and Child Health, 1983:4.

10. Missouri Division of Health. A community action approach for prevention of burn injuries. Washington, DC: U.S. Government Printing Office, 1969; DHEW publication no. (HSM)72-10008.

11. Dean D. Community outreach for health education programs, a practitioner's manual [Dissertation]. Boston: Statewide Childhood Injury Prevention Program, 1981.

12. Samms D. Benefits of a community needs assessment. Am J Public Health 1988;78:850–1.

13. Micik SH, Alpert JJ. The pediatrician as advocate. Pediatr Clin North Am 1985;31:243–9.

14. McGavran EG. What is public health? Can J Public Health 1953;44:446.

15. Cales RH. Concepts. In: Cales RH, Heilig RW, eds. Trauma care systems: a guide to planning, implementation, operation, and evaluation. Rockville, Maryland: Aspen Publishers, 1986:14.

16. Injury prevention in New England. Boston: New England Conference on Health Promotion and Illness Prevention, 1985.

17. Micik SH, Miclette M. Injury prevention in the community: a systems approach. Pediatr Clin North Am 32:251–65.

18. Statewide Comprehensive Injury Prevention Program. Report of the comprehensive injury prevention programs. Boston: Massachusetts Department of Public Health, 1987.

19. Rosenberg ML. 1987 conference on injury in America: a summary. Public Health Rep 1987;102.

20. U.S. Department of Health and Human Services, Office of Disease Prevention and Health Promotion. Locating funds for health promotion projects. 3rd ed. Washington, DC: U.S. Government Printing Office, 1988.

21. Breckon DJ, Harvey JR, Lancaster RB. Acquisition and management of grants. In: Community health education: settings, roles, and skills. Rockville, Maryland: Aspen Systems, 1985.

22. Breckon DJ, Harvey JR, Lancaster RB. Community fundraising. In: Community health education: setings, roles, and skills. Rockville, Maryland: Aspen Systems, 1985.

23. Gallagher SS, Messenger KP, Guyer B. State and local responses to children's injuries: the Massachusetts statewide childhood injury prevention program. J Social Issues 1987;43:149–62.

24. CDC recommendations for a community plan for the prevention and containment of suicide clusters. Morbid Mortal Weekly Rep 1988;37:S-6.

25. Educating professionals in injury control: a multi-disciplinary approach. Newton, Massachusetts: Education Development Center, 1987.

26. New England Network to Prevent Childhood Injuries. Injury prevention professionals: a national directory. Newton, Massachusetts: Education Development Center, 1988.

Chapter 2: Learning from Data

From June to October, tourists drive down the winding Newfound Gap Road across the Great Smoky Mountains from Gatlinburg, Tennessee, to Cherokee, North Carolina. The capital of the Cherokee Indian Reservation, Cherokee attracts a large share of these visitors. With the visitors comes a discernible increase in motor vehicle collisions, particularly those involving pedestrians.

Jackie Moore was an Indian Health Service Community Injury Control Coordinator based in Cherokee. "My role was to document the problem by collecting data," she said. "Using a simple map, with pins to locate each crash, I built up a picture of the collisions that had occurred during the period of a year. When I looked at the data, the problem was clear. It was 'the gap.' " As the road wound into Cherokee, motorists and pedestrians passed through a cleft in the rocks so narrow that there was no shoulder space where they could safely walk.

"I turned the data over to the Cherokee Tribal Planning Board," said Moore. "With that evidence and the cooperation of a number of state and federal agencies, we obtained funds to make improvements. We blasted away the side of the cliff to widen the gap and installed a mile of sidewalk and 22 street lights." "It's good to see that people finally have a safe place to walk," added Eddie Almond of the Cherokee Tribal Planning Office (J. Moore, personal communication).

Not all injury interventions involve moving a mountain. But Jackie Moore's experience demonstrates the progression from understanding an injury problem to developing and implementing an intervention. It also highlights the use of basic data collection techniques well within the scope of a local program or practitioner.

USES OF DATA

Injury prevention and control programs should be based on a solid foundation of data. This chapter and Chapter 3 focus on finding and using data to design, implement, and evaluate such programs. They explore the kinds of data that are useful for identifying and understanding an injury problem; existing sources for these data, their strengths and limitations; the circumstances under which it makes sense to collect data and the advantages and disadvantages of doing so; and the types of data analysis that can be helpful in identifying and understanding an injury problem and designing, implementing, and evaluating an injury prevention and control program.

It is possible for an important injury problem to escape detection if each injury event is separate enough in time or distance from other, similar incidents. Data collection and analysis can reveal these hidden problems. On the other hand, a sudden, highly publicized injury may suggest a trend that is not really there and fuel a public or political demand for action. The death of a college student falling from a hotel balcony during a party may provoke dramatic news stories and public interest. The more frequent injuries to the elderly resulting from falls in the home may not receive as much attention because they are more common and thus less "newsworthy." Injury data can help turn public attention (and public resources) from well-publicized but rare occurrences to more common injuries for which intervention is important.

Decisions about the focus of an injury prevention and control effort should not be based only on the number of injuries. Other factors, such as severity, should be considered. Despite their low frequency, serious injuries resulting in long-term disability may justify attention because of the profound consequences they have for those injured as well as the medical system. Weighing the relative importance of an injury problem in terms of both the number of people affected and the severity of the injuries is an important use of data.

Data can confirm, disprove, or refine analysis of an injury problem and aid in the design, implementation, and evaluation of an effective program. Thorough and systematic consideration of several basic questions is essential.

- Who is being injured (e.g., children, the elderly, low-income people, people who live in a particular neighborhood)?
- How are these people being injured (e.g., suicide with firearms, falls down stairs, pedestrian–motor vehicle collisions)?
- Where are these injuries taking place (e.g., home, work, school)?

• What are the circumstances under which these injuries occur (e.g., is alcohol involved; is a particular activity, such as hunting, or a particular consumer product involved)?

• How serious are these injuries (e.g., are they fatal; do they require treatment at home, in emergency rooms, or as hospital admissions)?

• How many of these injuries have occurred, and over what period of time? What proportion of the community is affected by them? Are they increasing or decreasing in frequency?

• Which of these injuries is most significant in its personal and social consequences (e.g., the time absent from school or work, permanent disability, the cost of emergency medical services, and Medicaid or Medicare expenditures)?

• Is the local injury rate from a particular type of injury higher, or lower, than the national (or state) rate? How does it compare with other health problems?

• What information will be needed to evaluate an intervention?

Data collected to answer these questions will identify areas in which prevention programs can be useful. Injury data can also be extremely valuable in educating the public and policymakers about the magnitude of a problem and the usefulness of injury control efforts as well as in redirecting misplaced concerns over relatively minor or rare problems. Data, especially local data, can help overcome practical or political barriers and gain support and funding for injury prevention.[1]

Sensational accounts of injury-producing events such as fires, motor vehicle collisions, suicides, and murders are the mainstay of any newspaper or news broadcast. Usually each is reported as a discrete event, with no connection made to similar incidents or to the "injury problem" and its consequences for the public's health. The judicious use of data can demonstrate that these incidents are part of a larger problem whose control deserves its share of public resources. A recent study used data on the consequences of injuries (including numbers of deaths and direct economic costs) to demonstrate that injury prevention and control are underfunded by the federal government compared to several other public health problems, including cancer and heart disease.[2]

Costs of injuries can provide dramatic evidence that prevention is needed. A 1986 National Center for Health Statistics report showed that childhood injuries accounted for 13% of acute medical costs for American children under the age of 17, or nearly $2 billion.[3] Data on the absolute or relative costs of a particular injury can also be valuable for gaining support for prevention. A 1981 study demonstrated that the costs of motor vehicle injuries alone equaled the costs of heart disease and exceeded those of stroke by more than two to one.[4]

The study of motorcycle injuries in Washington state referred to in the Introduction found that $1.9 million of the $2.7 million spent on care for injured motorcyclists in the state during 1985 came from public sources, especially Medicaid. This is because serious motorcycle injuries (especially head injuries) usually require medical care costly enough to make the patients eligible for state aid. These figures provide a strong economic argument in favor of helmet laws.[5]

Finally, data can be used to evaluate injury control programs. Injury prevention as a field has been severely hampered by inadequate—or more often nonexistent—program evaluation. Evaluation is essentially the task of collecting and analyzing data about the operation of the program as well as its results. Ongoing evaluation provides information to use in measuring program effectiveness and making changes to increase effectiveness during the course of the program. The whys and hows of program evaluation are further discussed in Chapter 4.

Injury control decisions should not be entirely data driven. In some cases, the existence of a low-cost and effective intervention (e.g., turning down the hot water heater to prevent tap water scald burns) will warrant its use even if the injury it addresses is relatively rare. Selecting a target for an injury prevention and control program is discussed in Chapter 4.

MORTALITY AND MORBIDITY DATA

Injury mortality data are the easiest to obtain because death records are maintained in every state and territory and statistics are aggregated annually at both the state and national levels. Because these data systematically include information on the cause of death, they can be used to demonstrate that injuries are, in fact, a serious public health problem. For example, the Maine State Department of Health conducted a study of childhood mortality using data from the years 1976–80. Although the overall children's death rate from both disease and injury was significantly lower in Maine than in the United States as a whole, the rate of children's deaths due to fires proved to be significantly higher. The study also found that the injury mortality rates for all injury categories except suicide were substantially higher for children in low-income families than for children in other income groups. Thus this study provided valuable information about the status of childhood injuries as a public health problem in Maine and highlighted the groups most in need of preventive efforts.

Mortality data do not always reflect the extent or severity of a particular injury problem. Most people who are injured do not die of their injuries. The extent of nonfatal injuries are reflected in injury morbidity data. The relationship between mortality and morbidity can be seen in Figure 1, the "injury pyramid."

The tip of the pyramid, deaths, represents only a fraction of the total number of injuries. One study of childhood injuries in Massachusetts showed that "for every death due to injuries among children 19 years of age and under, there are 45 hospitalizations and 1,300 visits to emergency rooms."[6] The same researchers estimated that the number of injuries treated at home and in physicians' offices may be double those treated in emergency medical facilities.[7]

Mortality data can be an important and sometimes dramatic indicator of a community's injury burden. At the same time, large numbers of people may be seriously harmed by injuries that rarely cause death but do result in long-term disability. Studies have shown that "the leading causes of nonfatal and fatal injuries are clearly different" and that mortality data are not a good guide to either overall injury incidence or the medical consequences of these injuries.[8] In Connecticut, for example, falls are ranked first among injuries as a cause of hospitalization for children 0–4 years of age but ninth as the cause of death. Within the 15–19-year age group, falls are ranked third as a cause of hospitalization but ninth as a cause of death.[9]

Thus a thorough examination of an injury problem must consider both mortality and morbidity data. Although they often are not as complete or accessible as mortality data, many sources exist for morbidity data, such as hospital discharge files and emergency room records. These are discussed later in this chapter.

USING AVAILABLE DATA SOURCES

Many government agencies, academic institutions, research programs, and other organizations collect, analyze, and disseminate injury data. It is important to learn as much as possible from these data before considering additional data collection activities. At the very least, existing data can define a context within which to analyze local problems. These data can also help target additional data collection activities. Using existing data sources generally takes less time and money than collecting new data, but the quality of data from these sources is uneven. The definition of an injury cause may not be standardized, or the data may differ in terms of geographic coverage and accessibility. The degree to which the combined data can be broken down by geographic location or particular population characteristics also varies.

NATIONAL DATA SOURCES

National data sources usually aggregate, or combine, the data they receive from individual states. The aggregation of large numbers of cases often reveals patterns that provide a new understanding of the injuries being examined. National data provide a general idea of the extent and patterns of a particular injury problem and allow a comparison of a community's injury rates with those of the nation as a whole. These comparisons can be useful in mobilizing political support and public resources, especially if they demonstrate that a community has a significantly greater than average occurrence of a particular type of injury. The National Electronic Injury Surveillance System (NEISS) of the U.S. Consumer Product Safety Commission (USCPSC), for example, gathers information about product-related injuries from the emergency room records of a national sample of hospitals. While no single community is likely to have a large number of injuries related to a specific toy, crib, or hairdryer, NEISS provides a national sample and supporting information. With this background, product-related injuries may turn out to be more important than their local numbers alone would suggest.

National data bases are often produced by sophisticated and comprehensive data collection systems with resources a local program could not

Figure 1. The injury pyramid.

Injuries resulting in deaths

Injuries resulting in hospitalization

Injuries resulting in treatment in emergency rooms, physician's offices, etc.

Injuries treated at home or not treated

hope to duplicate. The data are often available on computer tape, on floppy disk, and/or in annual or special reports issued by the data collection agency. Some agencies will handle specific data requests by phone or mail. The National Highway Traffic Safety Administration (NHTSA) oversees two such data bases: the Fatal Accident Reporting System (FARS), which contains data on all police-reported fatal traffic crashes, and the National Accident Sampling System (NASS), produced from a national sample of all police-reported motor vehicle crashes, including pedestrian- and bicycle-related injuries involving motor vehicles.

All data bases have limitations. They may lack information about the circumstances of an injury or descriptions of the victims that would be useful. They may also reflect geographic biases, since geographic differences exist in injury rates, circumstances, and outcomes. Rural areas, for instance, generally have higher death rates than urban areas for most types of unintentional injuries.[10]

There are often lengthy delays between data collection and availability. The data may be difficult and/or expensive to use. National data on injury outcome, rehabilitation, and long-term disability are almost nonexistent. And, because national data may not reflect a particular community's injury problem, community leaders and decision makers will need some local data to verify the existence of an injury problem before they commit themselves to participating in or funding local injury control activities.

STATE AND LOCAL DATA SOURCES

Sources of already-collected data can also be found in most states and many counties and communities. These data will often be more representative of a state's or community's injury problem than data aggregated at the national level. Many of these data are available from two sources: the local agency that actually collects the data (such as a hospital or a police department) and state collections of data compiled from all the communities in the state. Each of these sources has its own particular strengths and limitations.

Local data sources are valuable precisely because they are local and thus may better reflect the particular injury problems of the population a program will serve. Because these sources are collections of the original documents filled out by primary data collectors (such as hospital discharge forms or police accident reports), they often contain a wealth of detailed information that is lost once these data are computerized.

On the other hand, because local data are generally not computerized, retrieving this information from existing records can be a time-consuming and expensive task. Using local data can be very much like collecting new data, and much of the discussion in Chapter 3, "Working with Data," is also relevant to using data from local sources. In addition, unless the data have been collected in a fairly densely populated area, they may contain too few cases from which to draw statistically valid inferences about the nature of an injury problem. (See the sections "Time Considerations" and "Population Considerations" in Chapter 3 for a further discussion of this problem.)

State aggregations of local data are usually large enough to provide data more statistically representative of an injury problem. State data collection agencies often computerize their data and produce periodic reports on their findings, making the information easier and less expensive to obtain than local data.

Yet state sources are not without their limitations. Like national sources, data may be incomplete or of uneven quality. A considerable time lag can occur between the time the data are collected at the local level and the time they are available from the state. The patterns of injury delineated by state data sources may not be representative of the local population an injury control program serves. In Colorado, which has an average annual injury death rate of 65.6 per 100,000 inhabitants, the county death rates range from 36 to 234 deaths per 100,000.[11] Injury rates and patterns can vary between urban and rural areas in the same state and between high-income and low-income neighborhoods in the same city. Despite these limitations, both state and local data sources can provide a great deal of valuable information.

Vital Statistics Mortality Files and Death Certificates

State and county offices of vital statistics have computerized vital statistics mortality files, abstracted from death certificates, on every death that occurs in the state. These files include the date, place, and cause of the death and the age, sex, race, and residence of the deceased. Most states publish an annual report that summarizes this information. Data from all the states and territories are compiled by the National Center for Health Statistics and published in their annual Vital Statistics Series. The national report appears 3–4 years after the calendar year in which the deaths occurred. State data are usually available sooner, although there is often a 2-year lag.

County collections of death certificates provide information on deaths within a more accurately defined area, and provide it much sooner, than state or national vital statistics mortality files. As is often the case with local data sources, however, the files are generally not computerized. Physicians have no standardized system for filling out death certificates, and the certificates are often completed without an autopsy or before one is performed to establish a precise cause of death. Questions have been raised about the accuracy and comparability of the information on these certificates and the mortality files created from them.[12,13] A more uniform system is needed, and the National Center for Health Statistics has developed a new, standardized death certificate that is expected to be adopted by at least 47 (and possibly more) states during 1990.[14]

Medical Examiners' Reports

Medical examiners' and coroners' reports must be completed for all persons whose deaths resulted from an injury (intentional or unintentional) or an unexpected natural death. These reports, available in town halls or county halls of records, can be rich sources of injury mortality data. Some states maintain centralized (and occasionally computerized) files of county medical examiners' and coroners' injury–death investigations in a chief state medical examiner's office. However, these reports are usually presented as narratives, so locating injury causes and abstracting the relevant information can be time-consuming. The competence of the individuals who produce these reports varies widely, as does the quality of information and state policies on what types of injury deaths to include in these systems. The percentage of these reports based on full autopsies also varies dramatically among counties and states. During 1985, autopsies were performed for 51% of deaths caused by unintentional injuries, 52% of suicides, and 97% of homicides.[13] Some counties require a court order to release these reports, even to local or state health departments.

Hospital Discharge Data

Hospital discharge data provide extremely valuable information about injury morbidity. Records on every case admitted to a hospital, completed on discharge, contain information on the age and sex of the patient and the length of hospital stay, diagnosis, and medical costs. Some specify the causes of injuries (in the form of E codes, which are discussed later in this chapter). Many states require that all hospitals submit summary data on hospital discharges, which are then aggregated at the state

level in the form of Uniform Hospital Discharge Data Sets (UHDDS). In some areas, private organizations or companies, such as the Committee on Professional and Hospital Activities, aggregate and analyze hospital discharge data for client hospitals and hospital associations. The National Hospital Discharge Survey of the National Center for Health Statistics collects discharge data from a national sample of hospitals and issues a yearly report as Series 13 of Vital and Health Statistics.

Because discharge data also include information about medical treatment, they are a good source of information on the consequences and costs of injuries. Comparisons of these consequences and costs with those of other disease and medical problems can help convince policymakers of the importance of funding injury prevention activities.

Hospital discharge data sets also have some limitations. Because a patient may be admitted to a hospital more than once for the same injury, the data can overstate the incidence of an injury. And because discharge data are often used for reimbursement, they tend to focus on the nature of an injury and its medical treatment rather than its cause or the circumstances under which it occurred. Consistently including information on the causes of injuries would be a relatively inexpensive way to gather much information useful for injury prevention and control efforts.[15] One way of doing this is discussed under "Promoting the Use of E Codes" later in this chapter.

Aggregating hospital discharge data at the state level can take several years. Efforts are under way in New York State to computerize these records at hospitals and make the data directly available, online, to the state agency responsible for producing the annual summaries (S. Standfast, personal communication). This system would increase both the speed with which data become available and the possibilities of analysis at the state and local levels.

A study in Connecticut demonstrated the usefulness of uniform hospital discharge data. This study examined severe childhood injuries resulting in hospitalization or death. Statewide cause-specific injury rates were calculated by both age and sex. As a result, the researchers were able to develop—for the first time—a comprehensive picture of serious childhood injury morbidity in Connecticut and identify the most common injuries in each of eight age groups from 0 to 19 years. This project can be seen as a model for a relatively low-cost injury surveillance system that uses existing state hospital discharge data.[9]

A combination of information from vital statistics records and the uniform hospital discharge set can provide a very good picture of a state's injury problem. A project in Michigan used the computerized state vital statistics files as well as the state hospital discharge data set to produce *Childhood Injuries in Michigan: 1982 Deaths, Hospitalizations, and Costs of Hospitalization.*[16] This report presents in a clear and concise format a vivid picture of childhood injuries in Michigan, including how injury compares to other medical conditions as a cause of death and hospitalization, the major causes of childhood injury, breakdowns by injury type and patient age and race, and the hospital costs associated with injuries.

Emergency Room Data

Many injured persons treated in emergency rooms are not admitted to the hospital and consequently do not appear in hospital discharge data. Information on these patients and their injuries can be found in hospital emergency room data. Even a review of the emergency room log, which lists some minimal information on each patient treated, can provide some information on the characteristics of people injured and the types of injuries occurring in the community served by a hospital.

Emergency room data also have limitations. The Massachusetts Statewide Comprehensive Injury Prevention Program (SCIPP), a large, federally funded effort, conducted a 3-year study that gathered emergency room data from 23 hospitals on 87,000 children and adolescents. This project revealed some of the problems inherent in collecting and using emergency room data.

- The emergency room logs did not include information on the causes of the injuries.
- Because emergency room data were generally not computerized, collecting the information was very labor-intensive.
- Project staff had to work out a system of protecting patient confidentiality with each of the 23 participating hospitals.[6,17]

Further information on the use of emergency room data can be found in Chapter 3.

Emergency Medical Services Data

Emergency medical services (EMS) data, recorded in the form of "run sheets," provide additional information on the circumstances of an injury, such as time, place, and cause, that are not always recorded in the emergency room. EMS data can be correlated with emergency room data by matching the time and patient description on the run sheet with the same information in the emergency room log. EMS data can be especially useful when investigating in-juries resulting from interpersonal violence or suicide attempts. A description of the condition of a home and children in that home, for example, may indicate that an investigation into possible child abuse is warranted. Some states have agencies (such as a bureau of emergency medical services) that compile EMS statistics at the state level. Note, however, that only a minority of serious injury cases treated in emergency rooms are brought in by EMS.[18]

EMS and emergency room data can be combined with police, fire, and newspaper reports on the same incidents to provide descriptions of the circumstances and causes of some injuries. The value of combining data about the same incidents from different sources is further discussed under "Data Linkage" in Chapter 3.

Trauma Registries

Trauma registries are maintained by trauma centers, hospitals that specialize in or include a unit specializing in the care of victims of acute injury. Some trauma centers treat all types of serious injuries; others specialize in a particular type of injury —serious burns or acute spinal cord injury, for example. It is important to learn if people with serious injuries are being taken to a trauma center rather than the closest hospital. If so, any data collection must include that center to accurately represent the injury problems and patterns in the community.

Patients treated at a trauma center will not necessarily reflect the types of injuries occurring in the surrounding community, because most centers receive referrals from other hospitals and many centers specialize in particular types of injuries. Trauma centers also characteristically treat certain injuries and patients on an outpatient basis while admitting others. People with hand injuries, for example, are generally not admitted but are treated and sent home, while patients with injuries of the same severity to the legs or feet are hospitalized. Younger patients tend to be treated and released for the same injuries for which older patients are hospitalized.[19]

There are also registries of information gathered from more than one institution used to track serious injuries in a particular region (such as the Illinois State Trauma Registry) or to collect information regionally or nationally about a particular injury or risk group (such as Colorado's statewide spinal cord injury registry). The accuracy and completeness of the data in such registries vary greatly. Problems of underreporting and geographic variability also exist.[20,21] Trauma registries often are not computerized, so that using registry data can be a time-consuming job, but some registries (including some regional registries) have started to use computerized data processing systems or are expected to go on line in the future.[22] Most trauma registries use the Injury Severity Score (ISS)[23,24] or another system to measure the severity of injuries. Severity data can be helpful in assessing the need for an intervention effort. Information found in registries usually includes the nature of the injury and the patient's status, medical procedures, length of stay, and medical outcomes. The data collected and the format in which they are compiled are not consistent among registries. The Centers for Disease Control, in collaboration with a number of professional organizations, is developing a standard data collection procedure for all trauma registries.[25]

Physicians' Offices and Health Maintenance Organizations

Injuries are also treated in physicians' offices and health maintenance organizations (HMOs). The large number of private physicians and walk-in clinics, as well as issues of confidentiality, make collecting these data extremely difficult. In many cases, physicians will not permit access to their records. Fortunately, the number of serious injuries treated in such locations is relatively small. A study in Philadelphia estimated that only 8–16% of injuries in urban areas were treated in private physicians' offices and clinics, and most were not serious. Most serious injuries were referred to emergency rooms.[26]

State Motor Vehicle Data

Many state police, highway safety, and motor vehicle departments aggregate data on motor vehicle collisions, as well as on pedestrian and bicycle injuries involving motor vehicles (commonly referred to as crash or accident reports). The data can also be used to locate "hot spots" where many injuries take place and where an intervention, such as installing a traffic light or increased enforcement of speeding laws, may be appropriate. The Massachusetts Department of Public Works (DPW) has compiled lists of intersections where large numbers of collisions take place. Using a formula that assigns greater importance to collisions resulting in injury or the loss of life as opposed to property damage, DPW uses these data to identify the most dangerous intersections and set priorities for action and use of state resources. The data also are used in applying for federal funds earmarked for the correction of

roadway hazards.[27] Another project, in Queens, New York, used similar reports and reduced the rate of pedestrian injuries. This project is described in Chapter 3.

These data also have limitations. Only 55% of traffic injuries resulting in emergency room treatment or hospitalization are included in crash reports.[28] Information on safety belt use depends on self-reports of drivers and passengers and is not reliable. Information on the involvement of alcohol in an incident may depend only on the impression of the police officer at the scene unless accurate breath alcohol or blood alcohol content testing was done.

Poison Control Centers

Regional poison centers, located at selected hospitals, keep records of calls for help. These records often lack important information, such as medical outcomes, and are rarely computerized. They can, nevertheless, provide information on the extent, cause, and nature of poisonings for various age groups in a region. Although the American Association of Poison Control Centers has developed a standardized format for recording poison center data, not all centers use it. Available information usually includes the age of the victim, where the poisoning took place (residence, workplace, etc.), the substance involved, and the therapy recommended by the center.

There is movement toward greater accessibility. The New York State Department of Health, for example, is developing a uniform computerized data system for its eight regional poison control centers. The American Association of Poison Control Centers' National Database Collection System combines and analyzes data from 57 poison control centers. A report on this analysis is published annually in the September issue of the *American Journal of Emergency Medicine*.

Criminal Justice Data

State and local police, highway patrols, and other criminal justice agencies collect data on homicides, assaults, and other types of interpersonal violence. Much of this information is aggregated in the FBI's annual Uniform Crime Report (UCR). The UCR contains statistical material on homicides including the age, sex, and race of victims, victim–offender relationships, the weapons used, and the circumstances surrounding the events. It also contains information on rape and aggravated assault. Some states, such as Illinois and New York, issue their own annual compilations of these data. Anyone can obtain information on a state's uniform crime reporting activities from the Committees on Uniform Crime Reports of the International Association of Chiefs of Police and the National Sheriffs' Association.

Other State and Local Data Sources

Police, fire, school, and parks and recreation departments all keep records on injuries. Some of these are aggregated at the state level, while some are not. State departments of labor, state and local occupational safety and health administration offices, insurance companies, and unions may also be able to provide information on occupational injuries.

Table 1 shows the key features of a number of national, state, and local sources of injury data. Further information about each of the data bases is contained in Appendix A.

ADDITIONAL SOURCES OF INFORMATION

There are other sources of data that may be useful in designing or targeting an injury intervention program. Injury risk can sometimes be estimated by examining statistics bearing on risk factors. Because fires, for example, are correlated with overcrowded housing conditions (themselves an indication of a poorer neighborhood with older, less well maintained buildings), census data on population density may be helpful in targeting neighborhoods whose populations are at higher risk for fire-related injuries. Demographic data can help determine the size of certain groups that research has shown are at risk for particular injuries (e.g., falls among the elderly, burn injuries among children, automobile-related injuries among teenage boys). A wealth of demographic information is available from the U.S. Census Bureau. The Bureau of Labor Statistics of the Department of Labor generates a great deal of data on employment, income, and other economic indicators. Information produced by federal agencies and congressional hearings about injuries and injury prevention and control are located at federal depository document collections. There are at least two depositories in every congressional district (often as part of public or university libraries).

Research exists on the knowledge, attitudes, and behavior of people toward health in general and injuries in particular. This research may aid in the design of intervention programs and especially their educational components. A number of state health departments participate in the Behavioral Risk Factor Surveillance System (BRFSS) of the Centers for Disease Control (CDC), a series of telephone surveys on selected health behaviors. Some of the issues explored by these surveys are relevant to injury prevention and control. The BRFSS also includes questions on the demographic and socioeconomic status of the respondents. The resulting data may help clarify the relationship between health behavior and special population groups. Reports on these surveys are published periodically in the CDC publication *Morbidity and Mortality Weekly Report* (MMWR). Some states also conduct their own risk factor surveys. (Chapter 4 discusses risk factor surveys in more detail.)

There is also a wealth of program descriptions and research reports about injury prevention and control, many of which are published in medical journals. A number of newsletters also report on activities and developments in the field of injury prevention and control. Descriptions of some of these periodicals can be found in Appendixes B and C. Several computerized data bases can be used to generate guides to this literature, including MEDLINE (the on-line version of the National Library of Medicine's data base) and the CDC's Combined Health Information Database (CHID). On-line literature searches related to injury prevention and control using these data bases can usually be done, for a fee, at university libraries, especially medical libraries or libraries at universities that have a school of public health.

Newspapers can also provide information on injuries. Newspaper reports of the circumstances surrounding specific injuries can be a valuable tool for educating policymakers and others about the extent and consequences of injuries. In most states there are clipping services that will, for a fee, collect articles from newspapers about particular subjects (i.e., injuries in general or a particular type of injury such as burns). Because newspapers do not report on every injury that occurs in their communities, however, clues uncovered this way should always be confirmed by more systematically collected data. A single newspaper headline, no matter how dramatic, should not dictate program development.

IMPROVING INJURY DATA SOURCES

Promoting the Use of E Codes

External cause of injury codes (E codes) and nature of injury codes (N codes) are classifications developed by the World Health Organization to use with the International Classification of Disease (ICD) system. Currently in its ninth edition, the ICD is an internationally accepted listing of diseases used in hospital and vital statistics record keeping.

Table 1. Data sources

Data base	Source	Type of data	Level	Information on injury cause
National Center for Health Statistics Data Bases	National Center for Health Statistics, United States Public Health Service			
National Vital Statistics Mortality Data		Mortality	National, state	E codes
National Health Interview Survey		Knowledge, attitudes, behavior	National	Motor vehicle vs. non-motor vehicle
National Hospital Discharge Survey		Primarily morbidity	National	E codes when available
National Ambulatory Medical Care Survey		Morbidity	National	None
National Highway Traffic Safety Administration Data Bases	National Center for Statistics and Analysis, National Highway Traffic Safety Administration			
Fatal Accident Reporting System (FARS)		Mortality	National, state	Weather, speed, alcohol, etc.
National Accident Sampling System (NASS)				
NASS Crashworthiness System		Crashworthiness of different makes and models of cars	National	Not applicable
NASS General Estimates System		Morbidity only	National	Road conditions, alcohol use, etc.
National Electronic Injury Surveillance System (NEISS)	Consumer Product Safety Commission	Product-related injuries	National	By product
Accident investigations	Consumer Product Safety Commission	Product-related injuries	National, state, town	By product
Occupational Injury and Illness Statistics Program	Office of Occupational Safety and Health Statistics, Bureau of Labor Statistics			
Annual Survey of Occupational Injuries and Illness		Mortality and morbidity	National, state, injury, size of company	No
Supplementary Data System (SDS)		Mortality and morbidity	National, state, injury, size of company	"Accident type," associated object or substance
Work Injury Reports (WIR) Survey Program		Varies by study	Varies by study	Varies by study
National Traumatic Occupational Fatality Data Base (NTOF)	National Institute for Occupational Safety and Health, Division of Safety Research	Mortality	National, state, industry	Yes
Indian Health Service Data Base	Indian Health Service	Medical care at all IHS facilities	National	E codes
Behavioral Risk Factor Surveillance	Centers for Disease Control	Risk factors	National, state	Focus on causes (not outcomes)

Table 1. Continued

Data base	Source	Type of data	Level	Information on injury cause
National Fire Incident Reporting System (NFIRS)	National Fire Data Center	Fire injuries	National, state	Causes of fires
United States Department of Justice Data Bases				
Uniform Crime Reports	Federal Bureau of Investigation	Crime (better on mortality than morbidity)	National, state	Weapon, cause of event
National Crime Surveys	Bureau of Justice Statistics		National	Nature of attack
Criminal Justice Statistics Association	Same	State criminal justice statistical reports	State	Varies
Boating Accident Reporting System (BARS)	State Boating Law Administration, United States Coast Guard	Boating "accidents"	National	Cause of event, alcohol, etc.
Drug Abuse Warning Network (DAWN)	National Institute on Drug Abuse, Division of Epidemiology and Statistical Analysis	Emergency room episodes caused by drug abuse	National	Some
National Burn Registry	National Burn Information Exchange (NBIE)	Injuries treated at burn care facilities	National	Some
National Institute of Disability and Rehabilitation Research Pediatric Trauma Registry	Tufts–New England Medical Center	Children under 20 years of age admitted to participating trauma centers	United States and Canada	Circumstances and mechanism
National Study on Child Neglect and Abuse Reporting	American Association for Protecting Children	Child abuse	National, state	Varies by state
Big Ten Injury Surveillance	University of Iowa Hospital	Sports-related injuries	Big Ten Conference colleges	Some
National Head and Neck Injury Registry	University of Pennsylvania Sports Medicine Center	Football-related cervical spine and head injuries	National	Some
National High School Athletic Injury Registry	National Athletic Trainers' Association	High school football and girls' basketball injuries	National	Some
Regional Spinal Cord Injury Systems	National Spinal Cord Injury Statistical Center	Injuries treated at regional spinal cord centers	National, regional	Some
American Association of Poison Control Centers National Data Collection System (AAPCC)	National Capital Poison Center	Calls to regional poison centers	National, regional	Substance ingested
Vital statistics records	State offices of vital statistics	Mortality	State, town	E codes
Uniform Hospital Discharge Data Sets (UHDDS)	Varies by state	Primarily morbidity	State	E codes (extent varies)
Medical examiners'/coroners' reports	Coroners' offices, medical examiners' offices	Mortality	State, local	Varies, often rich source
Medicaid/Medicare data bases	Varies by state	Primarily morbidity	State	Focus on treatment

Table 1. Continued

Data base	Source	Type of data	Level	Information on injury cause
Workers' compensation records	Varies by state	Morbidity	State	Based on reports by the injured workers
Emergency room records	Hospital emergency rooms	Morbidity and mortality	Local	Varies
Ambulance and emergency medical service records	State bureaus or divisions of emergency medical services	Morbidity and mortality	Local, some state	Varies
Police reports	Local and state law enforcement agencies	Motor vehicle and pedestrian injuries, suicide, interpersonal violence	Local, state	Narrative
School health incident reports	Schools	Primarily morbidity	Local	Varies
Day-care incident reports	Day-care centers	Injuries occurring in day-care setting	Local	Varies
State fire marshal records	State fire marshal's office	Fire injuries	State, local	Varies
Trauma registers	Trauma centers	Morbidity and mortality	Local or regional	Varies

N codes describe the nature of an injury and the part of the body injured; they do not explain how the injury occurred.[29] That is the purpose of E codes. If a closed fracture of the ribs was sustained as the result of a fall, for example, E codes can be used to distinguish among a fall on or from stairs (E880), a fall from a ladder (E881), and a fall from a building or other structure (E882).[29] E codes also provide information on the location where the injury took place, such as home or work. Thus, E codes provide a critical link between the cause and nature of an injury.

E codes are mandatory on death records for all persons whose deaths are injury related. Including E codes in emergency medical service, emergency room, and hospital data would provide a wealth of information for injury prevention and control, information that would have direct applications to intervention efforts. Although E codes are included in some hospital discharge data, only New York, Virginia, Wisconsin, California, and Washington state have mandated the use of E codes in hospital discharge data. Several states have achieved impressive levels of E coding on a voluntary basis. Compliance in Rhode Island now stands at 80%.

Voluntary compliance clearly is not enough. Physicians, nurses, and hospital administrators are often unaware of the need for these data. Unlike N codes, E codes are not linked to the reimbursement process. So E codes are frequently dropped in abstracting records for transfer to computer files in favor of the codes used for reimbursement. The universal use of E codes in hospital discharge data is one of the most important goals for the injury prevention community. Consistent use of E codes would provide an extremely valuable resource for the study of injury rates and patterns. Injury control practitioners would be able to use this information to obtain a much more accurate view of a local injury problem. It would also allow much more thorough aggregation of injury morbidity data as well as comparisons of the data at the local, state, and national levels and among specialized studies. The Council of State and Territorial Epidemiologists (CSTE) has recently recommended that E codes be included in all hospital discharge data.[30] We recommend that every state and territory require the inclusion of E codes in hospital discharge data.

E codes, like all data, need to be recorded accurately and consistently. It is important that personnel with this responsibility be trained in the use of this system. The Division of Injury Epidemiology and Control of the CDC has developed EPIC, a software package for personal computers that facilitates the rapid and accurate assignment of E codes.

Making Acute Spinal Cord Trauma a Reportable Condition

An important step in understanding injury problems is the tracking of sentinel injuries, injuries so devastating in their personal and social consequences that preventing them should be an important national priority. A system for such data collection already exists: the United States National Notifiable Disease System. This system requires that physicians and health care facilities report all cases of certain diseases (e.g., infectious hepatitis, leprosy, and yellow fever) to their state health departments. States consolidate and forward the data to the Centers for Disease Control, which analyzes the data and issues periodic reports in MMWR.

In 1987 CSTE adopted a resolution (reprinted in Appendix D) recommending that acute traumatic spinal cord injuries be added to this system as a prototype for the future inclusion of other serious injuries in this system. CSTE and CDC have developed a case definition for use in the surveillance of acute spinal cord injuries (also reprinted in Appendix D). These activities represent an important step for injury data collection in this country. We recommend that all health practitioners cooperate fully in this important effort.

Improving Injury Surveillance at the State Level

As injury prevention and control have grown as a public health concern, there has been increased recognition of the need for more complete and more accurate injury data, especially injury morbidity data. *Promoting Health, Preventing Disease: Objectives for the Nation,*[31] included a recommendation for uniform reporting of injuries at the state level. The Select Panel for the Promotion of Child Health, in its 1981 report to the Congress and the Secretary of Health and Human Services, made a similar recommendation, as did the 1984 CDC Conference on the Prevention of Injuries and *Injury in America.*[32–34]

Systematic collection of injury data by all state and territorial health departments would provide much better information on most injuries than is currently available. Epidemiologic surveillance has been defined as

> the ongoing and systematic collection, analysis, and interpretation of health data in the process of describing and monitoring a health event. This information is used for planning, implementing, and evaluating public health interventions and programs. Surveillance data are used both to determine the need for public health activities and to assess the effectiveness of programs.[35]

When applied to injuries, such surveillance can identify emerging or recurrent injury problems; document the magnitude of injury problems; characterize populations at risk for injuries by using demographic, geographic, and environmental data;

and generate hypotheses of risk factors.[36]

At a minimum, every state and territory should make use of several readily available data sources to create an annual report on the extent and patterns of injuries that result in death and hospitalization. In June 1987 the Massachusetts Department of Public Health published the first issue of *Injury in Massachusetts: A Status Report.*[37] It forms the basis of the following description of a minimum state injury surveillance system.

Data on fatal injuries can be obtained from state vital statistics records. Because the annual number of the various types of fatal injuries in any one state will be relatively small, a 5-year aggregation of this information is recommended.

Information on nonfatal injuries can be collected from hospital discharge data, either from state aggregations, or, in states in which this information is not aggregated, from a representative sample of hospitals. The value of such surveillance for injury prevention is directly related to the systematic and accurate inclusion of E codes in hospital discharge data sets.

Injury morbidity and mortality rates should be calculated and these rates broken down by age, sex, and race. (These calculations are discussed further in Chapter 3.) The following age groups are recommended: Less than 1 year, 1–4 years, 5–9 years, 10–14 years, 15–19 years, 20–24 years, 25–29 years, 30–39 years, 40–49 years, 50–59 years, 60–69 years, 70–79 years, and 80 years and over. Such groupings are useful for analyzing the age-specific injury rates for various developmental stages and can be combined for comparison with census data.

Injury morbidity and mortality rates by selected causes should also be calculated. E codes related to some general categories of injuries are not necessarily numerically contiguous. A format developed by the Childhood Injury Prevention Resource Center at Harvard University groups E codes into general categories useful for injury surveillance. These groups are included in Appendix E.

Information produced by such state surveillance systems would be valuable in setting intervention priorities. One way of presenting this information is to create lists of the 10 leading causes of injury mortality and morbidity by age group. A further breakdown of injury rates (or injury rates for particular causes) by sex, race, income level, municipality, or a combination of two or more of these elements could also help identify high-risk groups and target specific interventions to them. Correlations between E codes and N codes or other information found in hospital discharge data (such as length of hospital stay) may also reveal patterns valuable to understanding an injury problem and setting priorities for intervention efforts. Comparing injury rates and costs with those of other medical conditions, such as cancer, cardiovascular diseases, and AIDS, can be useful for convincing public officials that injury prevention and control deserve their share of public health resources.

A more useful, although more complicated and costly, system would also sample emergency room records and link vital statistics and medical data sources to police, fire department, and EMS reports to provide a more elaborate description of an injury problem. However, even a minimum surveillance system would be extremely valuable for injury prevention and control. It would allow people to compare their state's injury problems with those of other states and regions, and it would provide valuable data for the evaluation of injury prevention and control programs. If every state and territory would institute such surveillance, the results could be aggregated to provide a comprehensive picture of the national injury problem. This information would provide a substantial base from which state and local injury control practitioners could begin to investigate their own community's injury problem.[35,38]

SUMMARY

The following are the key points on which this chapter focuses.

• Injury data can confirm, disprove, or refine an analysis of an injury problem and are essential for the design, implementation, and evaluation of an effective injury prevention and control program.

• Injury data are valuable for educating the public and policymakers about the magnitude of the injury problem and the usefulness of injury prevention and control efforts.

• Examination of an injury problem should consider both mortality and morbidity data because the causes of fatal and nonfatal injuries are sometimes different.

• There are many national, state, and local sources from which injury data can be obtained. Each of these sources has it own strengths and limitations.

• The inclusion of E codes in hospital discharge data should be required in every state and territory.

• Serious injuries, starting with acute spinal cord trauma, should be incorporated into the National Notifiable Disease System.

• Every state and territory should have an injury surveillance system.

INTERVIEW SOURCES

Jackie Moore, Indian Health Service, Cherokee, North Carolina, March 3, 1988.

Susan Standfast, Injury Control Program, Division of Epidemiology, State of New York Department of Health, Albany, February 10, 1988.

REFERENCES

1. Fawcett SB, Seekins T, Jason L. Policy research and child passenger safety legislation: a case study and experimental evaluation. J Social Issues 1987;43:133–48.

2. Hatziandrev E, Graham J, Stoto M. AIDS and biomedical research funding: comparative analysis. Rev Infect Dis 1988;10:159–67.

3. Parsons P, et al. Costs of illness, United States, 1980: national medical care utilization survey. Hyattsville, Maryland: National Center for Health Statistics, 1986; DHHS publication no. (PHS)86-20403.

4. Hartunian NS, Smart CN, Thompson MS. The incidence and economic costs of major health impairments: a comparative analysis of cancer, motor vehicle injuries, coronary heart disease, and stroke. Lexington, Massachusetts: Lexington Books, 1981.

5. Rivara F, et al. The public cost of motorcycle trauma. JAMA 1988;250:221–3.

6. Gallagher SS, Finison K, Guyer B, Goodenough SH. The incidence of injuries among 87,000 Massachusetts children and adolescents: results of the 1980–81 state-wide childhood injury prevention surveillance system. Am J Public Health 1984;74:1340–7.

7. Guyer B, Gallagher SS. An approach to the epidemiology of childhood injuries. Pediatr Clin North Am 1985;32:5–16.

8. Barancik JI, Cramer CF. Northeastern Ohio Trauma Study: overview and issues. Public Health Rep 1985;100: 563–5.

9. Spivak P, Smith G. Childhood injuries in Connecticut: development of new data sources. Newton, Massachusetts: New England Network to Prevent Childhood Injuries, 1986.

10. Baker S. Injury facts, risk groups, and injury determinants. Public Health Rep 1985;100:581–2.

11. Injury in Colorado. Denver: Colorado Department of Health, Injury Prevention Network, 1987.

12. Carter J. The problematic death certificate. N Engl J Med 1985;313:1285–6.

13. Autopsy frequency: United States, 1980–1985. Morbid Mortal Weekly Rep 1988;37:191–4.

14. Altman L. New certificate may ease criticism of death data. New York Times 1988 Oct 18:C3.

15. Estimated costs of state-wide E-code reporting using the Commission Hospital Abstract Reporting System. Olympia, Washington: Department of Social and Health Services, Disease Prevention and Control, 1986.

16. Anda R, Thar W. Childhood injuries in Michigan: 1982 deaths, hospitalizations, and costs of hospitalization. Lansing, Michigan: Michigan Department of Public Health (Undated).

17. Gallagher S. The Massachusetts experience in developing an injury surveillance system. Workshop presentation at the North Carolina Conference on Injury Prevention. Durham, North Carolina, Sep 9–10, 1986.

18. Cales R, Anderson P, Heilig R. Utilization of medical care in Orange County: the effect of implementation of a regional trauma system. Ann Emerg Med 1985;14:853–8.

19. Payne SR, Waller JA. Trauma registry and trauma center bias in injury research. J Trauma (in press).

20. Goldberg J, et al. An evaluation of the Illinois Trauma Registry. Med Care 1980;18:520–31.

21. Rossignol R, Locke J. An assessment of the completeness of the Massachusetts Burn Registry. Public Health Rep 1983;98:492–6.

22. Cales R, Bietz D, Heilig R. The trauma registry: a method for providing regional system audit using the microcomputer. J Trauma 1985;25:181–7.

23. Baker S, et al. The injury severity score: a method for describing patients with multiple injuries and evaluating emergency care. J Trauma 1974;14:187–96.

24. Baker S, O'Neill B. The injury severity score: an update. J Tramua 1976;16:882–5.

25. Trauma registry data set, format, and coding conventions. Atlanta, Georgia: Centers for Disease Control Trauma Registry Workshop, Jan 21–23, 1988 (fourth draft).

26. Philadelphia Injury Prevention Program: progress report and program plan. Philadelphia: Philadelphia Department of Health, 1988:30.

27. Howe PJ. Where accidents are waiting to happen. Boston Globe 1988 Feb 2:13.

28. Barancik J, Fife D. Northeastern Ohio Trauma Study IV: discrepancies in vehicular crash injury registry. Accident Anal Prev 1985;17:147–54.

29. International classification of diseases: 9th revision, clinical modification, vol. 1. Washington, DC: U.S. Department of Health and Human Services, 1982; DHHS publication no. (PHS)80-1260.

30. Council of State and Territorial Epidemiologists. CSTE position statement number 7. 1988 Aug 18.

31. U.S. Public Health Service. Promoting health, preventing disease: objectives for the nation. Washington, DC: U.S. Department of Health and Human Services, 1980:48.

32. Select Panel for the Promotion of Child Health. Better health for our children: a national strategy, vol. 1. Washington, DC: U.S. Department of Health and Human Services, 1981:88.

33. Wilson R, Barancik JI, Gallagher SS, Sattin RW, Smith GS, Sokal DC. Injury morbidity and mortality overview workshop. Public Health Rep 1985;100:560–1.

34. National Research Council. Injury in America. Washington, DC: National Academy Press, 1985:36.

35. Klaucke DN, et al. Guidelines for evaluating surveillance systems. Morbid Mortal Weekly Rep 1988;37(suppl 5):1–18.

36. Ing RT. Surveillance in injury prevention. Public Health Rep 1985;100:586–90.

37. Injury in Massachusetts: a status report. Boston: Massachusetts Department of Public Health, Statewide Comprehensive Injury Prevention Program, 1987.

38. Graitcer P. The development of state and local injury surveillance systems. J Safety Res 1987;18:191–8.

Chapter 3: Working with Data

Data collection and analysis can assist in an injury control program only if they are done well. Collecting poor-quality data or the wrong data, or drawing the wrong conclusions from the data, can have serious implications for an injury prevention and control program. Inaccurate data or data analysis can result in targeting the wrong injury or the wrong population or implementing the wrong intervention.

DEFINING DATA COLLECTION NEEDS

Collecting new data is more complex and expensive than using data from existing sources. It is always a sound practice to make as much use as possible of existing data before collecting additional data. It is possible a program's workers can learn what they need to know from data that already exist. The existing data at least indicate what is, and what is not, known about an injury problem. This allows a more precise definition of the additional data that are needed and provides a context for their analysis. An efficiently targeted data collection system will be more cost-effective than one that collects data that are unnecessary or available from another source.

For example, the Injury Prevention Unit of the Wisconsin Bureau of Health decided to investigate amputations, that is, injuries resulting in the loss of extremities, thumbs, or digits. Research indicated that these injuries are becoming more common with the increasing use of machinery. Wisconsin hospitals were asked to supply the records of patients with such injuries to the division of health, where they were abstracted and computerized. A pilot study revealed that hospital records usually lacked information on the circumstances of the injuries. Thus additional data collection (telephone interviews with amputation victims) was necessary to supplement the hospital record data.[1]

WHO CAN HELP COLLECT AND INTERPRET INJURY DATA?

Although some data can be collected and analyzed by nonspecialists, it is always useful to have expert advice when designing a data collection system and especially when analyzing data. Having an epidemiologist (especially with some experience with injuries) on staff or as a consultant is preferable, but there are other professionals who can assist in collecting and analyzing data. Health agencies, health care facilities, universities, and research institutions can provide assistance. Some private companies specialize in statistical analysis and epidemiology. The five Centers for Disease Control (CDC)-funded injury prevention research centers and eight collaborative centers (see Appendix F) can be valuable sources of expertise for programs. Another source for locating people with expertise in injury data is *Injury Prevention Professionals: A National Directory.*[2] Figure 1 summarizes information about some of the people who may be of help and where they can be found.

USING COMPUTERS FOR DATA COLLECTION AND ANALYSIS

Using computers to help collect and analyze data need not involve customized software and expensive mainframe systems. Commercial data base and statistical software packages are available for most types of personal computer. These programs store data as well as calculate various statistical measures. The capabilities of any software package (that is, the amount of information that can be entered on any one event or the types of calculations possible) should be fully explored before deciding whether it will be helpful.

It is essential to consult with someone who has experience with these types of computer applications before designing a data collection system. Knowing what types of statistical analyses are possible as well as the procedure for entering data may affect both the information collected and the manner in which it is collected.

TARGETING A DATA COLLECTION SYSTEM

One of the major purposes of collecting data on injuries is to define the problem by identifying the characteristic ways in which injuries occur, the types of people to whom they occur, and the typical consequences of these injuries. An important aspect of designing a system to define an injury

- **Epidemiologists** can be found in state and some local health departments, in academic and research institutions, and in consulting firms that specialize in epidemiology.

- **Statisticians and biostatisticians** can be found in state health and motor vehicle departments, academic and research institutions, and in corporate settings.

- **Nosologists** (experts in the classification of morbidity and mortality data) and medical records technicians can be found in hospitals and state health departments.

- **Medical records abstractors** (who are responsible for data abstraction of medical and other relevant records) can be found in academic, hospital, and clinical settings.

- **Individuals with expertise in economics, acute care, rehabilitation, and biomechanics** may also be helpful. They can be found in academic, hospital, and clinical settings as well as in private corporations and public service agencies.

- **Individuals familiar with the use of computers and statistical software** can help determine the feasibility of computerizing a data collection and analysis system. These persons can be found in the health departments of most states and large cities, as well as in many academic, hospital, and corporate settings.

Figure 1. People who can help with data collection and analysis.

problem accurately is effective targeting—that is, focusing on a specific type of injury (such as burns), injury-producing event (such as house fires), and/or population (such as people over the age of 65). The time period during which data are collected is also important, as collecting data over too short a period or at the wrong time can adversely effect the validity of the inferences drawn from the data (that is, the answer to the questions discussed under "Uses of Data" in Chapter 2).

Because most injury prevention and control programs have limited resources, data collection should be as cost-effective as possible. One method that can contribute to cost-effectiveness is to collect only the data needed to understand the problem under investigation. Often, however, the precise data useful for defining a problem or selecting an intervention, strategy, or target group are not obvious until at least some data have been collected and examined. Thus, in some cases, a program will begin with a broad data collection effort and later

narrow its focus. Again, the use of data from existing sources can often provide an important context for the design of a program's data collection activities.

Case Definition

A case definition specifies the injuries on which data will be collected. Such specification helps data collectors understand which cases should be included in or excluded from the system and lets data users know precisely which injuries the data represent. A case definition will usually specify injury type(s) (e.g., burns), severity (using the Injury Severity Score [ISS] or another severity rating system), population (e.g., children under the age of 5), geographic area (e.g., New York City), location (e.g., the home), mechanism and source of injury, and time frame (e.g., January 1, 1980, through December 31, 1985).

A case definition must be as specific as possible. The case definition used for a study of swimming pool drownings and near-drownings among children on the island of Oahu, Hawaii, for example, included the following elements.

- Injury type and severity: Serious immersion injuries in which, following immersion, a child was admitted to a hospital or died
- Population: Children under the age of 16
- Circumstance: Immersions in swimming pools, defined as "receptacles containing fresh water, 18 inches or more in depth, constructed and used primarily for swimming, dipping, or wading"; ornamental pools, fountain pools, and reflector pools were excluded
- Geographic area: The city and county of Honolulu
- Time frame: 1973–77[3]

Other case definitions may depend less on circumstance and more on injury type. A study of the incidence of acute brain injury and serious impairment in a defined population,[4] for instance, used a case definition that included

- Injury type and severity: "Physical damage to, or functional impairment of, the cranial contents from acute mechanical energy exchange . . . [including] penetrating and blunt forces that result in concussion, contusion, hemorrhage, or laceration of the brain or brain stem to the level of the first cervical vertebra." Excluded were "persons with fracture of the skull or facial bones, or injury to the soft tissue of the eye, ear, or face, without concurrent brain injury" or "brain damage from birth injuries, infections, chronic degenerative processes, or stroke. . . ."

• Geographic area: San Diego County
• Population: Residents, including active military personnel stationed in the county but excluding transients and those without identities. This definition of residency is consistent with the U.S. census eligibility rules for 1980. 1980 U.S. census data were used as denominators in this study.
• Circumstance: Patients dying from or hospitalized for new brain injuries
• Time frame: 1981

Sensitivity and Specificity

Sensitivity is the ability of a data collection system to include all of the cases of a particular injury. Specificity is the system's ability to exclude other phenomena that might be mistaken for the one under study. A data collection effort investigating adolescent suicide needs to be sensitive enough to identify all suicides within a specific area and time, including, for example, single-vehicle crashes that have been misidentified as accidental deaths. The system should also be able to exclude cases in which there is no evidence that the death was, in fact, intentional. The use of more than one data source can help improve sensitivity. An investigation of bicycle injuries, for instance, might include medical examiners' reports (for fatal injuries), emergency room records (for nonfatal injuries), and police records (for reports on incidents that involved motor vehicles though the injury was not treated in an emergency room). It is often necessary to weigh the sensitivity of a system against its cost-effectiveness and simplicity. The more sources from which data are collected, the more time it takes to collect data and match them with data from other sources. Any dramatic change of sensitivity must be taken into account. If the sensitivity of a system dramatically improves (perhaps because of new data recording procedures), then an increase in the calculated injury rates might be due not to an increase in injuries but to an increase in the percentage of injuries identified by the system.

Variables

A variable is a discrete piece of information that describes a specific aspect of the event or object being examined. A person investigating a homicide might want to know certain things about both the killer and the deceased (e.g., age, race, or presence of alcohol), the source of injury (e.g., knife, gun, or club), and the circumstances of the incident (e.g., family dispute, robbery, or gang fight). Comparing these variables across cases can reveal patterns im-

Time: Date and time of event

Place: Address (state; county; city; street address)
Description (factory; freeway; playground; home; etc.)

Injured person: Age; sex; race; occupation; place of residence; etc.

Injury: N code (if available)
Type (burn; concussion; poisoning; etc.)
Anatomic location (head; spinal cord; arm; etc.)
Severity

Cause: E code (if available)
Activity at time of injury (welding; skiing; driving automobile; etc.)
Intentionality (unintentional; assaultive; self-inflicted; etc.)

Other circumstances: Alcohol or drug involvement (medication; controlled substance; etc.)
Product involvement (automobile; handgun; toy; etc.)

Medical care: EMS; ER; hospital admission; trauma center; etc.

Outcome: Treated and released; number of days hospitalized; death; amputation; long-term disability; etc.

Figure 2. Examples of variables useful for injury data collection systems.

portant to understanding an injury problem and help in selecting interventions and strategies for reducing the number of these injuries. Figure 2 provides a summary of variables often useful in understanding injuries.

The variables on which a system focuses depend on the injury under investigation. A data collection system concerned with injuries from assaults could include information on the relationship among people involved in an assault (strangers, friends, family members, etc.). An investigation of drownings could include the type of body of water in which the immersion took place (bath tub, in-ground pool, lake, etc.). Both systems will require the age of the injured person as well as whether drugs or alcohol was involved.

Selecting Variables: An Example

An excellent example of both the selection of variables and the creation of a form for abstracting such variables from medical records is found in the Westchester County (New York) Health Department Injury Control Program. There, the Yonkers Emergency Room Surveillance System collects epidemiologic information on injuries in the five emergency rooms that serve the residents of Yonkers, a city of approximately 900,000. Data for a 25% random sample of emergency room injury cases are abstracted from emergency room logbooks and records and other hospital records onto a special surveillance form developed by the program. Based on this information, each case is assigned an E code. The data are then computerized, analyzed, and interpreted. Data collection procedures, as well as an explanation of the purpose of the program and how the data are computerized and analyzed, are described in the project's procedure manual. The person responsible for filling out the forms at each hospital was trained in assigning E codes to injury cases to ensure accurate and consistent E coding of all information.

Two variable lists were developed, one of demographic variables and the other describing the injury. These variables were abstracted from the emergency room and hospital records using the program's data collection form and later entered into a computer from these forms. This form also includes some narrative information not now computerized and is reprinted as Figure 3. The system uses a control number rather than the patient's name to preserve the confidentiality of the records.[5,6]

Time Considerations

Injury patterns and risk groups change over time. Therefore, data should be reasonably contemporary. For example, federal regulations on the flammability of fabrics combined with changes from loose- to tight-fitting clothing styles contributed to a dramatic decrease in clothing ignition deaths between 1968 and 1977.[7] Anyone seeking to prevent injuries of that kind would have been led astray if he or she had based programs on data from the early 1960s. Timeliness is even more important for evaluation purposes.

Some injuries, however, occur infrequently. Suicide, for example, although important and traumatic, is a relatively rare occurrence in any given community, so statistically meaningful results can be drawn only from data collected on a large population over a long period of time. Looking at too small a window of data can result in serious mistakes in analysis. If in one year a small town experiences an extremely large rise in the number of

deaths from poisonings, for example, the odds are great that the following year will show a marked reduction in the number of such deaths even if nothing is done to prevent them. By the same token, several instances of poisoning deaths after a long period in which none occurred does not necessarily indicate a sudden crisis but an averaging over time. Statisticians refer to this phenomenon as "regression to the mean." Its effect can be minimized by collecting comparative data (baseline data) for at least 5 years previous to the implementation of an intervention.

An adequate time span can also identify injury patterns that might otherwise go unnoticed. Some types of injuries are seasonal. House fire deaths tend to occur at night and primarily during the winter months.[7] Collecting data on injuries from house fires only during the summer would yield misleading information.

An adequate time span is also necessary to eliminate time effects, the consequences of unique circumstances that are restricted to a particular time period. Research done in Wisconsin, for example, indicated that lower water levels in lakes and streams during the drought of 1988 may have led to an increase in spinal cord injuries associated with diving because divers misjudged depths.[8]

Population Considerations

Few programs can collect data on all the injuries that occur. Most will select a particular type of injury and a particular group of people on which to focus. In some cases, a program's particular jurisdiction or funding source may direct the choice. An injury prevention and control program located in a state maternal and child health unit might collect data on injuries to the 0–19-year age group. An effort undertaken by a department of elderly affairs might look at the causes of injury among the elderly. A state highway department program might limit its investigation to injuries involving motor vehicles. In other cases, a more general data collection effort may be needed to help define the population most affected by a particular injury.

Choosing the group of people on whom data will be collected can have important consequences. One example of the way in which definition of a population can affect data analysis occurred in a study of children's deaths in Maine. When the Maine Department of Human Services defined the study population only by age, the data revealed death rates significantly lower than the national rates. When it analyzed the data by income level, injury death rates among children for all causes except

HEALTHY NEIGHBORHOODS
PREVENTIVE HEALTH CORNERSTONES
WESTCHESTER COUNTY DEPARTMENT OF HEALTH
INJURY CONTROL PROGRAM

Hospital Name _____ Code ☐☐

FILL IN FOR ALL INJURIES & ACCIDENTS:

A D D R E S S

BLD#	STREET NAME
APT	CITY
STATE	ZIP CODE

CENSUS TRACT	BLOCK #	RACE:

RACE:
1 ☐ Black   2 ☐ White
3 ☐ Amer. Indian   4 ☐ Oriental
5 ☐ Other   6 ☐ Unknown

SEX	BIRTHDATE		
	Month	Day	Year

Spanish Origin:
☐ 1 Yes ☐ 2 No ☐ 9 Unknown

PATIENT NUMBER:

PAYMENT SOURCE (check all that apply)
1 ☐ Medicare   2 ☐ Medicaid
3 ☐ Pvt Ins.   4 ☐ Self-Pay

Classification of Injury: (check one)
☐ 1 Unintentional   ☐ 2 Intentional   ☐ 3 Unknown

Injury Inflicted by (Check one):
☐ 1 Self   ☐ 2 Other   ☐ Unknown

Report date: ___/___/___ (month) (day) (year)   Date of Injury ___/___/___ (month) (day) (year)

Time of Injury ___:___ ☐ 1 -AM ☐ 2-PM

LOCATION OF INJURY EVENT: (CHECK ONE)

☐ 00 Home (specify):

☐ 01 Patient's own home
☐ 02 Other person's home
☐ 03 Unknown

☐ 10 Farm
☐ 20 Mine and quarry
☐ 30 Industrial place and premises
(e.g. garage, factory, workshop, warehouse)

☐ 40 Place of recreation and sport
☐ 50 Street and highway
☐ 60 Public building (if school, check below) (e.g. offices, commercial shops, stores, schools)
☐ 61 School

☐ 70 Residental institution (if hospital or nursing home, check below)
☐ 71 Hospital
☐ 72 Nursing home

☐ 88 Other location: _____
☐ 99 Unknown location

Did this event occur on the job?
☐ 1 yes
☐ 2 no
☐ 9 unknown

NATURE OF THE INJURY: (check all that apply)

☐ 01 Contusion/abrasion
☐ 02 Laceration/puncture
☐ 03 Foreign body in orifice
☐ 04 Other foreign body
☐ 05 Amputation/avulsion
☐ 06 Sprain/strain
☐ 07 Dislocation
☐ 08 Fracture

☐ 09 Concussion
☐ 10 Metabolic or toxic
☐ 11 Burn
☐ 12 Asphyxiation
☐ 13 Fetal/death injury
☐ 88 Other: _____
☐ 99 Unknown

DESCRIBE THE INJURY EVENT AND AGENT IN FULL DETAIL:

_____
_____
_____
_____
_____
_____
_____
_____

INJURY EVENT

FALL: (Check all that apply and specify)
☐ From height: From: _____
To: _____
☐ On same level
☐ From slipping, tripping or stumbling
☐ From collision, pushing, or shoving, by or with another person
☐ Other: _____
☐ From: bicycle
☐ Other: _____

FIRE/BURN (Check the one most applicable and describe)
☐ Fire in building (describe what happened below)
☐ Ignition of clothing
☐ Explosive material: _____
☐ Chemical: _____
☐ Tap water scald
☐ Other hot liquids/vapors: _____
☐ Hot object: _____
☐ Electric burn _____
☐ Sunburn
☐ Welding
☐ Other: _____

POISONING EVENTS (Check the one most applicable and describe agent)
BY:
☐ Drugs: _____
☐ Alcohol: _____
☐ Chemical: _____
☐ Gases: _____
☐ Food or Plants: _____
☐ Other: _____
☐ N/A

☐ **ACCIDENT INVOLVING A VEHICLE (Answer questions below)**

Describe specifically what happened _____
_____

Injured person:
☐ Operator   ☐ Bicyclist   ☐ Unknown
☐ Passenger   ☐ Motorcyclist
☐ Pedestrian   ☐ Other _____

Vehicle(s) involved:
☐ Automobile/Stationwagon/Van   ☐ Not applicable
☐ Truck   ☐ Other
☐ Motorcycle   ☐ Unknown
☐ Moped   ☐ N/A
☐ Bicycle

Place: ☐ Street or highway   ☐ Off street

Was injured person's vehicle in motion?
☐ Yes ☐ No ☐ Unknown ☐ N/A

OTHER EVENTS (Check all that apply)
☐ Struck by falling object
☐ Struck against, or by, object/person (describe event in detail below)
☐ Caught in or between objects

INJURY EVENTS caused by:
☐ Machinery, specify type: _____
☐ Cutting/piercing object/tool/appliance, splinter, specify object: _____
☐ Explosion, specify: _____
☐ Overexertion and strenuous movements (e.g. physical exercise/overexertion from lifting, pulling, pushing)

☐ Foreign body in eye
☐ Foreign body in other orifice, specify _____

☐ Human bite
☐ Animal bite, specify: _____
☐ Insect bite/sting specify: _____

☐ Drowning and submersion (describe in detail below)
☐ Obstruction of respiratory tract by object/food specify: _____
☐ Mechanical suffocation (e.g. plastic bag/pillow/refrigerator (detail below)
☐ Other: _____

DISPOSITON: ☐ 1 Released   ☐ 2 Admitted   ☐ 3 Deceased   E-Codes, if known: ☐☐☐☐☐☐ ☐☐☐☐☐☐

Figure 3. Sample data collection form.

suicide were found to be significantly higher for low-income children than for children of other income groups. This disparity in mortality rates could be used to target injury prevention efforts in a more effective way.[9]

DATA RECORDING

Data reflect more than facts about injuries; they also reflect the manner in which the facts were recorded. The less complicated a system is, the more likely it is that complete and accurate data will be collected. Simplicity is a characteristic that is desirable in any data collection system. Flexibility is also important. A data collection effort may need to be refined as more is discovered about the injury problem under investigation and/or the data sources being used.

Injury data are not always uniformly recorded. In some areas, a severely injured person may be transported to a trauma center that not only provides specialized treatment but has a recording system that can capture more data than other hospitals. The use of health care facilities by special population groups may also explain differences in data capture and the over- or underrepresentation of particular population groups. For instance, people with extremely low incomes often use emergency rooms for the same types of medical problems that middle- and upper-income people will take to HMOs or private physicians.

The type of injury may also affect reporting, particularly if the injury involves interpersonal violence or suicide. For reasons discussed in Chapter 16, suicide in the United States is substantially underreported. Child abuse, although reportable by law in every state, is also notoriously underreported.[10]

People who collect data at its source—the scene of an injury or the point at which an injury victim comes into contact with the medical or legal system —should also be consulted when evaluating the quality of data. Police officers, emergency medical technicians, and emergency room nurses can provide valuable information about the data they routinely collect. They can reveal what is systematically left off the records or point out that record keeping tends to decline during particularly busy periods, such as weekends.

Data collectors complete records and forms more carefully and meet the program's specific requirements more readily when they understand the purpose of their efforts. Every year, researchers at the University of North Carolina Highway Safety Research Center (HSRC) conduct traffic records workshops for police officers. "The first time we did this," said an HSRC official, "a police officer came up to one of my colleagues, put his arm around him, and said 'Son, I've been filling these forms out for 20 years and this is the first time anyone ever told me what they are used for' " (P. Waller, personal communication). Workshops given on the high rate of infant mortality and the importance of prenatal care for emergency medical technicians in the District of Columbia resulted in an increase in information on the ambulance run sheets of obstetrics patients. This information can help emergency room physicians determine whether they are dealing with a high-risk pregnancy (B. Coleman-Miller, personal communication).

A good data collection system is sensitive to the needs of the data collectors. Adding a complex and time-consuming data collection procedure to the already demanding job of emergency medical service technicians, police officers, or nurses is likely to yield data that are neither complete nor accurate. A data collection system must be nourished to remain effective. Feedback on the importance of their efforts is a major way to promote cooperation on the part of people whose efforts are essential to the success of a system. At the same time, it is a way to learn from the experience of data collectors what is working well and what is not.

Creating a clear and concise data protocol, that is, a written explanation and description of data collection procedures, can help ensure the quality of data. It can also be used to train new data collectors. Such protocols are especially important if the forms being filled out are computer-readable, a format that may not be familiar to everybody. Protocols are also useful for gaining the cooperation of the institutions in which one wants to collect data. This is discussed further under "Maintaining Confidentiality" later in the chapter.

DATA LINKAGE

Data linkage is the process of collecting data about the same cases from different sources. Because different sources collect different details about the same event, linking data can provide more information about an event than can data from any single source. Linking a program's data collection efforts with information available from existing sources can both limit the kinds of data a program needs to collect (thus simplifying the data collection system) and augment the information already available (thus customizing the information so it is directly relevant to the project).

The value of linking data from a number of sources can be seen in a study of child homicides.[11] The Connecticut Coalition to Prevent Childhood Injuries obtained death certificates from the office of health statistics for all homicide victims (identified by E codes) 19 years of age and under for the years 1980–85. The office of the chief medical examiner provided autopsy records for each victim. Local law enforcement agencies provided partial investigation records. From the information in the medical examiner's reports and law enforcement records, cases of justifiable or negligent homicide or killings resulting from legal intervention, such as a law enforcement officer killing someone in the line of duty, were excluded. Complete law enforcement records and birth certificates for all victims 4 years of age and younger were obtained and analyzed in order to provide information for a special substudy of the particularly tragic phenomenon of infant homicides.

The risk of child homicide in Connecticut proved to be greatest for those 15–19 years of age. Child homicide rates were found to be highest in the state's five largest cities. Firearms were used in 46% of these homicides. Over half of these firearms were handguns. The special substudy on homicides of children 4 years of age and younger revealed that the parent was the assailant in 80% of the cases. The parents' average age was about 22.

Using data that would not have been available from any one data source, this study showed that child homicide is a serious public health problem in Connecticut, identified those at greatest risk for such homicide, identified several important patterns for homicide that would be useful in targeting interventions, and provided baseline data and a model of a surveillance system for evaluation of intervention programs directed at this problem.

A number of states have projects that will permanently link several data bases to better track injury problems. The New York State Department of Health is creating a Highway Traffic Safety Coordinated Database by matching data from four information sources: the Department of Motor Vehicle files, death certificates, records kept by the state bureau of alcoholism, and information collected for the Fatal Accident Reporting System (S. Standfast, personal communication). The Maine Health Information Center has developed a system that links emergency room data, police crash reports, hospital discharge data, and death certificates. The system can trace an injury victim from the time the injury is reported to the police or emergency medical service through medical outcome (be that death or discharge from the hospital). It can reveal variations in the patterns of injury incidence and medical outcome by emergency medical service (EMS) region, county, and special population group. Because these data are also linked to the EMS Licensure Database, the system is also being used to evaluate the performance of the state's various emergency medical services. And because the resulting linked files can correlate the levels of training of the emergency medical technicians (EMTs) responding to an incident with other variables (such as medical outcome), they may yield valuable information on the relationship between the training levels of EMTs and medical outcomes.

Efforts to link data bases can be thwarted by the absence of names or other personal identifiers. There are techniques to match unidentified data.[12] One problem affecting data collection in general and linkage in particular is that facilities providing medical treatment consider information collected from patients to be confidential. Disclosing any identifying information may be seen as a violation of the patient–doctor relationship. For many injury prevention purposes, personal identification is not necessary. Sometimes, however, it may be useful to identify injured persons for follow-up interviews or to establish record linkage among several data bases.

MAINTAINING CONFIDENTIALITY

State laws on the confidentiality of hospital, police, and other official records vary. Some states have laws that allow health departments access to confidential medical records for the purpose of epidemiologic research. In some states, the state health department is allowed access to such records, while local health departments are not. Anyone starting a data collection program must be sure to investigate state laws and regulations concerning confidentiality.

Hospitals, trauma centers, and other institutions are often concerned about the confidentiality of their files. Institutional review boards (IRBs) protect the rights of subjects of biomedical and behavioral research. Most hospitals and health agencies have an IRB to fulfill U.S. Department of Health and Human Services requirements for such research. Among the rights they protect are those concerning the confidentiality of medical records. An institutional review board or data committee may have to approve any data collection based on its practical or research value and the ability of the system to protect the confidentiality of its records.

Several allies can help overcome these confidentiality concerns. Bringing a data collection and injury control program to the attention of the hospital commission or state medical association can be useful. Their support can be critical in gaining cooperation from individual hospitals or medical centers. Finding an ally within a hospital can be extremely valuable. Convincing a department head of the importance of the project or its usefulness to a particular department can create an important internal ally who will "run interference" with other hospital authorities. Data collection in an emergency room, for example, can help justify resource allocations such as increasing the number of staff allocated to the emergency room.

It will help to develop a protocol beforehand to demonstrate the ability to protect confidentiality. Such a protocol might include a written description of a project's purpose and methods, the responsibilities of all institutions and agencies involved in the project, and data collection and analysis procedures, with an emphasis on measures built in to protect the confidentiality of this information.

One technique that protects the confidentiality of records is assigning a code number to each case at the point of data collection. The code list linking names to code numbers can be left with the data provider (the hospital, trauma registry, etc.). This will preserve the confidentiality of the records yet retain the ability to return to the files to double-check the data or to update records. Using preexisting patient numbers, usually assigned when a patient is treated in an emergency room or admitted to a hospital, can simplify this process as well as make it easy to link emergency room and hospital discharge data. If two or more data sources on the same patient are being linked (e.g., those of a hospital emergency room and an ambulance service), a similar codebook can be created linking individuals' records in both sources by numbers instead of names.

Many states have laws governing the use of medical data. A familiarity with medical record laws is essential. Stating that one knows that any misuse of data by a project is not only unethical but illegal can help reassure cooperating institutions that the confidentiality of their data will be honored.

In fact, all data, even those accessible to the public, should be handled as if they were confidential. Assure sources that all names and codebooks matching names to code numbers will be kept under lock and key and that no names or other identifying information will ever appear in project reports or data summaries. And always seek the advice of an attorney on the requirements of a state's medical records laws.

DATA ANALYSIS

There are many ways to analyze data to reveal patterns useful for the targeting of injury prevention and control programs and the selection of appropriate interventions. Statistical analysis is a process that must be done accurately if it is to produce useful results. Improper analysis of even the highest-quality data can do immeasurable harm to the design and evaluation of a program.

Injury Rates and Years of Potential Life Lost

Two of the most common measures used in analyzing injury data are injury rates and years of potential life lost (YPLL). Injury rates describe the number of injuries that can be expected among a certain number of people (usually 100,000) over a period of time (usually 1 year), while YPLL is the difference between the length of time a person lived and the expected lifespan had he or she not died prematurely. Further information on the calculation of injury rates and YPLL can be found in "A Note on Calculations" at the end of this chapter.

Many injuries affect particular age groups disproportionately (e.g., pool drownings and toddlers, falls and the elderly, homicide and young males). For this reason, people often use age-specific injury rates, that is, injury rates for people within a particular age group, such as 1–4 years of age or over 65 years. Age-specific injury rates are also often combined with other population features, such as sex or race, to determine (or demonstrate) that a population is at high risk for a certain type of injury.

Age-adjusted injury rates reflect the differences in the average age of different populations. Suppose, for example, that a state has a much higher drowning rate than the nation as a whole. Does this mean that the citizens of that state are at greater risk for drowning than the citizens of the rest of the country? It might reflect the fact that the state contains a higher proportion of children than the country as a whole. More children drown than do adults. In fact, no resident of this state is any more at risk for drowning than if he or she lived anywhere else in the country.

Comparisons of non-age-specific injury rates should, if possible, be adjusted for age, using the national population as a standard. Because the average age of the population in a state (and the country as a whole) shifts over time, long-term comparisons of injury rates even in one place should also be adjusted for age.

The Problem of the Denominator

An injury rate is a fraction. The top half, or numerator, represents the number of cases under study, that is, the injured. The bottom half, or denominator, is the population from which the cases originate. The greater the uncertainty in defining the denominator or the inaccuracy in specifying its characteristics, the less accurate the resulting rate will be. If the number of homicides in a community (the numerator) doubled over a 10-year period, it would seem to indicate that the homicide problem has increased. But if the population of that community (the denominator) tripled in those 10 years, the homicide rate actually decreased. A similar effect can occur in communities with seasonally transient populations, such as university towns or vacation communities.

Another problem in calculating injury rates is specifying the size of the population whose injury problem is under investigation (the denominator). The larger the number of people included in a study (and the longer the time span for which data are collected), the more representative the data are likely to be. The degree of certainty that can be claimed for the statistical accuracy of an injury rate is called a "confidence interval." Determination of the number of cases necessary for a particular level of confidence in any calculation (or, conversely, the statistical confidence allowed by the number of cases a data collection system has revealed) is best left to persons with some statistical expertise.[13]

Exposure Rates

Another problem associated with the denominator is that of exposure rates. To assess accurately the injury rate associated with a particular activity, one must not only estimate accurately the number of people who take part in that activity but also consider the amount of time they engage in the activity. To compare the traffic injury rate of cab drivers with other automobile drivers would be misleading, because cab drivers, characteristically, spend more time behind the wheel than do people whose occupations do not involve driving. To compare these rates accurately, the differences in the time each group spends behind the wheel (that is, the time each is exposed to the risks inherent in that activity—the exposure rate) must be taken into consideration.

Suppose (this example is purely hypothetical) an investigation of traffic fatalities occurring among two groups, 10,000 cab drivers and 10,000 other drivers, revealed that one fatality occurred in each group over the course of a year. The injury rate for each group would be the same, 10/100,000. But suppose that investigation revealed that each cab driver drove an average of 100,000 miles a year, while the other drivers drove an average of 10,000 miles a year. A calculation of injury rates *per mile driven* would reveal that the other drivers had a mortality rate 10 times that of the cab drivers. Thus, while the injury rates based on population would be the same, the injury rates based on exposure would show the other drivers to be at considerably higher risk.

EVALUATING THE QUALITY OF THE DATA COLLECTION SYSTEM

Data provide information critical at every stage of injury prevention and control, from identifying a problem to demonstrating the efficacy of an intervention. Because the data a program collects help define a community's injury problem, inaccurate data collection and analysis can misdirect the selection of the program's target, interventions, or both. Inaccurate data collection can also affect the accuracy of a program's evaluation. It is thus critical periodically to evaluate the operation of the data collection system and its results. Several important questions to consider when evaluating a data collection system include the following:

• Are data being abstracted, coded, recorded, and aggregated consistently?
• Are desired standards of sensitivity and specificity being maintained?
• Do the data answer the questions that need to be answered?
• Are the data being used?
• Are the costs of data collection acceptable?
• Is the feedback to the data collectors adequate?
• Are the data being presented clearly?

Answers to these questions will show how well a data collection system is working as well as indicate the limitations of the collected data. How to use these data in the design, implementation, and evaluation of an injury prevention program is the subject of the next chapter.[14]

DISSEMINATING INJURY DATA

Information on the extent of injuries, the existence of specific injury problems, and what can be and is being done to address them can generate political, financial, and public support for a project as well as injury prevention and control efforts overall. A wide range of people may be interested in the details of a program and will benefit from having this information:

- Public health workers and officials
- Governors and state legislators
- Community groups, volunteer organizations, and official agencies interested in the health of particular populations (e.g., children, the elderly, workers, or minorities)
- Collaborating organizations and institutions, especially those that provided data
- The media
- The public

Data and the results of data analysis should be made available in an attractive and accessible format. Technical language should be avoided. Charts and other visual devices can help make statistical presentations more comprehensible to non-specialists. Data usually need to be interpreted. The audience should be told what the data indicate about the magnitude and nature of the injury problem being studied. (More ideas on dissemination are included in Chapter 5.)

COLLECTING NEW DATA: FOUR EXAMPLES

Data collection systems vary according to the information being collected, the number of people or injury events involved, the resources available, and so forth. No single program can serve as a model. Brief descriptions of three injury data collection efforts follow as examples of some of the possibilities that exist. A fourth example, the Queens Boulevard Pedestrian Safety Project, demonstrates how combining newly collected data with existing information played a major role in the design, implementation, and evaluation of a successful injury prevention and control program.

Oklahoma Spinal Cord Injury Surveillance Project

The Epidemiology Service of the Oklahoma State Department of Health created a spinal cord injury surveillance system in order better to understand both the extent and circumstances of this not infrequent and usually severe injury. Data are collected from neurologists and neurosurgeons throughout the state and at the state's two spinal cord rehabilitation centers. Since some spinal cord injury patients are also treated at smaller rehabilitation facilities in the state, as well as in neighboring states, these institutions also were included in the system.

Data on persons who die from spinal cord injuries before they arrive at a hospital are forwarded to the program by the state medical examiner's office.

Project staff at the state department of health negotiated written agreements with all of the participating institutions. After developing a case definition (that is, what specific types of injury would be included in this surveillance) and data collection forms, the project epidemiologist traveled to each of the sites to provide in-service training to the people who would fill out the forms.

The form solicits information on the date, day of the week, and time of the injury; the location of the incident; whether the injury was work related; whether the injury was intentional or unintentional; the hospital to which the injured person was taken; the admission and discharge dates; the circumstances of the injury; the possibility of alcohol involvement and the results of a blood alcohol content test, if one was administered; and whether the injury was motor vehicle related. If so, data on safety belt use, location of the injured person in the vehicle, speed at the time of the incident, and further questions on alcohol involvement are also asked. The form also asks for demographic information on income level and educational background.

The data collected by this system, as well as by two analogous state surveillance systems targeting burns and drownings, are expected to provide important information on the extent and patterns of these injuries in Oklahoma and be used to plan, implement, and evaluate prevention efforts.[15]

Philadelphia Injury Prevention Program Delay Studies

The Philadelphia Injury Prevention Program (PIPP), mentioned at the end of Chapter 2, is investigating the reasons why people delay seeking medical care for injuries, given that delays can increase both medical consequences and costs. The study focuses on falls among the elderly, falls among children, and stabbings. Interviews are conducted in participating hospital emergency rooms using an interview script specifically designed for each injury type. Information from these interviews is computerized and analyzed.[16] Obviously, an effort like this demands substantial staff time to identify cases, conduct interviews, code data, and so on, as well as specialized expertise to develop protocols and interview scripts and negotiate the use of identifiable medical records in participating hospitals. However, the quality of data generated by such an effort can be extremely valuable for planning interventions to decrease the number of delays in treatment.

Effects of Safety Seat Legislation on Pediatric Trauma

A number of studies have indicated that mandatory child passenger safety laws, if enforced, are effective in increasing safety seat use rates. However, the effects of these laws on injury severity, patterns, and utilization of the health care system have not been as widely addressed. Because of this, a project funded by the U.S. Department of Transportation at the University of California, Irvine, examined the effects of the California Child Passenger Safety Act. It studied changes in injury patterns and severity, changes in frequency and severity of head injuries, changes in the number of children injured in noncrash situations, and changes in the utilization of emergency rooms by young children.

Nine hospitals in Orange County, California, were selected to be part of an emergency room monitoring system, based on the large number of child motor vehicle injuries they treated. All child passengers were identified; data were collected through a standardized questionnaire with questions on (1) the injury event, (2) child restraint use, location in car, and so on, and (3) the medical consequences to and treatment of the child. Questionnaires were filled out by the parent or guardian when the child came to the emergency room. Project staff entered additional information from the child's medical records. Approximately 75% of the cases required a follow-up telephone interview with the parent or guardian because the forms had not been given to the parent or guardian by emergency room personnel or because this adult had also been injured. Data on fatal injuries were abstracted from documentation available at the Orange County Coroner's Office.

Data for the study were collected from the 2 years preceding enactment of safety seat legislation (1981–82) as well as the 2 subsequent years (1983–84). The findings of this study are discussed in more detail in Chapter 6. In brief, for children less than 4 years of age (those covered by the law), safety seat use increased from 25% to 50%. There was also a significant increase in the number of children involved in motor vehicle collisions who escaped injury. In particular, head injuries decreased by 16%, and there was a significant decrease in the number of children injured in non-collision situations (that is, by sudden swerves or stops of the vehicle).[17]

Queens Boulevard Pedestrian Safety Project

In 1985 the Traffic Safety Unit of the New York City Police Department began systematically to examine and intervene in the city's pedestrian injury problem. The unit reviewed police accident reports on fatal and severe pedestrian injuries from 1982–84. When these data were plotted on a street map of New York City, 39 severe and fatal injuries were found to have occurred in one 2.5-mile stretch of Queens Boulevard. Seventy-five percent of all pedestrian injuries in this area were to persons over the age of 65. Eighty-five percent of the pedestrians killed along this stretch were over the age of 65, while only 30% of the pedestrians killed in the city as a whole were in that age group.

The Traffic Safety Unit then undertook an indepth study to investigate why these injuries were occurring. Queens Boulevard, at this point, is a very wide roadway; pedestrians must cross distances of 150 feet. It carries a high volume of traffic, especially during the morning and evening rush hours. The study revealed the following:

• Much of the traffic exceeded the 30-mph speed limit.

• Elderly persons took an average of 50 seconds to cross the boulevard; however, the "WALK" sign allowed only 35 seconds.

• Because of the boulevard's width and vision loss among the elderly, many pedestrians could not read the "WALK/DON'T WALK" signs.

• Peripheral vision loss added to the confusion of elderly pedestrians about what direction in which to look for oncoming traffic.

• Vision loss obscuring the boundary between curb and street often caused elderly pedestrians to step off the curb and into the path of oncoming vehicles.

The unit designed and implemented a comprehensive intervention effort that included the following:

• The "WALK/DON'T WALK" signs were reset to allow more time to cross the street.

• The distance between pedestrians and the "WALK/DON'T WALK" signs was cut in half by placing additional pedestrian signals on median strips, so that vision-impaired elderly people could see them better.

• Large arrows were painted on the pavement near crosswalks to indicate the direction from which traffic was coming.

• Median island edge lines were painted to provide a highly visible distinction between the curb and the street.

• Oversized speed limit signs were installed at frequent intervals along the boulevard, and police increased enforcement of the speed limit in this area.

• An extensive public education campaign on pedestrian safety was conducted for senior citizens in the area.

A comparison of the injury data for this area for the 5 years before this intervention with the data for the 30 months afterward revealed a 44% decrease in the death rate (from 4.3 to 2.4 per 100,000), a 77% decrease in the rate of severe injuries (from 3.5 to 0.8 per 100,000); and a 60% overall reduction in the rate of fatal and severe injuries (from 7.8 to 3.2 per 100,000).

The Queens story holds lessons for other cities: data collected and analyzed by the National Traffic and Highway Safety Administration's Fatal Accident Reporting System (FARS) show that urban areas account for two-thirds of pedestrian fatalities in the United States. Its strategy can be used elsewhere:
• To identify hazards and injury-prone locations and establish priority sites for intervention
• To conduct in-depth analyses at priority sites to determine the circumstances in which injuries occur
• To select and implement interventions appropriate for those circumstances.

This project exemplifies the public health benefits that can result from a carefully designed and implemented injury prevention program.[18–20]

SUMMARY

Chapters 2 and 3 indicate the ways in which data can help identify and clarify an injury problem. The main points to remember in working with data are as follows:
• Experts in data collection and analysis can provide valuable technical assistance to an injury prevention and control program.
• It is efficient and practical to make as much use as possible of existing data before collecting new data.
• An appropriate case definition and the selection of relevant variables are important elements in the design of a data collection system.
• Data should be collected for a relevant population over an adequate period of time.
• A data collection system should be sensitive to the needs and constraints of the data collectors as well as to the confidentiality concerns of cooperating institutions.
• Linking data from a number of sources can reveal patterns that might not be apparent using data from a single source.
• Two of the most common, useful, and easy to calculate measures for the analysis of injury data are injury rates and years of potential life lost (YPLL).

• Evaluating a data collection system is essential. Inaccurate data collection can adversely affect a program's design, implementation, and evaluation.

A NOTE ON CALCULATIONS

The statistical measures reviewed in this section are frequently encountered in available data sources. Before using them in calculations and data analysis, it would be useful to review the more detailed, technical instructions available in textbooks such as *Statistics in Medicine* by Theodore Colton.[21]

Morbidity and Mortality Rates

Injury rates are calculated by dividing the number of injuries occurring in a population over a year by the total number of persons in that population and multiplying by 100,000. Injury mortality rates are calculated by dividing the number of injury deaths over the year by the population.

In either case, if more than 1 year of data is used, the result obtained by dividing the number of injuries by the population must then be divided by the number of years represented in the data. The result can then be multiplied by 100,000 to give the number of injuries (or deaths) that could be expected to occur to 100,000 people over the course of 1 year.

Age-Specific Morbidity and Mortality Rates

Age-specific rates are used to examine injury problems that affect a particular age group (such as falls among the elderly). Divide the number of injuries or deaths in the age group under consideration by the total number of people in that age group and then multiply by 100,000. For example:

$$\frac{\text{Injury deaths among males,}\ \text{age 1–4 years}}{\text{Total population of males,}\ \text{age 1–4 years}} \times 100,000$$
$$= \text{Injury mortality rate among}\ \text{males, age 1–4 years}$$

Again, if more than 1 year of data is used, the total number of injuries or deaths must be divided by the number of years.

Age-adjusted rates for a state are calculated using what is called the direct method, as follows:
1. Calculate age-specific rates covering the range of ages within the state's population (say, for those under 15 years, 15–44 years, 45–64 years, and 65 years and over).

2. Multiply each of these age-specific rates by the total number of people in that age group in the nation:

- State age-specific rate for those under 15 years × total number of persons under 15 years in the United States = A.
- State age-specific rate for those 15–44 years × total number of persons 15–44 years in the United States = B.
- State age-specific rate for those 45–64 years × total number of persons 45–64 years in the United States = C.
- State age-specific rate for those 65 years and over × total number of persons 65 years and over in the United States = D.

3. Add the resulting products together and divide by the total population of the United States.

$$\frac{A + B + C + D}{\text{Total population of the United States}} = \text{Age-adjusted rate}$$

Note: The CDC recommends the use of the 1980 census of the United States in determining a "standard population" for age-adjusted injury or death rates to ensure comparability.

Years of Potential Life Lost (YPLL)

Calculating YPLL is relatively straightforward:
- YPLL = Fixed age − Age at death
- *Fixed age* is an arbitrary age for potential years of life, typically 65 or 70 years of age, but any fixed age can be used
- *Age at death* is the person's age at the time of death

To calculate the YPLL for a specific population, total the YPLL for deaths under investigation (say firearms mortality) in that population over a time span (usually a year).

YPLL can be expressed as a rate by dividing the total YPLL for a population by the number of people in that population and multiplying the result by 100,000. Like other injury measures, YPLL is less prone to random fluctuation when based on more than 1 year of data. As a rule, 5 years of death data is recommended. As with the calculation of injury rates, the total YPLL should be divided by the number of years represented by the data to get an annualized rate (that is, the years of potential life lost per year).

For additional information about YPLL, see "Premature Mortality in the United States: Public Health Issues in the Use of Years of Potential Life Lost."[22]

INTERVIEW SOURCES

Beverly Coleman-Miller, District of Columbia Fire Department, September 27, 1988.

Susan Standfast, Injury Control Program, Division of Epidemiology, State of New York Department of Health, Albany, February 10, 1988.

Patricia Waller, PhD, Injury Prevention Research Center, University of North Carolina, Chapel Hill, April 28, 1987.

REFERENCES

1. Amputation Injury Surveillance System. Madison, Wisconsin: Wisconsin Department of Health, 1987.

2. Injury prevention professionals: a national directory. Newton, Massachusetts: New England Network to Prevent Childhood Injuries, 1988. Available from: Education Development Center, Inc., 55 Chapel Street, Newton, MA 02160.

3. Pearn J, et al. Swimming pool drownings and near-drownings involving children: a total population study from Hawaii. Milit Med 1980;145:15–18.

4. Kraus J, et al. The incidence of acute brain injury and serious impairment in a defined population. Am J Epidemiol 1984;119:186–201.

5. Procedure manual: Yonkers emergency room surveillance system. White Plains, New York: Westchester County Department of Health, 1988.

6. Program description, and goals and objectives. White Plains, New York: Westchester County Department of Health (undated).

7. McLoughlin E, Crawford JD. Burns. Pediatr Clin North Am 1985;32:61–75.

8. Maiman D, et al. Diving-associated spinal cord injuries during drought conditions: Wisconsin 1988. Morbid Mortal Weekly Rep 1988;37:453–4.

9. Children's deaths in Maine: 1976–1980, final report. Augusta, Maine: Maine Department of Human Services, 1983:9–10.

10. Reece RM, Grodin MA. Recognition of nonaccidental injury. Pediatr Clin North Am 1985;32:41–60.

11. Lapidus G. Child homicide in Connecticut. Hartford, Connecticut: Connecticut Coalition to Prevent Childhood Injuries, 1987.

12. Fife D. Matching fatal accident reporting system cases with National Center for Health Statistics motor vehicle deaths. Accident Anal Prev (in press).

13. Kraemer HC, Thielmann S. How many subjects?: statistical power analyses in research. Newbury Park, California: Sage, 1987.

14. Klauke D, et al. Guidelines for evaluating surveillance systems. Morbid Mortal Weekly Rep 1988;37(suppl 5): 1–18.

15. Oklahoma demonstration project on injury control. Oklahoma City, Oklahoma: Oklahoma State Department of Health (undated).

16. Philadelphia Injury Prevention Program: progress report and proposed plan. Philadelphia: Philadelphia Department of Public Health, 1988:31–42.

17. Agran P, Dunkle D, Winn D. The effects of safety seat legislation on pediatric trauma. Springfield, Virginia: National Technical Information Service, 1986.

18. Traffic safety unit. New York: City of New York Department of Transportation, 1986.

19. Retting R. Urban pedestrian safety. Presented at the New York Academy of Medicine Symposium on Motor Vehicle Injuries. New York: New York Academy of Medicine, 1987.

20. Retting R. Interventions need not be expensive: Queens Boulevard changes. Presented at the Centers for Disease Control/National Highway Traffic Safety Administration Second National Injury Control Conference. San Antonio, Texas: Centers for Disease Control/National Highway Traffic Safety Administration, 1988.

21. Colton T. Statistics in medicine. Boston: Little, Brown, 1974.

22. Centers for Disease Control. Premature mortality in the United States: public health issues in the use of years of potential life lost. Morbid Mortal Weekly Rep 1986;35 (suppl 2s).

Chapter 4: Program Design and Evaluation

Early one April morning in 1985, Emily and Susan, ages 10 and 8, respectively, were struck by a car as they started across an untended crosswalk on their way to the Madrona Elementary School in a Seattle, Washington, suburb. Severely injured, they were rushed to the Harborview Medical Center. Emily died on the operating table; her sister died the next day.

In June 1985 as 6-year-old Matthew left the Mount View Elementary School, also in suburban Seattle, for the day, he was hit by a car while crossing the street to join his mother and died.

Located in neighboring communities, the Madrona and Mount View Elementary Schools are part of the Highline School District in the State of Washington. The district serves a predominantly white, blue collar population of more than 100,000, of whom 16,000 are students. In April, after Emily and Susan died, there was growing concern for the children in the district. In June, with Matthew's death, concern gave way to anger and fear.

Highline Parent-Teacher Association (PTA) President Irene Jones organized a campaign to prevent further pedestrian injuries to local children. The PTA council surveyed the routes to school that children commonly used. Discovering that many had no sidewalks, the council requested that the school district take action.

At the same time, Drs. Frederick Rivara and Abraham Bergman were looking for neighborhoods where they could carry out interventions to prevent childhood pedestrian injuries. Both are pediatricians at the Harborview Medical Center in Seattle; Rivara, in fact, was part of the medical team that tried to save Emily and Susan. "We saw such injuries weekly," said Rivara. "So many of them resulted in death or serious head injury and long-term disability" (personal communication).

As long-time injury prevention researchers (at the University of Washington's Harborview Injury Prevention and Research Center), Rivara and Bergman were already investigating countermeasures against pedestrian injuries. The county engineering department put them in touch with PTA president Jones, and an injury prevention program was launched.

While PTA members informed parents about the community's pedestrian injury problem and enlisted the support of residents and officials, Rivara and Bergman focused on identifying the dimensions of the problem. Mortality data from the county medical examiner's office and morbidity data from the engineering department revealed that over a 5-year period, 65 Highline youths under the age of 15 had suffered a pedestrian injury. In 1985 Highline accounted for 30% of King County's child pedestrian fatalities but contained only 10% of the county's children. In terms of both morbidity and mortality, Highline's pedestrian injury rates were disproportionately higher than those of surrounding communities. In previous research, Rivara had discovered that 5- to 9-year-old children have double the pedestrian injury rate of all other age groups; low-income children are at particularly high risk, in part because they lack protected play areas.[1]

Based on a review of the literature, consultation with experts, and some group brainstorming, the PTA and Harborview team devised a multiyear effort to prevent pedestrian deaths to children. It focused on a four-pronged attack: improved police enforcement of laws to protect pedestrians, modification of environmental hazards (e.g., installing sidewalks), a school-based education program, and a mass media public education campaign. The group obtained a $400,000 3-year grant from the U.S. Public Health Service's Office of Maternal and Child Health to implement a Child Pedestrian Injury Prevention Program.

The project, launched in January 1988, accomplished several major tasks in its first 12 months: It developed an elementary school curriculum on pedestrian safety, implemented the curriculum in half of all Highline elementary schools, and supported the county engineering department in identifying and modifying hazardous areas. The project also surveyed parental attitudes about child pedestrian safety and established a formal Pedestrian Advisory Committee. And it developed two television public service announcements and initiated several newspaper articles about pedestrian safety. In its first year, the county council doubled the funds available to improve pedestrian areas, and the State House and Senate transportation committees began a review of legislation on pedestrian safety.

To monitor progress toward its objectives, the project is collecting evaluation data from several sources. Observations of child pedestrian behavior will be carried out before and at two points after the school program and mass media campaign. Parents of children who take part in the school program will be surveyed before and after the program to measure their attitudes and awareness of pedestrian risks. These data will be compared with data from surveys of parents whose children were not exposed to the program. Last, the impact of the program on pedestrian injury morbidity and mortality rates is being tracked through the Harborview trauma registry, county ambulance run records, police reports, and vital statistics records. In the second and third years of the program, the interventions will be expanded to several other communities across the state.

Unfortunately, in the field of injuries few programs are as well designed and evaluated as the pedestrian safety program. To address that need, this chapter examines how to develop program goals and objectives, select effective and appropriate interventions, and design and carry out evaluations. Because program design and evaluation are complex, resource-intensive activities, this chapter highlights methods for the development of partnerships among state and local agencies and community researchers and evaluators.

The examples used in this chapter are practical, real-world programs, each with strengths and limitations, and understanding both can assist practitioners in designing new injury prevention efforts. This chapter, unlike the other chapters in this section of the book, is divided into two distinct parts: program design and program evaluation. Because the two processes are so closely related and interdependent, we felt it better to cover them in one chapter rather than separating them.

PROGRAM DESIGN

No two injury prevention programs are precisely the same, but most share some features. By examining the key elements of one well-designed program, we can learn much about how to create others. First among these elements is the application of injury data and community assessment information to define the problem and to develop the goals that will guide the program's work.

Developing Goals

In Chapter 1, we discussed the importance of problem identification, the application of a systems approach to injury prevention, and the need to identify and work with the community and its leaders. Taken together, the first three chapters emphasized that community assessment information and new and existing injury data can help to identify and define both an injury problem and a population at risk. Deciding how to make use of this information to establish program goals is the first critical step toward prevention.

Goals, an essential component of an injury prevention program, determine the direction of the program and set forth what the program aims to achieve. Goals are written as broad, general statements of the long-term change the program is designed to bring about. Once the PTA members and Harborview staff had defined the local pedestrian injury problem and clarified some important contributing factors, they devised a program to "decrease child pedestrian morbidity and mortality by altering the manner in which our community thinks about and approaches pedestrian safety."[2]

Note that the goal is broad and its elements undefined. Goals are, as one writer phrased it, "timeless statements of aspiration."[3] Because they do not specify the program's approach, they must be accompanied by measurable objectives that detail how the program will achieve its goals and how it will measure its progress.

Setting Objectives

In the Highline School District, the program staff broke down their broad program goal into specific, time-limited, and measurable program objectives. Some objectives describe the desired outcome of the program in terms of injury rates; the target population's knowledge, attitudes, behavior, and physical environment; and public policy and practice related to injuries. These are called outcome objectives. Other objectives detail the level of activities designed to produce these outcomes and are called process objectives. It is important to remember that process objectives describe the relation between a project's activities and desired outcomes. Thus they are not simply lists of activities but rather quantifiable and measurable statements of what the program will have accomplished by certain dates. Table 1 provides a systematic overview of these elements of program design and examples from the pedestrian safety program.

A program's goals are stated in broad terms, but its objectives must be specific, time-limited, and measurable. For each of the pedestrian program's process and outcome objectives, staff have specified who will carry out the activity (process objectives only); who will be affected by it (outcome objectives

Table 1. The elements of an injury prevention program

	Definition[a]	Example from the Child Pedestrian Injury Prevention Program
Injury Problem Identified	Description of the injury problem and target population that the program seeks to address	Disproportionately high mortality and morbidity from pedestrian injuries to children 5 to 10 years of age in the Highline District and other parts of Seattle-King County
Program	A coordinated effort organized by a lead agency to reduce an injury problem among a target population	A pedestrian injury prevention program coordinated by the Harborview Injury Prevention and Research Center
Goal	A statement of changes sought in an injury problem	Decrease child pedestrian morbidity and mortality by altering the manner in which the community thinks about and approaches pedestrian safety
Objective	A statement of desired change in terms that are measurable, time-limited, and specific to a given target population	See below
Outcome objective	A statement of the desired impact of an intervention on injury morbidity and/or mortality, knowledge, attitudes, behavior, physical environments, or public policy and practice	

1. By December 1990 (after the program has expanded to other districts in the Seattle area), there will be a 40% reduction in the number of fatal and nonfatal pedestrian injuries in children under 10 years of age in Seattle-King County.

2. Compared to baseline data, the following changes in the pedestrian behavior of 5- to 10-year-old children will have occurred by December 1990:

    a. 50% reduction in the number of children under 6 years of age who cross streets without adult supervision

    b. 50% reduction in the number of children under 10 years of age who cross through streets or arterial highways without adult supervision

    c. 50% reduction in the number of children who cross streets when oncoming traffic is visible

    d. 50% reduction in the number of children who do not stop and search before entering the street

3. By December 1990, there will be a 50% increase in the number of driver citations for failing to yield the right-of-way to pedestrians, resulting from a higher degree of police enforcement of laws governing motor vehicle–pedestrian interactions.

4. By December 1990, there will be a 40% increase in the number of parents who report that they teach pedestrian skills to their 5- to 10-year-old children.

5. By December 1990, there will be a 40% increase in the number of parents with correct perceptions of children's developmental skills and how they relate to children's capabilities in traffic.

6. By December 1990, environmental modifications will be made to areas of the city defined as high-risk for pedestrian injury; there will be a 20% increase in the number of sidewalks installed in high-risk areas.

Table 1. Continued

	Definition[a]	Example from the Child Pedestrian Injury Prevention Program
Process objective	A statement of the desired level of achievement of program activities	(Sample of program process objectives) 1. By March 1988, the project coordinator will have developed a five-unit pedestrian safety curriculum for use in Highline elementary schools. 2. Beginning in March 1988 and October 1988, the project coordinator will have trained 20 teachers to use the curriculum, which will consist of in-classroom training, skill training in traffic environment, and reinforcement by parents. These teachers will then teach approximately 400 children using the curriculum. 3. By September 1988, the public information director will have produced and distributed two television and two radio public service announcements targeted to parents on the topic of pedestrian safety for children. 4. By October 1988, the public information director will have generated five articles in the print media targeted to parents on the topic of pedestrian safety for children. 5. By July 1989, the project coordinator will have worked with the county engineering department to increase the number of sidewalks available by 10%. 6. By July 1989, the project coordinator will have worked with the county engineering department to modify 20 areas hazardous to child pedestrians. 7. On an ongoing basis beginning in January 1989, the project director will monitor changes in prevalence, incidence, and severity of child pedestrian injuries by collecting and analyzing data from: trauma registry data on children treated in two hospital emergency rooms; county emergency medical services run report data; state department of transportation data on police-reported injuries; and county medical examiner's office logs on fatalities.
Intervention	A specific prevention measure or activity designed to meet a program objective. The three categories of intervention are: legislation/enforcement, education/behavior change, and engineering/technology	1. Improve police enforcement of pedestrian safety laws (legislation/enforcement) 2. Conduct a school-based educational program (education/behavior change) 3. Carry out a mass media educational campaign directed to the entire community (education/behavior change) 4. Make environmental modifications in high-risk areas (engineering/technology)
Strategy	An overall plan for meeting a program's goals and objectives that combines a set of interventions with the program's resources, a plan for the evaluation of its process and outcome, and a method for securing the necessary community and financial support to stay in operation	A multifaceted program to reduce child pedestrian morbidity and mortality, drawing on the strengths of the Harborview Medical Center and concerned citizens (More details are available in the program's proposal.)

[a] These are the working definitions of the National Committee for Injury Prevention and Control; other organizations or persons may define these terms differently.

only); what results are expected; how large an effect is necessary to demonstrate success; and how much time is required for the change. These elements should be made explicit in all process and outcome objectives.

Planning Interventions

The interventions adopted by the pedestrian safety program (i.e., the specific prevention measures or activities designed to meet a program objective) flow from its objectives. Interventions, as we saw in the Introduction, fall into three broad categories based on the approaches they reflect: legislation/enforcement, education/behavior change, and engineering/technology. Working with local police departments to enhance enforcement of existing laws, for example, is a critical legislation/enforcement intervention. The mass media and school-based efforts are education/behavior change interventions, designed to reach those most affected by pedestrian injuries—children and their parents. Engineering/technology interventions are embodied in the program's environmental modifications. Finally, deliberately including an evaluation component from the earliest stage ensured that the program's efforts would be fully documented and, if successful, could be replicated in other communities.

Identifying a Strategy

In their grant proposal to the Office of Maternal and Child Health, program staff delineated how the program's goals, objectives, resources, evaluation component, and long-term community support would coalesce into an overall program strategy. In essence, the program strategy is the comprehensive method for improving the target population's injury status.

Community Diagnosis: Using Injury Data and Community Assessment Information to Design a Program

In a world where staff time and other resources are always limited, planning is a wise investment and its payoffs significant. Setting goals and outcome and process objectives, selecting interventions, and developing a program strategy are the principal elements in the program design phase. Taking time to define each of these elements before initiating any activities can be crucial to success. The design then becomes a focus for future program activities. It is a tool through which staff can convey to others the program's goals and intentions, and it establishes

guidelines for evaluating the program's effect. But program managers need to be flexible about structure and activities throughout the life of a program. Flexibility enhances the program's ability to respond to obstacles and to take advantage of opportunities for improvement.

Community health programs, including injury prevention efforts, are not laboratory research enterprises. They are carried out in real communities with diverse populations. Examining and building on a community's values in the program design phase are important. Data can help to identify and define injury problems that are statistically important in a community. An intervention program is more effective when a community recognizes that the injury problem requires action. If community attitudes do not support the program, education about the injury problem may need to be carried out before an intervention program is launched.

Community members can play an active role in shaping program development and implementation. As one community organizer phrased it, "Citizens involved in the work of the program may serve as a source of continuous feedback to program staff. They may be consulted regarding the adequacy of the program—the degree to which it fulfills real community health needs. . . . In addition they may, and should, be asked for suggestions as to how the program might be improved . . ."[4] In some instances community involvement is crucial to program success: that is, when a program needs community volunteers in order to operate, or when a powerful public political constituency is required to keep a program funded. The PATCH program illustrates the role a community can play.

Local-Level Program Design: Holmes County, Ohio, PATCH Program

The 30,000-plus residents of Holmes County occupy 424 square miles of northeast central Ohio. A region of rolling farmland, oil and gas wells, coal strip mines, and small villages, Holmes County includes the largest Amish settlements in the world. One-third of the population of the county is Amish. The Holmes County Planned Approach to Community Health (PATCH) program, located at the county health department in Millersburg, is managed by Dr. Janet Fuchs, Director of Health Promotion for the agency.

PATCH is a model for program design and implementation in which a community moves through specific developmental stages. These stages are community mobilization (identifying people willing to participate), community diagnosis (examining morbidity and mortality data and conducting surveys of community opinion and behavioral risk factors), and community intervention (selecting, implementing, and evaluating interventions).[5] Although not formally included in the model, a review of what other community agencies are doing in injury prevention can also provide important information for program design.

Developed by the Centers for Disease Control (CDC), PATCH "provides a forum through which community health workers, health education professionals, and citizens plan, conduct, and evaluate health promotion programs at the community level," wrote CDC program coordinator Charles F. Nelson. "Working as a team, key representatives from the state health department, the local health agency, local community workers, citizens and CDC form an active partnership [to implement] health promotion programs designed to meet the priority health needs of the community."[6] Among those priority health needs is injury prevention.

The program organized an advisory group composed of staff from the state health department, the county health department, the county offices on education and aging, a local hospital, the local cooperative extension, CDC, and others. It also solicited advice from the local police and sheriff's office and members of the Amish community.

With direction from the advisory group, Fuchs began her community assessment by reviewing health department mortality data. The major causes of death in Holmes County were (by decreasing magnitude) heart disease, cancer, cerebrovascular disease, injuries, and pneumonia/influenza. When Fuchs calculated years of potential life lost (YPLL), however, injuries were far and away the greatest source of YPLL for the county as a whole. On average, each injury death resulted in 36.1 years lost, while each cancer death—the next greatest source of YPLL—resulted in 8.4 years lost.

The staff carried out two behavioral risk-factor surveys—one by telephone to gather data from the non-Amish, the other door to door to reach the Amish population. (Self-reported data on health and safety behaviors are less reliable then observational data but are much easier to collect.) These surveys considered the prevalence of obesity, sedentary lifestyle, smoking, use of safety belts, alcohol use, and hypertension among Holmes County residents. When the survey results were compared with statewide risk factor data, it was found that residents had much lower rates of safety

belt use and much higher rates of obesity and sedentary lifestyle. Fuchs reports that although the Amish rarely drive motor vehicles, they commonly ride as passengers—especially to and from work.

The staff supplemented this picture with additional data from the state department of highway safety and local sheriff's offices (data on motor vehicle crashes), local hospital emergency rooms (data on the location of unintentional injuries), and the state industrial commission (data on occupational injuries).

Taken together, these data revealed that motor vehicle occupant injuries were a major health problem in Holmes County. Averaging 800 to 900 crashes per year, Holmes had the highest per capita traffic collision rate in Ohio. Excessively high speeds and failure to yield were the primary causes of crashes. A surprisingly high 25% of all motor vehicle occupant fatalities in the county were of nonresidents. Most of these crashes occurred in the spring, summer, and fall months, during the peak of tourist traffic to and from the Amish settlements. Safety belt use was deplorably low throughout the county: 54% of the Amish and 50% of the non-Amish surveyed reported that they seldom or never wear a safety belt when they ride in a motor vehicle. (At the time of the surveys, Ohio had a mandatory child safety seat use law but no mandatory safety belt use law. A safety belt law went into effect in April 1987.)

"Our county is so small," commented Janet Fuchs, "and the sense of community so strong that it was fairly easy to collect the data and plan and carry out interventions. The fact that we were able to present our advisory group with *local* injury data that *we* had collected really convinced people that injury was a big problem—and one that we had to address" (personal communication).

In light of Fuchs's findings, the advisory group determined that one overall program goal would be to reduce motor vehicle occupant crashes and injuries in the county. Given data on the low rates of safety belt use throughout the county and an understanding of the correlation between nonuse of belts and higher occupant injury rates, the advisory group concluded that the program's primary focus should be to increase safety belt use. Based on county data for motor vehicle crashes and behavioral risk factors, the group identified adolescent and tourist drivers and local commercial truck drivers as high-risk target groups. They also selected a number of motor vehicle occupant–related program objectives and interventions, some of

Goal: Reduce motor vehicle occupant crashes and injuries in Holmes County

Outcome objectives

1. By December 1990, the program will have reduced the county motor vehicle occupant mortality rate from 22 to 18 per 100,000 persons.*

2. By December 1990, the program will have increased "always" and "nearly always" self-reported safety belt use among the Amish from 27% to 54% and among the non-Amish from 31% to 62%.*

3. By December 1990, the program will have decreased the annual number of Holmes motor vehicle crashes from 855 to 750.* (The crash rate should have been represented as the number of crashes per population or per miles driven.)

Process objectives and interventions

1. By December 1990, the program will have conducted a general education campaign informing the public about Ohio's mandatory use laws for child safety seats and safety belts and the importance of restraint use. Interventions: Various public awareness measures, from newspaper articles to notices on local bank statements.

2. By December 1990, the program will have conducted targeted educational sessions with small groups of children and adults. Interventions: A child safety seat curriculum in local preschools and safety belt curriculum in elementary schools, presentations to local 4-H clubs and other community groups.

3. By December 1990, the program will have conducted a safety belt use incentive campaign. Interventions: Distribution of raffle tickets to county fair attendees observed wearing a safety belt; distribution of prizes to safety belt wearers at drive-through banks.

4. By December 1990, the program will have conducted educational sessions for automobile and truck drivers on the topic of truck safety. Interventions: Special units in high school driver education courses, and seminars on truck maintenance, operation, and safety for local truck drivers and trucking employers.

---

* Note: The targeted reduced rates were arrived at after consultation with state health department and CDC staff, who suggested that a 15–20% improvement in rates of crashes and deaths and a 50% improvement in safety belt use rates (because of the new state law) would be appropriate objectives for this 5-year program.

Figure 1. Selected motor vehicle occupant-related objectives and interventions in Holmes County.

which are outlined in Figure 1. Program staff will measure the effectiveness of these interventions through pre- and post-intervention risk factor surveys, community opinion surveys, and county mortality and morbidity data.

Listing the program's objectives and interventions in tabular form highlights a number of points that require additional comment. Ideally, the process objectives and interventions should be stated in more quantifiable and measurable terms. For example, process objective 2 (educational sessions) should indicate the number of such sessions to be held, the expected audience, and the educational approaches to be employed.

Also, the program would have been strengthened by including a few legislation/enforcement and engineering/technology interventions along with those emphasizing education/behavior change. The education/behavior change focus was based on the advisory board's perception that the new state law was being vigorously enforced but needed a complementary effort to inform citizens about the law and the benefits of safety belt use. (It

is important to note that only secondary enforcement of the law is permitted—a motorist can be cited for violating the law only when he or she has been stopped for another violation.) More rigorous techniques for reaching tourist drivers are also needed in order to reduce their high occupant death rates.

As the Holmes County experience suggests, a community diagnosis comprising community assessment findings and morbidity and mortality data can provide a strong foundation for the development of program goals and objectives. Community diagnosis helps program developers document local health problems and community values, gives direction to program activities, and generates the baseline data against which progress toward program objectives can be measured.

Collaboration across agencies is critical for the collection and application of data. As Janet Fuchs noted, "A great deal of time and support must be committed to the effort. Agencies must be willing to work cooperatively. . . . From the onset of the project, participants should experience a sharing of activities and decision making, and old turf issues must be set aside."[7] The Holmes County PATCH program's advisory group clearly played an important role in moving the program from community diagnosis to intervention. As noted in Chapter 1, a lead state agency can design or implement a program itself, or it can facilitate community-based efforts by providing statewide injury morbidity and mortality data, technical assistance, and resources to local agencies, as shown in the following example.

The Lead State Agency and Program Design: CAHP

Founded in 1985, Colorado Action for Healthy People (CAHP) is a coalition composed of representatives from the state health department, public health association, health officers association, county nurses association, county extension agents, universities, medical societies, and a business coalition for health. Its mission is to motivate and assist communities to address the state's primary health problems: heart disease, tobacco use, substance abuse, teen pregnancy, and injuries.

In August 1987 CAHP, which is funded by the Kaiser Family Foundation and the Colorado Trust, formed a special Injury Prevention Task Force chaired by M. Patricia West, Director of Injury Prevention and Control at the Colorado Department of Health. The task force advised CAHP to allocate $40,000 to fund four community injury prevention programs. After a thorough analysis of statewide injury mortality data and a review of existing prevention programs, the task force specified that each funded program should focus its efforts on one or more of the following injury types: motor vehicle injuries, sports and recreation injuries, and intentional and unintentional home injuries.

CAHP's request for proposals (RFP), which was distributed to interested community agencies, suggested effective interventions in each of these areas.[8] The RFP was intended to motivate local agencies to develop programs that would address the community's injury problem and resources. CAHP monitors the funded projects, and the state health department provides technical assistance.

"We don't always have sufficient funds at the state level to stimulate the development of local prevention programs," said Patricia West. "But we do need to monitor injury fatalities and severe injuries. It's our responsibility to make these data understandable to people at the local level and to help communities make wise choices about where and how to intervene. We can educate communities and agencies and help reduce the barriers for injury

prevention practitioners" (personal communication).

Program Targeting

Whatever the characteristics of a program, it is important in the initial phases of program design to set goals and objectives that can reasonably be accomplished and avoid those that are vague, too high, too low, or too varied. Program targeting—the selection of feasible goals, objectives, and interventions and an appropriate, narrowly defined injury type and target population—is crucial to success. "Being successful in a reasonable time is extremely important," California injury expert Dr. Sylvia Micik and her colleagues pointed out, "because success will help you obtain community and political support, future funding and resources."[9]

Reducing injury morbidity and mortality can be a complex process; easy solutions are rarely available. Because a new program may take years to demonstrate a positive impact on community injury rates, defining and working toward attainable short-term program outcome objectives is an essential program strategy.

Hierarchy of Outcome Measures

A program's outcome can be measured in a variety of ways. Figure 2 outlines a hierarchy of injury outcome measures devised by Dr. Frederick Rivara.[10] The top four levels of the hierarchy represent measures of injury morbidity and mortality and thus are the most significant indicators of program outcome. Measures of knowledge and attitudes rank low because they do not *necessarily* lead to changed be-

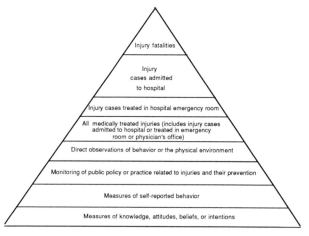

Figure 2. Hierarchy of injury outcome measures. Adapted from Frederick P. Rivara, MD, personal communication.

Data

1. Is there an existing system for collecting data on the host, agent, and environmental factors related to the community's injury problem, or will a new data collection system have to be designed and implemented?

2. Will the number of people exposed to the program be large enough to conduct a thorough outcome evaluation? If not, will the data gathered still be useful in evaluating the program's outcome?

3. In addition to measuring changes in injury morbidity and mortality, how many other ways are needed to measure the program's progress toward process and outcome objectives (e.g., self-report data; parents' reports of children's behavior; hospital, police, or school data; direct observation data; or a combination of these types of data)?

Target population

1. What is known about the target population (age, sex, race, socioeconomic status, educational level, location)?

2. What are the existing mechanisms for reaching the target population?

Intervention

1. What other injury prevention interventions are under way in the community?

2. What appears to be the most effective intervention or mix of interventions from the areas of legislation/enforcement, education/behavior change, and engineering/technology for the program to implement?

3. What other programs have implemented and evaluated similar intervention(s)? What successes and failures did they encounter? What can be learned from their evaluations about the effect of the intervention(s) on morbidity and mortality?

4. What type of materials, staff protocols, and feedback mechanisms will need to be developed? What is the best combination to use?

Resources and management

1. How much money will the budget allow for program implementation and evaluation? What outside funding sources are available to support the program?

2. What personnel will be needed? Will the implementation and evaluation be conducted by existing or new program staff or contracted to an outside agency? What other human and financial resources can be tapped? Can volunteers or a voluntary agency perform any program tasks?

3. How many and what level of agency staff are available to provide assistance?

4. What kind of training and how much will staff need?

5. Who are the major audiences for the program's evaluation findings? What do they want or need to know?

Figure 3. Issues to consider during program design.

havior and reduced injury rates, although they may predispose people to make behavioral changes.

Researcher David Sleet, Ph.D., points out that additional outcome measures may include an increase in the number of societal rewards for individuals who practice safe behavior (e.g., lower insurance premiums for people who have working smoke detectors in their homes) and an increase in the number of behavioral prompts present in the community (e.g., signs reminding people to wear their safety belts) (personal communication). (For additional references about developing goals and objectives for state and community injury prevention programs, see references 3, 6, 11, and 12.)

Issues to Consider During Program Design

After the injury problem and target population have been identified and defined, there are a number of questions to consider during the design phase of an injury prevention program. As outlined in Figure 3, these concern data, target population, intervention, and resources and management.

Selecting Program Interventions

Having used data and community assessment information to define the injury problem and establish realistic program goals and objectives, the next task is to select specific interventions. The following sections include both suggestions and examples from a variety of programs.

Cherokee Indian Hospital Pressure Cooker Intervention

Some years ago, Jackie Moore, the Indian Health Service Community Injury Control Coordinator in

Cherokee, North Carolina, worked as a radiologic technologist and volunteer emergency medical technician at the Cherokee Indian Hospital. There she developed an injury record system based on the hospital's ambulatory patient care reports. During a 1-year period, while following 1,500 cases through this system, Moore identified nearly a dozen women aged 40 to 50 who had suffered similar burns.

Struck by this odd injury cluster, she called each woman for additional information. Most cooperated, and she learned that each had been injured while canning foods with a faulty pressure cooker. In every case, the pressure gauge had stopped working, resulting in an explosion and burns.

Using these data, Moore designed a simple intervention. She arranged for a home extension agent to conduct a clinic to inspect pressure cooker gauges. Eighty-five people attended the clinic, which had been announced on cable TV and in the local paper. "We identified a large number of faulty gauges," said Moore, "including four or five that would have blown up had they continued in use" (personal communication). Although Moore did not continue her research beyond the intervention, her injury record system could also have been used to document changes in burn injury rates among the target population.

Small in scale and simple in its particulars, this is very nearly a paradigm of injury prevention design and implementation at the local level. Collection and analysis of data led to identifying a significant injury problem in a particular population. Additional data collection focused on the development and implementation of a targeted intervention. Further, the example illustrates a point as relevant to injury prevention as it is to advanced physics: the simplest, most direct solution that answers all of the relevant questions is the best. This should be borne in mind when selecting and evaluating injury prevention interventions.

A Multifaceted Approach to Injuries

Given the complexity of most injury problems and the paucity of well-evaluated, effective interventions, it is rare that a practitioner can attack an injury problem as directly as Jackie Moore did. It is often necessary to take a multifaceted approach.

As noted in the Introduction, interventions may be active or passive (sometimes also called "automatic"). Active interventions (e.g., wearing a safety belt or a bicycle helmet) require repeated actions by individuals to achieve the protective benefit. Passive interventions (e.g., installing air bags in motor vehicles or fire sprinkler systems in homes and workplaces) require little or no individual action yet provide automatic protection to everyone who is exposed to them.

Although some passive interventions (e.g., roadway design changes) are within the province of state or local injury prevention programs, most interventions that these programs employ are "active" to one degree or another. And because active interventions, by definition, require people to adopt and maintain a new behavior, a variety of complementary intervention approaches is often necessary.

An effort to pass and enforce a mandatory safety belt use law (a legislation/enforcement intervention) can benefit from an educational program to increase compliance among the public (through education/behavior change interventions). Because it is rare that a single intervention will significantly reduce a complex injury problem, program designers should carefully consider a mix of legislation/enforcement, education/behavior change, and engineering/technology interventions that complement each other and increase the likelihood of success. Consider, for example, the Children Can't Fly program in New York and the North Carolina Child Passenger Safety effort.

Children Can't Fly. In 1972 the New York City Department of Health launched Children Can't Fly,[13] a pilot prevention program to address the high incidence of childhood injuries and deaths resulting from falls from upper-floor apartment windows. During a 4-year period, such falls had claimed the lives of 123 children under the age of 15.

Program interventions included distribution, and in some cases installation, of free window guards (engineering/technology), education about the hazards of unguarded windows for families with young children living in high-rise apartments (education/behavior change), and follow-up home visits to the families of fall victims to provide information, referral, and window guards and to collect demographic and sociological information (engineering/ technology and education/behavior change). There were also community outreach and media activities.

Citywide mortality data revealed that from 1973 to 1975 the program reduced children's deaths caused by falls from heights by 35%. A similar decline in morbidity was recorded by the program's data collection system, which was based on voluntary reporting of children's falls from windows by hospital emergency rooms and police precincts.

Legislative intervention in the form of a 1976 amendment to the city health code further boosted the program's effectiveness; it mandated that landlords provide window guards in apartment buildings that housed children under 10 years old. At that point the focus of education/behavior change shifted; the program began to educate property owners about compliance with the new regulation and to inform parents of children under the age of 10 about their right to have window guards installed in their homes.

The basis for this legislation was the program's ability to demonstrate the effectiveness of installing window guards and educating parents about fall hazards. By employing a judicious mix of education/behavior change, legislation/enforcement, and engineering/technology interventions, Children Can't Fly made a significant impact on childhood deaths from falls in a relatively short time.

Budget cuts forced the program's elimination in 1981, and there was a resurgence of falls from windows among New York City children. In 1986, following several highly publicized deaths, the window guard regulation was strengthened, and enforcement and public education efforts were reinstated. Not surprisingly, window fall rates have begun to drop sharply. As program director Charlotte Spiegel notes, "There is an ongoing need for continued . . . intervention to control and effectuate a decline in [falls]" (personal communication).

North Carolina Child Passenger Safety Program. Several statewide child passenger safety programs have been successful in applying a multifaceted approach to preventing injuries. One of the best examples is based at the University of North Carolina Highway Safety Research Center (HSRC).

In 1977 the North Carolina Governor's Highway Safety Program began allocating to the HSRC funds to provide public education about motor vehicle safety and how to select and use child safety seats (engineering/technology and education/behavior change). These early efforts were correlated with a slight decrease in child motor vehicle occupant death rates and a 6% increase in safety seat usage rates (from 5% to 11%) for children involved in crashes from 1978 to 1981.

In 1980 the HSRC began receiving additional state funds to organize safety seat rental programs in North Carolina (engineering/technology and education/behavior change). In 1980 there were only 10 such rental programs; by 1984 there were approximately 125 in the state. The average rental fee to low-income parents was just 50 cents a month. The HSRC also established a toll-free telephone line to answer questions about child safety seats and to refer parents to local rental programs (education/behavior change). Written materials were also distributed throughout the state (education/behavior change).

Because data showed that the program's mix of engineering/technology and education/behavior change interventions was having a limited impact on safety seat use rates, the need for a legislation/enforcement intervention became apparent. In early 1982 advocacy efforts by an informal statewide citizens' network of child passenger safety supporters —the North Carolina Child Passenger Safety Association (NCCPSA)—resulted in passage of a statewide mandatory child passenger safety bill. North Carolina became the 10th state to enact a child passenger safety law.

The law went into effect in July 1982 and was narrow in scope; it covered children only up to the age of 2 and only when the parents were the drivers. As with other North Carolina laws, it was subject to a "sunset clause," which required legislative review and repassage in three years. The sunset clause became the law's saving grace.

Spurred on by the 3-year deadline, HSRC and NCCPSA stepped up their public education efforts. HSRC conducted full-scale observational surveys in eight cities each summer. With added data from telephone interviews of drivers in crashes that involved children and statistics from the state division of motor vehicles, the surveys allowed HSRC staff to monitor the combined impact of education, rental programs, and the new law on safety seat use rates.

The results were startling. Between July 1982 and December 1984, usage rates for children covered by the law jumped from 30% to 70%. Children in safety seats were 88% less likely to be killed and 56% less likely to be seriously injured than unrestrained children.[14]

In 1985, as the legislature reviewed the law's impact, all the public education and advocacy work bore fruit. According to one witness, "Dozens of parents from across the state, with their young children in tow, flocked to the state capitol in Raleigh, each to tell the Highway Safety Committee that their child had survived a motor vehicle crash unharmed because he or she was restrained. Most of them added that it would never have occurred to them to put their child in a safety seat if it weren't for the state law requiring them to do so."[15]

The pleas of these concerned parents, coupled with the convincing data compiled by the HSRC, led the North Carolina state legislature to make the law permanent, to expand its coverage to all children up to the age of 6, and to make all noncompliant drivers, not only parents, liable to fines. HSRC staff now work closely with law enforcement officials across the state to improve enforcement of the law.

"The key to our program's success," said Forrest Council, former program director, "is the combination of public education, local rental programs, and passage and enforcement of the law. None of these elements in isolation would have been nearly as effective as all three in combination" (personal communication).

Two Frameworks for Identifying Interventions

Identifying interventions is a deliberate process and not one to be performed in isolation. Injury prevention has a history from which we can learn. Thus, the first step in selecting program interventions is to identify effective interventions that have been used elsewhere. (Effectiveness here includes both the *efficacy* of an intervention [e.g., evidence that an adequately designed and tested motorcycle helmet will protect the wearer's brain in the event of a crash] and its *application,* or the extent to which it is used [e.g., even a superbly designed and tested helmet will protect only those motorcycle riders who wear it].)

Chapters 6–17 identify effective and promising interventions. Other useful sources of information include national, state, and local experts in the injury field and research findings reported in medical and public health journals. See Chapter 3 and Appendix B for further suggestions.

Conducting such a search gives a sense of the possibilities while conserving program resources. The next step is to expand the list by identifying *all* of the possible interventions for addressing the problem. Advisory boards or coalitions can be helpful in developing a list and then paring it down. "A group can be more creative," noted Sylvia Micik et al., "generate a greater number of options, and propose unique options not previously considered. Even more important is the capacity of the group to evaluate the options, develop implementation strategies and tactics, and set priorities for implementation."[9] In addition, professionals from the local community may be better equipped to judge the feasibility of implementing a given intervention than are outsiders unfamiliar with the target population and setting. Several tools and techniques can be used to expand the list of potential interventions, including the Haddon Countermeasures and the PRECEDE model.

The Haddon Countermeasures. The 10 countermeasures developed by Dr. William Haddon, Jr.,[16] to supplement his phase-factor matrix were described in detail in the Introduction. A thorough examination of these strategies can help program designers to avoid neglecting whole classes of potential interventions. Table 2 illustrates how the list can be used to generate interventions to prevent kitchen scald burns to young children.[9] The 10 countermeasures can help identify a variety of possible interventions for a selected injury type, but clearly some interventions will be more practical than others. As Haddon himself said, "The analysis is not per se a means for choosing policy . . . but is rather an aid for identifying, considering, and choosing the various means by which policy might be implemented."[16]

The PRECEDE Model and Maryland's Project KISS. PRECEDE is a diagnostic health promotion model[17] developed by a group of researchers led by Lawrence Green, now vice-president of the Kaiser Family Foundation. It focuses on *predisposing factors* that provide the rationale or motivation for behavior (including knowledge, attitudes, beliefs, values, and perceptions), *enabling factors* that allow a motivation or aspiration to be realized (including personal skills and availability and cost of health resources), and *reinforcing factors* subsequent to behavior that provide reward, incentive, or punishment for a behavior and contribute to its persistence or extinction (such as social support or tangible rewards).

PRECEDE is widely used to develop education/ behavior change interventions in health promotion. The Maryland Department of Health and Mental Hygiene's Project Kids in Safety Seats (KISS)[18] provides a good example of the process.

Begun in 1980, Project KISS involves the design, implementation, and evaluation of statewide child safety seat efforts. It seeks to address the behavioral factors that influence parents either not to use safety seats at all or to use them improperly. Following review of the literature, project staff identified the predisposing, enabling, and reinforcing factors and the corresponding interventions summarized in Figure 4.

Observational surveys at three Maryland sites in 1986 showed that safety seats were used among 70% of children under the age of 5—and were correctly used for 68% of those children. This represents a dramatic shift from the 22% overall use rate in 1982 (L. Bernstein, personal communication). As in North Carolina, however, these gains were the result of combining more than one approach to intervention: education/behavior change (presentations to the public), engineering/technology (loaner programs through Project KISS beginning in 1980),

Table 2. Haddon's countermeasures: scald burns in the kitchen to children under 5 years of age

Countermeasure	Intervention	Intervention type
Prevent the creation of the hazard.	Do not cook food (not practical). Eliminate hot water (not practical).	Education/behavior change
Reduce the amount of the hazard brought into being.	Lower temperature of water heaters.	Education/behavior change
	Manufacture hot water heaters to heat water only to safe temperatures.	Engineering/technology and legislation/ enforcement
Prevent the release of the hazard.	Create nontip pots, short-cord appliances.	Engineering/technology
Modify the rate of release of the hazard from its source.	Install small spouts for hot water.	Engineering/technology
Separate the hazard from that which is to be protected by time and space.	Cook when children are out of kitchen. Do not use hot liquids near children.	Education/behavior change
Separate the hazard from that which is to be protected by a physical barrier.	Use gates to kitchen to prevent entry. Use counter guards and pot handles that prevent spills.	Education/behavior change and engineering/ technology
Modify relevant basic qualities of the hazard.	Manufacture appliances that are low conductors of heat.	Engineering/technology
Make what is to be protected resistant.	Have children wear protective clothing (not practical).	Education/behavior change and engineering/ technology
Begin to counter damage done by the hazard.	Apply cold water to burns.	Education/behavior change
Stabilize, repair, and rehabilitate the object of the damage.	Develop a regional treatment system (pre-hospital, burn center, and rehabilitation care).	Engineering/technology and legislation/ enforcement

Adapted from reference 9.

and legislation/enforcement (starting in January 1984, when Maryland's mandatory child passenger safety law went into effect).

For more information on applying the PRECEDE model to injury problems, see reference 19. For a description of the Multilevel Approaches Toward Community Health (MATCH) model, which builds on the PRECEDE framework to conduct needs assessments and plan interventions, see reference 20.

After a list of potential interventions is created using the Haddon countermeasures, Green's PRE-

CEDE model, or another method, these criteria should be considered for each intervention.

• Has the efficacy of the intervention been demonstrated through research?

• Has the intervention been properly implemented in any other setting?

• Has it significantly reduced injuries or other risk factors related to injuries?

• Was there public acceptance of the intervention?

• How flexible was its implementation in the face

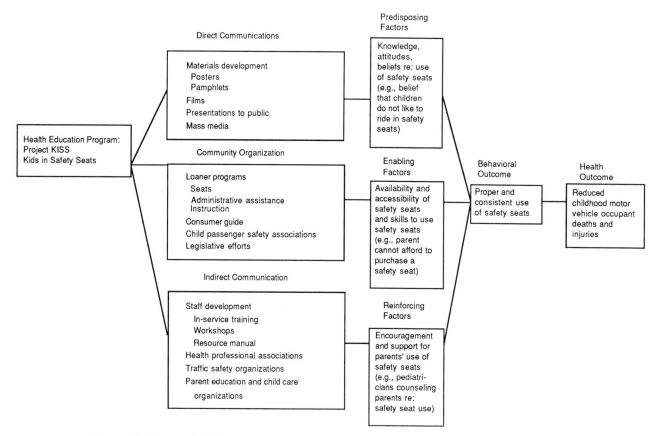

Figure 4. Project KISS application of the PRECEDE model. Adapted from reference 18.

of political, economic, or logistical barriers?

• How much commitment did it require by the target population?

• What did it cost to implement?

• How similar to the original intervention setting are your population, geography, and available resources?[21]

The answer to these questions may have to come in part from other programs that have carried out the intervention. Reviewing research on the role of contributing factors and injury antecedents can help in selecting and implementing program interventions. (For a review of some factors contributing to childhood injuries, see references 22, 23, and 24.)

Integrating Interventions into Existing Program Activities

Another important criterion is how well the proposed intervention can be integrated into the ongoing work of the lead agency or participating agencies. Interventions that can be integrated are more easily implemented and sustained.

Concerned that the eventual end of short-term state funds will signal the disappearance of injury prevention activities at the community level, sev-

eral state health departments have begun to require that contracting agencies integrate injury prevention activities into their ongoing direct care and preventive services.[25,26] As one state health department RFP phrased it, "Funds may *not* be used to establish separate programs that will end when the funding period is over."[27]

Recent evidence from the field of traffic safety reinforces the wisdom of this policy. Two strategies for safety belt promotion campaigns were tested in 1986 in Albany and Greece, New York.[28] In Albany, periods of increased enforcement of the state safety belt law were accompanied by a media blitz announcing the program. In Greece, police officers incorporated increased secondary enforcement of the law (ticketing violators only when they were stopped for other offenses) into their regular duties. Four short-term public information and education campaigns were conducted, many by the Greece police officers themselves. The Greece program was much less expensive than the one in Albany.

Safety belt use rose comparably in both cities. But 8 months after the Albany program had ended, safety belt use had returned to the lower, precampaign level. In Greece, however, many of the activities undertaken by the police continued and the increased level of safety belt use remained steady. Once interventions have been selected, the next critical question to ask—and answer—is, how will the program's effects be measured?

PROGRAM EVALUATION

Nearly every city and many smaller towns have some form of alcohol safety education program. Persons convicted of driving under the influence are sentenced to 10- to 13-hour educational courses to reduce the likelihood of their repeating impaired-driving offenses. The problem is that these short-term programs, in the absence of other interventions, appear to have little or no effect on subsequent drinking and driving by individuals who complete the courses. (See Chapter 6 for more information.)

When evaluation is done properly and consistently, it can determine whether and how a program is effective. From the evaluation of other programs, practitioners can learn which interventions might work in their local community and avoid those that are likely to be ineffective. The alcohol safety education experience represents failure on two counts: too few programs have been evaluated and too many programs have been implemented in the face of the generally negative evaluation findings from earlier ones. This ineffective intervention continues to be replicated on a large scale, thereby draining precious resources that would be better spent on more effective interventions. In a time of scarce resources, practitioners must be especially careful to use injury prevention resources wisely.

Every program should be evaluated, but not necessarily in the same way or at the same level of methodological rigor. An intervention that has been subjected to exacting evaluations and found effective in a variety of settings does not require intensive evaluation every time it is implemented. When funding is limited, programs can select interventions known to be effective with a similar population and then limit the outcome measures evaluated.

Programs that implement untried interventions or ones that have yielded conflicting evaluation findings, however, should include extensive outcome and process evaluation measures. There is an important distinction to be made between program evaluation, as described in this chapter, and "evaluative research." Evaluative research refers to the "utilization of scientific research methods" for the evaluation of public service and social action programs.[29] Marked by the use of research designs that "almost always [attempt] to conform to the experimental model," this long-term study of the effectiveness of new interventions is more appropriately housed in an academic setting where evaluation resources are sufficient.[29] Community-based programs frequently lack the funding and expertise to carry out such efforts.

Practitioners have a tendency to be intimidated by the notion of program evaluation. According to one community health planning manual, the very term "evaluation" carries a connotation of "judgment ('you didn't do this right') rather than assessment ('what happened, what lessons are in it, what steps can we take?')." The manual concluded that "the sting is taken out of the evaluation process when it is seen as just an extension of more familiar forms of decision making."[30]

Researcher Carol Weiss and others suggested that often evaluation findings have little impact on program and policy decisions because of "costs, ideology, self–interest, public reaction, [and] the rules and standard operating procedures of the institution."[31,32] Nevertheless, outcome and process evaluation is the only way to determine both if and how a program has reached its stated goals and objectives. Without evaluation, practitioners will be unsure of the effects of their labors and unaccountable to themselves, their funders, and their communities. "It is clearly irresponsible to undertake a new or major program without examining results at each stage, for it is only by discovering which strategies work that we can begin to build successful programs in injury control," according to several researchers.[33]

The discussion of the hierarchy of objectives earlier in this chapter indicated that there is a continuum along which a program can affect its target audience. Some level of evaluation is always possible. However, one level of evaluation should not be confused with another. A program may be able to measure a change in the knowledge, attitudes, or behavior of its participants, but demonstrating that such changes directly correlate with reductions in injury mortality and morbidity requires a more intensive and long-term evaluation process. Thus, it is important to bear in mind not only the level of an evaluation, but what that evaluation is capable of demonstrating about the injuries under consideration.

Why and When to Evaluate Programs

The most significant reasons for conducting both process and outcome evaluations of injury prevention programs are the following:

• Clarification of the impact of the program on the target population

• Identification and correction of implementation problems, thereby improving the program's future efficiency and effectiveness

• Development of program data for use in marketing the program

• Facilitation of replication of the successful aspects of the program

• Justification of the program's costs

Evaluation is an essential tool for program management; it clarifies if (and how much) the program is off track and helps identify necessary modifications. Attkisson and Broskowski suggest that "program evaluation is foremost a practical matter, of use to others primarily to the extent that it informs their current decision making, fulfills environmental demands for accountability, and points to future program improvement and efficiency."[34] (For a thorough discussion of program management, see Chapter 5.)

The time to plan an evaluation is in the very earliest stages of program design: "planning and evaluation should be thought of as interrelated and dependent processes working together at varying levels of emphasis throughout the life of a program."[11] Yet more times than not, unfortunately, evaluation is an afterthought. The intervention has been launched, some data may have been collected, and suddenly it is time to prove whether the program has worked or explain why it has not.

Such ex post facto evaluation simply does not work. "Once a program has begun, valuable baseline data are lost, compliance to data collection protocols is more difficult to obtain, and certain useful evaluation methods cannot be used," wrote one expert.[9] Accurate baseline data on the target population and the community are of paramount importance in assessing the program's performance. Before planning an evaluation, practitioners should consider again all of the questions raised in Figure 3 of this chapter.

Challenges to Injury Prevention Program Evaluation

Regrettably, real world practice often falls far short of the ideal. The number of long-standing intervention programs whose effectiveness has simply been presumed—without evidence or in the face of negative evidence—is legion. Why? Reasons include target populations too small to demonstrate statistically significant effects, inadequate funds for evaluation, lack of program staff well trained in evaluation, lack of understanding of the importance of de-

veloping evaluation components, focus on service delivery rather than evaluation, concern among program managers that negative findings may doom a program, and so on.

Practitioners and researchers must work with policymakers at the local, state, and national levels to ensure that more resources are allocated for evaluation of ongoing programs. Even in small, resource-scarce programs, managers should design their efforts so that essential process and outcome data are collected along the way.

Evaluation costs, though substantial, are unavoidable. In fact, they can be seen as analogous to the costs of navigation capability on a commercial airliner. Navigation not only gives location and progress toward the destination but also the corrections to be made to reach that destination. Evaluations provide the same kind of information.

Injury prevention efforts should be evaluated in terms of both process and outcomes. Process evaluation measures how the program conducted its interventions, what portion of the target audience was reached, and the cost of the program. Outcome evaluation examines progress toward the program's outcome objectives: Have injuries been reduced? Are there improvements in the target group's knowledge, attitudes, and behavior or the physical environment? Has the program had an impact on public policy or practice related to injuries? There may be other (perhaps unplanned) effects on the community worth noting: Is the community more aware of the injury problem? Have more local agencies become involved in injury prevention? Does the community's satisfaction with the program bode well for future efforts?

Generally, state and local health departments are experienced in conducting process evaluations of health promotion and disease prevention efforts. This type of evaluation usually translates well to process evaluations in the injury field.

Outcome evaluations, however, often require a substantial investment of time, money, and expertise. Not surprisingly, a recent national survey of injury prevention programs in state health departments revealed that 23 states conduct process evaluations of their injury prevention programs, but only 13 states report carrying out outcome evaluations.[35]

The experience of those who have attempted to measure the outcomes of other health promotion interventions sheds some light on why it is so difficult to demonstrate the impact of interventions on injury morbidity and mortality. The National High Blood Pressure Education Program (NHBPEP) was launched in 1972 by the National Heart, Lung and Blood Institute to conduct research and provide education to reduce hypertension among Americans. Seven years into the multimillion-dollar campaign, researchers were able to document slight increases in the percentage of the public who had had their blood pressure checked within the previous year and the percentage of hypertensives who were still taking prescribed medication. During this same period, stroke-mortality rates—which are closely associated with hypertension—dipped dramatically.

Researchers point out that there are simply too many extraneous variables involved to claim a direct cause–effect relationship between the NHBPEP interventions and the outcome of improved behavior and reduced mortality. Researcher Graham Ward suggested, "Although the temporal relationship is strong, one cannot accurately determine how much hypertension control has contributed to this decline. Our citizens also smoke less, have changed their eating habits, are more physically active, and receive better medical care. All these factors doubtless contribute as well."[36]

Given that this large-scale research effort was unable to demonstrate a direct correlation between its interventions and changes in health behavior and health outcome, it is not surprising that small-scale community-based programs experience the same difficulty. Establishing an association between interventions and reduced morbidity or mortality generally requires that a very large sample of the population be exposed to the program, a comprehensive evaluation be conducted to collect and analyze process and outcome data from a wide variety of sources, and data be collected for a fairly long period of time following the intervention. All of these conditions hold true for injury outcome evaluations as well.[37]

Collaborating with Outside Experts on Program Evaluation

As noted in Chapter 3, injury prevention practitioners sometimes need expert assistance, particularly with data collection and analysis and program evaluation. In designing an evaluation, a public health/social science evaluator (often found in an academic or research institution) or an epidemiologist (frequently based at a state health department or an academic or research institution) can be helpful. Chapter 3 gives advice about locating outside experts. If there is no one available in a local or state health department or university, CDC's Division of Injury Epidemiology and Control or the U.S.

Public Health Service's Office of Maternal and Child Health may have suggestions. For evaluating motor vehicle safety programs, the state office of highway safety may be able to help.

It is wise to begin working with an experienced evaluator as early as possible, even before planning a program. Some common obstacles—each with a workable solution—can arise when public health practitioners collaborate with academicians. Dr. Richard Biery, Director of Health of the Kansas City Health Department, phrased the potential difficulties this way: "Refining data to the level of academic or research desirability is not generally a health department priority. . . . [Health department personnel] frequently expect conclusions from the research community that the researcher is unwilling to make, or, on the other hand, so many contingencies are placed on the conclusions that the policymaker is frustrated."[38]

Learning to work together is a matter of compromise and maintaining respect for one another's disciplines. Clarifying discrepancies in the vocabulary of research and practice as they arise can move the focus from differences in perspective to what each side has to offer the other.

Another common barrier to collaboration is financial: Community and state agencies with few resources often cannot afford to hire evaluators from private institutions. Some injury prevention program directors overcome this obstacle by working with university researchers to develop internships. Graduate students knowledgeable about program evaluation give technical assistance to the agency as part of a thesis project or in order to receive academic credit.

The major issue for both sides to address is this: With the given resources—and in the real, uncontrollable world, with small sample populations—what kind of program evaluation is most appropriate and at what level of detail? With the money and staff to carry out a process evaluation but not a full-scale outcome evaluation, what is the best procedure?

One planning manual makes this suggestion: Rather than beginning program evaluation by asking the question, "What are we doing that we can measure?" program managers and evaluators should ask, "What do we need to measure in order to know what we are doing?"[11]

The next two sections explore some of the essentials of process and outcome evaluations. Before launching into a process or outcome evaluation, however, it is essential to prepare a brief written description of the program. A program description should include a summary of the magnitude and characteristics of the injury problem(s) to be addressed; the program's goals, process and outcome objectives, interventions, program strategy, and evaluation measures; and, perhaps most important, the rationale for selecting this particular approach to reduce the injury problem among the target population. A clearly written program description will go far to define the questions that process and outcome evaluations seek to answer.

Process Evaluation

An evaluation of a program's process objectives examines such questions as: Who was reached by the program? To what extent? Did implementation occur as planned? "A thorough description of what happened during program implementation can provide program staff and other interested parties information about which program features worked and which did not," suggested one group of experts. "At the same time such a description creates an historical record of the program that may be of value to others who want to implement it or a similar program."[39] (The terms "process evaluation" and "program monitoring" are used interchangeably throughout this section.)

Although an outcome evaluation can tell whether a suicide prevention program has changed teenagers' knowledge and attitudes about suicide and the actual incidence of suicide attempts and completions, process evaluation reveals which program components were successfully implemented and therefore may have been instrumental in bringing about those changes. In addition, process evaluation findings made available during the course of the program can help identify where alterations or "midcourse corrections" could improve the program's effectiveness. To measure progress toward each of the program's process objectives, staff will need a method to keep track of all program activities related to each of these objectives. Activities to be monitored may include, for example, the number of people exposed to each intervention, the number and type of materials distributed to the target population, the number of staff who receive special training, and the community's response to program materials and outreach activities. In designing the process evaluation, it may be necessary to modify existing agency record-keeping forms or develop new ones to keep track of each of these activities.

All process evaluations should carefully document any changes in program management as well as modifications to program goals, objectives, and interventions. Any changes at this level will eventually affect the program outcome. Monitoring alterations in the program's management and implementation systems simultaneously with outcome data will clarify how process influences outcome.

Three common methods of collecting data for a process evaluation exist: tabulating and analyzing records on program activities, interviewing or surveying the program participants and program staff, and observing the program in action. All are usually needed to form a complete process evaluation. Program records provide quantitative data on activities, and interviews and observations clarify the qualitative aspects of those activities. Program observation is much more time-consuming and expensive than interviews or surveys, but it can yield more accurate and useful information. Creating simple forms or logs can make record keeping easier and encourage consistency in data collection.

After the program is underway, a program manager can facilitate the process evaluation by making sure that all staff keep accurate records of materials distributed, outreach activities, and contacts with the target population and media by requesting that staff submit regular written progress reports and by holding regular staff meetings to discuss implementation problems. Lead agencies that fund community-based programs often find it valuable to organize regular meetings for staff from the various programs to discuss common problems. Providing feedback on data collection and evaluation findings can increase the motivation and accuracy of staff.

The REACT Project

The Rapid Evaluation and Counseling in Trauma Project (REACT), based in Boston at the Children's Hospital Division of Ambulatory Medicine, conducted a particularly thorough process evaluation of its first 12 months of activity, July 1986–June 1987. REACT, which was funded by the Massachusetts Statewide Comprehensive Injury Prevention Program (SCIPP), carried out an evaluation of the process used to implement each of three major program objectives: public education, clinic-based safety counseling, and home visits.

The project produced public service announcements and wrote feature articles, and staff appeared on radio talk shows to promote awareness of childhood injuries and their prevention. Project staff kept careful records on use of the public service announcements and on where newspaper articles were reprinted. They also conducted nine seminars on childhood safety for parents of young children that reached more than 350 families during the year. The seminar agenda, the number of participants, the types of materials distributed, staff time to prepare for and conduct the seminars, and the cost of materials were documented for each meeting (the cost for nine seminars was $470.00). Participant evaluation forms were not used, but they will be in the future.

REACT included child safety counseling in the well-child checkups conducted by nurse practitioners and physicians in the ambulatory care clinic. Eight in-service training sessions were held for the pediatric and house staff. Large quantities of home safety supplies were provided to the medical staff for distribution to parents who received child safety counseling. Staff also held telephone interviews with a sample of parents who had received the counseling. Along with collecting outcome data on how parents used the safety supplies, they questioned parents about such process issues as the number and type of safety supplies and pamphlets they had received and their reaction to the pamphlets. An additional process measure, a random audit of patient charts to determine how much safety counseling was documented by medical staff, revealed that over a 4-week period 40% of 385 eligible families were counseled. The total cost for safety supplies, the development of new forms, and project staff time—to train medical staff, order and renew supplies, and carry out the process and outcome evaluation—was estimated at $7,700.

Collaboration with a visiting nurse association to provide home safety inspections and safety supplies to parents of young children was the third component REACT evaluated. Staff recorded the number of inspections (70) and the number and type of supplies distributed. The estimated cost of training, supplies, travel, and staff time for the inspections was $8,000.

Remember that this process evaluation was conducted not in lieu of an outcome evaluation, but rather as a complement to it. An analysis of outcome data on the safety counseling and home inspection portions of the REACT project is now underway. (For more resources on process evaluation, see references 9, 33, 39, and 40.)

OUTCOME EVALUATION

Outcome evaluations are conducted to measure progress toward improving injury rates, the knowledge, attitudes, behavior, or physical environment of the target population, or public policy or practice related to injury. Outcome evaluation is markedly more complex than process evaluation. Figure 2 illustrates the hierarchy of outcome measures, or ways to determine the program's success in meeting its outcome objectives. The program budget, staffing, and other factors enable practitioners to decide which of these measures to use, but it is essential that every program include some form of outcome evaluation.

In general, it is better to conduct an outcome evaluation that measures the program's impact on injury rates rather than to rely on measurements of knowledge, attitudes, or behavior. But that may not be necessary if the intervention has already been used and evaluated with a similar target population and shown to reduce injury rates by a statistically significant amount.

Sometimes, too, the cost of evaluating the impact on injury rates is beyond the budget of the program. (The outcome evaluation budget should be lower than the budget for implementing the intervention.) This type of outcome evaluation is also inadvisable when the sample size is too small or the time frame is too short for the impact to reach statistical significance. An evaluation consultant can help determine what sample size and time frame are needed for the program impact to reach statistical significance.

Two situations require evaluation of program impact on injury morbidity and mortality rates: implementation of a previously untested intervention or application of an existing intervention to a new target population. Injury outcome data from these efforts are sorely needed because they can help other program planners make informed choices and move the field forward.

After outcome objectives are selected, the next task is to develop an evaluation design to track progress toward the objectives. Issues to consider include collection of baseline and outcome data, use of control and experimental groups, and assignment of individuals to groups.

When will evaluation data be collected? If data on the target population are not available before the intervention, there is no way to demonstrate the effect of the intervention. Therefore all outcome evaluations should include a method of collecting and analyzing both baseline and outcome data.

From whom will data be collected? Outcome evaluation usually requires data on an experimental group (those exposed to the intervention) and a control group (those not exposed). "Without any comparison group it is hard to know how good the results are, whether the results would have been as good with some other program, and even whether the program had any effect on the results at all," cautioned two researchers.[41] It is suggested that program directors applying untested interventions on a large sample size employ a control group. As an example, for a clinic-based project a similar clinic in another town or another neighborhood may provide a good control.

Using a control group is not always feasible. In an intervention to provide treatment for suicide attempters, for example, it could constitute withholding treatment, an unethical practice. Or in the case of a home safety inspection program, doing preintervention inspections in a group of control homes may itself improve the families' home safety practices, thereby invalidating outcome data.

A program using a control group must decide how to assign individuals to groups. Evaluations that randomly assign individuals to groups are termed experimental; nonrandom assignment is called "quasi-experimental." An experimental design maximizes the probability that the experimental and control groups will be alike initially, which helps form accurate conclusions about program outcome data. In the real world, however,

> "it is rarely feasible to assign individuals randomly to program participation or comparison groups, [and] most public health programs follow some type of quasi-experimental design. . . . To maximize both the internal and external validity of a quasi-experiment, evaluation planners try to ensure that the program participants and the comparison group are as similar as possible with respect to characteristics that might affect program results."[33]

(For more resources on outcome evaluations, see references 33 and 41–46.)

Evaluating Changes in Injury Rates: Two Examples

Documenting program impact on injury rates is both more significant and more challenging than documenting changes in knowledge, attitudes, behavior, or the environment, as these examples illustrate.

Massachusetts Statewide Childhood Injury Prevention Program. The Massachusetts Statewide Childhood Injury Prevention Program (SCIPP) evaluated the 3-year trial of five interventions[47] aimed at reducing childhood injuries: child motor vehicle passenger safety, home safety, poisoning prevention, burn prevention, and pediatric safety counseling. (The full program evaluation also measured changes in knowledge and behavior.)

The impact of the interventions on injuries of young children was measured through an injury surveillance system carried out in 14 communities. The system recorded injuries to experimental and control community residents who appeared at an emergency room or were admitted to a hospital as well as those that resulted in death. Outcome data showed that changes in injury rates were similar in both experimental and control communities for all types of injuries except for a statistically significant decrease in motor vehicle passenger injuries among young children in the experimental communities. This reduction cannot be attributed solely to the SCIPP intervention; concurrent debate over a proposed child passenger safety law was receiving a great deal of media attention and probably reinforced the message of the SCIPP intervention. A process evaluation telephone survey confirmed that a wide variety of passenger safety activities were underway in the experimental communities.[47]

In general, it is extremely difficult to determine people's level of exposure to a program's interventions by relying on morbidity and mortality data from a surveillance system. The surveillance system alone does not clarify which persons have not, for whatever reasons, been exposed to the interventions. However, a retrospective case–control evaluation design can overcome this barrier. In this type of outcome evaluation, data are collected only from individuals who have been treated for an injury. Then, by differentiating between those who have and have not been exposed to the intervention, evaluators can more accurately assess program impact.

Colorado evaluation of safety belt effectiveness. Retrospective case–control studies can be complex and expensive and are thus rare in the injury field. A model study was carried out by evaluators from the University of Colorado School of Medicine,[48] who were seeking to measure the effectiveness of safety belts in one- and two-car crashes.

From a pool of more than 24,000 eligible vehicles involved in crashes in 1984, evaluators identified 256 matched pairs of belted and unbelted front seat occupants, one or both of whom had been injured. Individuals were matched for crash severity, impact mode (front, side, rear, overturn), and size of car. The average age difference between belted and unbelted partners was less than 1 year.

The findings were conclusive: Although 84% of the unbelted partners received an injury, fewer than 50% of their belted partners were injured. Further analysis of safety belt effectiveness by type of impact "showed statistically significant reductions for frontal, overturn, and side impacts, with the best efficacy for frontal impacts."[48] Thus, this evaluation was able to show a clear correlation between safety belt use and reduced morbidity from motor vehicle crashes.

Evaluating Other Changes

Documenting the impact of program interventions on knowledge, attitudes, behavior, physical environment, or public policy and practice is one step removed from demonstrating changes in injury rates, but it serves an important purpose. In fact, some behavioral changes and modifications to state or local policies can serve as "proxy" measures for a reduction in injury. Proxy measures are alternative or substitute outcomes that have been proven by research or are generally accepted to be associated with reduced injury morbidity or mortality. For example, the correct use of child safety seats can function as a proxy measure for reduced childhood injury in motor vehicle crashes, given the clear relationship between proper use and decreased injury and death rates (see Chapter 6). Likewise, improving law enforcement officers' methods for responding to spouse abuse cases has led to a reduction in injuries sustained by women (see Chapter 13).

Positively influencing knowledge or attitudes does not automatically lead to changed behavior and reduced injury. Other factors such as having the skill to perform the behavior, prevailing social norms, and available reinforcement also promote behavioral change.

Attenuation of effect, or the decreasing impact of the intervention as it is carried out through the population, can greatly influence the outcome of education/behavior change interventions. Researchers from Project Burn Prevention, which sought to increase awareness about burn hazards and reduce the incidence of burn injuries in two communities in the greater Boston area, believe that the effectiveness of their interventions may have suffered greatly from attenuation of effect. Figure 5 represents the researchers' estimation of the impact of attenuation during the course of a hypothetical education/behavior change intervention targeted at 100 people. At each stage, only two-thirds of the remaining population move ahead to the next step. The net effect is that the intervention produces behavioral change leading to an actual injury being prevented among only 13 of the original 100 persons in the target population.[37]

Attenuation of effect may be influenced by the type of education/behavior change intervention employed. It has been suggested, for example, that untargeted information and awareness campaigns are likely to be less successful than behavioral interventions targeted at a single type of behavioral change (D. A. Sleet, personal communication).

Instruments to measure knowledge, attitudes, and behavior. Table 3 is a partial listing of instruments that measure some combination of injury risk, knowledge, attitude, behavior, or injury incidence. Although only two of the instruments listed have been scientifically validated, the others can provide a helpful point of reference. Practitioners who want to develop a new instrument should consult with professionals experienced in instrument design.

Direct observation to measure changes. The best way to measure changes in behavior or physical environments is through direct observation. Observation surveys of safety belt and child safety seat use are quite popular, and, when done by trained observers, tend to yield valid data. Home visits can determine major home injury risks and general safety practices among members of the target population. Questionnaires and telephone surveys can be used to supplement data gathered through observation surveys and home visits.

Evaluating changes in knowledge and behavior: an example. With funding from the Wisconsin Electric Power Company, pediatrician Dr. Murray Katcher directed a recent study[49] to evaluate the effectiveness of a mass media campaign on knowledge and behavior to prevent tap water scald burns. The program used newspaper, radio, and television messages to advertise a free liquid crystal thermometer for tap water testing and an accompanying educational brochure. Some 140,000 Greater Milwaukee citizens requested the thermometer.

Random telephone surveys of Milwaukee households before and after the campaign (with separate samples of 337 and 318, respectively) indicated an increase in awareness but no behavior change in tap water testing or lowering of water temperatures. However, a random telephone survey of 325 households that had requested the thermometer (those

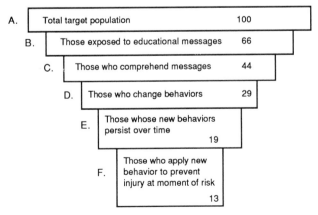

Figure 5. Hypothetical example of attenuation of effect. Adapted from reference 37.

Table 3. Selected list of injury prevention instruments

Name of instrument	What it measures	No. items	Where available
National Health Interview Survey: Health Promotion and Disease Prevention Section Questions on Injury Control and Child Safety and Health (1985)	Parents' knowledge and behavior; incidence of childhood poisonings, fire, and scald burns; use of child safety seats and safety belts	9	National Health Interview Survey Division of Health Interview Statistics National Center for Health Statistics 3700 East-West Highway Hyattsville, MD 20782
National Adolescent Student Health Survey (Office of Disease Prevention and Health Promotion, USPHS, 1987)	Form 1: Eighth and tenth graders' knowledge, attitudes, intentions, and behaviors concerning suicide and unintentional injuries	24	National Adolescent Student Health Survey Office of Disease Prevention and Health Promotion U.S. Public Health Service Room 2132 Switzer Building 330 C Street, SW Washington, DC 20201
	Form 2: Eighth and tenth graders' knowledge, attitudes, and behavior concerning violence	12	
Statewide Childhood Injury Prevention Program Telephone Survey (Massachusetts, 1982)[a]	Parents' knowledge, attitudes, and behavior and incidence of childhood poisonings, burns, falls, and motor vehicle occupant injuries	95	Statewide Comprehensive Injury Prevention Program Bureau of Parent, Child and Adolescent Health Massachusetts Department of Public Health 150 Tremont Street, 3rd Floor Boston, MA 02111
Childhood Accident Prevention Project Telephone Survey (California, 1981)[a]	Parents' knowledge, attitudes, and behavior and incidence of childhood burns, poisonings, and head injuries	400 +	Childhood Injury Prevention Project Maternal and Child Health North County Health Services 348 Rancheros Drive San Marcos, CA 92069
North Carolina Childhood Injury Prevention Project Telephone Survey (1987)[a]	Parents' knowledge, attitudes, and behavior and incidence of childhood falls, burns, choking, poisonings, and motor vehicle occupant injuries	53	Center for Health Statistics Division of Health Services North Carolina Department of Human Resources P.O. Box 2091 Cotton Classing Building Raleigh, NC 27602
New York State Bureau of Child Health and American Red Cross Safety Survey (1986)[a]	Parents' knowledge, attitudes, and behavior concerning childhood poisonings, burns, and motor vehicle occupant safety	10	Injury Control Program New York State Department of Health Empire State Plaza Corning Tower, Room 621 Albany, NY 12237
Injuries and Accidents: A Survey of Key Massachusetts Citizens Concerning a Critical Health Issue (1985)[a]	Knowledge and attitudes of legislators, health care, public health, and public safety practitioners concerning childhood injuries	31	Statewide Comprehensive Injury Prevention Program Bureau of Parent, Child and Adolescent Health Massachusetts Department of Public Health 150 Tremont Street, 3rd Floor Boston, MA 02111
Teaching Safety: A Survey of Massachusetts Elementary School Teachers (1985)[a]	Elementary school teachers' attitudes and practices concerning teaching safety to children in the classroom	28	Statewide Comprehensive Injury Prevention Program Bureau of Parent, Child and Adolescent Health Massachusetts Department of Public Health 150 Tremont Street, 3rd Floor Boston, MA 02111

Table 3. Continued

Name of instrument	What it measures	No. items	Where available
Survey of Parents' Attitudes Toward Child Safety (Safe Kids Campaign, 1987)[a]	Parents' knowledge, attitudes, and behavior concerning general child safety issues	37	National Coalition to Prevent Childhood Injury % Children's Hospital National Medical Center 111 Michigan Avenue, NW Washington, DC 20010
Violence Prevention Project (Boston) Telephone Survey (1987)[a]	Adolescents' and adults' knowledge, attitudes, and behavior concerning interpersonal violence	34 (pre) 35 (post)	Violence Prevention Project Health Promotion Program for Urban Youth Department of Health and Hospitals 818 Harrison Avenue, N.E.B. Room 112 Boston, MA 02118

[a] Has not been scientifically validated.

known to have been exposed to the intervention) revealed that 54% had used the thermometer and could report the test results. More than half of those who tested the water and found it higher than the safe limit of 130°F lowered their water heater thermostats.

Extrapolating to the target population at large, Katcher estimated that the program had prompted approximately 20,000 households to lower their hot water temperatures to 130°F or below. This relatively simple evaluation cost only $10,000 of the project's $210,000 budget.

Measuring changes in the physical environment: an example. In the late 1970s, the New York State Department of Health received two contracts totaling $15,000 from the U.S. Consumer Product Safety Commission (CPSC) to conduct and evaluate a pilot playground injury prevention program.[50] In cooperation with three county health departments and a university health education class, project staff sponsored 30 40-minute workshops for 1,500 community leaders involved in purchasing, installing, and maintaining public playgrounds. The workshops reviewed how to correct common playground equipment hazards. In addition, more than 60 community volunteers ("consumer deputies") attended a half-day training seminar to learn how to identify and remedy 12 observable, easily corrected hazards. The consumer deputies were charged with recording equipment hazards at their local playgrounds and working to convince community leaders and playground supervisors to rectify the hazards voluntarily.

Using a CPSC checklist, project staff did pre- and postintervention random surveys of 110 playgrounds in the three participating counties. The counties experienced an average drop of 35% to 51% in observed number of hazards per playground site. One county replaced hard playing surfaces (associated with severe injuries from falls) in 15 out of 24 playgrounds. Two large hospitals in one of the counties experienced a statistically significant 22% drop in playground injuries treated during the project period. This brief, well-designed program was thus able to evaluate the effectiveness of educating adults to make modifications in playground environments hazardous to children's safety.

Combining process and outcome evaluation: an example. An evaluation component that collects process and outcome data simultaneously can be designed. A good example is the recent evaluation of a Baltimore, Maryland, program[51] that distributed 3,720 free smoke detectors to city residents at their request. (The presence of working smoke detectors in residences is considered a proxy measure for reduced mortality and morbidity from smoke inhalation and flame burns.) Program evaluators measured an important process question (i.e., was the population reached by the program at high risk for fire injury or death?) and a crucial outcome factor (i.e., did the households that received a smoke detector install it?).

A researcher and a fire department official visited a random sample of 231 homes that had received a free smoke detector. They noted whether the detector was installed and operational. Baltimore Fire Department data were then used to compute fire death and injury rates for groupings of census tracts across the city. This information was correlated with data from a questionnaire administered during the home visit, which gathered demographic and personal risk factor information from each of the 231 households.

Evaluation findings revealed that the program

had achieved both process and outcome objectives: 81% of the inspected homes had an installed, operational smoke detector, and those who received free detectors were in census tracts at higher risk for fire than other parts of the city. This small-scale evaluation was thus able to measure program process and outcome through brief home visits to a sample of the target population.

USING THE FINDINGS FROM A PROGRAM EVALUATION

Once program evaluation data have been analyzed, they need to be translated into nontechnical terms more easily understood by a wide range of people —from program funders and local and state legislators to the media and general public. Communicating evaluation results is something few people take the time to do well. As one researcher/evaluator lamented, "Most evaluation reports seem to wind up as 40 mimeographed copies, 400 pages long, submitted to a program or a funding agency, and piled on a shelf."[42]

Following are guidelines for preparing a nontechnical version of an evaluation report.

• Above all, the report should be simple, clear, and brief.

• The executive summary is the most important part of the report.

• Statistical tables are often difficult for the lay reader to interpret; use them sparingly. Figures, bar graphs, and pie charts are easier to comprehend.

• Do not present data without analysis and conclusions. Data *do not* speak for themselves and must be interpreted for the audience. Describe the limitations of the data and the data analysis.

• State clearly what about the program works and does not work.

• Use the findings to make some recommendations, especially when there are clear policy implications that arise from the data.

Evaluation findings should be disseminated to everyone involved or interested in the program, including funders, program developers, program staff, all advisory committee or coalition members, local businesses and industries, the city council, state legislators, the media, and other injury prevention professionals in the state and across the nation. Policymakers at the local, state, and national level are an especially important audience. Staff should keep a list of people they contact during the course of the program and send them copies of the findings.

The findings can be disseminated through oral presentations, full-length written reports, press releases, legislative testimony, articles in agency or program newsletters, articles in professional newsletters and journals, and as background for interviews on radio or television or in newspapers. Chapter 5 includes a more detailed discussion of disseminating program information.

SUMMARY

Effective program design and evaluation involve understanding and applying the following concepts:

• Relevant injury data and community assessment information must drive program design.

• Selecting feasible goals, objectives, and interventions and a narrowly defined injury type and target population are crucial to a program's success.

• Programs should select the most effective mix of available legislation/enforcement, education/behavior change, and engineering/technology interventions.

• Integrating interventions into existing program activities is frequently the most effective way to ensure their continuity.

• All injury prevention programs need process evaluation and some form of outcome evaluation. These processes determine whether and how a program has reached its goals and objectives.

• Program evaluation must be planned at the initial stages of program design.

• Collaboration with experienced evaluators can greatly facilitate evaluation of state and local agency-based programs.

• Evaluations should collect both baseline and outcome data. Experimental and control groups should be used when appropriate, especially for new or untested interventions.

• Evaluation findings must be translated into clear, nontechnical formats and then disseminated to a wide audience.

INTERVIEW SOURCES

Larraine Bernstein, Director, Project KISS, Maryland Department of Health and Mental Hygiene, Baltimore, March 21, 1988.

Forrest Council, MS, Deputy Director, University of North Carolina Highway Safety Research Center, Chapel Hill, March 22, 1988.

Janet Fuchs, PhD, Director, Health Promotion, Holmes County General Health District, Millersburg, Ohio, March 15, 1988.

Jackie Moore, Nashville Area Community Injury Control Coordinator, Indian Health Service, Cherokee, North Carolina, March 4, 1988.

Frederick P. Rivara, MD, MPH, Director, Harborview Injury Prevention and Research Center, Harborview Medical Center, Seattle, Washington, August 10, 1988.

David A. Sleet, PhD, Professor, Department of Health Science and Graduate School of Public Health, San Diego State University, California, May 19, 1988.

Charlotte N. Spiegel, MA, Director, Window Falls Prevention, City of New York Department of Health, New York, August 15, 1988.

M. Patricia West, MSW, Director, Injury Prevention and Control Program, Colorado Department of Health, Denver, March 18, 1988.

REFERENCES

1. Rivara FP, Barber M. Demographic analysis of childhood pedestrian injuries. Pediatrics 1985;76:375–81.

2. Rivara FP. Child Pedestrian Injury Prevention Project (unpublished ms). Seattle, Washington: Harborview Injury Prevention and Research Center, 1987.

3. Model standards: a guide for community preventive health services. Washington, DC: American Public Health Association, 1985.

4. Dean D. Community outreach for health education programs: a practitioner's manual (unpublished manuscript). Boston: Statewide Childhood Injury Prevention Program, 1981.

5. Planned approach to community health. Atlanta, Georgia: Centers for Disease Control (undated).

6. Nelson CF, Kreuter MW, Watkins NB, Stoddard RR. A partnership between the community, state and federal government: rhetoric or reality. Int J Health Educ 1986;5: 27–31.

7. Fuchs JA. Planning for community health promotion: a rural example (unpublished ms). Millersburg, Ohio: Holmes County General Health District, 1988.

8. Request for proposals: community health grants. Denver, Colorado: Colorado Action for Healthy People, 1988.

9. Micik S, Yuwiler J, Walker C. Preventing childhood injuries: a guide for public health agencies. 2nd ed. San Marcos, California: North County Health Services, 1987.

10. Rivara FP. Injury prevention: how to tell if it works (unpublished ms). Seattle, Washington: Harborview Injury Prevention and Research Center, 1988.

11. Program management: a guide for improving program decisions. Atlanta, Georgia: Centers for Disease Control (undated).

12. U.S. Public Health Service. Promoting health/preventing disease: objectives for the nation. Washington, DC: U.S. Department of Health and Human Services, 1980.

13. Spiegel CN, Lindaman FC. Children Can't Fly: a program to prevent childhood morbidity and mortality from window falls. Am J Public Health 1988;67:1143–7.

14. Hall WL. Evaluation of the effects of the North Carolina Child Passenger Protection Law: final report to the general assembly. Chapel Hill, North Carolina: University of North Carolina Highway Safety Research Center, 1985.

15. Waller P. Using state injury data to identify injury problems, design interventions, and evaluate programs. Remarks at Current Issues in Injury Data Conference. Boston, Massachusetts: 1987.

16. Haddon W Jr. Advances in the epidemiology of injuries as a basis for public policy. Public Health Rep 1980; 95:411–21.

17. Green LW, Kreuter MW, Deeds SG, Partridge KB. Health education planning: a diagnostic approach. Palo Alto, California: Mayfield Publishing, 1980.

18. Eriksen MP, Gielen AC. The application of health education principles to automobile child restraint programs. Health Educ Q 1983;10:30–55.

19. Sleet DA. Health education approaches to motor vehicle injury prevention. Public Health Rep 1987;102: 606–8.

20. Simons-Morton BG, Brink S, Simons-Morton DG, et al. An ecological approach to the prevention of injuries due to drinking and driving. Health Educ Q (in press).

21. Calonge N. Objectives for injury control intervention: the Department of Health and Human Services model. Public Health Rep 1987;102:602–5.

22. Rivara FP, Mueller BA. The epidemiology and causes of childhood injuries. J Soc Issues 1987;43:13–31.

23. Zuckerman BS, Duby JC. Developmental approach to injury prevention. Pediatr Clin North Am 1985;32:17–29.

24. Spivak H, Prothrow-Stith D, Hausman AJ. Dying is no accident: adolescents, violence, and intentional injury. Pediatr Clin North Am (in press).

25. Request for proposals FY 1988: prenatal (MIC) programs and pediatric (C&Y) programs. Boston: Division of Family Health Services, Massachusetts Department of Public Health, 1986.

26. Request for proposals: home hazard control and program evaluation in a high-risk pediatric population. Madison, Wisconsin: Wisconsin Division of Health; 1985.

27. Bid application package for Comprehenisve Injury Prevention Programs. Boston: Division of Family Health Services, Massachusetts Department of Public Health, 1985.

28. Rood DH, Kraichy PP, Carman JA. Selective traffic enforcement programs for occupant restraints. Washington, DC: National Highway Traffic Safety Administration, 1987; DOT publication no. HS807.

29. Suchman EA. Evaluative research. New York: Russell Sage Foundation, 1967.

30. An Indiana guide to community health planning. Indianapolis: Indiana State Board of Health, 1987.

31. Weiss CH. Evaluating social programs: what have we learned? Society 1987(Nov):40–45.

32. Attkisson CC, Brown TR, Hargreaves WH. Roles and functions of evaluation in human service programs. In:

Attkisson CC, Hargreaves WA, Horowitz MJ, eds. Evaluation of human service programs. New York: Academic, 1978.

33. Lee AM, Vince CJ, McLoughlin E, et al. Guidelines for state and local agencies to conduct fire and burn injury control programs. Newton, Massachusetts: Education Development Center, 1981.

34. Attkisson CC, Broskowski A. Evaluation and the emerging human service concept. In: Attkisson CC, Hargreaves WA, Horowitz MJ, eds. Evaluation of human service programs. New York: Academic, 1978.

35. Harrington C, Gallagher SS, Burgess LL, Guyer B. Injury prevention programs in state health departments: a national survey. Boston: Harvard School of Public Health, 1988.

36. Ward GW. The National High Blood Pressure Education Program: an example of social marketing in action. In: Frederickson LW, Solomon LJ, Brehony KA, eds. Marketing health behavior: principles, techniques, and applications. New York: Plenum, 1984.

37. McLoughlin E, Vince CJ, Lee AM, Crawford JD. Project Burn Prevention: outcome and implications. Am J Public Health 1982;72:241–7.

38. Biery RM. Collaborative opportunities: the view of the US Conference of Local Health Officers. Public Health Rep 1985;100:606–7.

39. King JA, Morris LL, Fitz-Gibbon CT. How to assess program implementation. Beverly Hills, California: Sage, 1987.

40. Attkisson CC, Hargreaves WA, Horowitz MJ, eds. Evaluation of human service programs. New York: Academic, 1978.

41. Fitz-Gibbon CT, Morris LL. How to design a program evaluation. Newbury Park, California: Sage, 1987.

42. Weiss CH. Evaluation research: methods for assessing program effectiveness. Englewood Cliffs, New Jersey: Prentice-Hall, 1972

43. MacMahon B, Pugh TF. Epidemiology: principles and methods. Boston: Little, Brown, 1970.

44. Campbell DT, Stanley JC. Experimental and quasi-experimental designs for research. Chicago: Rand McNally, 1963.

45. Issac S, Michael WB. Handbook in research and evaluation. San Diego, California: Robert R. Knapp, 1971.

46. Kiresuk TJ, Lund SH. Goal attainment scaling. In: Attkisson CC, Hargreaves WA, Horowitz MJ, eds. Evaluation of human service programs. New York: Academic, 1978.

47. Guyer B, Gallagher SS, Chang BH, Azzara CV, Cupples A, Colton T. Prevention of childhood injuries: evaluation of the Statewide Comprehensive Injury Prevention Program (SCIPP) (unpublished ms). Boston: SCIPP, 1988.

48. Kerwin EM, Marine WM, Lezotte DC, Baron AC. Seat belt effectiveness in injury-producing accidents: the Colorado matched pairs study. Paper presented at the annual meeting of the American Public Health Association. Washington, D.C.: American Public Health Association, 1985.

49. Katcher ML. Prevention of tap water scald burns: evaluation of a multimedia injury control program. Am J Public Health 1987;77:1195–7.

50. Fisher L, Harris VG, VanBuren J, Quinn J, DeMaio A. Assessment of a pilot child playground injury prevention project in New York state. Am J Public Health 1980;70:1000–2.

51. Gorman RL, Charney E, Holtzman NA, Roberts KB. A successful city-wide smoke detector giveaway program. Pediatrics 1985;75:14–18.

Chapter 5: Program Implementation

Most residential hot water heaters are set at 140°F–150°F to provide a large reservoir of very hot water for taking showers and washing dishes. But only 2–5 seconds' exposure to water this hot can cause deep second- or third-degree burns in adults. Children burn in a quarter of that time. Each year an estimated 1,300 children under the age of 5 are hospitalized as a result of tap water scalds.[1]

In the late 1970s Wisconsin pediatrician Murray Katcher, having read work by Feldman et al.,[2] became increasingly concerned about the number of infants and children with tap water scald burns he was seeing in his Madison practice. Hoping to learn more about the incidence and etiology of these burns, Katcher began a series of studies. His data were conclusive: tap water scalds were an important cause of preventable burn injury; those most at risk were "the physically or mentally disabled, children younger than 5 years, and adults older than 65 years."[1] These early activities in surveillance and epidemiology would become the cornerstone of Katcher's efforts to prevent tap water scalds.

From the outset, Katcher believed the most effective intervention would be state or national regulations that required manufacturers to set all new hot water heaters at a maximum temperature of 120°F–130°F. However, there was very little support for such a mandate either in Wisconsin or at the federal level.

So Katcher began to focus his efforts on educational interventions. Following the positive outcome evaluation of the Greater Milwaukee mass media scald prevention program described in Chapter 4, Katcher targeted his educational efforts to protect young children. From 1982 to 1985, he used his contacts with Wisconsin's medical professional associations and worked with teams of pediatricians in clinical[3] and hospital settings to convey his message: the importance of educating parents about the dangers of excessively hot tap water.

By 1985 Katcher felt the climate was right to initiate state legislation to reduce hot water heater settings. But, believing the idea would be received enthusiastically, Katcher neglected to organize support for the bill, and the effort failed. Within 2 years he had put together a strong lobbying group of his associates from state medical and nursing associa-

tions and advocacy groups and had enlisted the support of several state legislators. This time the assembly and senate passed the bill.

The law, enacted in November 1987, requires that new water heaters sold in the state be preset at 125°F or lower, that new heaters bear a label describing the danger of hot water, that public utility companies mail annual notices recommending that water heaters be set at less than 125°F, and that landlords of dwellings in which the water heater serves a single unit reset the heater at a safe temperature before a new tenant moves in.

Similar laws were passed in Florida and Washington State. Katcher then approached water heater industry leaders to urge adoption of voluntary standards for safe temperature settings and labeling on all water heaters sold in the United States. It now appears likely that the industry will adopt such standards rather than face the need to comply with a series of different and possibly conflicting state laws.

Katcher's efforts offer a model for implementing an injury prevention program. At each stage of this 10-year process, moving from problem identification and data collection to educational interventions and then to legislation, he was careful to concentrate his activities in the areas in which success was most likely. When the political climate would not support a legislative intervention, he focused on educational interventions for medical professionals and parents. When education had helped to build sufficient support, he convinced his friends and colleagues to advocate passage of the water heater law. And when the opportunity arose to use the Wisconsin legislation as leverage for national voluntary standards, Katcher pursued that task.

"In retrospect," he notes, "it has been a logical progression from one step to the next. I do feel a need, though, to be ever alert to what lies ahead. For example, now that it looks likely all new hot water heaters will be installed at safe temperatures, we must educate all those people whose heaters aren't preset to test their water temperature and lower the setting if it's too high" (personal communication).

As this example illustrates, program implementation—which rests on such diverse components as program management, coalition building, and injury prevention advocacy—can be the most rewarding and most challenging aspect of injury prevention work. Although each program follows a unique course, common elements persist: program management, program evolution, coalition building, media and public relations activities, injury prevention advocacy, and program institutionalization. This chapter treats each of these topics in turn.

PROGRAM MANAGEMENT

Effective program management is one of the most crucial—and most often overlooked—factors. The leadership of a concerned, motivated individual can be the catalyst for getting a fledgling program off the ground, but once start-up funding has been secured, longer-term management issues come into play. In this section, we examine two particularly important management tasks: securing a staff and moving the program from design to implementation.

Program management is a skill that is developed over time and with practice. Anyone new to it may benefit from drawing on the experience of longtime managers. Just as we recommend that novices to data collection and analysis consult with experts in that field, so too can new program managers learn from their more seasoned colleagues. There is also a large, specialized literature on management, and most universities and colleges offer management courses.

Program Staffing

"You won't have a successful program without good staff," commented Wisconsin injury control coordinator Hank Weiss (personal communication). The program director is the pivotal staff position. If the director is a staff member from the lead agency who was instrumental in getting the program funded, it is simply a question of reassigning this individual to the new position.

Although agencies prefer to make use of existing personnel, this is not the best method if already overburdened staff are assigned injury prevention as an additional responsibility. On the other hand, Joan Ascheim, injury prevention program manager at the New Hampshire Division of Public Health Services, commented that several of her staff prefer to split their time among injury prevention and a number of related child health programs because it allows them to develop professionally in more than one area (personal communication).

Staff are sometimes hired to implement a program that has been designed by administrators who will be only minimally involved in its implementation. The new program director and staff must put into effect plans that others laid out. As a result, they may need to modify the program's goals and objectives based on their skills and their understanding of the community.

If a program involves community outreach, it can be essential to obtain staff who share the language and culture of the target population. When administrators at the Violence Prevention Project of Boston's Health Promotion Program for Urban Youth decided to hire community development coordinators who had grown up in the project's inner city intervention sites, interactions between the project and community organizations were greatly facilitated. Noted former codirector Dr. Howard Spivak, "Having staff who represent the community you're working with can make an enormous difference to the success of a program" (personal communication).

Staff training. Program staff new to the field of injury prevention or to the lead agency need training. Susan Gallagher, former Massachusetts injury prevention director, suggests that every program develop standard procedures for training new staff. "In my experience," Gallagher noted, "helping staff become comfortable with program design and the application of injury data are the two most important objectives of training" (personal communication). There are other topics to review in training sessions: the magnitude and characteristics of the injury problem in the community, state, and nation; injury prevention strategies; the history of the agency's involvement in prevention; and the existing program. Depending on staff expertise, training may also include program evaluation, coalition building, and media/public relations activities.

Staff training generally is the director's responsibility, but whoever conducts training should be familiar with the injury literature and have experience in training so that information is presented to new staff at the right level and in the appropriate amounts. Refresher sessions and discussions of recent developments in the injury field can help to maintain staff morale.

Injury prevention seminars or courses in schools of medicine or public health are another vehicle for training, as are national injury conferences, such as the annual meeting sponsored by the Centers for Disease Control (CDC), or injury prevention sessions at other national meetings (e.g., American Public Health Association). (See references 4–6 for notes on injury prevention training manuals that have been developed to familiarize practitioners with the field.)

Beyond training itself, continued contact with other injury prevention professionals can provide staff with a tremendously rich source of information. The New England Network to Prevent Childhood Injuries has developed a computerized national directory of injury prevention professionals. (For more information, see reference 7.)

Keeping abreast of the literature on injuries and injury prevention can be both a training method and a management tool. Reading books on injuries and skimming relevant journals and newsletters for new information about injuries, though time-consuming, is fruitful. "In conversations with fellow practitioners and researchers," said Gallagher, "I regularly hear about new articles or papers. We seem to help each other keep informed" (personal communication). (See Appendixes B and C for a list of relevant journals and a partial list of injury prevention newsletters.)

Until there are well-established injury prevention training programs at the undergraduate and graduate levels, staff training will remain a challenge for most programs. However, as Patty Molloy of the New England Network to Prevent Childhood Injuries noted, "So many of us come to the field of injury prevention from related public health or safety programs. This type of background is extremely valuable when it comes to developing expertise in injury prevention" (personal communication).

Long-term staffing issues. Maintaining a well-trained, competent staff can be difficult in a field where salaries and job security are comparatively low. Practitioners interested in the field can become discouraged when they realize they will have no "natural home" or clear-cut career path to follow. Program managers, who should be sensitive to both the aims of the program and the personal needs and career goals of staff, must recognize that the two may not always be in accord.

As a program expands, there is a natural transition when the program manager needs to turn over control of day-to-day activities to staff. This can be uncomfortable for manager and staff alike. Weiss, of the Wisconsin Division of Health, recalled that as he assumed additional managerial responsibilities he found it difficult "*not* to dive into the data. It's frustrating to lose some of your research skills because you're too busy with management. But time demands make these two priorities incompatible" (personal communication).

Along with regular staff meetings, it is useful to provide a forum once or twice a year for staff to re-flect on recent accomplishments and plan for the future. "Holding a half-day or all-day retreat is a good technique for reviewing how far the program has come in the previous year and where it's headed in the next year," suggested Molloy (personal communication).

Working with volunteers. Some programs rely on community volunteers to assist staff. Volunteers can help a program accomplish more, extend limited staff resources, and establish the program's credibility in the community. As a group of vocal supporters, they can be advocates for a program's continuation. The disadvantage of relying on volunteers is that they may lack the training and commitment to continue their participation in the program over a long period of time.

Before recruiting volunteers from the community, their tasks should be determined. The activities of volunteers should complement staff efforts, not replace them. Deborah Dean, a former injury prevention community organizer, noted that volunteer roles usually fall into four categories: outreach assistants, service provision assistants, administrative assistants, and fund-raisers. She also suggested that "the type of volunteer who seems to work out best is the person for whom service to the program fits in with or enhances his/her regular duties or job description."[8]

Volunteers can be trained through a simplified version of the procedures used for staff members. As with program staff, it is necessary to have a system for supervising volunteers' work and documenting their activities. Developing a protocol for volunteers to use as they perform their work (e.g., when collecting data or conducting safety checks) can be particularly helpful. When volunteers perceive their work as interesting and rewarding, they are much more likely to remain committed to the program.

Moving from Program Design to Program Implementation

The program design phase outlined in Chapter 4 provides the manager with a useful blueprint for program implementation—a blueprint that must be updated and modified as the program develops. Finding a program focus during the design phase and holding on to it—while still allowing room for flexibility—is perhaps the greatest challenge facing injury prevention program managers.

"Having a long-range vision of where the program is headed is essential," suggested Ascheim. "Without this vision it's easy to get sidetracked by initiatives that seem to make sense at the time but don't lead anywhere. It's better to do one thing well than ten things poorly" (personal communication).

Reflecting on her many years of experience with community-based and state injury prevention programs, Gallagher concluded, "The program planning process is an extremely critical aspect of program management. Making sure you've done your homework—by collecting local injury morbidity and mortality data, doing a needs assessment in the community, and building a strong coalition—will make program management and implementation much easier" (personal communication).

Defining staff roles and translating process objectives into tasks. After program staff are on board and the process of identifying the injury problem and designing the program has been completed, the next step is to clarify each staff member's responsibilities. A labor loading chart can help allocate staff time. The chart estimates the number of days each person will devote to each process objective during a given period, usually 1 year. (Process objectives are quantifiable and measurable statements of what the program will have accomplished by specific dates.) Table 1 shows a labor loading chart from a playground safety project conducted by the Statewide Comprehensive Injury Prevention Program (SCIPP) at the Massachusetts Department of Public Health.

After the chart is completed, each of the process objectives should be broken down into a series of tasks, with each staff person's role clearly defined. In a large program, it can be useful for the program manager and staff to work together to define these tasks and understand how each task fits into the program's goals and objectives.

A timeline should then be constructed. Putting all of the tasks on paper clarifies the general sequence of events and highlights particularly busy periods. It also provides an opportunity to redefine or re-order tasks, if necessary. Program managers should make realistic estimates of how long each task will take and allow time for the inevitable delays of any project.

Table 2 shows a portion of the timeline from the playground safety project. It illustrates how program staff transformed two process objectives—playground safety inspections and a survey of children's playground injuries—into a series of concrete tasks. Note that the timeline describes each task and the staff member(s) responsible for completing it. In the interest of efficiency, large programs may choose to use a timeline to monitor only a few major tasks.

A timeline is a valuable management tool. It clarifies at a glance which staff members should be engaged in what activities at any time. Managers should be especially sensitive to periods during which many tasks are being performed simultaneously or a particularly crucial task seems to be moving too slowly, which often occurs when the

Table 1. Labor loading chart from the playground safety project conducted by the Massachusetts Statewide Comprehensive Injury Prevention Program (SCIPP)

Process objectives	Program director (person/days)	Health educator (person/days)	Assistant director of research (person/days)	Training coordinator (person/days)	Conference coordinator (person/days)	Administrative assistant (person/days)
1. By July 1988 inspections of all selected playgrounds in the three communities will be completed.	5	14	24	—	—	8
2. By May 1988 parents of elementary school–aged children in three communities will have completed a questionnaire regarding the incidence and circumstances of playground injuries incurred by their children.	1	5	18	—	—	2
3. By mid-June 1988 SCIPP will have sponsored a statewide conference on playground safety.	5	15	8	5	25	5
4. By August 1988 SCIPP will have produced a report summarizing the project's findings and recommendations.	7	7	15	—	—	6
Total days	18	41	65	5	25	21

Table 2. Portion of the timeline from the playground safety project conducted by the Massachusetts Statewide Comprehensive Injury Prevention Program (SCIPP)

Program goal: to develop recommendations for the reduction of playground hazards by pilot-testing an epidemiological investigation of playground hazards in three communities
Process objective 1: By July 1988, inspections of all selected playgrounds in the three communities will be completed.

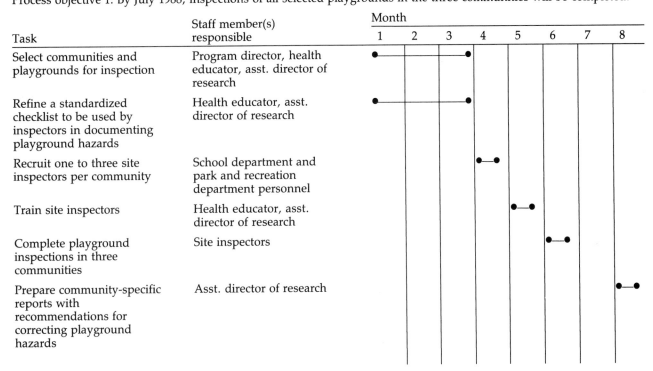

Task	Staff member(s) responsible	Month 1	2	3	4	5	6	7	8
Select communities and playgrounds for inspection	Program director, health educator, asst. director of research	●━━━━━━●							
Refine a standardized checklist to be used by inspectors in documenting playground hazards	Health educator, asst. director of research	●━━━━━━●							
Recruit one to three site inspectors per community	School department and park and recreation department personnel				●━●				
Train site inspectors	Health educator, asst. director of research					●━●			
Complete playground inspections in three communities	Site inspectors						●━●		
Prepare community-specific reports with recommendations for correcting playground hazards	Asst. director of research								●━●

Process objective 2: By May 1988, parents of elementary school–aged children in three communities will have completed a questionnaire regarding the incidence and circumstances of playground injuries sustained by their children.

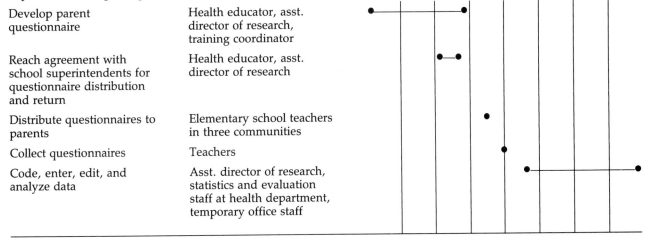

Task	Staff member(s) responsible	Month 1	2	3	4	5	6	7	8
Develop parent questionnaire	Health educator, asst. director of research, training coordinator	●━━━━━━●							
Reach agreement with school superintendents for questionnaire distribution and return	Health educator, asst. director of research				●━●				
Distribute questionnaires to parents	Elementary school teachers in three communities					●			
Collect questionnaires	Teachers						●		
Code, enter, edit, and analyze data	Asst. director of research, statistics and evaluation staff at health department, temporary office staff						●━━━━━━━●		

timeline does not accurately reflect the amount of time needed.

If the program is falling behind schedule because staff, volunteers, or participating agencies or individuals have lost their commitment, it is the manager's responsibility to get them back on track. One group of health planners offered the following tips: reinforce "the importance of each step in the project; make participants aware that others are counting on them; identify the participants' problem and help solve it; . . . be willing to compromise —few activities are as we anticipate them, so flexibility is a part of creative leadership."[9]

Coordinating with other agencies in the community and state. It is almost impossible to imagine an effective injury prevention program operating in isolation from other local and state agencies. Injury prevention programs, multidisciplinary by nature, require collaboration to accomplish objectives, reduce duplication of effort, and give injury prevention a higher profile in the community.

The most traditional method of establishing ties with a broad range of agencies is to invite them to participate in a state or community injury prevention coalition. An alternative technique, and one that can lead to closer working partnerships, is to join forces with one or more related agencies in the community to carry out interventions.

Implementing interventions with other agencies. Wisconsin's La Crosse County health department built on its long-standing relationship with the local Hmong Mutual Assistance Association to obtain state health department funding for a home safety program targeted at Hmong refugees from Laos who have settled in La Crosse. The program, mentioned in Chapter 1, is described by Denis Lee Tucker of the association as "a truly beneficial partnership for both agencies" (personal communication).

Program funds enable association staff to conduct home safety inspections and provide families with information about home safety. When staff cannot convince a landlord to correct a hazardous condition, they refer the case to the county health department, which draws upon its housing code enforcement authority to ensure that the situation is remedied. The association submits quarterly reports to the county health department, which acts as the program monitor.

This type of working relationship is most appropriate when the participating agencies have similar objectives and complementary resources. One group of injury prevention professionals has developed criteria for selecting outside agencies or individuals to participate in implementation of interventions: they must have credibility with the target population, perceive that injury prevention activities are compatible with their primary role, and have an organizational structure and ongoing activities that facilitate integration of the intervention.[6]

Drawing on the resources of other agencies to help carry out interventions is a sound practice, especially when budgets are limited. Pooling resources lets each of the participating agencies accomplish much more than it could alone. Reciprocal relationships seem to work best: if one agency assists another with a certain aspect of program implementation, the second agency should repay the favor—perhaps by offering technical assistance or training on a special topic.

Turf issues among agencies. Cooperation and co-ordination with other agencies can greatly enhance programs. There are, however, some natural barriers to interagency coordination. The most pervasive are conflicts over the boundaries that separate one agency's domain from another's.

Wanting to protect its own turf, an agency can neglect to form ties with related institutions and thereby become isolated from cooperative injury prevention initiatives in the community. In the extreme, turf issues can embroil agencies in lengthy disputes over who has the prerogative to offer which injury prevention services to the community.

Understanding and addressing a community's or state's injury problem requires participation by a wide variety of groups, such as health departments, hospitals, highway safety agencies, public safety agencies, academic researchers, and consumer advocacy groups. Each is eager to stake out its own territory. Indeed, some jockeying for position is natural during the start-up phase of new programs. Arriving at a mutually acceptable division of responsibilities among agencies is the key to resolving turf issues. Dr. Kathleen Acree, Assistant Chief at the California Department of Health Services, noted the importance of fostering mutual respect among agencies: when it is inappropriate for the health agency to assume "the oversight role, a coordinative one may substitute. The health agency then bears responsibility to assure that the primacy of other agencies is respected in their areas of authority" (personal communication).

The best time to avert potential disagreements over turf is at the very inception of a program. Some sure ways to avoid problems are to ask all would-be competitors to join an advisory board, invite them to participate in a community injury prevention coalition, explore how programs might share staff and/or funding, or look into cosponsoring public education materials or events.

Building and maintaining strong links with other agencies is perhaps the most effective way to prevent the development of turf conflicts. As community organizer Deborah Dean commented, "perhaps the greatest benefit deriving from liaison efforts is the opportunity to resolve potential disputes over turf or community resources before they evolve into conflicts disruptive to the smooth functioning of the program."[8]

Experienced program managers deal with interagency conflicts before they erupt into territorial disputes. Susan Gallagher of Massachusetts advises managers to "approach other organizations in a nonthreatening manner and explain to them 'I need to learn from you. Let's try to work together'" (personal communication). If one agency is duplicating the efforts of another, both groups can increase their effectiveness through collaboration. Gallagher noted that opportunities to apply for new local, state, or national funds for injury prevention can provide a good incentive for organizations to break down turf barriers (personal communication). In addition, demonstrating a willingness to collaborate with other organizations often increases the strength of a proposal.

Increasingly, state health departments and highway safety agencies are making great strides toward coordinating their passenger safety efforts and defining complementary roles. Many state partnerships were launched when each agreed to serve on the other's advisory board or planning committee. In others, collaboration grew as one agency was willing to see the other as a resource rather than a competitor. M. Patricia West, Director of the Injury Prevention and Control Program at the Colorado Department of Health, suggested, "Read your state's highway safety plan. I find some of my best data are in that plan" (personal communication).

PROGRAM EVOLUTION

One measure of a program's success is its ability to expand and improve over time. Programs that use their ongoing evaluation data to refine operations and respond creatively to new opportunities for implementation tend to have the greatest impact and survive the longest.

Using Data to Modify Programs: The Massachusetts Passenger Safety Program

One program in Massachusetts illustrates well how efforts can evolve as new data emerge. In response to a growing concern among pediatricians and parents, the Child Passenger Safety Resource Center (CPSRC) was established in 1979 at the Massachusetts Department of Public Health. CPSRC's original goal was to "decrease the incidence of premature death and disability in children under 14 through the increased safe and consistent use of child automobile restraint devices."[10]

During its first year, the program focused on increasing safety seat use among Massachusetts children under the age of 5, in particular newborn infants leaving the hospital obstetrics ward. This decision was based on national data and the assumption that children so protected will progress to regular use of safety belts later in life. Activities centered on establishing safety seat loaner programs (especially in hospitals with obstetrics wards) and training loaner program staff to inform new parents about child safety seat use.

In 1980 new data from a statewide observational survey became available, and CPSRC began to expand its focus. The data indicated a 50% safety seat use rate among children under the age of 1, an 18% rate for ages 1–4, and a 5% safety belt use rate among 5- to 9-year-olds.[11] Former CPSRC assistant director Cindy Rodgers commented, "We learned that infant seat use was fairly commonplace. Adults seemed to be a lot more hesitant, however, to use toddler seats for older preschoolers and to convince their school-age children to wear safety belts" (personal communication). CPSRC refined its activity to focus on children between the ages of 1 and 10 and their parents.

The program's efforts were given a boost in January 1982 by the state's new child passenger safety law, which covered children from birth to the age of 5. Concerned that the law was not being fully enforced, Rodgers surveyed local police departments and learned that enforcement of the law was minimal.[12] So CPSRC developed and disseminated materials to familiarize law enforcement officers with the law and provide suggestions for enforcement.

In 1984 CPSRC evolved into the Massachusetts Passenger Safety Program (MPSP), a comprehensive program focusing on all age groups. New developments since then have been based on data from outcome and/or process evaluations. For example, 1984 statewide observational data on high rates of child safety seat misuse led to the provision of special training sessions and educational materials for parents and service providers. In 1985 observational data on safety belt use and a new mandatory safety belt use law caused the program to focus on teenagers, the elderly, and blue collar workers.

Although the safety belt law was repealed in late 1986, 6 months later the state's child passenger safety law was expanded to cover children up to the age of 12. This has provided a new focus for MPSP's public education efforts, as well as outreach to police departments about enforcement. Thus we see how the ongoing use of outcome and process data enabled this program to modify its goals, objectives, and interventions in order to address the issue of motor vehicle safety more effectively.

Responding to Opportunities in the Community: The Liberty City Injury Prevention Program

The other principal way in which programs grow and change over time is in response to new opportunities. Liberty City, a primarily black Miami neighborhood of approximately 85,000 persons (1980 census), is one of the poorest sections in Florida's Dade County. In 1978, while implementing a large-scale rodent control program, environmental health workers from the county health department identified environmental hazards that put residents at risk for a variety of injuries, from flame and scald burns to poisonings and falls.

As injury prevention program director and Assistant Environmental Health Director Al Ros pointed out, "That was the late 1970s. Very little work on home injuries was being done in any part of the country, so there were few people for us to consult with. Added to that, we had almost no local data to base our program on" (personal communication). In 1982, when federal rodent control funds were turned over to the states, Ros allocated approximately $200,000 a year to carry out a home injury prevention program in Liberty City. The program lasted until 1985, when federal funds were eliminated.

The first step was to collect data to clarify where to intervene. Based on a review of the literature and their understanding of the hazards in Liberty City homes, the staff devised a 47-item survey form. Environmental health workers trained in its use conducted a door-to-door survey in a random sample of 500 homes. Forming the basis for the program, the findings indicated a high proportion of environmental hazards that could lead to home fires, tap water scalds, and falls.

Because of budget limitations, Ros and his staff initially targeted the most prevalent hazard (home fires) and the population most at risk (young children and the elderly). Using computerized home hazard survey data, the staff identified homes with one or more of the following risk factors for a death or injury from a home fire: the presence of a child under the age of 5, an adult over the age of 65, or an invalid; the presence of a smoker; or residence in a wood-frame housing unit.

More than 10,000 homes in two especially high-risk target areas within Liberty City were selected as the intervention group. Complementing the survey data, medical examiner records indicated that these sites accounted for almost 10% of all Dade County home injury deaths, although less than 2% of the county's population lived there. In addition, at least 30% of the families in the two target areas were living below the poverty level.[13] A third target area, contiguous to the first two, served as the control site.

With assistance from the county fire department, the county mayor's office, and several large businesses in Miami, the program purchased and installed over 3,000 smoke detectors in high-risk homes. Staff used the home visits as an opportunity to inform residents how to prevent scald burns and injuries from home fires and how to provide first aid to fire and burn victims.

Program efforts also focused on another prevalent home hazard—residential hot water heaters set above 130°F. By giving informal educational presentations on scald burns at local churches and in public housing projects, outreach workers were able to convince residents to allow them to lower their settings to 120°F. "Our staff [many of whom are minority members] were in the neighborhood on a daily basis—that really helped to form a bond of trust between the community and the health department staff," commented Ros (personal communication).

In addition, outreach workers addressed childhood poisonings by distributing bottles of syrup of ipecac, poison control center telephone stickers, and cabinet locks. Outlet plugs (to prevent electrical burns) were given to families with young children, and nonslip strips were installed in the bathtubs and showers of homes with one or more elderly occupants.

Two follow-up surveys were conducted in each of the intervention homes at 1-year intervals, while control homes were followed up once. Whereas the preintervention survey showed that only 25% of all homes had a working smoke detector, that figure increased to 66% among the intervention homes just 1 year later. Likewise, the incidence of homes with tap water temperatures above 130°F was reduced from 45% preintervention to 20% postintervention. The control group showed no statistically significant change on either measure (personal communication). Overall, an average of two injury hazards (25% of all hazards) were eliminated in each intervention home.[13]

Program staff convinced the Department of Housing and Urban Development (HUD) to replace the hot water heaters in several Liberty City housing projects with new, preset electric heaters rather than gas ones. Because the temperature settings on electric hot water heaters are much harder to readjust than on gas heaters, staff believe this simple engineering/technology intervention will help to reduce tap water scald burns in these housing projects for years to come.

Figure 1 delineates key events in the Liberty City Injury Prevention Program. Although envisioned as a 3-year effort, activities spanned a 6-year period. As former Massachusetts injury prevention program director Susan Gallagher commented, "In this

1982	1983	1984	1985	1986	1987
• Staff training in home hazard survey techniques • Sample home hazard surveys conducted in target areas 1 and 2	• Initial comprehensive home hazard surveys conducted in target areas 1 and 2 • Interventions initiated in target area 2	• Initial comprehensive home hazard survey conducted in random sample in control area • Injury surveillance initiated in emergency room of hospital 1 • Interventions initiated in target area 1 • Follow-up comprehensive home hazard surveys conducted in target area 2 • Replacement of smoke detectors in target area 2	• Follow-up comprehensive home hazard surveys conducted in target area 1 • Injury surveillance initiated in emergency room of hospital 2 • Replacement of batteries in smoke detectors in target areas 1 and 2	• Second follow-up comprehensive home hazard surveys conducted in target areas 1 and 2	• Follow-up comprehensive home hazard surveys conducted in control area

Figure 1. Key events in the Liberty City Injury Prevention Program.

field, it takes time to do something right. Program successes don't come about overnight" (personal communication).

The pace and process of the Liberty City program, as it progressed from its beginning stages to implementing a series of different interventions, reflected the program's setting and funding level. Program administrators took advantage of new opportunities as they arose and employed a variety of interventions to address the neighborhood's major home injury hazards.

Just as the passenger safety and home injury prevention programs followed dramatically different developmental paths, managers can expect their programs to evolve in a way that reflects the community served, the agency in which the program is based, and the personalities and skills of program staff. The secret is to manage the program with a light hand, always alert to new opportunities for improving the program's effectiveness.

As programs evolve, they often seek to become more strongly connected to the wider community. Forming these community ties can help gain communitywide support for the program, prevent it

from becoming too narrowly focused, and minimize turf problems. Establishing a coalition is one way to forge such ties.

ESTABLISHING AND WORKING WITH A COALITION

"In the space of a few minutes on a July afternoon in 1978 my life was turned completely upside-down," said mother of four Nadina Riggsbee. After falling into the family's backyard swimming pool, Riggsbee's 2-year-old daughter Samira drowned and 1-year-old Jayjay became a quadriplegic with severe brain damage. "I came away from that experience determined to do something to prevent other children from drowning or nearly drowning," remembered Riggsbee (personal communication).

Riggsbee learned from state vital statistics that drowning in home pools was the leading cause of death for California children 1–4 years of age. Yet she found that state and local health officials were unaware of the magnitude of the problem, and almost no data were available on nonfatal drowning incidents.

I convinced the intensive care unit director of the hospital where my son was treated to begin compiling statistics for me on near-drowning victims treated in the hospital. I also looked into interventions for preventing drownings, but it soon became clear that working alone I wouldn't be able to accomplish much. I needed to band together with others who were also concerned with this issue (personal communication).

The partnerships developed during the problem identification phase can be the foundation for a long-lasting coalition that, because of its broad base, has the potential to exert great influence on injury prevention policy and programming in a community. Depending on the design of a program, some of its objectives may be well served by forming a coalition: when an intervention requires that the program achieve a high degree of recognition and credibility in the community quickly, when materials need to be disseminated or individuals trained as inexpensively and as comprehensively as possible, when attempting to improve communication and coordination with related organizations in the community or state, when the program requires a powerful and diverse constituency to support legislative or policy change, or when it needs vocal support from lay people and professionals to obtain start-up or long-term funding.

Coalition membership can be composed entirely of professionals or represent both volunteers and professionals. Every injury prevention coalition has its own mission and history. Nevertheless, each can be described as

an organization of individuals representing a variety of interest groups who come together to share resources and effect change. The purpose of a coalition is to try to achieve greater impact by having member groups plan and work together. Coalitions increase the "critical mass" behind a community effort, help groups to trust one another, and reduce the likelihood of resource squandering through unnecessary competition among groups.[14]

Some coalitions are launched by agencies to support their common injury prevention objectives (such as the Alternatives to Violence and Abuse Coalition, described below); other coalitions, like the Drowning Prevention Foundation, have been organized by a group of concerned private citizens with the broader aim of promoting the cause of injury prevention within the community.

Grassroots Coalitions: The Drowning Prevention Foundation

Nadina Riggsbee began to contact parents of other victims, local pediatricians, and public health pro-

fessionals and invited them to join with her to create the San Francisco Bay Area Coalition on Drowning Prevention. Its first initiative was to develop and lobby for a new county ordinance mandating safety measures for backyard pools.

Despite opposition from pool manufacturers, the efforts of the coalition were successful. In August 1984 Contra Costa County required that all backyard pools have at least one of three safety measures: perimeter fencing over 4½ ft high with a self-closing and self-latching gate, a pool safety cover/dome, or an audible alarm on all home exits leading to the pool. (A discussion of the coalition's advocacy techniques appears later in this chapter.)

The ordinance, covering pools installed after 1984, is enforced by staff from the county's building inspection department. Preliminary death and hospital admission data indicate a statistically significant drop in childhood drownings and near-drownings since the ordinance went into effect (Riggsbee, personal communication).

The most stringent pool safety law in the country when it was passed, the ordinance became a catalyst in generating support for the coalition. Incorporated in 1985 as the Drowning Prevention Foundation, the group has flourished. Today its activities include working to strengthen local, state, and national legislation on pool fencing; providing technical assistance to other state and local coalitions and agencies; producing and distributing educational materials on drowning prevention; advising parents on improving the safety of backyard pools, spas, and hot tubs; sponsoring a speakers' bureau on drowning prevention; and working with the pool industry to design voluntary standards for pool safety covers.

Interagency Coalitions: The Alternatives to Violence and Abuse Coalition

Frustrated by a lack of communication with colleagues and eager to establish a forum for coordinating educational services and brainstorming solutions to common problems, a handful of service providers from northern California's Contra Costa County formed the Abuse Prevention Training Committee in 1982. It soon evolved from a forum for open discussion into a coalition focusing on education and advocacy. Larry Cohen, head of the county health department's prevention program, was selected coalition leader because of the credibility and perceived neutrality of his agency. In 1985 the group changed its title to the Alternatives to Violence and Abuse Coalition (AVAC) and expanded membership to include service-providing agencies (like rape crisis and elder abuse treatment programs) and alternative programs (for example, a conflict resolution group).

Over the next 2 years, the coalition launched a school-based pilot project, organized a speakers' bureau, and sponsored two conferences. The conferences, attended by county supervisors, state legislators, and agency heads, highlighted the need for increased funding for preventive health services throughout the county. AVAC members also helped draft a countywide policy statement on the prevention of substance abuse and violence. As a direct result of these efforts, county supervisors voted to increase the funds available to local health and human service agencies for primary prevention activities.

In 1987 the health department's prevention program, which staffs the 25-member coalition, received a 3-year grant of $125,000 a year from the federal Office of Maternal and Child Health to expand the coalition's adolescent violence prevention efforts. Additional coalition staff have been hired, and some participating agencies have been reimbursed for conducting educational sessions on violence prevention for parents and high school students. Along with a dramatic increase in the number of adolescents and adults reached by AVAC educational sessions, coalition staffers count among their biggest accomplishments increased media attention to violence prevention, revision/development of local school policies on violence, and a strengthened partnership with the United Way.

Coalition staffer Nancy Baer noted that AVAC's early decision to be as inclusive as possible has greatly contributed to its success: "In most coalitions, it's a good idea to include a broad representation of disciplines, perhaps even those people you think might disagree with you. Bring them into the discussion right from the beginning" (personal communication).

Not every community may want or need to form a coalition. Lead agencies coordinating coalitions report that the groups can consume an inordinate amount of staff time and financial resources, and a coalition in itself is no panacea for the challenges that face injury prevention programs. Yet in the right setting a strong coalition can act as a powerful agent for change.

Steps to Effective Coalition Building

Larry Cohen, Nancy Baer, and Pam Satterwhite of the Contra Costa County Health Services Department in California have devised an approach to effective coalition building for lead agencies. Summarized in Figure 2, it is based on a series of steps.

It is important first to explore whether there are

1. Review the program's injury prevention objectives and decide if a coalition is needed.

2. Identify key participants in the coalition.

3. Devise a set of preliminary objectives for the coalition.

4. Invite members to join the coalition. Schedule the first coalition meeting to review the group's objectives and structure.

5. Operate with as much flexibility as possible; continue to modify the coalition's objectives as the need arises.

Adapted from reference 15.

Figure 2. Steps for effective coalition building at the state and local levels.

any coalitions in the community or state dealing with similar injury issues. If so, it is probably more productive to join that coalition than to organize a new one.

Coalition budgets are generally limited to preparation of materials and mailings to members. However, a sizable amount of staff time is always necessary. Usually the lead agency provides staff to accomplish clerical duties; planning, preparation, and facilitation of meetings; membership recruitment, orientation, and ongoing communication; research and fact gathering; and coordination of special activities and projects. The planning and follow-up of coalition meetings are particularly time-consuming.

"Organizations," commented two experts, "should be considered as potential coalition members if they possess needed resources, have relevant programs, or if their staff have special expertise to help address the issue."[14] Figure 3 provides a list of potential members of state and local injury prevention coalitions.

After the agency has created a preliminary list of possible organizations for the coalition, it must decide who from each organization should be invited to join. High-level officials, although often too busy to attend meetings and more likely to get caught up in turf battles with other members, can lend credibility to a newly forming coalition. Second- or third-level employees, despite their lack of influence, may be more committed and enthusiastic, and "it may be important for the top leadership to appoint staff from this level to represent their organization."[15] It is helpful to meet with each member individually before the coalition is launched to clarify what he or she wants the coalition to accomplish.

Size is an important factor. Coalitions larger than 20 people tend to create staffing and logistical

Medical
  Professional organizations
    American Academy of Pediatrics
    American College of Surgeons
    American Association of Poison Control Centers
    Medical societies
    Community clinics

Allied health
  Hospital associations
  Pharmacist associations
  Nursing associations
  Paramedic associations
  Dental associations

Voluntary organizations
  Red Cross
  Spinal cord injury associations
  Head injury associations
  Sports injury associations
  National Safety Council
  Child passenger safety associations

Consumer
  Parent–teacher associations
  Public interest research groups
  Consumer Product Safety Commission
  YMCAs
  Mothers' groups
  Service clubs
  Youth and senior associations
  American Automobile Association clubs

Media
  Newspapers
  Radio and television stations

Advocacy groups
  Children's lobbies

State health department
  Maternal and child health
  Preventive health

Environmental health
Occupational health
Emergency medical services
Food and drug
Licensing
Epidemiology
Vital statistics
Health education

Local health departments

State departments on aging

Public safety
  Highway patrol–state police
  Local law enforcement
  Fire services
  Lifeguard services
  Municipal engineering departments

Educational
  Public and private schools
  Day-care centers
  Head Start programs
  WIC programs

Academia
  Schools of medicine
  Schools of public health
  Schools of law
  Community colleges
  State colleges and universities
  Private colleges and universities
  Departments of community medicine and
    occupational health

Business organizations
  Insurance companies
  Banks
  Sporting goods manufacturers
  Supermarkets
  Employee assistance programs

Adapted from Micik S, Miclette M. Injury Prevention in the Community: A Systems Approach. *Pediatr Clin N Am* 1985;32:258.

Figure 3. Potential members of state and local coalitions on injury prevention.

strains and make consensus building more difficult. Small coalitions seeking to expand membership can sponsor a workshop or conference to attract new members. Or, if a coalition becomes too large, it can be broken down into smaller subcommittees or working groups.

In the start-up phase of a coalition, as in the start-up phase of a program, it is better to focus on objectives that can result in short-term successes. The coalition's objectives will undoubtedly grow and change over time, but throughout the group's life the objectives should "remain clear, focussed, and relevant to all members."[15] Objectives should be chosen that will avoid turf battles among participating organizations.

In its first meeting the group should formulate a shared vision of its mission. It should also address target dates for reaching the objectives, the length and frequency of meetings, the coalition's decision-making process, and how much time members can spend on coalition work between meetings.

The common pitfalls of coalition work fall into five major areas: group dynamics, problems with hard-to-please members, frequent turnover in representatives, poor planning by the lead agency, and changes in the issue the coalition is designed to address.[15] Members' commitment to the coalition can be heightened by sharing its benefits (e.g., media attention, opportunities for staff training) and by ensuring that members and their organizations receive credit for the group's successes.

Coalitions, like injury prevention programs, benefit from process and outcome evaluation. Evaluation of coalition effectiveness tends to focus on such process measures as increases in skills and commitment of coalition members, formation of new working relationships among member agencies, and participant and community satisfaction with coalition activities and products. This information can be gathered by conducting mail, telephone, or in-person surveys of coalition members and community residents. It is also important to measure the coalition's achievement of its formally stated objectives. Putting the evaluation findings to work *for* the coalition "is a way of assuring that the community and the participants grow from the experience, regardless of the outcome."[15] For additional resources on coalition building, see references 16 and 17.

A counterpart to injury prevention coalitions, advisory committees are sometimes set up to make injury prevention recommendations to a state agency or state legislature. Members of legislative advisory committees are frequently political appointees who may or may not have technical expertise in injury prevention. Also, state agency directors and, to a lesser extent, local agency directors sometimes establish special committees to act in an advisory capacity on the formation of injury prevention policies or programs. Staffing for both types of advisory committees is usually provided by the state agency.

MEDIA AND PUBLIC RELATIONS ACTIVITIES

Backyard tents would appear to offer children a safe, entertaining summer play space. Before industrial or governmental flame-retardancy standards, however, these tents were likely to burst into flames on contact with fire or high temperatures. In the summer of 1973, in separate incidents, two New York state children received fatal third-degree burns when their backyard tents caught on fire.

Leslie Fisher, then director of the New York State Health Department's burn prevention program, responded by conducting investigations of the circumstances of the deaths and surveying 24 styles of tents on sale in stores throughout the Albany area. Only one of the 24 tents was labeled flame retardant. Shortly thereafter, several major newspapers in the state published interviews with Fisher in which he discussed the dangers of backyard tents.

By 1975, having arranged with the news director of the Albany NBC television affiliate to produce a weekly 5-minute spot on home safety issues, Fisher decided to devote one session to tents. To reinforce the life-threatening danger of the tents, he convinced the news director to insert into the live interview a previously filmed segment of himself setting fire to a popular brand of tent, which was reduced to cinders in 90 seconds.

The media coverage generated an outcry of concern among parents and legislators, culminating in 1978 with the passage of a new regulation which required that all tents sold in the state be manufactured with flame-retardant materials and clearly labeled as such.[18] Said Fisher, "The newspaper coverage and television spot were the catalyst for everything that followed—support from consumers and elected officials and passage of the flame-retardancy regulation" (personal communication).

Efforts to inform the community through media and public relations can go a long way toward influencing how the public and professionals respond to the injury problem. In most communities injuries are already recognized as newsworthy events by local reporters and editors. Child abuse, homicides, suicides, car crashes, drownings, and fires appear frequently on the front pages of daily newspapers and are given much attention in radio and television news reporting.

Unfortunately, most news stories focus on the circumstances and aftermath of a single injury. What is missing is the depiction of injuries as understandable, predictable, and in many cases preventable events. As a result, the vast majority of legislators, governors, commissioners, physicians, nurses, lawyers, and the general public continue to view injuries as acts of fate. And as Cheryl Vince of Education Development Center, Inc., noted, "Until those of us in the injury field try to shake those beliefs to their very foundations, we will continue to have an uphill battle in trying to achieve significant reductions in injury death and disability" (personal communication).

On its own, media coverage is unlikely to effect change. But media attention to the injury problem and to prevention measures that work can be essential in reinforcing and supplementing the range of interventions discussed in this book. Information given to a reporter may not always appear exactly as it was given. Providing the reporter with a written summary of key points can help ensure accuracy. One guidebook suggests that where the mass media are concerned, "Loss of control over how the information is communicated may be a trade-off for broad and rapid transmission."[19] Figure 4 provides a summary of the types of issues for which a program may decide to seek media coverage.

A program can increase its chances of coverage by getting to know key staff at local newspapers and radio and television stations and contacting reporters and editors on the phone or in writing with well-timed, relevant information on injury and the community's efforts to attack the problem. The people to contact at weekly newspapers are the news editor and the reporters who cover injuries, health, or other pertinent areas; at dailies they are the city editor and a few key reporters. Reporters are usually assigned to "beats," so it is best to focus on those who cover health, consumer affairs, transportation, or crime. Editors and reporters also face deadlines for receiving information and preparing articles for their papers.

One particularly effective strategy for getting media coverage is to offer additional information or statistics related to an injury story that the newspaper or station has just covered. Timing is critical,

1. **The scope of the injury problem in the community and/or state**

2. **Stories about people who have been involved in fatal or severe injuries (with permission of the victim or his/her family, of course)**

3. **The start-up of the program or coalition**

4. **The story of an interesting individual (e.g., Nadina Riggsbee, head of the Drowning Prevention Foundation)**

5. **Special events and activities sponsored by the program**

6. **Program outcome**

7. **Information about preventing injuries, especially following a well-publicized injury or during a season with predictable injuries**

8. **Pending state or local legislation related to injuries or injury prevention**

9. **Recognition for community sponsors who have provided financial support or donated materials to the program**

Figure 4. Issues for which a program may decide to seek media coverage.

and immediate response is most likely to get coverage.

At some state and local agencies, program staff are prohibited from contact with the media; this function is performed by senior-level administrative staff or the agency's public relations office. In this case, efforts should be made to enlist that person in the cause of injury prevention. Also, one program manager suggested "asking others who are not government employees to do the footwork with the media or 'arranging' for the media to contact the head of your agency about the issue you want covered."

A program looking for assistance in developing a media strategy might want to contact a university communications studies department, a public service advertising council, or an advertising agency. Any one of them may have interns available who can help to augment program staff. The following sections examine the process of working with the print and broadcast media.

Print Media

In this country, approximately 600 children die each year and over 300,000 receive emergency room treatment for bicycle-related injuries. Four-fifths of these deaths and injuries are the result of head trauma. The most effective intervention to prevent bicycle head injuries is a simple one: wearing a bicycle helmet. Yet less than 1% of children in the nation wear helmets. In 1986 a handful of concerned organizations in Washington State formed a coalition to launch the Children's Bicycle Helmet Campaign.

Along with facilitating parents' purchase of low-cost helmets and distributing free helmets to children from low-income families, the coalition is working to make helmet wearing an acceptable behavior among children as well as educating parents about bicycle injuries. Print and broadcast media have been the focus of campaign efforts to reach children and parents. Newspaper coverage of the campaign has been extensive. The campaign's full-time public information director, Lisa Rogers, based at Harborview Injury Prevention and Research Center in Seattle, has relied on personal contact with editors and reporters from papers across the state in convincing them to cover the campaign. She has found newspaper health reporters particularly receptive to covering the issue.

"Perhaps most successful in reaching parents has been the proverbial 'victim' story appearing in local community newspapers detailing the injury of a young child not wearing a helmet," explained

Rogers. "These stories have had a tremendous effect on other parents . . . in terms of both helmet buying and general public awareness of the problem."[20]

Recently, the campaign received national coverage in a *Parade* magazine story.[21] Using personal contacts, Rogers and other staff members also have placed articles in the newsletters of professional organizations (e.g., the Washington State Medical Association and the American Academy of Pediatrics) and civic groups (e.g., the Washington State PTA and the Boy Scouts). She feels the campaign has been able to capture media attention because "bike riding is such a common activity and bike injuries affect an enormous number of children. Most important, the solution—wearing a helmet—is a relatively easy one to implement" (personal communication).

The major advantage of using print media is that it allows more detailed coverage of an issue than does television reporting. A newspaper may use editorials, feature stories, and regular news stories as well as reader-contributed columns (called "op-ed" pieces) and letters to present the issue of injuries.

A program that decides to inform local newspapers of an injury problem and current prevention efforts can request a meeting with the papers' editorial boards or news editors which may result in an article or an editorial. Such a meeting can also sensitize the newspeople to the issue and help assure continuing coverage, with the program as a source of information.

Another route to newspaper coverage is the press conference, especially for a particularly newsworthy event. The press conference should be announced through a press release accompanied by a press kit with statistics on the injury problem, a description of the prevention program, and copies of public information and education materials used by the program. A press release should attempt to establish a link between the program and community concern about injuries. For example, a press release describing a new program to reduce interpersonal violence among adolescents might include quotes from concerned teenagers, parents, and teachers. Figure 5 provides additional tips on preparing a press release.

A program staffer can also submit an op-ed piece to a local newspaper. These are articles written by readers that appear on the page opposite the editorials.

If a local newspaper prints a story or editorial about injuries or injury prevention that is inaccurate, incomplete, or misleading, the program di-

1. **Prepare the release in double-spaced type on the organization's letterhead.**

2. **Include the date when the story can be aired. Use either "For immediate release" or "For release after (date)."**

3. **Provide a "kill date," the date when the program or event ends.**

4. **Give the name and phone number of a contact person who can supply additional information.**

5. **Write a headline that describes the event and will capture the reader's attention.**

6. **Summarize the who, what, where, when, and why of the story in the lead paragraph.**

7. **Use quotes, and make sure that they are completely accurate.**

8. **Keep the release as brief as possible—two pages at most.**

9. **When using two pages, type "more" at the bottom of page one.**

10. **Use the "#" symbol to indicate the end of the release.**

Figure 5. Tips on preparing a press release.

rector should write a letter to the editor correcting the facts. Pending legislation provides another opportunity to send a letter to the editor advocating or defending the legislation. And when a newspaper publishes a story that accurately describes the injury problem or injury prevention efforts, complimentary letters to the reporter and editor can be effective in cementing a good relationship.

Broadcast Media

Radio and television stations offer four options for injury coverage: news programs, talk shows and interview programs, public service announcements, and purchased air time. The visual nature of television and the aural character of radio affect how topics are covered. Television coverage must include visual displays (charts, demonstrations of safety techniques), and radio coverage requires an engaging style for presenting the information.

It is important to select appropriate stations. According to one guidebook, "Radio stations and independent TV stations (including UHF and cable) have more of their own time to fill than ABC, CBS, NBC, and PBS affiliates."[22]

News programs. Soliciting news program coverage of injury problems or prevention efforts is advisable only when the story can be told concisely; the average television news story is slightly less than 2 minutes long. To capture the attention of a news director, assignment editor, or reporter, the program should present a brief synopsis of the story idea and a list of facts that make it particularly newsworthy. A press release for newspapers might contain still photographs of featured individuals; a release for radio and television should suggest possible live footage.

Anyone whose story idea is turned down should ask what would make it more appealing, then refine and resubmit it. The best time to reach a television or radio reporter is in the morning; in the afternoon they are busy trying to meet deadlines.

Talk shows and interview programs. Talk shows and interview programs can devote time to exploring a subject in depth. The producer or assistant producer is the person to contact. The program should send a letter describing the topic, its newsworthiness, and any controversy it has stirred up in the community.

Anyone invited to appear should be familiar with the show's general format and tone and should go prepared with an outline of the points to be made. It is wise to focus on no more than three main points and to repeat those points several times in different ways throughout the interview. Remarks should be brief, concrete, and engaging, explaining the injury problem in simple terms and giving the viewers or listeners a sense of the impact it has on every member of society. It helps if the interviewer has a list of questions to ask to bring out specific points.

Public service announcements. Public service announcements (PSAs), typically 10, 30, or 60 seconds long, use air time donated by the sponsoring station. Unfortunately, PSAs often run late at night when the audience is at its smallest. PSAs can convey only a limited amount of information on any given topic. They are most useful "for creating an awareness or heightening the public's sensitivity to a health problem or issue, or reinforcing a newly established behavior."[19]

In 1987 Boston's Violence Prevention Project was selected by the Advertising Club of Greater Boston to receive $300,000 worth of concept and product development and advertising free of charge from a major advertising agency. The campaign produced two 30-second television and radio PSAs on the topic of acquaintance violence among adolescents. According to former project codirector Dr. Howard Spivak, the PSAs have received a fair amount of airplay because they were well designed and expertly produced (personal communication).

National organizations such as the National Highway Traffic Safety Administration have prepackaged PSAs available. Or the program can develop its own videotaped or audiotaped message and submit it to the station, reviewing it, if possible, with the station's public service or public affairs director. Short—10- or 30-second—PSAs are more favorably received than longer ones. The program can also submit brief written copy to be read on the air. The message should be simple and to the point, in a conversational style free of jargon.

Purchased air time. Purchased air time, although more costly, provides more freedom in selecting when and how frequently the message will be aired. The station selected and the time of day during which the message is played determine the type of listening audience, especially with radio and cable television stations. (For additional resources on working with the print and broadcast media, see references 19 and 22–25.)

Public Relations Activities

Print and broadcast media reach a broad, general audience. There may be times, however, when the program needs to target a specialized audience, such as key decision makers. Program funding frequently hinges on how the program is perceived by the agency head and related agencies, legislators and other elected state and city officials, health care practitioners, and other local service providers. The program may then decide to draw on public relations techniques.

Often program managers face a need to counter misconceptions about injuries. Methods depend in large part on program funding for public relations activities and the skills of program staff. Many programs develop a simple brochure outlining the toll of injury morbidity and mortality in the city or state they serve and include a description of the program's prevention efforts. The brochure invites people to contact program staff for further information. The brochure should reach a wide range of people: members of the target population served by the program, legislators and city councillors, directors of hospitals and community health centers, professionals in local agencies whose work touches on injury, representatives of the media—anyone who should be aware of the program's existence.

With sufficient funding, a variety of attractive public relations materials can be developed to raise the awareness of decision makers. In a six-state region, the New England Network to Prevent Childhood Injuries produced state-specific briefing packets to inform legislators, health professionals, and the general public about injuries and efforts to prevent them.[26] The packets form part of an overall plan to gain recognition and additional resources for the injury prevention programs housed in the six state health departments. Audiovisual materials also can be developed to make injury data appealing and understandable.

Not only can public officials be targets of a public relations campaign, they also can lend credibility and visibility to a program. City councils can be asked to issue proclamations; the water department might be willing to enclose a message about hot water temperature and tap water scalds with its bills. Mayors or governors may not have the time or inclination to delve into the day-to-day concerns of an injury prevention initiative, but they are often happy to appear at publicity events or make introductory remarks at conferences and workshops.

Any agency that produces a periodic newsletter or bulletin that is distributed widely throughout the city or state should be sure to include regular updates on the local injury problem and program activities. Some programs have launched injury prevention newsletters of their own, but producing a newsletter can be a large drain on staff time.

If program staff members regularly participate in health fairs and conferences or make presentations before business and community leaders, it may be useful to prepare a visually appealing portable display that describes the program's efforts for use in these settings.

INJURY PREVENTION ADVOCACY

Almost all injury prevention practitioners become involved in advocacy work. Advocacy can range from building a statewide coalition to testifying before a legislative committee to convincing a professional organization to include injury prevention on the agenda of its annual meeting or conference.

Advocacy at the National Level: Dr. Garen Wintemute and Plastic Handguns

In early 1987, University of California assistant professor Dr. Garen Wintemute read about one firearms manufacturer's plans to market a plastic handgun in designer colors—a handgun that could easily be mistaken for a toy but would be as deadly as its metal counterparts. Having earlier reviewed 7 years of data on California children who were unintentionally shot and killed by themselves or their siblings or peers, Wintemute became gravely concerned that proliferation of the proposed plastic handguns would lead to even higher rates of unin-

tentional firearm deaths among children.

When he heard that a bill was before Congress to block production and distribution of the plastic handgun, Wintemute volunteered to testify. Although the bill was premised on the fact that the gun would be undetectable by airport security systems, Wintemute's May 1987 testimony focused on the likelihood of children mistaking the gun for a toy. In preparing his remarks for any testimony, Wintemute said, "I read up as much as I can on the topic, so when questions come my way they won't throw me. It's very important not to act flustered or defensive. . . . There's nothing wrong with saying 'I don't know, but I will find out and let you know.' "[27]

Five months later, shortly before the bill was about to come up for a vote, Wintemute did a mass mailing to over 500 injury prevention professionals across the country, asking them to voice their support of the bill to their representatives in Congress. He also helped convince several national organizations of health professionals to lend their support.

Wintemute's advocacy, coupled with that of dozens of others, paid off. The bill, which passed both branches of Congress by overwhelming majorities, prohibits all manufacture and distribution of undetectable handguns in the United States.

Advocacy at the Local Level: Nadina Riggsbee and Pool Safety

Much advocacy takes place on the local level. The activities of Nadina Riggsbee and her drowning prevention coalition, described earlier in this chapter, were instrumental in gaining passage of a Contra Costa County, California, ordinance that mandates safety measures for backyard pools.

In 1983, at the first board of supervisors meeting at which the proposed ordinance was discussed, Riggsbee arranged for pediatricians and relatives of drowning victims to testify. She herself cited statistics on drownings and near-drownings and recounted the story of her daughter's death and her son's near-death. The majority of supervisors remained unmoved.

Soon after, the pool builders' industry, which objected strongly to any regulations related to pool fencing, contacted county pool owners, urging them to oppose the proposed ordinance. Riggsbee and the coalition convinced a few local television news reporters to cover the board's hearings on pool safety, and the coalition was able to put pressure on the supervisors to come to a decision quickly. Behind the scenes, coalition members lobbied individual board members intensively.

To win passage of the ordinance, the coalition found it necessary to make some compromises. As we saw, the ordinance, which finally received the approval of the board of supervisors in August 1984, requires that all new pools have either perimeter fencing with self-closing and self-latching gates or one of two less effective safety measures—a pool cover/dome or alarms on home exits leading to the pool. Riggsbee noted that although the ordinance is "not the sweeping solution to drownings"[28] that she had hoped for, it has set an important precedent for legislative interventions to prevent backyard drownings.

The experiences of Wintemute and Riggsbee illustrate that while becoming an advocate for injury prevention can be time-consuming and frustrating, those who stick with it are often able to bring about substantial changes.

Much advocacy work in the field of injury prevention focuses on trying to influence the opinions of policymakers or legislators at the city, county, state, or national level regarding proposed legislation or regulations. State and municipal employees prohibited from actively engaging in this type of lobbying may find it beneficial to form links with nonprofit organizations that can lobby (e.g., American Academy of Pediatrics, child passenger safety associations) and citizen's advocacy groups (e.g., Mothers Against Drunk Driving [MADD], children's lobbies).

In building new relationships or strengthening existing ones with local, state, and national representatives, there are a few essential steps to take. The first is to determine at what level the issue will be decided and then identify the individuals who will be involved in the decision. Before making contact with representatives, learn about their positions on related public health issues.

Having identified a few sympathetic individuals, an advocate must work to develop personal relationships with them by writing or meeting to discuss the topic under consideration. If the representative is too busy to meet, a meeting with one of his or her staff can be helpful. Legislative aides are often a legislator's primary source of information about "new" issues, so convincing an aide of the merits of a position is a significant accomplishment.

In any meeting, an advocate's pitch should be brief and compelling, presenting data that are accurate and easy to understand and relating the topic to the district in question (e.g., "In our city last year, five adolescents were killed as a result of peer violence"). At the end of the briefing, the representative (or aides) should understand the problem *and* the action needed: for example, sponsorship of a proposed piece of legislation or a call for the establishment of a task force to make recommendations about a particularly severe injury problem.

Whether the initiative passes or not, the advocate should be sure to express appreciation to the representatives who supported it and to their staff. If the bill passes, the advocate might encourage press coverage highlighting the efforts of the key representative(s) or arrange for a local or state organization to present an award to the representative.

Clearly, the combination of legislative initiatives and a lot of elbow grease goes a long way in shaping the local and national agenda for injury prevention. As Judith Stone, the National Safety Council's Director of Federal Affairs, noted, "We influence the agenda by creating it."[29] (For additional references on advocacy and lobbying, see references 6 and 30.)

INSTITUTIONALIZING INJURY PREVENTION PROGRAMS

A program is *institutionalized* when it has achieved ongoing financial support and commitment from the agency and the community in which it is based. Institutionalization is crucial; without comprehensive, long-term responses to the injury problem, programs will never significantly reduce injury death and disability and their associated costs. Programs that are in operation for only a year or two and then disappear (usually because of a loss of funding) are unlikely to bring about real change.

Designing, implementing, and evaluating one or more interventions usually bring program staff a renewed sense of commitment to addressing the injury problem and a better understanding of which interventions are most likely to succeed. After learning some valuable lessons, the staff has, as one planning manual suggested, "new skills and insights to offer. You will have undoubtedly turned up new aspects of the problem you can deal with or you could move on to another phase of the problem."[8]

The New England Network to Prevent Childhood Injuries, housed at Education Development Center, Inc. (EDC) in Newton, Massachusetts, is a collaborative effort between the six New England state health departments and EDC staff to enhance and institutionalize each state's capacity to respond to the injury problem. Institutionalization activities in the states have two major emphases: to educate relevant health department personnel (commissioners and deputy commissioners) about injuries and prevention efforts and to raise awareness about injuries among members of the media and key decision makers using a variety of methods (press releases, educational materials, and events).[26]

Network staff see these as ongoing processes that must be tailored to the individual state's needs and political realities. "Each state comes to the task of institutionalization with a unique history, set of goals, and political makeup," wrote Daphne Northrop, the network's Associate Director for Communications.[26]

Program administrators should strive for institutionalization from the very first day of a program's existence. The key is to identify the factors that are most likely to convince the public and decision makers of the worthiness of the program and then to highlight those elements in contacts with agency heads, the media, and the general public.

Institutionalization at the State Level: The Wisconsin Comprehensive Child Injury Prevention Program

The road from program inception to institutionalization can be a long and rocky one, as Wisconsin injury control coordinator Hank Weiss knows. In 1979, when Weiss was hired as an environmental epidemiologist by the state division of health, Wisconsin had no state-level staff working on injury prevention, and community-based activities were virtually nonexistent.

As Weiss reviewed the literature on injury, he became convinced of the inconsistency of his agency's attending to "related lifetime risks from toxics on the order of one in a million while ignoring severe injury risks on the order of one in a hundred."[31] Urged on by his section supervisor, Weiss prepared a position paper describing injuries as a public health priority and suggesting how state and local injury prevention activities might be designed, funded, and carried out. He then received permission to spend a small portion of his time writing grant applications to obtain funds for an injury prevention program.

In the process of learning about the theory and practice of injury prevention through his contacts with the Centers for Disease Control and by meeting with injury prevention staff from other states at national conferences, Weiss developed an expertise in injury prevention. Agency administrators noted this new expertise. Thus, Weiss noted, "the stage was set for establishing a full-time presence" responsible for defining—and defending—the agency's role in preventing injuries.[31]

Weiss was eventually successful in getting injury

prevention included as a priority area for the expenditure of Wisconsin Maternal and Child Health (MCH) block grant funds beginning in 1984, and the state continues to provide funding for about eight community-based injury prevention projects a year.

In 1985 Weiss and his colleagues applied for and received a 3-year federal MCH grant to establish the Wisconsin Comprehensive Child Injury Prevention Program (WCCIPP). Through the grant, which totaled about $200,000 a year, they implemented targeted injury surveillance strategies in the state and trained public health and health care professionals in injury prevention. The grant enabled Weiss to hire several injury prevention staff members and to devote himself full time to the position of statewide injury control coordinator.

By definition, the federal funding was short-term. Recognizing the need to obtain state funding, Weiss worked with his immediate superior to develop a proposal to fund the state's injury prevention and emergency medical services programs through a $1-addition to the state's motor vehicle registration fee. In 1986 this so-called "one dollar for life" proposal was approved by the state health agency, but it was less well received by the state department of transportation, the agency that traditionally was awarded all funds raised through vehicle registration fees. The proposal was subsequently withdrawn.

In early 1988, with federal funding soon to expire, Weiss made "my last attempt at pushing a request for injury prevention funding through the department's long and involved planning process" (personal communication). When state planning analysts reviewed his budget request, they asked how much the state would save in health care costs by funding the program. Weiss explained that because of the poor quality of Wisconsin injury morbidity data and the lack of funding for a comprehensive injury surveillance system he was unable to answer the question.

As a result, the planning analysts did not disapprove the funding request. Meanwhile, senior-level administrators in the Wisconsin Department of Health and Social Services finally decided to fund the program either through new state funds or by reallocating existing funds. Although the funding level may be modified, Weiss and his colleagues remain cautiously optimistic that the state will support the 2.5 injury staff members and much of the program operating costs now paid for with federal funds.

Given his 8-year battle to obtain a stable source of

injury prevention funding, Weiss is keenly sensitive to the need for a full-time injury prevention coordinator in every state. He sees this person's most important function as continuing to carry the case for injury prevention to decision makers. "If you want to succeed at [institutionalizing your program], you have to play giraffe and stick your neck out," said Weiss (personal communication). As Weiss's experience exemplifies, with determination —and a lot of patience—injury prevention managers at the state and local level can institutionalize a coordinated response to the injury problem.

Hank Weiss's program is located in a state agency, but the process of institutionalization at the local level is often similar. No matter what the setting, most programs find that some combination of the following elements is helpful in achieving program institutionalization:

• A designated lead agency with an injury prevention coordinator (see Chapter 1)

• Accurate and timely injury mortality and morbidity data from the community or state (see Chapter 2)

• Clearly defined program goals, objectives, and interventions (see Chapter 4)

• Evidence that the program does or will be able to reduce injury rates or injury risk (see Chapter 4)

• Media and public relations efforts to increase community awareness of injuries and the need for action (see this chapter)

• A coalition of committed people working together to coordinate their injury prevention efforts (see this chapter)

(For additional references on program institutionalization, see references 32 and 33.)

SUMMARY

In summary, program implementation involves applying the following concepts.

• All program staff should receive training in the theory and practice of injury prevention. Training should focus on the magnitude and characteristics of the injury problem, injury prevention strategies, program design, working with injury data, the history of the agency's involvement in prevention, and an overview of the existing program.

• Building and maintaining strong links with other agencies and organizations is the most effective way to prevent the development of turf conflicts.

• The two primary ways in which programs evolve over time is by applying outcome and process evaluation data and responding to new opportunities in the community or state.

• Broad-based injury prevention coalitions have the potential to exert great influence on injury prevention policy and programming.

• Media attention to the injury problem and to prevention measures that work can be essential in reinforcing and supplementing a program's interventions.

• Institutionalization of injury prevention programs is essential: without comprehensive, long-term responses to the injury problem, we will never significantly reduce injury death and disability.

INTERVIEW SOURCES

Kathleen Acree, MD, MPH, JD, Assistant Chief, California Department of Health Services, Sacramento, October 21, 1988.

Joan Ascheim, MSN, Child Health Program Chief, New Hampshire Division of Public Health Services, Concord, October 21, 1988.

Nancy Baer, MSW, Project Coordinator, Prevention Program, Contra Costa County Health Services Department, Pleasant Hill, California, October 11, 1988.

Leslie Fisher, MPH, Manager, Preventive Services, Injury Control Program, New York State Department of Health, Albany, August 18, 1988.

Susan Gallagher, MPH, Director, Childhood Injury Prevention Resource Center, Harvard School of Public Health, Boston, Massachusetts, October 3 and 21, 1988.

Murray Katcher, MD, PhD, Section Chief, Family and Community Health, Wisconsin Division of Health, Madison, August 8, 1988.

Patty Molloy, MSW, Director, New England Network to Prevent Childhood Injuries, Education Development Center, Inc., Newton, Massachusetts, October 19 and 20, 1988.

Nadina Riggsbee, President/Executive Director, Drowning Prevention Foundation, Alamo, California, October 13, 1988.

Cindy Rodgers, MS, Director, Statewide Comprehensive Injury Prevention Program, Massachusetts Department of Public Health, Boston, October 21, 1988.

Lisa Rogers, Public Information Director, Harborview Injury Prevention Research Center, Harborview Medical Center, Seattle, Washington, August 22, 1988.

Al Ros, MS, Assistant Environmental Health Director, Dade County Department of Public Health, Miami, Florida, August 24 and September 8, 1988.

Howard Spivak, MD, Deputy Commissioner, Health Promotion Sciences, Massachusetts Department of Public Health, Boston, December 6, 1988.

Denis Lee Tucker, Executive Associate Director, Hmong Mutual Assistance Association, La Crosse, Wisconsin, October 11, 1988.

Cheryl Vince, EdM, Vice President, Education Development Center, Inc., Newton, Massachusetts, July 11, 1988.

Hank Weiss, MPH, MS, Supervisor, Injury Prevention Unit, Wisconsin Division of Health, Madison, October 7 and 20, 1988.

M. Patricia West, MSW, Director, Injury Prevention and Control Program, Colorado Department of Health, Denver, September 14, 1988.

REFERENCES

1. Katcher M. Scald burns from hot tap water. JAMA 1981;246:1219–22.

2. Feldman KW, Schaller TS, Feldman JA, McMillon M. Tap water scald burns in children. Pediatrics 1978;62:1–7.

3. Katcher ML, Landry GL, Shapiro MM. Use of a liquid-crystal thermometer in pediatric office counseling about tap water burn prevention. Pediatrics (in press).

4. Statewide Comprehensive Injury Prevention Program. Safestate: safehome, safechild, safe daycare, and safeschool. Boston: Massachusetts Department of Public Health, 1987.

5. Moore JD, Gerken EA, eds. You can make a difference: preventing injuries in your community. Chapel Hill, North Carolina: University of North Carolina Injury Prevention Research Center, 1988.

6. Micik S, Yuwiler J, Walker C. Preventing childhood injuries: a guide for public health agencies. 2nd ed. San Marcos, California: North County Health Services, 1987.

7. New England Network to Prevent Childhood Injuries: Injury prevention professionals: a national directory. Newton, Massachusetts: Education Development Center, 1988. Available from: Education Development Center, Inc., 55 Chapel Street, Newton, MA 02160.

8. Dean D. Community outreach for health education programs: a practitioner's manual [Dissertation]. Boston: Statewide Childhood Injury Prevention Program, 1981.

9. An Indiana guide to community health planning. Indianapolis, Indiana: State Board of Health, 1987.

10. A child passenger safety resource center: a proposal. Boston: Massachusetts Department of Public Health; 1979.

11. Summary of child automobile safety observation study. Boston: Massachusetts Department of Public Health (undated).

12. Rodgers CN, Rosen L. Increasing enforcement of the Massachusetts child restraint law. Paper presented at the annual meeting of the American Public Health Association. Washington, DC: American Public Health Association; 1985.

13. DeVito CA, Ros A, Strait R, et al. Effectiveness of injury control interventions in the Liberty City area (un-

published ms). Miami: Dade County Public Health Unit, 1988.

14. Rogers T, Houston Miller N. Building and maintaining coalitions. In: Childhood injury control in California: coming of age. Second annual conference. North County Health Services, 1988.

15. Cohen L, Baer N, Satterwhite P. Developing effective coalitions: a how-to guide for injury prevention professionals (unpublished ms). Pleasant Hill, California: Contra Costa County Health Services Department, 1988.

16. Brown CR. The art of coalition building: a guide for community leaders. New York: American Jewish Committee, 1984.

17. Mulford CL, Klonglan GE. Creating coordination among organizations: an orientation and planning guide. Ames, Iowa: Cooperative Extension Service, Iowa State University, 1982; North Central Regional Extension Publication 80.

18. Fisher L. Childhood injuries: causes, preventive theories and case studies: an overview on the role for sanitarians and other health professionals. J Environ Health 1988;50:355–60.

19. Making health communications programs work. Bethesda, Maryland: National Cancer Institute (in press).

20. Rogers L. Bicycle helmet campaign update (unpublished ms). Seattle, Washington: Harborview Injury Prevention and Research Center, 1988.

21. Berger D. Wear a helmet, save a life. Parade 1988 May 29:22–3.

22. How to plan a comprehensive community occupant protection program. Washington, DC: National Highway Traffic Safety Administration (undated).

23. Publicity and advocacy for injury prevention: proceedings of a conference sponsored by the New England Network to Prevent Childhood Injuries. Newton, Massachusetts: Education Development Center, 1988.

24. Public relations guide. Elk Grove Village, Illinois: American Academy of Pediatrics, 1987.

25. National Highway Traffic Safety Administration. Public information, education and relations for EMS providers: your second most important service. Washington, DC: U.S. Department of Transportation, 1986.

26. Northrop D. Progress report: New England Network to Prevent Childhood Injuries implementation incentive grant. Newton, Massachusetts: New England Network to Prevent Childhood Injuries, Education Development Center, 1988.

27. The plastic handgun: visible and invisible. In: EPIC: Educating Professionals in Injury Control: pilot core curriculum. Newton, Massachusetts: Education Development Center, 1988.

28. Riggsbee N. Biography of a parent–organizer and founder of the Drowning Prevention Foundation (unpublished ms). Alamo, California: Drowning Prevention Foundation, 1986.

29. Stone JL. Advocacy: how to influence the agenda. Paper presented at the Second National Injury Control

Conference of the Centers for Disease Control. San Antonio, Texas: Centers for Disease Control, 1988.

30. Redman E. The dance of legislation. New York: Simon and Schuster, 1979.

31. Weiss H, Schmidt W. Developing a statewide injury prevention program. The Wisconsin experience—past, present and future. Paper presented at the annual meeting of the American Public Health Association. Las Vegas, Nevada: American Public Health Association, 1986.

32. Capwell EM. Institutionalization of health promotion programs: going beyond implementation. Paper presented at the annual meeting of the American Public Health Association. New Orleans, Louisiana: American Public Health Association, 1987.

33. Yin RK. Changing urban bureaucracies. Santa Monica, California: Rand Corporation, 1979.

Part 2: The State of the Art of Injury Prevention and Control

Injury Interventions: A Guide To Chapters 6–17

The next twelve chapters are intended to be a resource for injury prevention and control practitioners, for the directors of new injury prevention programs, for decision makers trying to respond to this major health problem, and for concerned citizens seeking to reduce the injury burdens of their communities. From traffic injury to residential, recreational, and occupational injuries to the injuries that result from interpersonal violence and suicide, these chapters document what is known about the injuries themselves and about who is at greatest risk and explore the state of the art in interventions. Before reading the chapters, however, it is important to consider the answers to several questions: How were the chapters created? What are their limitations? How do they relate to the earlier chapters? How they can best be used?

At its first formal session, in April 1987, the 31 members of the National Committee for Injury Prevention and Control were organized into the seven working groups whose membership is listed on page 000. Three groups investigated the process of injury prevention and control, and their work resulted in Part I of this book. Another three working groups—traffic injury; residential, recreational, and occupational injuries; and interpersonal violence and suicide—were charged with identifying, reviewing, and appraising interventions in each broad area. The focus of the final working group was on trauma care systems.

Each working group began by identifying points of consensus or controversy within its field. Then drawing upon the expertise of its members, other specialists, and a wide-ranging literature search by the staff, the working group collected as many reports (both published and unpublished) of specific intervention projects as possible, with an emphasis on those that had been subjected to *some* form of evaluation. These documents, drawn from peer-reviewed and non-peer-reviewed journals, from individual project reports, from published and private literature reviews, and from federal and state program reports, formed the basis for each working group's deliberations about and categorization of the interventions.

Interventions were assigned to one of four categories based on what was known about their efficacy—did they reduce injuries or have some other discernible positive effect? The categories used were "Proven effective," "Promising," "Ineffective," and "Unknown or insufficiently studied."

The groups' recommendations included both the statement of efficacy and suggestions for further use of the intervention. These reflect the discussion of program monitoring and evaluation in Chapters 4 and 5. Thus:

• An intervention that is proven effective (e.g., bicycle helmets) should be used and the program should be routinely monitored.

• An intervention that is promising (e.g., raising alcohol excise taxes to reduce availability) should be used, monitored, and its outcomes should be rigorously evaluated.

• An intervention that is ineffective or counterproductive (e.g., painted crosswalks) should not be used.

• An intervention whose efficacy is unknown or insufficiently studied (e.g., designated driver and safe-ride programs) should be the subject of further research.

Several things should be pointed out about this scheme. It is not the only way in which the interventions could have been categorized. Nor is it complete. In many cases individual projects are described as representative of a type. More important, it is not intended to imply scientific precision in the assignment of interventions to categories. It was selected by the committee as a flexible way of organizing a large amount of information that was the best available but nonetheless of distinctly uneven quality. And primarily it was adopted because it reflected the current realities of injury prevention.

The state of the art in injury prevention can be summarized thus: There are too few interventions that have been proven effective, and there are many promising interventions about which too little is known. There are a smaller number of ineffective or counterproductive interventions that should be abandoned and a great many ideas for new programs and future research. That description, while accurate in general, requires some modification as it applies to each of the three main intervention areas.

Traffic injury, which has the longest history of public and governmental interest and well-organized, multidisciplinary research, also has the greatest number of proven effective interventions. However, even here there are a great many promising countermeasures for which sufficient evaluative data are not yet available. In residential, recreational, and occupational injuries, there are fewer proven effective measures and more that fall into the promising category. Finally, in interpersonal violence and suicide, the paucity of effective interventions highlights the need to make these injuries a national priority.

Any attempt to describe the state of the art is always and foremost a description of the "state of the past." The following 12 chapters delineate the recent past of injury prevention and control activities. This is a selected past, a representative past digested and organized by experts from many disciplines. It is, we believe, a reasonable guide for the future. How, then, should it be used?

First, we must emphasize what it is *not:* It is not a cookbook designed to show that success in injury prevention can be achieved by applying intervention *A* to problem *B* in community *C*. That "cookie cutter" approach to social problems is more likely to result in frustration, the wasting of scarce resources, and more difficulty in beginning another program when the first one fails. Also, no matter how successful an intervention has been in its original setting, it will likely require some modification to maximize the chances of success in another community. Indeed, the most effective interventions described in these chapters are those that most clearly reflect the needs and realities of the communities in which they were implemented.

It is critical, then, that these chapters be read and used in the context provided by the first part of this book. The understanding of a community, the use of data and other techniques to identify its injury problems, and the skills required to design, implement, and evaluate a program are the prerequisites to making effective choices about the interventions reviewed here.

The 12 chapters of this section were designed to be self-contained. Each contains descriptive material about a group (or type) of injuries, the magnitude of their occurrence, risk factors, and the role of prevention, followed by the interventions themselves. We suggest that the section be read as a whole, however, as interventions in one area may suggest new ideas to experts in another. This is particularly true of the chapters on interpersonal violence and suicide. Only by considering these together is it possible to comprehend fully the magnitude of the injury problem related to violence and the critical need for imaginative new programs.

Finally, the practical development and successful implementation of the interventions discussed in these chapters require the collaboration of state and local health officials, traffic safety and public safety experts, public health specialists, physicians, emergency medical services personnel and other health care workers, behavioral science and health education specialists, injury researchers, and others. Public officials, policymakers, and the media have key roles to play as well. And as injury prevention practitioners have discovered countless times, the understanding and support of the local community can make an enormous difference in the success or failure of injury prevention efforts.

Chapter 6: Traffic Injuries

On Wednesday, September 13, 1899, Mr. Henry H. Bliss of New York stepped off a streetcar at the corner of 74th Street and Central Park West. As he did, the 68-year-old real estate dealer was struck and killed by a passing motorcar, thereby becoming the first recorded casualty in America's love affair with the automobile.[1] But if Mr. Bliss was unique on that September day, he would soon be one statistic among millions.

Within a decade, the automobile had surpassed the horse and buggy as an agent of death and injury. From 1913 to 1919, each year saw an average of 6,800 auto-related deaths. And by 1924 there were 23,600 deaths (including 10,000 children) and more than 700,000 nonfatal injuries.[1,2]

Since Mr. Bliss's death, there have been enormous numbers of such incidents. Indeed, if every war since 1776 is taken together, no instrument of death—flintlock, repeating rifle, machine gun, tank, plane, or bomb—has resulted in as many American fatalities as has the motor vehicle. From 1910 to 1985, there were more than 2.5 *million* traffic fatalities; only New York City, Chicago, and Los Angeles have populations greater than that.[3]

How our understanding of these traffic injuries and deaths has changed in the last several decades is a subject of this chapter. What has been done with that understanding to reduce traffic injuries is its primary focus. What can be done in the future is the message of its recommendations.

THE MAGNITUDE OF THE PROBLEM

Among all fatal injuries, motor vehicles are the most common cause of death for individuals 1–34 years of age, and a significant cause of death at all ages. According to the National Highway Traffic Safety Administration (NHTSA), there were 46,056 motor vehicle–related deaths in 1986.[4] The National Safety Council estimates that there were 1.8 million nonfatal, disabling traffic injuries in 1987.[5] Traffic injuries cause the majority of cases of paraplegia and quadriplegia and are the single leading cause of severe brain injury and severe facial lacerations and fractures.[6]

Population-based rates for traffic injury death vary widely by age and sex, peaking in the late teenage years and early twenties. Death rates then decline, increasing again at about the age of 65. Males, especially those in their twenties, are more often fatally injured than are females.[7,8]

As with fatalities, the rates for nonfatal injuries are highest for males 15–24 years of age. For example, a study of crash-related nonfatal injuries revealed much higher rates in the teenage and young adult groups, particularly among males.[9] Annually, about one of every 21 American males 15–24 years of age is injured in or by a motor vehicle in a crash severe enough to require that a vehicle be towed from the scene; one of every 565 males between 15 and 24 years of age sustains a life-threatening motor vehicle–related injury annually, excluding those who die.[10]

It is clear from such statistics that young male drivers constitute a special category when it comes to designing and implementing traffic injury prevention efforts. In both emotional and economic terms (particularly when one calculates years of potential life lost), their deaths and severe injuries cost society dearly.

Traffic injuries also vary by race and income. The highest traffic injury death rates are found among Native Americans (51 deaths per 100,000, compared to 24 for whites, 19 for blacks, and 9 for Asians). And while whites have the highest motorcycle death rates, Native Americans have higher death rates as motor vehicle occupants and as pedestrians.[7]

A portion of the Native American traffic injury death rate is attributable to a high incidence of alcohol use, but there are other important factors. A 1983 population-based survey of Hopi Indian injuries suggested that Native Americans may be disproportionately likely to walk on roads without pedestrian areas or to ride in open pickup trucks.[11]

Per capita income varies inversely with traffic injury death rates. One study found that for all races, the 1977–79 death rate in countries where the annual per capita income was less than $3,000 was more than double the rate in counties where the average income was $6,000 or more.[7] Most of this difference is attributable to variations in the occupant death rate, which is almost three times higher in low-income areas. Poor roads, older and poorly maintained vehicles, different driving practices (e.g., less frequent use of safety belts), and lack of efficient emergency and medical care and transport services play a significant role in this difference. Some of these conditions may also help to explain why traffic injury death rates are more than twice as high in rural areas of the U.S. as in major urban centers.[7]

These statistics, however, both highlight and obscure the problem they describe. As Mr. Bliss's experience reveals, every statistic has its beginning in a single event. And each event has its specific consequences for the victim, for family and friends, and for society at large.

Robert D. was 23 years old in 1980, when the head injury he suffered in a single-car collision put him into a coma. An accountant, Robert was employed by one of Chicago's major firms and had, by all reports, a bright future ahead. Within a few months, he was able to leave the acute care hospital. But head injuries are insidious and unpredictable in their consequences. He moved in with his family. His insurance ran out. His parents spent more than $25,000 in medical expenses. At one point Robert left home and was picked up by the police and institutionalized. Later he lived in his car for a time. Now 31, he still cannot work and is supported by his parents, Medicare, and other government programs. Robert is a traffic injury survivor whose many predictable years of productive life are lost to himself, his family, and society (M. P. Spivack, personal communication).

THE COST OF TRAFFIC INJURIES

The nearly 45,000 traffic fatalities and 3.5 million nonfatal injuries that occurred in 1984 cost the citizens of this country more than $69 billion. That was the message of an important NHTSA report, *The Economic Cost to Society of Motor Vehicle Accidents.* "Both the individual . . . victims and society as a whole, are affected in numerous ways that, unlike pain and suffering, lend themselves to valuation in quantitative terms."[12]

Medical costs, for example, are paid by the individual through payments for uninsured expenses and by society through higher insurance premiums and the diversion of medical resources away from other vital needs such as disease or medical research. Even higher costs are associated with the productivity that is lost when an individual is disabled or killed at an early age.

Those dependent on a victim suffer the immediate economic hardship from lost income, but society also suffers through efforts to support the victim's dependents and eventually through contributions not made to the nation's productivity. The costs associated with human suffering and loss are inestimable.

The NHTSA researchers considered property losses, insurance expense, productivity losses, legal and court costs, medical costs, emergency costs, and other expenses. The $69.52 billion total for 1984 included the following expenditures (in billions of dollars): property losses, $27.54; insurance expenses, $17.7; productivity losses, $15.19; legal and court costs, $4.13; medical costs, $3.78; emergency costs, $0.71; other costs, $0.44.[12]

Property losses make up a relatively large share of the total because most collisions involve vehicle or related property damage but not injuries. The figure for insurance expenses includes both motor vehicle and health insurance. However, as the authors point out, almost all of this cost represents motor vehicle insurance because of the preponderance of property damage claims and high overhead.

An earlier version of this report considered the costs sustained by federal, state, and local governments as a result of traffic injuries.

"In 1980, those portions of public assistance programs that accrue because of motor vehicle accidents, the immediate expenses relating to government workers involved in motor vehicle accidents, tax losses, and the cost of government programs related to motor vehicle safety, are estimated to have cost the federal government $7.5 billion in revenue and expenditures. Similar expenditures cost state and local government about $3.4 billion."[12]

A significant share of the annual cost of motor vehicle crashes is borne by employers. They pay as much as one-third of the cost of off-the-job motor vehicle–related trauma, some $10 billion annually.[13,14] Businesses bear a large portion of insurance costs and the costs of lost productivity, including some 45 million days of work lost annually from traffic injuries.[15] And, to the extent that some of these costs are reflected in higher prices, they become part of the general traffic injury–related costs borne by society as a whole.

It has been estimated that the cost of all crashes averages $308 annually for every man, woman, and child in the United States, that total medical costs per critical injury average $138,000, and that every traffic-related fatality costs society more than $330,000.[12] A more recent analysis suggests that traffic fatalities cost society an average of $425,000 per fatality.[16]

Thus are we all diminished by any person's death or injury: first as the relative or friend of a victim and second as a member of a society whose resources must be expended on this problem. Society is diminished by the lost economic and social productivity of traffic victims as well as by the increased costs of insurance and medical care. And to the extent that providing income support and Medicare services requires the expenditure of state and federal revenues, traffic injuries affect the ways in which tax dollars are spent.

There are less tangible, if no less real, losses as well. Sometimes the lost productivity associated with a person's death or long-term disability is of a kind that affects the national shared culture. There are all too many examples of talent lost in traffic crashes: James Dean (actor), Ernie Kovacs (comedian), Margaret Mitchell (author of *Gone with the Wind*), Jackson Pollock (painter), Bessie Smith (blues singer and early recording star), David Smith (sculptor), and Nathanael West (author of *Miss Lonelyhearts* and other novels), among others.

In any consideration of traffic injuries in America two facts stand out: their cost and their prevalence. In comparison with the costs associated with cancer, coronary heart disease, and stroke, for example, traffic injuries are second only to cancer in their economic cost.[9]

And, as to prevalence, NHTSA provides the following chilling timetable. "Every ten minutes, 24 hours a day, 365 days a year, another American is killed in a motor vehicle crash, and every ten seconds another person is injured, more than the casualties of a major airline crash every day of the year."[17]

That traffic injury is a public health problem of enormous proportions is unmistakable. Nonetheless, progress in moving traffic injury prevention and control into the mainstream of public health has been slow. Not until the publication of *Healthy People: The Surgeon General's Report on Health Promotion and Disease Prevention* in 1979 did traffic injuries take their place on the nation's public health agenda. "The clear message," proclaimed the report, "is that much of today's premature death and disability can be avoided."[18]

RISK GROUPS

Apart from the general considerations discussed earlier, each type of traffic injury (e.g., motor vehicle occupant, pedestrian) involves a different pattern of risk groups and other significant factors.

Motor Vehicle Occupants

A total of 38,211 motor vehicle occupants (both drivers and passengers) died in the United States in 1986.[4] Motor vehicle occupant death rates per billion miles traveled peak at ages 16–19.[7] And among males 15–19 years of age, injuries sustained as a motor vehicle occupant account for one-third of the deaths from *all* causes. The rates for females of similar age are substantially lower, although 1986 Fatal Accident Reporting System (FARS) data indicated a 16% increase in crashes involving 15- to 17-year-old female drivers.[4]

These deaths are most likely to result from single-vehicle or nighttime collisions. The typical fatal crash involves a single vehicle on a rural road at night; the typical nonfatal crash occurs on an urban street during the day.[6]

Geographically, occupant death rates are highest in the southern and western United States. Fatalities are most common in summer and least common in winter. And more than one-third of the deaths occur between 10:00 P.M. and 4:00 A.M., although only 17% of all crashes occur during these hours. Rural roads have higher death rates than urban roads; interstate highways have relatively low crash rates.[7] Most important, alcohol consumption is a leading factor in those crashes resulting in serious injuries and fatalities. Nearly 40% of all fatally injured drivers have sufficient alcohol in their systems to be legally intoxicated.[4]

The demographic shift under way in the United States—the steady aging of the general population—will have a substantial effect on the driving population as well. Approximately 22% of the drivers in 1985 were 55 years of age or older; older drivers will constitute 39% of drivers by the year 2050. Many are careful and experienced, but there is growing evidence that the skills necessary for safe driving begin to deteriorate at the age of 55 or thereabouts, perhaps decreasing dramatically at the age of 75 and beyond.[19,20]

Pedestrians

Pedestrian injuries, which affect primarily young children, the elderly, and the intoxicated, claimed the lives of 6,771 Americans in 1986.[4] Males account for 70% of pedestrian fatalities across all ages. Among children, pedestrian injury deaths are highest in the 5- to 9-year-old age group. It has long been established that the elderly experience the highest population-based pedestrian death rates, and people over the age of 70 are the most frequent victims.[8,21]

Pedestrian injuries are essentially an urban problem. The percentage of pedestrian deaths is fairly constant from month to month, apart from peaks in the fall and winter. Two-thirds of fatalities happen between sunset and dawn, and most occur away from intersections.[4] Although pedestrian fatalities to older children are often the result of "dart-outs" into traffic,[22] fatalities to children younger than 5 years tend to occur when a child is backed over in the home driveway by a vehicle driven by a parent.[23]

Alcohol plays a role in nearly half of all adult pedestrian fatalities. Among all fatally injured pedestrians in 1986, more than 20% were legally intoxicated.[4] And if all drivers involved in pedestrian deaths were to be tested for alcohol consumption, the role it plays might become even more apparent.[7]

The physical vulnerability of pedestrians (and bicyclists) is a major factor in their injury. While motor vehicle drivers and occupants are protected by the vehicle's interior, and perhaps by a safety belt, most pedestrians and bicyclists travel without protective devices. A 50-pound child or even a 175-pound adult is no match for a 2,000- to 3,000-pound moving vehicle.

Motorcyclists

Among 15- to 34-year-olds, one out of every 10 traffic fatalities involves a motorcycle driver or passenger. More than 4,500 motorcyclists were fatally injured in 1986.[4] For every death there are approximately 37 reported injuries and many that go unreported to the police.[7] Death rates are highest between the ages of 18–24 and are 10 times as high for males as for females. Male victims are much more likely to be driving; female victims are much more likely to be passengers.[7] A California population-based study indicates that nonfatal injuries peak sharply at the age of 18.[24]

Motorcyclist deaths are highest during the spring and summer months; most occur on weekends and during the evening and night hours. State-by-state comparisons reveal two factors as being critical in determining death rates: the amount of travel by motorcycle and the existence or absence of helmet-use laws.[7]

Fatally injured motorcyclists are most likely to have illegally high blood alcohol concentrations if they are involved in nighttime, single-vehicle collisions. Nearly 80% of 25- to 29-year-old motorcyclists killed under such conditions had used excessive amounts of alcohol.[7]

A recent investigation of motorcyclists hospitalized at a major trauma center revealed that the average total direct cost for care per patient was over $25,000. More than 60% of these costs were paid by public funds.[25]

Bicyclists

About 90% of bicyclist deaths result from collisions with motor vehicles;[7] according to FARS data, 941 such deaths occurred in 1986.[4] Most nonfatal injuries, however, result from falls. The U.S. Consumer Product Safety Commission estimates that in 1985 there were 574,000 emergency room visits for bicycle-related injuries.[26]

Bicycle death rates rise rapidly from the age of 4 and peak for 13-year-old males. Unlike other forms of traffic injury, bicycle fatalities are less subject to regional variation. Nor do the degree of urbanization or per capita income reveal specific patterns.[7]

Bicycle deaths, like motorcycle fatalities, occur more frequently during the spring and summer months. The peak time is different, however, lasting through the after-school and early evening hours, from 3:00 P.M. to 8:00 P.M. As half of fatally injured bicyclists are children, alcohol consumption (on the part of the bicyclist, not necessarily the driver) is a less important factor.[7]

Helmet use can play a critical role in reducing bicycle injuries. Field studies demonstrate that few bicyclists wear helmets, and a Florida study reported that five out of six fatally injured bicyclists died as a result of head or neck injuries.[27] Frequently those bicyclists whose deaths resulted from serious head injuries did not suffer other life-threatening or potentially disabling injuries. Thus, if bicyclists used helmets, many fatalities and serious head injuries would not occur.[28,29]

ATTITUDES AND TRAFFIC INJURIES

During each of the 10 years from 1964 to 1973, there were more than 50,000 traffic-related fatalities in the United States. There were also the years culminating in the heaviest American combat role in Vietnam, where more than 50,000 U.S. servicemen and women were killed. In *each* year from 1964 to 1973, as many Americans died on the highway as died during the *entire* Vietnam War.

Throughout the whole quarter-century from 1962 to 1986 the annual traffic fatality toll generally ranged between the mid-40,000s and mid-50,000s. How and why our society accepts that annual level of carnage are questions that are easy to pose but hard to answer. What is it about us and our relationship to the automobile that permits, and even encourages, the continuing epidemic?

Jack Kerouac's *On the Road* is a novel that has long been popular, particularly with males in the most vulnerable traffic injury age groups:

I wasn't frightened at all that night; it was perfectly

legitimate to go 110 and talk and have all the Nebraska towns—Ogallala, Gothenburg, Kearney, Grand Island, Columbus—unreel with dreamlike rapidity as we roared ahead and talked. It was a magnificent car; it could hold the road like a boat holds on water. Gradual curves were its singing ease. "Ah, man, what a dreamboat," sighed Dean. . . . "You and I, Sal, we'd dig the whole world with a car like this because, man, the road must eventually lead to the whole world."[30]

Speed, the limitless vista of the highway, escape, freedom, status, and even sexuality are so bound up in this one machine, this "dreamboat" of a car, that they are almost impossible to disentangle. But we can try.

Standing on a corner in a strange town, watching the cars go by on a 1950s night, reporter Eric Larrabee experienced "a sense of private destinies, of each making his own choice, of being independent of everything but statistics."[31] That sense of private destinies, of independence, helps in part to explain the American ability to accept with seeming equanimity 40,000–50,000 highway deaths each year. Several factors would seem to be at play.

First, there is the appearance of randomness and the geographic dispersion of collisions: "There may be a traffic fatality every ten minutes, but not on my corner." To this is added denial—the general human inability to believe that harm will really befall oneself. Finally, as the automobile is an inanimate machine ostensibly under the operator's complete control, it hardly seems reasonable to "blame" it for collisions. Therefore, each crash must be the result of some driver's error; a skillful driver can avoid the errors and hence the collisions. As race car driver Sterling Moss observed, there are two areas in which no one will admit inadequacy: driving and lovemaking.[31]

Moss's observation leads to the last of the elements entwined with the automobile: sexuality. "Perhaps the greatest marketing coup of the twentieth century," wrote Stephen Bayley, "was to relate the automobile to sex."[31] No need to belabor the point. One need only look at advertisements, read books, visit the movies, watch TV, or listen to popular songs. The automobile is "the stuff dreams are made of." Nightmares, too.

One of the strongest scenes in Alfred Hitchcock's dark classic *Notorious* (1946), for example, is played out at night in the front seat of a convertible speeding precariously down a Florida road. Cary Grant and Ingrid Bergman are the protagonists, and the film has barely begun. Each is trying to exert power over the other. Bergman pushes the car faster and faster to frighten and thereby control

Grant; the sexual tension between them is almost unbearable. The scene is resolved by the sudden appearance of a police officer.

It's all there—the automobile as a vehicle for exhilarating speed, as a catalyst for passion, desperation, power, and control. But there's one other element that makes the scene worthy of consideration. Bergman, the driver, is drunk.

ALCOHOL AND TRAFFIC INJURIES

Alcohol is thought to be the most abused drug in America. Many studies document that crashes in which alcohol plays a role tend to be much more severe than other crashes. The more severe the crash, the higher the percentage in which alcohol contributes. Nationally, alcohol plays a role in about 20% of crashes involving serious injury to a driver or passenger, about 50% of all fatal crashes, and about 60% of single-vehicle fatal crashes.[32]

In 1987 an estimated 23,630 Americans were killed in alcohol-related motor vehicle crashes, accounting for a total of 783,304 years of potential life lost.[33] And a recent analysis of North Carolina crash data on more than 1 million drivers indicates that in crashes of equal severity, "the drinking driver is more likely to suffer serious injury or death compared with the nondrinking driver."[34] The connection is irrefutable; eliminating the mixture of alcohol consumption and automobiles would lead to dramatic decreases in death, injury, disability, and costs for all age groups.

In 1983 some $60 billion was spent on alcoholic beverages in the United States. In the country as a whole, 27% of men and 42% of women do not drink at all. On the other hand, only one-third of the adult U.S. population drinks 95% of all of the alcohol consumed in 1 year, and, even more startling, a mere 5% of the people drink 50% of the total.[35] According to the National Research Council (NRC), alcoholism implies "at a minimum, a loss of control over the intake of alcohol or an inability to stop drinking."[36] What alcohol use implies for the driver is illustrated by the following facts.

One 12-ounce can of most beers, one 4-ounce serving of most wines, and one cocktail made with 1.2 ounces of 80-proof spirits contain identical amounts of alcohol. Alcohol is metabolized by the liver, on average, at the rate of one drink per hour (this rate varies, however, depending upon an individual's weight and other factors).

Blood alcohol concentration (BAC) is the standard measure of intoxication. A BAC of 0.10% means a level of 0.10 g of pure alcohol per 100 mL of the person's blood. An increasing BAC correlates

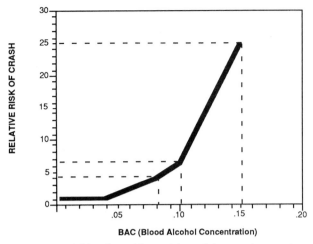

Figure 1. Odds of crashing. Adapted from reference 37.

with decreased coordination, reasoning, and balance. Many laws on impaired driving use 0.10% as the definition of intoxication. However, as the NRC report indicates, "At a BAC of 0.05 percent, which for most people requires three or more drinks within an hour, physical skills begin to deteriorate."[36] As Figure 1 illustrates, the rate of motor vehicle crashes also increases at this point.[37]

An NHTSA statement of 1981 concluded that solutions to reducing impaired driving are best implemented at the state and local level. Furthermore, "state and local law governs in this area and state and local courts are the only forum for this case. . . . The crux of the drunk driver problem in most states is not lack of adequate laws . . . but the lack of consistent, convincing enforcement of those laws by state and local officials."[38]

A BRIEF HISTORY OF TRAFFIC INJURY PREVENTION

The automobile was barely out of its infancy when concern over increasing traffic injuries gave rise to the belief that nearly all collisions resulted from poor driving and human error. By 1914 that belief had found an organizational home in the recently established National Safety Council. The council and other similarly focused groups adopted an approach embodied in the phrase, "The Three E's: Engineering, Enforcement, and Education."

In practice, however, this theoretically comprehensive approach was directed almost entirely at education. And education, in those boom days of American advertising, meant public relations, outdoor advertising, and simple messages. "Success," according to historian Joel Eastman, "tended to be measured in the number of posters or pamphlets distributed, the number of safety campaigns con-

ducted, and the number of column inches of newspaper space or radio . . . time acquired."[2]

In the 1920s President Herbert Hoover convened the National Conference for Street and Highway Safety, the first in a series of presidential initiatives. The conference's goal was to agree on a body of uniform traffic laws that would prevent collisions. The 1936 Accident Prevention Conference, convened by President Roosevelt, briefly shifted the focus of traffic safety to the vehicle by calling for slower speeds, better lighting, and stronger and safer body construction. However, these recommendations were put on hold by wartime priorities and the postwar auto boom. President Truman maintained an interest in highway safety, although to no greater effect than his predecessors. In 1954, by executive order, President Eisenhower created the President's Action Committee for Traffic Safety.

During these same years, critics of the highway safety movement, and even some of its members, began calling for more research, citing the inadequate basis upon which earlier judgments had been made. In testimony before Congress in 1956, Professor Ross A. McFarland of the Harvard School of Public Health decried "overgeneralizations from inadequate or improperly handled data [that] have been taken to constitute the facts of accident causation."[2]

THE SCIENCE OF TRAFFIC INJURY PREVENTION

The science of traffic injury prevention had already taken its first significant steps in the midst of World War II. "In the 14 years between 1942 and 1956," wrote Dr. William Haddon, Jr., et al. in the landmark volume *Accident Research: Methods and Approaches*, "a new engineering field has been created, namely, that of crash-survival design engineering."[39] The modern understanding of injury prevention and control began in 1942 with Hugh De Haven's brief but seminal *War Medicine* study, "Mechanical Analysis of Survival in Falls from Heights of Fifty to One Hundred and Fifty Feet."[40]

Both De Haven and his pioneering work have already been discussed in some detail in the introduction to this book. However, in the context of traffic injuries, it is useful to remember the conclusion of his *War Medicine* article: "It is reasonable to assume that structural provisions to reduce impact and distribute pressure can enhance survival and modify injury within wide limits in aircraft and automobile accidents."[40]

Additional pressure for increased safety came from the medical profession. As early as the 1930s, Dr. Claire Straith had installed safety belts and padding in his own car. Straith, a prominent Detroit plastic surgeon, had seen the consequences of poor interior vehicle design on the faces and bodies of his patients. He met with industry representatives and strongly urged redesign of the passenger compartment to eliminate protruding knobs, unyielding dashboards, easily shattered window glass, poorly placed window cranks, and other injury-producing elements.

Through the prewar years and the early 1940s, Straith had limited success. In 1948 he joined with the Accident Prevention Bureau of the Detroit Police Department to conduct a study of crashes, which revealed that during a 1-month period 70% of those injured were right front seat passengers who were thrown into the instrument panel.

Meanwhile, the pace of research—and the broadening of the safety research community—was quickening. John Stapp, like Hugh De Haven, was fascinated by the effects of mechanical forces on human tissue. A physician, biophysicist, and Air Force colonel, Stapp conducted a 12-year series of experiments in which he used a rocket-propelled sled to simulate crash forces and to test safety belts and other restraints, sometimes serving as his own subject.

In 1955 the Society of Automotive Engineers began working with Stapp and others involved in crash research. The Annual Automotive Crash Research Field Demonstration and Conference was the result.

Nor did traffic safety research stop with the physicians and the engineers. "Since 1950," said Frank J. Crandell, vice-president and chief engineer for Liberty Mutual Insurance Company of Boston, "we have been working on this project in order to try to design the proper packaging principles so that we could have a crash and still reduce the injuries."[2] In 1951 Liberty Mutual began research studies in conjunction with Cornell University.

Following on the efforts of some of its members, organized medicine took an increasingly larger role in the automobile safety debate. Both the American Medical Association and the American College of Surgeons passed strong resolutions calling for tougher safety standards and particularly for the mandatory installation of safety belts. And 1957 witnessed the founding of the American Association for Automotive Medicine (now the Association for the Advancement of Automotive Medicine), which was to play a major role in the passage of safety belt legislation and in many other areas of traffic safety.

By the mid-1950s the new field of traffic safety, drawing on the expertise of engineering, bio-

physics, biomechanics, medicine, public safety, and design was off the ground. It had documented, wrote Dr. Julian Waller, "that body impact against hard, sharp objects within the vehicle resulted in more serious injury than did impact against broader, more flexible surfaces. In particular, the unyielding steering column penetrates the driver's chest and jagged windshield fragments and sharp dashboard edges lacerate and macerate faces and heads of front seat passengers."[41]

Many engineers, researchers, public safety officials, and public health officials had now concluded that the machine and not the driver was at fault for much of the carnage. But if the safety constituency was broadening and the research agenda was prospering, the numbers of injuries and fatalities were still climbing. The result was an increasing pressure on the government to act.

TRAFFIC INJURY: THE FEDERAL RESPONSE

By the mid-1960s, strong currents were propelling the likelihood of federal regulation of the automobile industry. In 1965, under the leadership of then-Senator Abraham Ribicoff, a subcommittee of the Senate Government Operations Committee began a highly publicized investigation of automotive safety. Testimony revealed, wrote automotive historian James J. Flink, that "since 1960 over eight million cars (about one out of every five manufactured) had been recalled for defects, many of them safety related."[1]

Public pressure for legislation increased dramatically in late 1965, when Ralph Nader, a consultant to the Ribicoff committee, published *Unsafe at Any Speed: The Designed-in Dangers of the American Automobile*. Finally, through the efforts of Senators Ribicoff, Gaylord Nelson, and Warren Magnuson, and with the support of President Lyndon Johnson and his special assistant Joseph Califano, the National Traffic and Motor Vehicle Safety Act of 1966 became law. The act has been described as a "revolutionary statute,"[42] and it has been suggested that there are probably no other examples of the adoption of a federal initiative in medicine or the social sciences "where the process has occurred with such unanimity and essential support in the space of nine months, at the initiative of the President, by the unanimous vote of the Congress, and with the wholly honorable acquiescence of the industries involved."[43]

The 1966 act established the National Highway Traffic Safety Administration (originally called the National Highway Safety Bureau) and empowered it to set safety standards for new cars, beginning with 1968 models. In a move that symbolized the importance of uniting public health and the traditional traffic safety community, Dr. William Haddon, Jr., was appointed NHTSA's first administrator. The new agency immediately began "issuing rudimentary federal standards that have made cars safer ever since: laminated windshields, collapsible steering assemblies, dashboard padding, improved door locks, dual braking systems, and many other automatic safety features."[17]

The changes mandated by the first regulations required automobile manufacturers to adopt current state-of-the-art engineering to resolve long-recognized problems. They did not guarantee that change would continue as knowledge increased. "The state of the art has advanced dramatically since the 1960s," noted a 1981 publication from the Insurance Institute for Highway Safety, "but government foot-dragging and regulatory uncertainty, coupled with industry lethargy, have frustrated the translation of engineering advances into new-car design realities. No new cars presently embody all the modern-day technologies that for some time have been available to substantially reduce motor vehicle crash injuries."[44]

Chief among the unused technologies was the inflatable, energy-absorbing air bag. However, perhaps because it was a genuine, exciting innovation or because the public/legislative/bureaucratic/legal controversy about it had assumed almost mythic proportions, there was a danger that the air bag would be seen as *the* solution to the traffic injury problem. To accept that would have been to make a mistake very similar to that of the early highway safety advocates, who saw in each collision the workings of human error, stupidity, or laziness—when research has now shown that each collision is a complex interaction of driver, vehicle, and roadway.

A NEW UNDERSTANDING OF TRAFFIC INJURY

As Haddon, Edward Suchman, and David Klein wrote in 1964, "The need for a balanced approach" is critical; those seeking to understand and prevent injuries must guard against "attempts to explain phenomena in narrow terms"[39] or from the perspective of a single discipline. Haddon's contributions to injury prevention and control include his identification (following on the work of J. J. Gibson) of injury with the harmful transfer of energy to human tissue, the Haddon matrix, and the 10 countermeasure strategies. These are discussed in the Introduction, but they should be borne in mind as the context within which many of the intervention strategies discussed in this chapter were developed.

Highway safety has been a crusade that required the contributions of many experts from many disciplines. And so, more than 70 years after the birth of the highway safety movement, we have come to understand the importance of combining the best elements of the three major preventive strategies (legislation/enforcement, education/behavior change, and engineering/technology) to reduce the tremendous toll of traffic injuries. For example, the success of laws on lower speed limits, a minimum legal drinking age, safety belts, child safety seats, and motorcycle helmets still requires a personal decision to comply as well as active enforcement on the part of society. We must find, wrote psychologist Michael Roberts, "the most comprehensive, effective, acceptable, and implementable strategy to achieve . . . health benefits through a multifaceted approach targeted at numerous levels of intervention involving both active and passive and both individual and population approaches."[45] If we have learned anything since Henry Bliss's death, it is that traffic injuries are the result of a complex process, a mixture of vector (machine), host (driver), and environment, conditioned by history, social practices, and individual attitudes, desires, and capabilities.

PREVENTING TRAFFIC INJURY: THE STATE OF THE ART

The lessons of this consideration of the traffic injury problem are unambiguous. By any measure, traffic injuries are a public health problem of enormous magnitude. They affect everyone, every day, in ways both apparent and subtle. That is the bad news. The good news is that we possess a broad array of strategies and techniques that can be brought to bear to ameliorate the problem.

There are, however, several points about implementation that should be repeated here. It is helpful to divide interventions into the categories of legislation/enforcement, education/behavior change, and engineering/technology. The distinction is apparent, if not always real. For example, the lifesaving technological changes implemented during NHTSA's early days were adopted by automakers under federal mandate. And safety belts and child safety seats, each an extraordinarily effective technological advance, have required sustained legal and educational approaches to fully achieve their promise.

The modern understanding of traffic injuries requires a broad, multidisciplinary effort. It is unreasonable to assume that a problem understood only through the combined expertise of highway safety specialists, physicians, researchers in biomechanics, engineers, public safety officials, public health scientists, and health care practitioners—among others—could be solved by a less multipronged assault.

One promising strategy is to combine a broad range of approaches to establish a community-based traffic safety program. Under such a program, a community would use both public and private resources to understand and attack all of its significant traffic problems (e.g., impaired driving, occupant protection, pedestrian injuries).[46] This would require making use of the full array of available interventions, both those described here and adaptations based on local need. The concept of the comprehensive community program is an innovative one for which there are no prescribed models.

The history of traffic injury prevention programs teaches a deceptively simple but critical lesson. Since the pioneering work of Haddon and others, collisions have been understood as singular events reflecting the interaction of a variety of separate factors related to the vehicle, the driver, and the external environment that coalesce at a precise moment. Each event, then, can be seen as the culmination of a chain of causation. The old saying that even the strongest chain can be broken at its weakest link can teach us a valuable lesson. No matter what "link" an intervention breaks, it will break the chain of causation. And the important, although often overlooked, corollary is that an intervention need not be targeted at the strongest or most obvious link in order to succeed.

The problem of impaired driving suggests a practical example. The general approach to reducing impaired driving has been the passage of increasingly strict laws directed at the driver, the most obvious link. Although the laws have important short-term effects, these effects degrade over time.[47] It is also possible to conceive of countermeasures directed not against the driver but at the vehicle. An ignition interlock, for example, can prevent the vehicle from starting if the driver is drunk. This is not to suggest that such devices are *the* solution, merely to indicate that approaches directed at the motor vehicle are important elements in a comprehensive effort.

Finally, no one who seeks to reduce the traffic injury problem—whether on the local, state, or national level—can afford to overlook the political dimensions of injury prevention and control. It should be humbling to remember that legislation mandating the use of two of the most effective traffic injury interventions—safety belts and motorcycle helmets—has been enacted, only to be overturned later in some (safety belts) or many (helmets) states. In addition, although all 50 states now have a minimum legal drinking age of 21, some states are considering lowering the age. There can exist no stronger argument for the importance of combining legislation/enforcement, education/behavior change, and engineering/technology approaches, drawing upon what is best in each, in the service of what is critically necessary to society.

The interventions discussed below are grouped into the following categories: impaired driving, occupant protection systems, teenage drivers, pedestrians, motorcyclists (including those on mopeds), bicyclists, enforcement of speed limits, and vehicle design.

INTERVENTIONS

Impaired Driving

Legislative/enforcement interventions. Using a blood alcohol concentration (BAC) level of 0.05 g/mL or above as per se evidence of impaired driving. Scientific evidence gathered over the last 50 years indicates the direct relationship between an increasing blood alcohol concentration (BAC) and the risk of collision and documents that driving skills begin to deteriorate markedly at 0.05 BAC. The implementation of a per se law makes it illegal to drive with a BAC at or above the specified level whether or not there are observable behavioral effects.

Congress has directed NHTSA to determine what BAC level is most appropriate for a legal limit. The American Medical Association and many experts have called for lowering the legal definition of impaired driving from 0.10 to 0.05.[48-50] In some states such a reduction may require more than one step, such as going from 0.10 to 0.08 and then to 0.05.

Instituting a lower BAC for teenage drivers. In June 1983 Maine implemented an automatic 1-year license suspension for any person under the age of 20 identified by police as driving with a BAC above 0.02. In comparing the first 2 years after passage with the year before the law was passed, one study noted continued increases in injuries and fatal crashes among both teenagers and adults. However, while adult crashes and injuries increased 13%, the increase for teenagers was only 4%.[51]

There is evidence, though inconclusive, that the combination of inexperience in driving and inexperience in drinking results in lower levels of alcohol having a greater impact on the driving of teenagers. In effect, these are two learning curves that overlap dangerously (P. Waller, personal communication).

Two crucial issues must not be overlooked with regard to impaired driving among teenagers. First, despite a minimum legal drinking age of 21 in every state, society does not currently present a clear message that abstinence, i.e., *no* use of alcohol, is a viable and respectable choice for teens. Second, society encourages drinking, particularly among young males, through marketing, advertising, television, and movies. In this, as in other areas, public policy creates predictable problems for which we then blame individuals.[52]

Implementing compulsory BAC tests in traffic injury cases. In April 1974 the government of Victoria, Australia, mandated compulsory BAC tests on all road crash casualties of persons 15 years of age or older. The legislation requires the physician of first attendance to take a blood sample. Any sample taken within 2 hours of the crash and indicating a BAC of 0.05 or greater is accepted as prima facie evidence of impaired driving under the law. While researchers have discussed the law's positive effects in providing increased data about impaired driving (particularly in nonfatal collisions) and in stimulating prosecutions, they have not assessed its impact on fatality or injury reduction.[53]

According to a panel of the 1988 Surgeon General's Workshop on Drunk Driving, the public health uses of BAC information on fatally and seriously injured individuals include "patient diagnosis and clinical management, aiding in the diagnosis of alcohol abuse, and providing data to document the epidemiology of alcohol in . . . injury."[54] The Australian statutory approach, which requires physicians to measure BACs as part of the legal process, would be less practical in the United States, where the focus is on physicians reporting BACs for public health rather than law enforcement purposes.

Administrative per se license suspension. In Oklahoma, as in other jurisdictions that have adopted this intervention, legislation requires the immediate surrender of a license if the driver refuses to submit to a chemical test or if such a test records a BAC of greater than 0.10 (refusal to be tested results in a 6-month suspension; a BAC of greater than 0.10 results in a 90-day suspension). Paul Reed, Jr., Oklahoma's Commissioner of Public Safety, reported that only 2 years after the law went into effect it had had a substantial impact on traffic fatalities.[55]

Recently a team of researchers compared the impact of several types of impaired driving laws on drivers involved in fatal crashes. They found that only administrative per se license suspension laws were associated with a statistically significant decrease (5%) in fatal crashes.[56]

Statewide action is necessary to enact and enforce this legislative intervention. The support of state public safety officials, including the Governor's Highway Safety Representative, can be of help, as can local political support.

Enforcement of minimum legal drinking age laws. From 1970 to 1975 many U.S. states and Canadian provinces reduced their minimum age for alcohol purchase. Research conducted in both countries demonstrated an increase in the crash involvement of drivers under the age of 21. Beginning in 1976, the trend to increasing the minimum age for alcohol purchase gained momentum and at present all 50 states have 21-year-old purchase laws.

Based on a review of many of the studies carried out since the early 1980s, researchers concluded that "raising the legal minimum age for purchasing alcoholic beverages reduces fatal crash involvement among youthful drivers."[57] Their analysis of 87,153 nighttime crashes between 1975 and 1984 indicated that raising the alcohol purchase age to 21 "was estimated to produce a 13% reduction in nighttime driver fatal crash involvements."[57] A more recent analysis estimates that minimum drinking age laws "saved between 977 to 1,071 lives in 1987 and 7,433 to 8,142 lives cumulatively between 1975–1987."[58]

However, as is the case with any traffic safety law, its effectiveness depends in large measure on its public support and enforcement by law enforcement officials. One group of researchers maintains that single-vehicle nighttime fatal crashes among teenagers (a common measure of impaired driving) actually increased 17% in 1986 after having been on the decline for several years. A crucial factor in reversing this upward trend, they maintain, will be improved police enforcement and public support for enforcement laws to combat impaired driving.[59]

Dram shop laws (civil liability of servers of alcoholic beverages.) Dram shop laws make liable those who serve alcoholic beverages to minors or to persons already "obviously" or "visibly" intoxicated. Some 38 states have now enacted dram shop laws. One study examined dram shop laws as a method of "reshaping the drinking environment" and of preventing alcohol-related problems. "Server intervention, with supporting dram shop legislation," the study concludes, "is a practical prevention strategy that can reduce the incidence of impaired driving and other alcohol-related problems. It also provides a means to increase public awareness of the responsibility of providers, both private and commercial, to serve alcohol responsibly, and of the need for a focused prevention effort."[60] Therefore, we recommend adoption of the Model Alcoholic Beverage Retail Licensee Liability Act of 1985.

Recommendations:

• The following legislative interventions have been proven effective: (1) administrative per se license suspension; (2) enforcement of minimum legal drinking age laws; and (3) dram shop laws. States and municipalities should adopt these legislative interventions and monitor their enforcement.

• Implementing compulsory BAC tests in traffic injury cases is clearly beneficial from a public health perspective. This intervention should be used and monitored.

• Using a BAC of 0.05 g/mL or above as per se evidence of impaired driving is a promising intervention. It should be implemented and monitored and its outcomes evaluated.

• Instituting a lower BAC for teenage drivers is an intervention that has not been sufficiently studied. Further research should be conducted in this area.

Other interventions. Sobriety checkpoints. After reviewing several U.S. and Canadian studies of sobriety checkpoint implementation, one researcher concluded that checkpoints are valuable because they increase public awareness of enforcement efforts and heighten the perception that impaired driving will be detected and punished. Further, "there is convincing evidence that the use of checkpoints has a marked, dramatic effect on reducing alcohol-related crashes in a community. This effect [is enhanced] by the existence of concerted public information efforts."[61] The evaluation of a New South Wales, Australia, campaign that includes extensive use of sobriety checkpoints (approximately one in three drivers is tested annually) and widespread media attention indicated a 22% drop in fatal crashes per week.[62]

Some researchers are pessimistic, seeing any positive impact as a result of the publicity itself.[47,63–65] In addition, it is noted that the positive effect of such activities falls off over time.[63]

Recommendation: The use of sobriety checkpoints has not been sufficiently evaluated. Additional research should be conducted. Because of the potential civil rights and civil liberties issues surrounding the implementation of checkpoints, their use should be carefully monitored to assure fair and minimally intrusive implementation.

Alcohol safety education schools for convicted impaired drivers. Alcohol safety education schools typically require convicted first offenders to pay for and attend 10–13 hours of classroom instruction on alcohol and its effects on the body and driving, impaired driving laws, and penalties. In return, they receive limited driving privileges and a faster reinstatement to full driving privileges than would otherwise be the case. Such programs are often used by courts in lieu of complete license suspension or revocation.

"Each year," wrote three highway safety researchers, "thousands of persons convicted of driving under the influence are sentenced to short-term educational courses directed at modifying their drinking driving behavior. Do such courses really impact drinking driving recidivism?"[66] They analyzed the North Carolina program, which is similar to many others in use around the country.

Course attendees, the authors found, demonstrated a highly significant increase in knowledge about alcohol and driving. However, "the salient feature of the recidivism analysis is that for every measure taken and for every time frame examined the study group fared worse than the comparison group."[66] However, a more recent evaluation suggests that when alcohol safety education schools are used in conjunction with license suspension, the combination affects recidivism rates more than license suspension alone.[67] The 1988 Surgeon General's Workshop on Drunk Driving noted that alcohol safety education schools "are insufficiently based in theory" and concluded that the schools might be improved by infusion of "current knowledge in the field of social psychology, mass communication and organizational change."[54]

Thus, in their current form, interventions such as the North Carolina program have not demonstrated their effectiveness. It would be counterproductive for a community to use such programs *in place of* license suspension, diagnosis, screening, and treatment for impaired drivers. Their use as a primary approach to impaired driving can create a false sense of security in a community. (For a review of progress in the field to date, see reference 68.)

Recommendation: In their current form, alcohol safety education schools have not demonstrated their effectiveness. Further research is necessary to design effective educational efforts for impaired drivers.

Raising state and federal alcohol excise taxes to reduce alcohol availability. For a long time, the real price of alcoholic beverages has been declining relative to other commodities. This has been due primarily to the stability of federal excise taxes and the modest increases in state and local taxes. Between 1960 and 1980, for example, the real price of liquor fell by

48%, the real price of beer by 27%, and the real price of wine by 20%.

One study explored, through econometric methods, the effects on youth motor vehicle fatalities of an increase in the price of beer. "Simulations suggest," the authors concluded, "that the lives of 1,022 youths between the ages of 18 and 20 would have been saved in a typical year during the sample period if the federal excise tax rate on beer, which has been fixed in nominal terms since 1951, had been indexed to the rate of inflation since 1951. This represents a 15% decline in the number of lives lost in fatal crashes."[69-71]

Recommendation: Raising state and federal alcohol excise taxes to reduce alcohol availability is a promising intervention. It should be implemented and monitored and its outcomes evaluated.

Server training. Server training seeks to teach waiters, waitresses, and bartenders to employ techniques to avoid serving already intoxicated customers and to prevent customers from becoming intoxicated. A team of researchers evaluated the effectiveness of a program titled "Training for Intervention Procedures by Servers of Alcohol" (TIPS) by comparing trained and untrained servers' interactions with alcohol-impaired bar patrons. They found that trained servers intervened more often than did untrained personnel, and that the customers served by trained servers reached substantially lower BACs than those served by untrained servers. The conclude, "If implemented on a large scale, server intervention programs have the potential of reducing drunken driving by helping to decrease the exit BACs of bar patrons."[72]

In addition, NHTSA is developing and field testing two programs for training individuals in the responsible serving of alcohol. One focuses on commercial servers, the other on social hosts. The 6-hour modular program is designed "to provide servers with a set of strategies they can employ to prevent patrons from becoming intoxicated."[73] Preliminary findings from the field test indicate that although the program had a statistically significant impact on server training at one site, at another site it appeared to have no impact.[74]

A variation on server training is embodied in a program called Techniques for Effective Alcohol Management (TEAM), which was designed for use in large sports facilities. Unlike the tavern setting, in which only servers are trained, TEAM provides training for the entire arena staff from ticket takers to managers. "The ultimate purposes of the training," a reviewer wrote, "are to promote effec-

tive crowd control and to address the issue of safety en route to and from and while attending sporting or entertainment events in public assembly facilities."[75] TEAM also works to stimulate communities to carry out comprehensive efforts to prevent drinking and driving. The program is a collaboration among NHTSA, the National Basketball Association, the National Safety Council, the International Association of Auditorium Managers, the Motor Vehicle Manufacturers Association, the National Automobile Dealers Association, the Government Employees Insurance Company, and CBS, Inc.

Recommendations:

• Server intervention programs directed at waiters, waitresses, and bartenders are a promising intervention. They should be used and monitored and their outcomes evaluated.

• TEAM is an intervention that has not been sufficiently studied. Further research should be conducted.

Educational programs to prevent impaired driving among youth and young adults. A variety of educational programs targeted at youth are in use in schools, youth groups and clubs, religious groups, etc. In order to be successful, these programs must go beyond simply providing information and increasing awareness. Rather, they must be directed at equipping youth with skills to resist peer pressure and at changing attitudes and social environments in support of sober driving. These educational programs should not be used as an isolated intervention but should be incorporated with other proven legislation/enforcement and engineering/technology interventions. (For more information about these primarily local-level programs, see references 76–78).

Recommendation: Educational programs to prevent impaired driving among youths and young adults are a promising intervention. They should be used and monitored and their outcomes evaluated.

Designated driver and safe ride programs. In the last few years, a great number of designated driver and safe ride programs have been initiated under public and private auspices.[79] They have been useful in raising public consciousness about impaired driving. It is possible that they have also had an effect upon traffic injuries; however, this has not been demonstrated.

Recommendation: Designated driver and safe ride programs are interventions that have been insufficiently studied. Further research is needed.

Ignition interlocks. An ignition interlock device prevents a driver from starting a vehicle unless he or she passes an alcohol-detecting breath test. Washington is one of several states that have passed ignition interlock legislation or have established pilot testing programs. In Washington, "The court may order any person convicted of an offense involving the use, consumption, or possession of alcohol while operating a motor vehicle to drive only a motor vehicle equipped with a functioning ignition interlock device, and the restriction shall be for a period of not less than six months."[80] The offender pays for installation and rental of the device.

Additional work is under way to render the devices immune to cheating (for example, one major manufacturer has introduced a complex "breath code" to prevent someone else "passing" the test for the offender). Pilot testing and evaluation of the ignition interlock systems can be implemented through state courts.

Recommendation: Ignition interlock systems are an intervention that has not been sufficiently studied. Further research is needed on their effectiveness in reducing impaired driving. Ignition interlock systems should be used in addition to, not in lieu of, proven effective measures like license suspension.

Roadway countermeasures to reduce impaired driving. "Many alcohol-related crashes," reported a joint working group that explored initiatives to combat impaired driving through roadway engineering, "involve driver errors which result in the vehicles running off the road or crossing the centerline, particularly along curved stretches of roadway. Some of these behavioral errors might be reduced through implementation of roadway countermeasures like improved roadway marking, signing, signals, etc."[81]

The group considered a broad range of countermeasures, including edge lines, raised lane delineators (raised lane markers embedded in the road surface), herringbone patterns (to make the road appear narrower, thus compelling drivers to reduce speed), rumble strips cut into the roadway surface as tactile cues, warning signs, and other devices. The report concludes that edge lines and wrong-way signs appear helpful in preventing collisions in general and alcohol-related crashes in particular. They suggest that standard 4-in edge lines be used on all multilane and high–traffic volume systems and that wrong-way signs continue to be installed

at all off ramps and entry points to one-way roads. They further suggest that raised lane delineators and rumble strips may be beneficial.

Recommendations:

- Edge lines and wrong-way signs are promising interventions to prevent injuries from impaired driving. They should be used and monitored and their outcomes evaluated.
- Other roadway countermeasures, including raised lane delineators, rumble strips, and herringbone patterns, have not been sufficiently studied. Further research should be conducted.

Occupant Protection

Safety belt use. The effectiveness of safety belts in preventing injury and death in motor vehicle crashes is well established; they are estimated to reduce motor vehicle fatalities by 40% to 50% and serious injuries by 45% to 55%.[82] In addition, one study found that safety belt use reduced hospital admissions from crashes by 65% and hospital charges by 67%.[83]

Mandatory safety belt use laws. The movement to mandate safety belt use gained momentum in 1970, when the state of Victoria, Australia, enacted the first safety belt use law. Many other foreign countries[84] and 32 U.S. states and the District of Columbia have since enacted similar legislation. (Massachusetts and Nebraska enacted laws that were subsequently repealed by voters.[85])

The efficacy of these laws (and of the enforcement and public education efforts that complement them) has been demonstrated in many studies.[86–88] A recent analysis of the impact of the laws in 24 states and the District of Columbia revealed a 7% drop in fatalities. The same researchers found that in five states with belt laws, nonfatal injuries decreased by 10%.[89]

Safety belt use has not had a greater impact on injuries and deaths in part because many people covered by the laws do not use safety belts, most belt use laws have gaps in coverage (i.e., not all vehicles and individuals are required to comply with the law), and safety belts are not 100% effective in reducing death and disability—as stated earlier, their effectiveness is estimated to range from 40%–55%.[84]

Enforcement of safety belt laws clearly affects compliance. However, as has been noted, "That does not mean there is a one-to-one relationship between changes in enforcement and changes in belt usage." But there is evidence that "if the public believes there will be *no* consequences of ignoring the belt law, compliance is apt to be lower."[84]

Safety belt use laws are implemented either through primary enforcement (i.e., a law enforcement official may stop a driver solely on the basis of a safety belt law violation) or through secondary enforcement (i.e., only after having stopped the driver for some other purpose may the officer address a safety belt law violation). Not suprisingly, states with primary enforcement policies have had the highest increases in belt use rates.[88]

In Elmira, New York, a law enforcement and publicity campaign related to the state belt (primary enforcement) use law increased use from 49% to 77%. Four months after the campaign, belt use dropped to 66%, only to be raised to 80% by a reminder campaign. Eight months after the second campaign, use had declined to 60%. Throughout this period, belt use in the comparison community of Glens Falls hovered around 41%. The researchers concluded that "initial and reminder programs are necessary to sustain high use rates over time."[90]

Recommendations:

• Enactment and enforcement of safety belt use laws have been proven effective in reducing motor vehicle injury and death. All states should enact and enforce a primary enforcement safety belt use law. States with secondary enforcement policies should amend the laws to allow for primary enforcement.

• Requiring safety belt use by employees who drive or ride in federal, state, municipal, or private fleet motor vehicles is a promising intervention.[91] Managers of fleets should implement this intervention, and its use should be monitored and the outcomes evaluated.

• Local ordinances requiring taxicabs to have accessible and usable safety belts are a promising intervention. Communities should enact such ordinances, monitor their implementation, and evaluate their outcomes.

Safety belts in school buses. The question of whether school buses should be equipped with lap belts is a controversial one. In any given year, fewer than 20 of the more than 40,000 traffic fatalities are school bus passengers. Most school bus fatalities occur to pedestrians—as students get on or off the bus or as they cross the street.[92]

Advocates argue that if children can be protected with lap belts they should be, and further, that the wearing of belts in school buses educates children and "carries over" to belt use in automobiles and other vehicles. Opponents point to the difficulty of retrofitting older buses and to the possibility that lap belts may cause injury in frontal crashes by forcing a child to double over, thereby striking his or her head on the seat in front.

A 1987 study by the National Transportation Safety Board, *Crashworthiness of Larger Poststandard Schoolbuses*, concludes that current federal standards that require higher-backed, better-padded seats and other modifications work well to protect passengers.[93] Moreover, a 1986 NHTSA study found the evidence of carryover effects inconclusive.[94] Most recently, the National Academy of Sciences has been charged by Congress with conducting a 2-year study of school bus safety.

Recommendation: The use of safety belts in school buses is being examined by the National Academy of Sciences. We make no recommendation on this intervention, pending the completion of this study.

Research to improve safety belt systems for children under the age of 14 and the elderly. A recent review of motor vehicle–related injuries sustained by children wearing a safety belt revealed that adult safety belts may not provide optimal protection for child occupants. Of particular note is the fact that 12% of injured 4- to 9-year-old children suffered abdominal injuries from straining against a safety belt, and one-third of 10- to 14-year-olds experienced a whiplash injury.[95]

Similarly, in a discussion of crash protection for older persons, one researcher noted that "the comfort and convenience of many current active and passive belts, as well as their crash performance, are not optimal."[96] He suggested that safety belt systems be redesigned to accommodate the limited joint movements of older persons.

Recommendation: Research is needed on how safety belt systems can be improved to provide optimal protection and comfort for children under the age of 14 and the elderly.

The Convincer. The Convincer is a popular device in widespread use by public safety agencies and others to educate the public about the importance of wearing safety belts. Sitting in a carlike device that runs down an incline, an individual is exposed to forces similar to those generated by a low-impact speed crash. Although the Convincer is highly regarded by many who use it, there are little formal data about its effectiveness. In addition, questions have been raised about its safety (at least one lawsuit by a plaintiff who alleged injury by the device was settled out of court; P. Waller, personal communication).

Recommendation: The Convincer is an intervention that has been tried but not sufficiently evaluated.

Pending evaluation and given the potential problems of injury and liability, use of the Convincer should be substantially curtailed. It should be used only with extreme care by trained public safety officials.

Child safety seat use. Child safety seats are designed to protect infants and young children from injury during a collision or a sudden stop. It has been estimated that the correct use of safety seats reduces fatality risk by 71% and serious injury risk by 67%.[97] It has also been suggested that safety seats are effective in reducing childhood injuries that occur during noncrash events (e.g., sudden stops, turns, opening the door of a moving vehicle).[98,99]

Mandatory child safety seat use laws. In the past several years, all 50 states and the District of Columbia have passed child safety seat use laws that generally cover children up to the age of 4, although there are differences in coverage from state to state. A study of Michigan's child safety seat law demonstrated a 25% reduction in injuries[100]; a California study that investigated injury patterns showed not only a 13% decrease in injuries but, most significant, a 17% decrease in head injuries among children in safety seats.[101]

Unfortunately, many states' child safety seat laws provide exemptions for children according to their age, the type of vehicle traveled in, whether the driver is or is not the child's parent or guardian, presence of an out-of-state driver or vehicle, and others. Researchers who analyzed the 50 states' child safety seat laws concluded not only that gaps in coverage are counterproductive but also that adopting comprehensive, uniform laws "would reduce ambiguities, facilitate compliance with the laws, and enhance protection."[102] NHTSA has developed model uniform legislation for this and other motor vehicle ordinances.

Recommendation: Child safety seat laws have been proven effective in reducing childhood injury morbidity and mortality. To close gaps in coverage in existing laws, uniform, comprehensive laws should be adopted in all 50 states. The laws should require safety seat use for all children up to the child's 5th birthday. (State mandatory safety belt use laws should cover all children over the age of 5.)

Child safety seat loaner programs. Since the late 1970s, many child safety seat loaner programs have sprung up, largely in response to financial constraints that prevent parents from purchasing safety seats. The programs, many of which are voluntary and are based in local hospitals, health departments, or service clubs, lend safety seats to parents for a small deposit and/or a minimal rental fee.[103,104]

An evaluation of a hospital-based loaner program in Kansas City showed that the program (which included pediatricians' providing regular reinforcement to parents for seat use) led to better than 90% correct safety seat use rates when the mothers were discharged from the hospital with their infants and more than 80% correct use rates 12 months later.[105] Safety seat usage rates are on the rise across the nation, but it is difficult to determine precisely the role that loaner programs have played in this upswing. However, loaner programs, in concert with child safety seat legislation, improved seat design, and public education efforts, are generally credited as being major contributors to improved safety seat usage rates among the general population.[106]

Recommendations:

• Comprehensive child safety seat loaner programs have been proven effective in increasing proper and consistent use of safety seats. Loaner programs should continue to be implemented and monitored. They should be targeted particularly at low-income parents unable to purchase a safety seat.

• Parents with young children who rent a car while away from home are often faced with the problem of transporting children without safety seats. Requiring rental car companies to provide loaner child safety seats as well as instruction on their proper installation and use is a promising intervention that should be used and monitored and its outcomes evaluated.

Child safety seat use for low-birthweight infants. A recent study noted that advances in health care have made it possible for many premature infants weighing less than 2.2 kg (5 lb) to be discharged from the hospital. However, current standards do not specify the minimum weight at which an infant is to be protected by a safety seat. The Automotive Safety for Children Program of Indiana State University suggests modifications of existing safety seats as well as the need for continued research.[107] Anyone interested should contact the Automotive Safety for Children Program (702 Barnhill Drive, P-121, Indianapolis, IN 46223) or the American Academy of Pediatrics (P.O. Box 927, Elk Grove Village, IL 60009) for the most up-to-date information.

Recommendation: Further research should be conducted on the development and use of safety seats for low-birth-weight infants.

Child safety seats and children with special needs. Children with disabilities who are not able to sit comfortably in an approved child safety seat are often held on an adult's lap, left unrestrained in a motor vehicle, or placed in seating systems that are not crashworthy.[108] None of these alternatives prevents injury in a serious collision. In fact, a child held on an adult's lap may become a "cushion" for the adult in a crash, with the adult's body crushing the child against the dashboard or windshield.

A wide variety of commercially produced restraint systems—including special safety seats, convertible wheelchairs, and harnesses and vests—are available for use by children with special needs.[109] Information on these systems is available through qualified health care professionals and medical equipment vendors. Unusual seating problems for children can be resolved by developing customized seating systems with the assistance of adaptive equipment programs, special equipment clinics, or equipment vendors. Additional sources of information on this topic include the Automotive Safety for Children Program at Indiana State University and the American Academy of Pediatrics.

Recommendation: Only the use of properly designed seating systems for children with special needs can prevent or reduce the severity of motor vehicle injuries in this population. All professionals who provide services to parents of children with special needs should give up-to-date information about the availability and use of such equipment.

Air bags. The technology behind the air bag has been available for over 20 years. Dr. William Haddon, Jr., NHTSA administrator in 1968, was responsible for initiating the process by which these highly effective passive devices should have been expeditiously approved. The effectiveness of the air bag—when used in conjunction with a safety belt—has been proved both through crash testing and through road experience with an increasing number of air bag–equipped cars.[110,111]

Yet only now, after years of regulatory and judicial entanglements,[112–114] are auto manufacturers producing air bags. For the 1989 model year, approximately 480,000 U.S. cars were produced with air bags; by 1990 that figure is expected to increase to 3.3 million.[115]

Public reaction to air bags has been mixed. A telephone survey of a nationally representative sample of over 1,000 persons revealed that while the public recognizes the extra protection offered by air bags, they are concerned about the reliability of the device (e.g., inflating by mistake, not knowing if it would work).[116]

One common misconception about air bags is that their use obviates the need for other types of occupant protection. To counter this misunderstanding, NHTSA's public information and education materials stress, "For maximum protection in all types of crashes—side, rear, and rollover collisions—drivers and passengers in air bag–equipped cars should wear safety belts at all times."[117] The need for safety belt use in cars with air bags is highlighted by recent findings that, during deployment, the bag can be a potential source of severe injury to an *unbelted* occupant—most commonly a child standing in the footwell in front on the passenger's side.[118,119]

Recommendations:

• The installation of air bags in motor vehicles has been proven effective in increasing occupant safety, particularly when air bags are used in conjunction with safety belts. Automobile manufacturers should continue to expand the installation of both driver and right front passenger air bags in all new models.

• Managers of federal, state, municipal, and private fleets should purchase air bag–equipped vehicles for use by employees.

Proper use of occupant protection systems. *Proper use of safety belts.* To be most effective, safety belts should be worn over the shoulder, across the chest, and low on the lap. Unfortunately, occupants frequently wear safety belts in the wrong position, such as under the arm, behind the back, or across the stomach.

Research has documented the incidence of safety belt misuse[120] and clarified how misuse reduces protection and often leads to additional injury or even death in the event of a crash.[121] A recent investigation in Japan correlated the crash deaths or serious injuries of safety-belted occupants with their incorrect belt use.[122] Another group of researchers presented case reports of six persons whose fatal injuries were caused by wearing a shoulder belt under the arm during a crash—all six crashes were determined to be "otherwise survivable."[123]

A question has been raised about the use of safety belts by women during pregnancy. Their use has been recommended by the American College of Obstetricians and Gynecologists.[124] All pregnant women should wear safety belts (and wear them correctly—with the shoulder belt across the chest and the lap belt under the abdomen). On a related issue, researchers from California who have studied cases of fetal death secondary to maternal involvement in a motor vehicle collision (and in which the women were not wearing safety belts) stress the need for prolonged continuous fetal monitoring for all pregnant women involved in collisions.[125]

Another potentially dangerous situation involves the use of a fully reclined bucket seat while a vehicle is in motion. A safety belt worn in a reclining seat during a frontal crash will not be optimally effective and may lead to severe abdominal and spinal cord injuries (A. I. King, personal communication). Therefore it is recommended that front seat passengers use the reclining feature only when the vehicle is not in motion.

Recommendation: Safety belts can be maximally effective in preventing or reducing the severity of injury in motor vehicle crashes only if they are worn correctly. All health professionals should inform the public about the importance of proper use of safety belts.

Correct installation and use of child safety seats. Like safety belts, child safety seat misuse is common. One recent study found a 65% misuse rate for safety seats.[126] Common errors include seats facing in the wrong direction, safety belts improperly routed around the safety seat, and misuse of toddler seat harnesses or shields.[126]

In a detailed review of 53 crashes involving restrained and unrestrained children, the National Transportation Safety Board identified safety seat misuse as a significant problem. The authors concluded that "it is absolutely essential that safety seat design and instructions for use of safety seats be as simple, clear, and precise as possible."[127]

Recommendations:

• Child safety seats can provide optimal protection in motor vehicle crashes only if they are installed and used correctly. All health professionals who work with parents of young children should inform them about the importance of proper installation and use of child safety seats.

• Research is needed to explore how design modifications can decrease child safety seat misuse and increase comfort and vehicle compatibility.

Interventions to increase safety belt and child safety seat use. Educational interventions. Education about motor vehicle–related risk and the importance of occupant protection is a gradual process. It moves from a person's being informed through understanding the message, to believing it and finally consistently modifying behavior. The effects of educational campaigns are not seen overnight. Thus practitioners should not expect to observe reductions in the incidence of injuries from educational programs alone.

Educational programs that rest solely on information as a means for change rarely succeed.[106] All programs should take into account target group characteristics, appropriate message content, and educational technology. The combined use of health education and behavioral science should make attempts to increase the use of safety belts and child safety seats more effective.[128] Further, motor vehicle injury prevention education should be presented in the context of other major health problems—heart disease and cancer prevention, for example.

Health risk appraisals (HRAs), which are used in a variety of primary health care and employee wellness settings, are a method of estimating one's risk for a particular cause of death and for assessing the prospects for future good or ill health. They can be used to stimulate healthful behavior and discourage health risks. HRAs have been widely used and positively evaluated in health care settings.[129] Two researchers studied the modification of HRAs to include information related to safety belt use. Their findings support the hypothesis that HRAs that include safety belt information can increase safety belt use.[129]

Recommendations:

• Educational interventions are a promising approach for increasing safety belt and child safety seat use. They should be used and monitored and their outcomes evaluated; ways to combine educational programs with other approaches to accelerate behavior change should be investigated.

• Integrating traffic safety information into health risk appraisals is a promising intervention. It should be used and monitored and its outcome on safety belt use evaluated.

Behavioral change interventions. Behavioral science has much to offer the field of injury prevention in understanding the circumstances that produce injuries and in developing effective strategies for behavioral change. Although psychological research and applications in injury prevention have been slow to develop, more recent attention has been given to the science of behavioral change. This reflects a recognition that injury prevention involves individual actions related to protection, caution, and risk.

Behaviorally based interventions seek to motivate changes in behavior through reward and punishment, modeling, prompting and feedback, skills development, and guided practice.[128] These techniques have been directed at a variety of health problems among children, teenagers, and adults. Traffic injury prevention programs must include components to encourage safe behavior because there are few purely automatic strategies to protect occupants in a crash.

The research on behavioral change interventions to increase the use of child safety seats reveals several promising approaches: prenatal counseling,[130] hospital-based education by physicians and nurses,[131,132] physician counseling during hospitalization,[133] loaner programs,[105] rewards and positive incentives,[134] and health education.[135]

Some of the promising approaches described in the behavioral change literature on increasing the use of safety belts among older children and adults include safety belt use policies,[106,136,137] participative risk education,[138] incentives,[139-143] and behavioral feedback.[144] Behavioral and educational approaches that modify attitudes of specific target groups are suggested, because research shows that attitudes toward belt use predict usage behavior better than any other factor.[145] More research is needed on the timing of reinforcement for belt use, the influence of social support, and methods for increasing compliance with safety belt use laws.

Recommendation: Behavioral change interventions show promise in increasing safety belt and child safety seat use. They should be implemented and monitored and their outcomes evaluated.

Teenage Drivers

Driving is a complex activity. It takes years to become a proficient driver. Therefore, although age of licensure is important, it is a less critical factor than how people learn to drive. All of the available evidence indicates that beginning drivers should learn under the least hazardous conditions, including a long period of daylight-only driving, preferably with supervision, before moving on to more complex driving conditions.[146]

Increasing the age of licensure. The minimum age of licensure varies from state to state, although most have adopted the age of 16. (At one time, many states had lower licensure ages; some, like Maine, still allow teens to become fully licensed drivers at age 15.) The authors of one study compared the fatal crash experience in states with different minimum licensing ages (New Jersey, age 17; Massachusetts, age 16.5; Connecticut, age 16). They concluded that, "New Jersey's 17-year-old minimum licensing age is associated with reduced fatal crash involvement among 16-year-olds." They note, however, that "it is possible that the reduced crash involvement of 16-year-old drivers in New Jersey may be partially offset at age 17." Further, "the

higher fatal crash involvement of 17-year-old drivers in New Jersey is compatible" with a hypothesis of driver inexperience.[147]

Recommendation: Increasing the age of licensure is an intervention that has not been sufficiently studied. It needs further research.

Graduated licensure. "Young people are going to learn to drive, and the question remains as to the best way to instruct them," wrote Patricia Waller.[146] Recognizing that existing driver training programs provide only a bare minimum of the practice that beginning drivers need, she argued that formal driver training could be started much earlier than it is now but without full licensure occurring any earlier. Initial driving would be limited to daylight hours, with a designated supervisor (generally a parent) in the passenger seat. After a specified amount of driving time (or number of miles driven), the hours of driving would be extended, but still with a supervisor. Finally, after all of the requirements had been met, full licensure would occur. "The difference would be that at the time of full licensure the novice driver would have far more supervised practice acquired at minimal cost to the taxpayer," Waller noted.[148]

Recommendation: Research should be conducted to develop, implement, and evaluate imaginative programs such as graduated licensure to meet beginning drivers' need for supervised on-the-road practice.

Curfew laws to restrict teenage driving. Researchers viewed the experience of several of the 12 states that had curfew laws as of 1984.[149] In general, they pointed out, many crashes involving teens occur during nighttime hours and (according to another study) "45 percent of fatal crashes of drivers under 18 occur from 8:01 p.m. to 4:00 a.m."[150]

Although the effectiveness of curfew laws varied somewhat among the states (in part as a function of the actual number of hours involved), the authors concluded that "curfew laws can be effective in reducing the high crash rates of teenage drivers which result in large numbers of injuries to themselves and others."[149] However, in a separate study, the same researchers found that many teenagers do not respect curfew laws, in part because they perceive that police enforcement of the laws is low.[151]

Recommendation: Curfew laws to restrict teenage driving have been proven effective. States without a curfew law should enact and enforce one, and existing curfew laws should be enforced.

Driver education programs. The first driver education programs were developed in the United States in the mid-1930s. Currently, most driver education courses include 30 hours of classroom instruction and 6 hours on the road. Possibly no single traffic safety intervention has been the subject of so much controversy. To supporters, the programs are responsible for decreasing highway carnage. To some opponents, they are responsible for *increasing* death and injury, because they encourage more drivers in the highest-risk years to be on the road.[10,152,153]

Other opponents argue that given what we know of the complexity of the task of driving and the limitations (in time, resources, and conceptual understanding) of all driver education programs, there is no reason why they should work.[146] The final view holds the possibility that by building on human factors research in such areas as pilot and athlete training, an effective driver education program could be developed.[146] (For additional information, see references 154–157.)

Recommendation: Because no current driver education course can perform effectively the tasks required of it, research should be conducted to document the skills involved in driving and the most appropriate strategies for teaching the skills to beginning drivers.

Pedestrians

Roadway countermeasures. One researcher has analyzed and summarized data on the effectiveness of more than a dozen roadway countermeasures in reducing pedestrian injuries.[158] These countermeasures can be implemented by state highway departments and, in some cases, local traffic and parking departments. Our recommendations, presented by countermeasure type, are consistent with the author's conclusions.

It is important to note the author's comment that "the use of a comprehensive program of public education and police enforcement must complement roadway improvements to help insure more widespread benefits to pedestrian safety."[158]

Recommendations:

• *One-way street* networks and the conversion of two-way to one-way streets have been effective in increasing pedestrian safety. These interventions should be used where feasible and monitored.
• Adequate *roadway lighting* is effective in increasing pedestrian safety. It should be installed and monitored.
• The moving of a transit or school *bus stop loca-*

tion from the near side to the far side of an intersection is a promising intervention that provides some level of improved pedestrian safety at certain locations. This intervention should be used and monitored and its outcomes evaluated.
• Although universal in urban areas, *sidewalks* are not normally constructed in rural areas and often are not provided in suburban neighborhoods. They are an effective intervention that should be used wherever feasible and monitored. Their installation should be required in new housing subdivisions.
• *Roadway barriers*, including chains, fences, and other devices to physically separate pedestrians from vehicles, are of proven effectiveness. They are most likely to be feasible at sites with a high incidence of midblock crossings and will vary widely in cost. Barrier use should be explored wherever possible and monitored.
• The use of *pedestrian crossing signs* in unusually hazardous locations is of proven effectiveness. They should be used and monitored. (The effectiveness of regulatory and informational signs is less clear and requires additional research.)
• Certain *school-zone measures* (e.g., use of crossing guards and police enforcement of vehicle speeds) have been found effective. (Other measures, including signs and pavement markings, are of undetermined efficacy.)
• Studies of marked versus unmarked *crosswalks* strongly indicate that pedestrian safety is not improved and may be adversely affected by crosswalk marking. This is probably because marked crosswalks give pedestrians a false sense of security. Crosswalk marking should be halted pending additional research.
• *Curb parking regulations* (e.g., diagonal parking, one-side-of-the-street parking, removal of on-street parking, and restrictions on the distance that one can park from a corner or crosswalk) have been tried but not adequately evaluated. Further research is needed in these areas.
• Although the positive effect of *traffic signals and pedestrian indicator lights* has long been assumed, there is little evidence to support the assumption. Research should be conducted on the effectiveness of these interventions.

Reduce hazards of right turn on red. In all jurisdictions, with the exception of New York City, drivers are permitted to make a right turn at a red light *unless* a sign specifically prohibits them from doing so. A study carried out by a team of researchers (whose findings have since been confirmed by others) demonstrated significant increases in pedestrian and bicyclist collisions involving right-turning vehicles at red signals following introduction of the rule.[159] The vast majority of crashes involve a victim coming from the driver's right who is not seen because the driver's attention is focused to the left while searching for a gap in oncoming traffic.

Right-on-red regulations spread rapidly in the 1970s in response to the need for both fuel-conservation and antipollution measures. In some western states, such as Colorado, right-on-red is still seen as a critically necessary air-pollution control device. Thus, while the injury toll is clear, rescinding such laws may not be practical. In the meantime, designing, implementing, and evaluating educational interventions to reduce the hazards of right-on-red laws seem appropriate. Such education should begin with state driver instruction materials.

Recommendation: Research should be conducted to develop, implement, and evaluate driver, pedestrian, and bicyclist education programs and other countermeasures to minimize the adverse effects of right-on-red laws.

Conspicuity. Knowing that nighttime pedestrian and bicycle activity is inherently dangerous and to be avoided when possible, a group of researchers tested various commercially available reflective materials and devices (flashlights, retroreflective disks worn near waist level, joggers' fluorescent vests, and retroreflective headbands, wrist bands, belts, and anklebands). They concluded that each of these devices produced significantly better detection distances than did a pedestrian control wearing a white T-shirt and blue jeans. Among the devices, the flashlight (or bicyclist's leg lamp) yielded the best results. The retroreflective disks ("dangle tags") were the least effective.[160]

Recommendation: Conspicuity-enhancement devices and materials are interventions that have been proven effective. They should be used by all nighttime pedestrians and bicyclists.

NHTSA model legislation to reduce pedestrian injuries. Based on an analysis of the various situations that result in pedestrian injuries (the six most common of which are described as "dart-out," "intersection dash," "vehicle-turn merge," "multiple threat," "bus stop related," "ice cream vendor related," and "backing up"), NHTSA has developed several pieces of model legislation.

The Model Ice Cream Truck Ordinance (which has been adopted in New Jersey and Detroit and is under consideration elsewhere), for example, requires drivers to come to a full stop at a working ice cream vending truck and then to proceed at a pru-

dent speed, yielding the right of way to pedestrians. A field test of this ordinance was conducted in Detroit. In the first partial vending season, vendor-related child pedestrian injuries decreased 54%. During the first full vending season, the ordinance resulted in a 77% reduction in injuries from the previous 3-year preordinance average.[161] Other model legislation includes commercial bus stops, school bus pedestrians, and a model vehicle overtaking ordinance (governing the right of way at crosswalks).[162,163]

Recommendation: The NHTSA model legislation is an intervention that has not been sufficiently studied. Further research should be conducted to determine the relative effectiveness of various types of legislation to reduce pedestrian injuries.

Pedestrian safety education for children. NHTSA developed a primary school pedestrian safety program to address child pedestrian injuries stemming from two leading causes, midblock dart-outs and intersection dashes. A demonstration project conducted from 1977 to 1982 found significant reductions (ranging from 13% to 57%) in both injury types.[164]

One study presents an analysis of a pedestrian safety program that was field tested in Los Angeles, Milwaukee, and Columbus, Ohio, and which reduced dart-out injuries by about 20%. The program, conducted during 1976 and 1977, employed "Willy Whistle," an original animated character. It included a 6- to 7-minute classroom film, reinforced by three 30-second and three 60-second TV spots and a poster. Tests showed high visibility of the program's character and messages (which stressed simple pedestrian safety techniques such as stopping at the corner and searching "left-right-left" before crossing) among children after it had been running for a year.[165] A complementary pedestrian safety education film for 9- to 12-year-old children has been found to improve knowledge and behavior and reduce pedestrian injuries by more than 20% in comparison with control sites.[166]

Many pedestrian safety educational materials (both publicly funded and commercially produced) are available for use in schools and other community settings. NHTSA recommends that pedestrian safety be taught in all preschool and primary school systems and that the pedestrian safety problem be addressed through "coordinated, comprehensive approaches after specific problem areas have been identified" (J. Hedlund, personal communication).

It has also been suggested that public information and education efforts should be directed at drivers to alert them of the need to be aware of pedestrians and bicyclists, especially when near residential, playground, senior citizen, and school areas and when turning or backing up (V. Litres, personal communication). (For additional information on pedestrian safety education for children, see references 167–169.)

Recommendation: Pedestrian safety education for children is a promising intervention. It should be used and monitored and its outcomes evaluated.

Protecting elderly pedestrians. As discussed earlier, the highest population-based pedestrian death rates are found among the elderly. Many of the pedestrian injury countermeasures already reviewed are as applicable to the elderly pedestrian as to the younger. However, the overrepresentation of the elderly among pedestrian injury victims clearly indicates the need for special consideration.

The National Research Council (NRC) recently released a detailed study of the problems of elderly drivers, passengers, and pedestrians.[20] In part, it examines what consequences of the aging process (e.g., loss of visual acuity, slower reflexes) affect driver and pedestrian activity and to what extent. It now remains to put this knowledge to work and to develop effective countermeasures. One study indicated that resetting "WALK/DON'T WALK" signals to allow more time for street crossing, when combined with additional engineering and education interventions, can significantly reduce pedestrian injury to and death among to the elderly.[170]

Recommendation: There are few interventions in the area of elderly pedestrian injury. Interventions should be designed with specific, measurable objectives and then implemented and evaluated and the results disseminated widely.

Motorcyclists

Helmet use laws. The use of a motorcycle helmet is an intervention that protects users, as shown by the result of a tragic "natural experiment." Only three states required the use of motorcycle helmets before 1967, when a federal standard was issued requiring that all states pass a helmet use law to qualify for safety and highway funds. By 1975 motorcyclists in all but three states were covered.

In 1976, despite evidence that nearly all motorcyclists in states with helmet laws were complying, Congressional pressure was lifted. Within 2 years, 26 states had rescinded their laws. The result was predictable and overwhelming: "the repeals or weakening of motorcyclist helmet use laws were typically followed by almost 40 percent increases in the numbers of fatally injured motorcyclists."[171]

Additional evidence was supplied by Louisiana, which reenacted a mandatory helmet use law in 1981. An analysis revealed that following reenactment fatalities fell from 3.63 per 100 collisions to 1.07 per 100 collisions. Crashes resulting in reportable, serious injuries fell from 84% to 74%. Also, there was a substantial reduction in the average medical cost per injury ($2,071 before reenactment, $835 after).[172]

Recommendation: Motorcycle helmet use laws are an intervention of proven effectiveness. Existing helmet laws should be retained and enforced, and all states without a helmet law should enact one. Concerted coalition building and political action will be needed to regain the level of implementation that existed before 1976.

Conspicuity. Since 1979 nearly all motorcycles sold in the United States have been wired so that the headlight automatically stays on while the engine is running. Because the average life expectancy of a motorcycle is approximately six years, the vast majority of motorcycles now in use have this feature. However, although one evaluation of the effect of daytime headlamp usage laws was inconclusive,[173] another found that the laws led to a 13% reduction in fatal daytime crashes.[174]

One team of researchers conducted a field evaluation of a variety of conspicuity measures for motorcycle riders. They concluded that "daytime conspicuity can most effectively be improved by use of fluorescent garments or steady or modulating lights. Nighttime conspicuity seems to be aided by use of retroreflective garments and running lights."[175] NHTSA also has conducted research in this area and continues to support the use of such conspicuity measures.[176]

Recommendation: Measures to enhance the conspicuity of motorcycle riders are a promising intervention that should be used and monitored and the outcomes evaluated.

Motorcycle rider education. Based on comprehensive, in-depth investigations of 900 motorcycle crashes and a review of 3,600 traffic reports of motorcycle crashes, one study concluded that riders involved in crashes "are essentially without training; 92 percent were self-taught or learned from family or friends." Furthermore, the researchers note, "Motorcycle rider training experience reduces accident involvement and is related to reduced injuries in the event of accidents."[177]

In 1983 Victoria, Australia, introduced a new motorcyclist licensing/training system that included rider training and written and skills tests. The program led to a statistically significant (and sustained) reduction in crashes involving novice riders. "Unfortunately," the authors note, "it is not possible to entirely disentangle the effects of the separate components of written test, skills test, and training."[178]

Recommendation: Motorcycle rider education is a promising intervention. It should be used and monitored and its outcomes evaluated.

Moped legislation. Mopeds are popular motor vehicles that offer economical transportation. During the past decade, however, there has been an increase in moped crash-related orthopedic and neurologic injuries, primarily among children and young males.[179]

One researcher reported on moped legislation passed by Ohio in 1984. The law required that riders be 14 years of age or older, wear an approved helmet, carry no passengers, ride within 3 feet of the right side of the road (where possible), and be equipped with a rearview mirror and turning signals. The author found postlaw decreases in injuries.[180] (For additional information, see references 181–183.)

Recommendation: Moped legislation is a promising intervention that should be implemented and monitored and its outcomes evaluated.

Bicyclists

Bicycle helmets. There is little argument over the fact that a well-made helmet bearing a seal of approval from the Snell Foundation or the American National Standards Institute (ANSI) can reduce the extent of head injuries sustained by a bicyclist in many falls or collisions, regardless of the involvement of a motor vehicle.[184–188] Yet few cyclists take advantage of this protection.[189,190]

Apart from regulatory action, there are other strategies to promote the use of helmets. A comprehensive bicycle helmet program in Seattle, Washington, including a media campaign directed at children and parents, school, community, and corporate participation, and a helmet discount program, has proved successful in raising awareness of the issue and promoting the use of bicycle helmets by children.[191] Other strategies include educational efforts in schools; mandatory provision of educational information on helmets at point of purchase of bicycles; school department regulations requiring the use of helmets by children riding bicycles to and from school; and public-, foundation-, or corporate-funded rebate offers for the purchase of approved helmets or the provision of low-cost or free helmets to low-income families.

Recommendation: The above-mentioned interventions (including municipal ordinances requiring bicycle helmet use) are all promising methods of increasing helmet use. They should be used and monitored and their outcomes evaluated.

Conspicuity. As with pedestrian activity, nighttime bicycle riding greatly increases one's risk for injury. See the earlier discussion of conspicuity-enhancement devices for pedestrians for a review of the literature on this topic.

Recommendation: Conspicuity-enhancement devices and materials are interventions that have been proven effective. They should be used by all nighttime bicyclists and pedestrians.

Bicycle safety programs. Comprehensive educational programs, especially for children and their parents, may be a way to address safety-related riding habits. Such programs typically include information on helmets, safe riding habits, and bicycle maintenance. Bicycle checks by qualified mechanics, designed to reduce injuries caused by equipment failures, may be an additional program component.

NHTSA has produced a technical report examining a broad range of bicycle safety programs, including community bicycle programs, bicycle rodeos, safety towns, bicycle monitors, and master plans (each of which the report characterizes as effective). Also included are "programs which promise effectiveness," such as in-school and on-bike training and education, safety messages, model regulations, protective clothing for cyclists, and bicycle lanes.[192]

Recommendation: The bicycle safety programs described above have been insufficiently studied. Further research should be conducted to determine their effectiveness.

Bicycle paths and lanes. In recent years, many towns and cities have developed bicycle paths and lanes in order to separate motor vehicle and bicycle traffic and to provide cyclists with a road surface free from obstructions and potholes.[193] Long-term maintenance issues are an important consideration in the development of these paths.

Recommendation: Bicycle paths and lanes are promising interventions for preventing bicycle injuries. Communities should construct and maintain such paths, monitor their use, and evaluate their outcomes on bicycle injuries.

Bicycle selection. The American Academy of Pediatrics has warned that oversized bicycles can present a danger to children. They recommend that parents choose appropriately sized bicycles for their children, not ones that the child will "grow into." "The child should be able to place the balls of both feet on the ground when sitting on the seat with the hands on the handlebars. . . . Straddling the center bar should be possible with both feet flat on the ground."[194] Children should be able to reach the handlebars easily. Because young children do not have the strength and coordination to use hand brakes, their bicycles should be equipped with coaster brakes.

Recommendation: The American Academy of Pediatrics' guidelines for selecting a child's bicycle are a promising intervention. The guidelines should be implemented and monitored and their outcomes evaluated.

Speed Limits

Congress enacted the 55-mph speed limit in 1974 as a fuel conservation measure. Evaluations of its effects have repeatedly demonstrated reductions in death and injury (an annual savings of 2,000–4,000 lives and 2,500–4,500 serious to critical injuries) and increases in fuel economy (a yearly savings of approximately $2 billion). Now that the states have the authority to increase speed limits to 65 mph on rural interstate highways, higher speed limits will be paid for with increased injuries and fatalities and higher fuel consumption. In fact, fatalities on rural interstates rose an average of 19% in 1987 in those states that enacted higher speed limits.[195]

The 55-mph limit not only reduced speeds; it also reduced speed variability among cars. All available evidence indicates that increased speed variability increases collisions.[196]

Most important, when the speed limit is increased on major highways, it affects other roadways profoundly. As early as 1954, the Highway Research Board described the process (then called "velocitization" and now termed "speed adaptation"). After driving at highway speeds, a driver will slow down less on secondary roads. After driving at 65 or 70 mph, 55 mph will "feel" like a safer speed (although it may be well above the secondary road's limits). The 55-mph limit affected speed adaptation significantly; most of the lives saved as a result of the 55-mph limit were on roads other than interstates.[197]

Recommendations:

• The 55-mph speed limit is a proven intervention for reducing traffic injury and death. All states should retain or reinstate the 55-mph speed limit on relevant state roads and highways.

• All states and municipalities should also enforce maximum speed limits on all roads.

Automated Enforcement Devices

A number of automated enforcement devices (e.g., "red light cameras," drone radar systems) have been developed in recent years to facilitate enforcement of traffic laws. One example is the implementation of a red light camera in New York City, where red light offenses are of increasing concern. According to one recent estimate, as many as 1 million such offenses occur daily. It is further estimated that during the 5 years from 1981 to 1985, more than 13,000 pedestrians and nearly 400,000 motorists and bicyclists were killed or injured as a result of such infractions.[198]

From July 1985 to March 1986, New York experimented with the installation of a red light camera. Mounted at an intersection, the camera is triggered by electromagnetic loops embedded in the crosswalk. When the light is red, any car entering the intersection triggers the mechanism and pictures are made documenting the violation. (The use of such devices also requires a change in the law such that red light violations, like parking tickets, accrue to the vehicle, no matter who is driving.)

Red light cameras offer a variety of advantages to communities, including law enforcement, deterrence, and revenue production. The New York pilot test was successful; however, the evaluation was conducted by the camera's distributor and should therefore be confirmed by independent investigators.[198] (For more information, see reference 199.)

Drone (or unmanned) radar systems are targeted at drivers who use radar detectors to exceed the speed limit, which leads to a potentially dangerous variance between their speed and the speed of nearby vehicles. Researchers evaluating the effectiveness of drone radar in the state of Kentucky concluded that it was "an effective means of reducing the numbers of vehicles traveling at excessive speeds."[200]

Recommendation: Automated enforcement devices are an intervention that has not been sufficiently studied. Further research should be conducted regarding their effectiveness.

Vehicle Design

Headrests. A study of nearly 3,400 drivers and front seat passengers involved in rearend collisions evaluated the injury rates for those whose vehicles had integral headrests, adjustable headrests, or no headrests. In crashes severe enough to require that the vehicle be towed from the scene, integral headrests were highly effective (31.6%) in reducing injuries. Adjustable headrests also demonstrated a positive effect, but not at a statistically significant level.[201] (To be fully effective, a headrest must be adjusted so that the center of the headrest is even with the user's ears.)

Recommendations:

• Integral headrests have been proven effective in reducing injuries. All motor vehicles should be equipped with integral rather than adjustable headrests whenever possible.

• The correct use of adjustable headrests should be emphasized in occupant protection materials and programs and in manufacturers' information supplied with the vehicle.

Lowering of bumper height. Serious injuries of automobile occupants often result from side impact, which occurs when the side door intrudes into the passenger compartment. It is the bumper of the striking vehicle that causes the door to intrude. If the bumper height were lowered, injuries to both occupants and vehicle could be reduced (A. I. King, personal communication). The lowering needs to take place gradually over a period of about 10 years so that there would not be a mismatch of bumper heights among old and new vehicles. Because there are currently no cars with lowered bumpers, there are no data related to their beneficial effects. However, it has been suggested that the reduced severity of side impacts can be deduced from basic principles of mechanical engineering (A. I. King, personal communication).

Recommendation: Lowering of bumper heights is a promising intervention that should be implemented and monitored and its outcomes evaluated.

INTERVIEW SOURCES

Jim Hedlund, Office of Driver and Pedestrian Research, National Highway Traffic Safety Administration, Washington, DC, January 27, 1989.

Albert I. King, PhD, Professor of Engineering, Wayne State University College of Engineering, Detroit, Michigan, January 11, 1989.

Virginia Litres, Safety Countermeasures Division, National Highway Traffic Safety Administration, Washington, DC, November 6, 1987.

Marilyn Price Spivack, President, National Head Injury Foundation, Framingham, Massachusetts, June 1987.

Patricia Waller, PhD, Director, University of Michigan Transportation Research Institute, Ann Arbor, May 21, 1987 and June 23, 1988.

REFERENCES

1. Flink JJ. The car culture. Cambridge, Massachusetts: MIT Press, 1975.

2. Eastman JW. Styling vs. safety: the American automobile industry and the development of automotive safety, 1900–1966. New York: University Press of America, 1984.

3. The figure of 2.5 million fatalities was derived from data from the National Center for Health Statistics and the National Highway Traffic Safety Administration for the period 1910–1974, and from the Fatal Accident Reporting System (FARS) for the period 1975–1985.

4. National Highway Traffic Safety Administration. Fatal accident reporting system, 1986: a review of information on fatal traffic accidents in the United States in 1986. Washington, DC: U.S. Department of Transportation, 1988; DOT publication no. (DOT HS)807 245.

5. Accident facts, 1988 edition. Chicago: National Safety Council, 1988.

6. Holden JA, Christoffel T. A course on motor vehicle trauma: instructor's guide—final report [users manual]. Washington, DC: U.S. Department of Transportation, 1986; DOT publication no. DOT/OST/P-34/86-050.

7. Baker SP, O'Neill B, Karpf RS. The injury fact book. Lexington, Massachusetts: Lexington Books, 1984.

8. Waller JA. Injury control: a guide to the causes and prevention of trauma. Lexington, Massachusetts: Lexington Books, 1985.

9. Hartunian NS, Smart CN, Thompson MS. The incidence and economic costs of major health impairments: a comparative analysis of cancer, motor vehicle injuries, coronary heart disease, and stroke. Lexington, Massachusetts: Lexington Books, 1981.

10. Robertson LS. Injuries: causes, control strategies, and public policy. Lexington, Massachusetts: Lexington Books, 1983.

11. Simpson SG, Reid R, Baker SP, Teret S. Injuries among the Hopi Indians: a population-based survey. JAMA 1983;249:1873–6.

12. National Highway Traffic Safety Administration. The economic cost to society of motor vehicle accidents. Washington, DC: U.S. Department of Transportation, 1983 and 1985; update; DOT publication no. (DOT HS)806-342.

13. National Highway Traffic Safety Administration. The profit in safety belts: an introduction to an employer's program. Washington, DC: U.S. Department of Transportation, 1984; DOT publication no. (DOT HS)806-492.

14. Rosenfield HM. Drunk driving and occupant restraints. Traffic Safety 1985;85:6.

15. National Highway Traffic Safety Administration. Loss prevention through safety belt use: a handbook for employers. Washington, DC: U.S. Department of Transportation, 1986.

16. Miller TR, Luchter S, Brinkman CP. Crash costs and safety investment. Assoc Adv Automotive Med 1988:69–88.

17. Claybrook J and the staff of Public Citizen. Retreat from safety. New York: Pantheon Books, 1984.

18. Healthy people: the Surgeon General's report on health promotion and disease prevention. Washington, DC: U.S. Government Printing Office, 1979; DHEW publication no. (PHS)79-55071.

19. Malfetti JL, ed. Drivers 55+: needs and problems of older drivers. Falls Church, Virginia: AAA Foundation for Traffic Safety, 1985.

20. Transportation in an aging society: improving mobility and safety for older persons, vols. 1, 2. Washington, DC: National Research Council, 1988.

21. Haddon W, Valien P, McCarroll JR, et al. A controlled investigation of the characteristics of adult pedestrians fatally injured by motor vehicles in Manhattan. J Chron Dis 1961;14:655–78.

22. Rivara FP, Barber M. Demographic analysis of childhood pedestrian injuries. Pediatrics 1985;76:375–81.

23. Brison RJ, Wicklund K, Mueller BA. Fatal pedestrian injuries to young children: a different pattern of injury. Am J Public Health 1988;78:793–5.

24. Kraus JF, Riggins RS, Franti CE. Some epidemiologic features of motorcycle collision injuries: I: introduction, methods and factors associated with incidence. Am J Epidemiol 1975;102:74–98.

25. Rivara FP, Dicker BG, Bergman AB, Dacey R, Herman C. The public cost of motorcycle trauma. JAMA. 1988;260:221–3.

26. Bicycle-related injuries: data from the National Electronic Injury Surveillance System. Morbid Mortal Weekly Rep 1987;36:269–71.

27. Fife D, Davis J, Tate L, Wells JK, Mohan D, Williams A. Fatal injuries to bicyclists: the experience of Dade County, Florida. J Trauma 1983;23:745–55.

28. Weiss BD. Bicycle helmet use in children. Pediatrics 1986;77:677–9.

29. Williams AF. Factors in the initiation of bicycle-motor vehicle collisions. Am J Dis Child 1976;130:370–7.

30. Kerouac J. On the road. New York: New American Library, 1955.

31. Bayley S, ed. Sex, drink and fast cars. New York: Pantheon, 1986.

32. Haddon W, Blumenthal M. Foreword. In: Ross HL, ed. Deterring the drinking driver: legal policy and social control. Lexington, Massachusetts: Lexington Books, 1984.

33. Premature mortality due to alcohol-related motor vehicle traffic fatalities: United States, 1987. Morbid Mortal Weekly Rep 1988;37:753–5.

34. Waller PF, Stewart JR, Hansen AR, Stutts JC, Lederhaus Popkin C, Rodgman EA. The potentiating effects of alcohol on driver injury. JAMA 1986;256:1461–6.

35. Moore MH, Gerstein D. Alcohol and public policy: beyond the shadow of prohibition–a report by the National Research Council of the National Academy of Sciences. Washington, DC: National Academy Press, 1981.

36. Olson S, Gerstein DR. Alcohol in America. Washington, DC: National Academy Press, 1985.

37. Borkenstein RF, Crowther RF, Shumate RP, Ziel WB, Zylman R. The role of the drinking driver in traffic accidents (the Grand Rapids study). Alcohol Drugs Behav 1974;2(suppl 1):8–32.

38. Valle S, ed. Drunk driving in America. New York: Haworth, 1987.

39. Haddon W Jr, Suchman E, Klein D. Accident research: methods and approaches. New York: Harper and Row, 1964.

40. De Haven H. Mechanical analysis of survival in falls from heights of fifty to one hundred and fifty feet. War Med 1942;2:586–96.

41. Waller J. Injury as a public health problem. In: Last JM, ed. Maxcy-Rosenau Public Health and Preventive Medicine. 11th ed. New York: Appleton-Century-Crofts; 1980.

42. Mashaw JL, Harfst DL. Regulation and legal culture: the case of motor vehicle safety. Yale J Regul 1987;4:257–316.

43. Boyd DR. The history of emergency medical services (EMS) systems in the United States of America. In: Boyd DR, Edlich RF, Micik SH, eds. Systems approach to emergency medical care. Norwalk, Connecticut: Appleton-Century-Crofts, 1983.

44. Insurance Institute for Highway Safety. Policy options for reducing the motor vehicle crash injury cost burden. Washington, DC: Insurance Institute for Highway Safety, 1981.

45. Roberts M. Public health and health psychology: two cats of Kilkenny? Prof Psychol Res Pract 1987;18:145–9.

46. Community traffic safety programs. Washington, DC: US Conference of Mayors, 1988.

47. Ross HL. Deterring the drinking driver: legal policy and social control. Lexington, Massachusetts: Lexington Books, 1984.

48. Sixth special report to the U.S. Congress on alcohol and health from the Secretary of Health and Human Services. Rockville, Maryland: U.S. Department of Health and Human Services, 1987; DHHS publication no. (ADM)87-1519.

49. Council report—automobile-related injuries: components, trends, prevention. JAMA 983;249:3216–22.

50. AMA Council on Scientific Affairs. Alcohol and the driver. JAMA 1986;255:522–7.

51. Hingson R, Hereen T, Morelock S. Preliminary effects of Maine's 1982 .02 law to reduce teenage driving after drinking. Paper prepared for the International Symposium on Young Drivers, Alcohol, and Drug Impairment, Amsterdam, 1986 (in press).

52. Waller PF, Waller MB. The young drinking driver: cause or effect? In: Benjamin T, ed. Young drivers impaired by alcohol and drugs. London: International Congress and Symposium series no. 116, Royal Society of Medicine, 1987.

53. McDermott F, Strang P. Compulsory blood alcohol testing of road crash casualties in Victoria. Med J Aust 1978:612–5.

54. Preliminary recommendations from the Surgeon

General's Workshop on Drunk Driving, 1988 Dec 14–16. Washington, DC: U.S. Department of Health and Human Services (in press).

55. Reducing highway crashes through administrative license revocation. Washington, DC: National Highway Traffic Safety Administration, 1986; DOT publication no. (DOT HS)806 921.

56. Zador PL, Lund AK, Fields M, Weinberg K. Fatal crash involvement and laws against alcohol-impaired driving. Washington, DC: Insurance Institute for Highway Safety, 1988.

57. Dumochel W, Williams AF, Zador P. Raising the alcohol purchase age: its effects on fatal motor vehicle crashes in twenty-six states. J Legal Stud 1987(Jan):249–66.

58. Womble KB. The impact of minimum drinking age laws on fatal crash involvements: an update of the NHTSA analysis (research notes). Washington, DC: National Highway Traffic Safety Administration, 1988.

59. Hingson RW, Howland J, Levenson S. Effects of legislative reform to reduce drunken driving and alcohol-related traffic fatalities. Public Health Rep 1988;103:659–67.

60. Mosher JF, Colman VJ. Prevention research: the Model Dram Shop Act of 1985. Alcohol Health Res World 1986:4–11.

61. Baker Dickman F. Sobriety checkpoints for DWI enforcement: a review of current research. Washington, DC: National Highway Traffic Safety Administration, 1987.

62. Alcohol-related fatalities continue to drop in New South Wales. Insurance Inst for Highway Safety Status Rep, 1988;23:1–2.

63. Mercer GW. The relationships among driving while impaired charges, police drinking-driving roadcheck activity, media coverage, and alcohol-related casualty traffic accidents. Accident Anal Prev 1985;17:467–74.

64. Epperlein T. The use of sobriety checkpoints as a deterrent: an impact assessment. Phoenix: Arizona Department of Public Safety, 1985.

65. Epperlein T. An evaluation of Arizona's crackdown on drinking drivers: the impact after three years. Phoenix: Arizona Department of Public Safety, 1986.

66. Popkin CL, Lacey JH, Stewart JR. Are alcohol safety schools effective? Proc Am Assoc Automotive Med 1985(Oct):45–58.

67. Popkin CL, Stewart JR, Lacey JH. A follow-up evaluation of North Carolina's alcohol and drug education traffic schools and mandatory substance abuse assessments: final report. Chapel Hill, North Carolina: University of North Carolina Highway Safety Research Center, 1988.

68. Triton Corporation. Treating the drinking driver, vols. 1, 2. Washington, DC: National Highway Traffic Safety Administration (undated).

69. Safer H, Grossman M. Beer taxes, the legal drinking age, and youth motor vehicle fatalities. Cambridge, Massachusetts: National Bureau of Economic Research, 1986 (May); working paper no. 1914.

70. Cook PJ. The effect of liquor taxes on drinking, cir-

rhosis, and auto accidents. In: Moore MH, Gerstein DR, eds. Alcohol and public policy: beyond the shadow of prohibition. Washington, DC: National Academy Press, 1981.

71. Phelps CE. Alcohol taxes and highway safety. In: Graham JD, ed. Preventing automobile injury: new findings from evaluation research. Dover, Massachusetts: Auburn House, 1988.

72. Russ NW, Geller ES. Training bar personnel to prevent drunken driving: a field evaluation. Am J Public Health 1987;77:952–4.

73. Vegega ME. NHTSA responsible beverage service research and evaluation project. Alcohol Health Res World 1986:20–3.

74. Responsible beverage service training: preliminary field test results. Nat Highway Traf Safety Admin Res Notes 1987(May):1–2.

75. Prugh T. Arena managers project, a TEAM approach. Alcohol Health Res World 1986:30–31.

76. Join the celebration: project graduation. Washington, DC: National Highway Traffic Safety Administration, 1987; DOT publication no. (DOT HS)807 063.

77. Klitzner M, Balsinskt M, Marshall K, Paquet U. Determinants of youth attitudes and skills towards which drinking/driving prevention programs should be directed. Vol. 1: The state of the art in youth DWI prevention programs. Washington, DC: National Highway Traffic Safety Administration, 1985; DOT publication no. (DOT HS)806-903.

78. National Commission Against Drunk Driving. Youth driving without impairment: report on the youth impaired driving public hearings. Washington, DC: National Highway Traffic Safety Administration, 1988; DOT publication no. (DOT HS)807-347.

79. Idea sampler to promote observance of National Drunk and Drugged Driver Awareness Week. Washington, DC: National Highway Traffic Safety Administration, 1986.

80. State of Washington, House Bill 663: an act relating to breath alcohol testing, 1987.

81. Finkelstein M, Smolkin H. Feasibility of roadway countermeasures to reduce alcohol-related crashes (a joint NHTSA/FHWA review paper) (unpublished ms), 1987.

82. Final rule, FMVSS 208: occupant crash protection, 49 CPR, part 571. Washington, DC: National Highway Traffic Safety Administration, 1984.

83. Mueller Orsay E, Turnbull TL, Dunne M, Barrett JA, Langenberg P, Orsay CP. Efficacy of mandatory seat belt use legislation. JAMA 1988;260:3593–7.

84. Campbell BJ, Campbell FA. Seat belt law experience in four foreign countries compared to the United States. Falls Church, Virginia: American Automobile Association Foundation for Traffic Safety, 1986.

85. Hingson R, Morelock Levenson S, Hereen T, et al. Repeal of the Massachusetts seat belt law. Am J Public Health 1988;78:548–52.

86. Latimer EA, Lave LB. Initial effects of the New York

State auto safety belt law. Am J Public Health 1987;77:183–6.

87. Petrucelli E. Seat belt laws: the New York experience–preliminary data and some observations. J Trauma 1987;27:706–10.

88. Williams AF, Lund AK. Mandatory seat belt use laws and occupant crash protection in the United States: present status and future prospects. In: Graham JD, ed. Preventing automobile injury: new findings from evaluation research. Dover, Massachusetts: Auburn House, 1988.

89. Campbell BJ, Campbell FA. Injury reduction and belt use associated with occupant restraint laws. In: Graham JD, ed. Preventing automobile injury: new findings from evaluation research. Dover, Massachusetts: Auburn House, 1988.

90. Williams AF, Preusser DF, Blomberg RD, Lund AK. Seat belt use law enforcement and publicity in Elmira, New York: a reminder campaign. Am J Public Health 1987;77:1450–1.

91. Geller ES, Lehman GR, Rudd JR. Long-term effects of employer-based programs to motivate safety belt use. Washington, D.C.: National Highway Traffic Safety Administration, 1987.

92. A special issue: school buses and seat belts. Ins Inst Highway Safety Status Rep, 1985;20:1–12.

93. Crashworthiness of larger poststandard schoolbuses. Washington, DC: National Transportation Safety Board, 1987.

94. School bus safety belts: their use, carryover effects and administrative issues. Washington, DC: National Highway Traffic Safety Administration, 1986.

95. Agran PF, Dunkle DE, Winn DG. Injuries to a sample of seatbelted children evaluated and treated in a hospital emergency room. J Trauma 1987;27:58–64.

96. Mackay M. Crash protection for older persons. In: National Research Council, Transportation Research Board. Transportation in an aging society: improving mobility and safety for older persons. Vol. 2: Technical papers. Washington, DC: National Research Council, 1988.

97. Kahane CJ. An evaluation of child passenger safety. The effectiveness and benefits of safety seats (summary). Washington, DC: National Highway Traffic Safety Administration, 1986; DOT publication no. (DOT HS)806-889.

98. Agran PA, Dunkle DE. Motor vehicle occupant injuries to children in crash and noncrash events. Pediatrics 1982;70:993–6.

99. Agran PF, Dunkle DE, Winn DG. Motor vehicle childhood injuries caused by noncrash falls and ejections. JAMA 1985;253:2530–3.

100. Wagenaar AC, Webster DW. Preventing injuries to children through compulsory automobile safety seat use. Pediatrics 1986;78:662–72.

101. Agran PF, Dunkle DE, Winn DG. Effects of legislation on motor vehicle injuries to children. Am J Dis Child 1987;141:959–64.

102. Teret SP, Jones AS, Williams AF, Wells JK. Child restraint laws: an analysis of gaps in coverage. Am J Public Health 1986;76:31–4.

103. Orr BT, Hall WL, Woodward AR, Desper LP. A guide for establishing a car safety seat rental program. Chapel Hill, North Carolina: University of North Carolina Highway Safety Research Center, 1985.

104. Orr BT, Hall WL, Marchetti LM, Woodward AR, Suttles DT. Increasing child restraint usage through local education and distribution efforts: final report. Chapel Hill, North Carolina: University of North Carolina Highway Safety Research Center, 1986.

105. Christopherson ER, Sosland-Edelman D, LeClaire S. Evaluation of two comprehensive infant car seat loaner programs with 1-year follow-up. Pediatrics 1985;76:36–42.

106. Nichols JL. Effectiveness and efficiency of safety belt and child restraint usage programs. Washington, D.C.: National Highway Traffic Safety Administration, 1982; DOT publication no. (DOT HS)806-142.

107. Bull MJ, Bruner Stroup K. Premature infants in car seats. Pediatrics 1985;75:336–9.

108. Feller N, Bull MJ. A multidisciplinary approach to developing safe transportation for children with special needs. Orthopaed Nurs 1986;5:25–7.

109. Hauser Holland S. Car safety for special children. Except Parent 1983(Oct):15–20.

110. Final regulatory impact analysis: amendment to Federal Motor Vehicle Safety Standard 208—passenger car front seat occupant protection. Washington, DC: National Highway Traffic Safety Administration, 1984; DOT publication no. (DOT HS)806-572.

111. O'Neill B. A note on air bag effectiveness. Washington, DC: Insurance Institute for Highway Safety, 1983.

112. Fleming A, ed. Air bags: a chronological history of delay. Washington, DC: Insurance Institute for Highway Safety, 1984.

113. Tolchin SJ. Air bags and regulatory delay. Issues Sci Technol 1984;1:66–83.

114. Graham JD. Auto safety: assessing America's performance. Dover, Massachusetts: Auburn House, 1989.

115. Automakers gear up to produce millions of air-bag-equipped cars. Ins Inst Highway Safety Status Rep 1988(Dec):1–3.

116. Loux S, Hersey J, Greenfield L, Sundberg E. National understanding and acceptance of occupant protection systems. Washington, DC: National Highway Traffic Safety Administration, 1986; DOT publication no. (DOT HS)807-025.

117. Protecting yourself automatically. Washington, DC: National Highway Traffic Safety Administration, 1986; DOT publication no. (DOT HS)806-866.

118. Mertz HJ, Driscoll GD, Lenox JB, Nyquist GW, Weber DA. Responses of animals exposed to deployment of various passenger inflatable restraint system concepts for a variety of collision severities and animal positions. In: Proceedings of the 9th International Technical Conference on Experimental Safety Vehicles, Kyoto, Japan, 1982.

119. Prasad P, Daniel R. A biomechanical analysis of head, neck, and torso injuries to child surrogates due to sudden torso acceleration. In: Proceedings of the 28th annual Stapp Car Crash Conference. Warrendale, Pennsylvania: Society of Automotive Engineers. 24–40; SAE paper no. 841656.

120. Goryl ME, Bowman BL. Restraint system usage in the traffic population. Washington, DC: National Highway Traffic Safety Administration, 1987; DOT publication no. (DOT HS)807-080.

121. Green RN, German A, Gorski ZM, Tryphonopoulos JP. Improper use of occupant restraints: case studies from real-world collisions. Proc Am Assoc Automotive Med, 1986:423–38.

122. Sato TB. Effects of seat belts and injuries resulting from improper use. J Trauma 1987;27:754–8.

123. States JD, Huelke DF, Dance M, Green RN. Fatal injuries caused by underarm use of shoulder belts. J Trauma 1987;27:740–5.

124. American College of Obstetricians and Gynecologists. Automobile passenger restraints for children and pregnant women. ACOG Tech Bull 1983(Dec):1–3.

125. Agran PF, Dunkle DE, Winn DG, Kent D. Fetal death in motor vehicle accidents. Ann Emerg Med (in press).

126. Ziegler PN. Child safety seat misuse. Washington, DC: National Highway Traffic Safety Administration Research Note, 1985.

127. National Transportation Safety Board. Child passenger protection against death, disability, and disfigurement in motor vehicle accidents. Washington, DC: National Highway Traffic Safety Administration, 1983; NTIS publication no. NTSB/SS-83/01.

128. Sleet DA Hollenbach K, Hovell M. Applying behavioral principles to motor vehicle occupant protection. Educ Treat Child 1986;9:320–33.

129. Perkins DD, Dunton SM. Health risk appraisal and safety belt use. Washington, DC: National Highway Traffic Safety Administration, 1987.

130. Greenberg LW, Coleman AR. A prenatal and postpartum safety education program. J Dev Behav Pediatr 1982;3:32–4.

131. Christophersen ER, Sullivan M. Increasing the protection of newborn infants in cars. Pediatrics 1982;70:21–5.

132. Colletti RB. A statewide hospital-based program to improve child passenger safety. Health Educ Q 1984;11:207–13.

133. Reisinger KS, Williams AF, Wells JK, John CE, Roberts TR, Podgainy HJ. Effects of pediatricians' counseling on infant restraint use. Pediatrics 1981;67:201–6.

134. Roberts MC, Layfield DA. Promoting child passenger safety: a comparison of two positive methods. J Pediatr Psych 1987;12:257–71.

135. Christophersen ER, Guylay J. Parental compliance with car seat usage: a positive approach with long-term follow-up. J Pediatr Psych 1981;6:301–12.

136. Report to regional NHTSA administration. Des Moines, Iowa: Department of Transportation, 1981.

137. Pabon, Sims, Smith, and Associates. Motivation of employers to encourage their employees to use safety belts. Washington, DC: National Highway Traffic Safety Administration, 1983.

138. Weinstein ND, Grubb PD, Vautier JS. Increasing automobile seat belt use. J Appl Psych 1986;71:285–90.

139. Geller ES. Motivating safety belt use with incentives: a critical review of the past and a look to the future. In: Advances in belt restraint systems: design, performance and usage (no. 141). Warrendale, Pennsylvania: Society of Automotive Engineers, 1984:127–52.

140. Streff FM, Geller ES. Strategies for motivating safety belt use: the application of applied behavior analysis. Health Educ Res Theory Pract 1986;1:47–59.

141. Sleet D, Geller ES. Do incentive programs for safety belt use work? Health Educ Focal Points 1986;3:2–4.

142. Gemming AG, Runyan CW, Hunter WW, Campbell BJ. A community health education approach to occupant protection. Health Educ Q 1984;11:147–57.

143. Campbell BJ. The use of economic incentives and education to modify safety belt use behavior of high school students. Health Educ Q 1984;15:33–4.

144. Van Houten R, Nau PA. Feedback interventions and driving speed: a parametric and comparative analysis. J Appl Behav Anal 1983;16:253–81.

145. Mayas JMB, Boyd NK, Collins MA, Harris BI. A study of demographic, situational and motivational factors affecting restraint usage in automobiles. Washington, DC: National Highway Traffic Safety Administration, 1983; DOT publication no. (DOT HS)806-402.

146. Waller PF. Young drivers: reckless or unprepared? In: Mayhew DR, Simpson HM, Donelson AC, eds. Young driver accidents: in search of solutions. Canada: Traffic Injury Research Foundation of Canada, 1985:103–46.

147. Williams AF, Karpf RS, Zador PL. Variations in minimum licensing age and fatal motor vehicle crashes. Am J Public Health 1983;73:1401–2.

148. Waller P. A graduated licensing system for beginning drivers. Chapel Hill, North Carolina: University of North Carolina Highway Safety Research Center, 1986.

149. Preusser DF, Williams AF, Zador PL, Blomberg RD. The effect of curfew laws on motor vehicle crashes. Law Policy 1984;6:115–28.

150. Robertson LS. Patterns of teenaged driver involvement in fatal motor vehicle crashes: implications for policy choice. J Health Politics Policy Law. 1981;6:303–14.

151. Williams AF, Lund AK, Preusser DF. Night driving curfews in New York and Louisiana: results of a questionnaire survey. Accident Anal Prev 1985;17:461–6.

152. Robertson LS. Crash involvement of teenaged drivers when driver education is eliminated from high school. Am J Public Health 1980;70:599–603.

153. Robertson LS, Zador PL. Driver education and fatal crash involvement of teenaged drivers. Am J Public Health 1978;68:959–65.

154. Stock JR, Weaver JK, Ray HW, Brink JR, Sadof MG. Evaluation of Safe Performance Secondary Driver Education Curriculum Demonstration Project. Washington, DC: National Highway Traffic Safety Administration, 1983.

155. Lund AK, Williams AF, Zador P. High school driver education: further evaluation of the DeKalb County study. Accident Anal Prev 1986;18:349–57.

156. Smith MF. Summary of preliminary results— follow-up evaluation: Safe Performance Curriculum Driver Education Project. Paper presented at annual conference of the American Driver and Traffic Safety Education Association, Spokane, Washington. Aug 10, 1987.

157. Potvin LP, Champagne F, Laberge-Nadeau C. Mandatory driver training and road safety: the Quebec experience. Am J Public Health 1988;78:1206–9.

158. Zegeer CV. Feasibility of roadway countermeasures for pedestrian accident experience. Warrendale, Pennsylvania: Society of Automotive Engineers, 1984:104–14.

159. Preusser DF, Leaf WA, DeBartolo KB, Blomberg RD, Levy MM. The effect of right-turn-on-red on pedestrian and bicyclist accidents. J Safety Res 1982;13:45–55.

160. Blomberg RD, Hale A, Preusser DF. Experimental evaluation of alternative conspicuity-enhancement techniques for pedestrians and bicyclists. J Safety Res 1986;17:1–12.

161. Hale A, Blomberg RD, Preusser DF. Experimental field test of the model ice cream truck ordinance in Detroit. Washington, DC: National Highway Traffic Safety Administration, 1978; DOT publication no. (DOT HS)803-410.

162. Pedestrian accident reduction guide. Washington, DC: National Highway Traffic Safety Administration, 1981; DOT publication no. (DOT HS)805-850.

163. State and community area report: pedestrian safety, 1985–1986. Washington, DC: National Highway Traffic Safety Administration, 1986.

164. Madiedo EC, Goodman R, Thompson D, Sabates C, Zimmerman M. Urban pedestrian safety demonstration project: K-3 Safe Street Crossing Program analytic study. Washington, DC: National Highway Traffic Safety Administration, DOT publication no. (DOT HS)7-01808.

165. Blomberg RD, Preusser DA, Hale A, Leaf W. Experimental field test of proposed pedestrian safety messages, vols. 1–3. Washington, DC: National Highway Traffic Safety Administration, 1983.

166. Preusser DF, Lund AK. And keep on looking: a film to reduce pedestrian crashes among 9 to 12 year olds. J Safety Res 1988;19:177–85.

167. Yeaton WH, Bailey JS. Teaching pedestrian safety skills to young children: an analysis and one-year follow-up. J Appl Behav Anal 1978;11:315–29.

168. Fortenberry JC, Brown DB. Problem identification, implementation and evaluation of a pedestrian safety program. Accident Anal Prev 1982;14:315–22.

169. Race KEH. Evaluating pedestrian safety education materials for children ages five to none. J School Health. 1988;58:277–81.

170. Retting R. Interventions need not be expensive: Queens Boulevard changes. Presented at the Centers for Disease Control Second National Injury Control Conference, San Antonio, Texas, Sep 1988.

171. Watson GS, Zador PL, Wilks A. The repeal of helmet use laws and increased motorcyclist mortality in the United States, 1975–1978. Am J Public Health 1980; 70:579–92.

172. McSwain NE, Willey J. The impact of re-enactment of the motorcycle helmet law in Louisiana. Proc Am Assoc Automotive Med 1985:425–46.

173. Robertson LS. An instance of effective legal regulation: motorcyclist helmet and daytime headlamp laws. Law Society Rev 1976:467–77.

174. Zador PL. Motorcycle headlight-use laws and fatal motorcycle crashes in the US, 1975–83. Am J Public Health 1985;75:543–6.

175. Olson PL, Halstead-Nussloch R, Sivak M. Enhancing motorcycle and moped conspicuity. In: proceedings of the International Motorcycle Safety Conference, vol. 3; Washington, DC, May 18–23, 1980:1029–57.

176. Planning for safety priorities. Washington, DC: National Highway Traffic Safety Administration, 1983.

177. Hurt HH, Ouellet JV, Thom DR. Motorcycle accident cause factors and identification of countermeasures. Vol. 1: Technical report. Washington, DC: National Highway Traffic Safety Administration, 1981; DOT publication no. (DOT HS)805-862.

178. Wood T, Bowen R. Evaluation of the revised motorcycle learner permit scheme, July 1983 to December 1985: summary report. Hawthorn, Victoria, Australia: Road Traffic Authority, 1987.

179. Westman JA, Morrow G. Moped injuries in children. Pediatrics 1984;74:820–2.

180. Westman JA. Effectiveness of moped legislation. Clin Perinatol 1985;12:331.

181. Hunter WW, Stutts JC. Mopeds: an analysis of 1976–1978 North Carolina accidents. Chapel Hill, North Carolina: University of North Carolina Highway Safety Research Center, 1979.

182. Hunter WW, Stutts JC. Exposure characteristics of North Carolina moped riders. Chapel Hill, North Carolina: University of North Carolina Highway Safety Research Center, 1980.

183. Moped safety: guidelines for comprehensive state programs. Washington, DC: National Highway Traffic Safety Administration; DOT publication no. (DOT HS)805-558.

184. Dorsch MM, Woodward AJ, Somers RL. Effect of helmet use in reducing head injury in bicycle accidents. Proc Am Assoc Automotive Med, 1984.

185. Dorsch M, Woodward A, Somers R. Do bicycle safety helmets reduce severity of head injury in real crashes? Accident Anal Prev 1987;19:183–97.

186. Bishop P, Briard B. Impact performance of bicycle helmets. Can J Appl Sport Sci 1984;9:94–101.

187. Thompson RS, Rivara FP, Thompson DC. Preven-

tion of head injury by bicycle helmets. Paper presented at the American Public Health Association Annual Meeting, Boston, Nov 1988.

188. Wasserman RC, Waller JA, Monty MJ, Emery AB, Robinson DR. Bicyclists, helmets and head injuries: a rider-based study of helmet use and effectiveness. Am J Public Health 1988;78:1220–1.

189. Selbst S, Alexander D, Ruddy R. Bicycle-related injuries. Am J Dis Child 1987;141:140–4.

190. Weiss B. Bicycle helmet use by children. Pediatrics 1986;77:677–9.

191. Developing a children's bicycle helmet safety program: a guide for local communities. Seattle, Washington: Harborview Injury Prevention and Research Center, 1987.

192. Erlich P, Farina A, Pavlinski L, Tarrants WE. Effectiveness paper: bicyclist safety programs. Washington, DC: National Highway Traffic Safety Administration, 1982; DOT publication no. (DOT HS)806-132.

193. Lott DF, Lott DY. Effect of bike lanes on 10 classes of bicycle-automobile accidents in Davis, California. J Safety Res 1976;8:171–9.

194. Injury control for children and youth. Elk Grove Village, Illinois: American Academy of Pediatrics, 1987.

195. Report to Congress on the effects of the 65 mph speed limit during 1987. Washington, DC: National Highway Traffic Safety Administration, 1989.

196. 55: A decade of experience. Washington, DC: National Research Council, 1984.

197. Waller P. The 55 mph national maximum speed limit. Testimony presented to the Subcommittee on Surface Transportation of the U.S. House of Representatives Committee on Public Works and Transportation, Mar 19, 1987.

198. Comptroller's report. New York: City of New York, 1986.

199. Cooper P. Evaluation of the effectiveness of the Multanova photo-exposure radar unit in reducing traffic speed. North Vancouver, British Columbia: Insurance Corporation of British Columbia (undated).

200. Pigman JG, Agent KR, Deacon JA, Kryscio RJ. Evaluation of unmanned radar installations. Paper presented at the 68th annual meeting of the Transportation Research Board, Washington, DC, Jan 22–26, 1989.

201. Stewart JR. Statistical evaluation of the effectiveness of FMVSS 202: head restraints. Washington, DC: National Highway Traffic Safety Administration, 1980.

Chapter 7: Residential Injuries

Residential injuries—injuries that occur in and around homes, day-care centers, nursing homes, and other institutional settings—stem from a broad range of host, agent, and environmental factors. The results are often tragic.

• An East Boston couple was killed and their 18-month-old son critically injured when a three-alarm fire ripped through their two-story home. The toddler had stopped breathing and suffered from burns and smoke inhalation. Careless disposal of smoking materials caused the fire.[1]

• A 75-year-old man suffered five falls in one year. An assessment of his home revealed numerous throw rugs and several staircases that were difficult to negotiate.[2]

• A 2-year-old girl suffered a poisoning from trisodium phosphate. Ironically, this strong household cleaning agent was being used to clean her home following the removal of lead paint (M. Bond, personal communication).

These incidents are especially tragic because knowledge and technology are available to prevent them. Over the past 30 years researchers, architects, environmental health workers, and legislators have helped to reduce risks once common in the home. The rate of fatal home injuries has fallen accordingly. Despite this progress, residential injuries continue to occur, taking a toll that is unacceptably high.

DEFINITIONS

The term "home injuries" refers to injuries that occur in the home (including apartments and boardinghouses) and the immediate surroundings, such as garage and yard, to occupants and guests. It does not include suicides, homicides, or deaths of undetermined intent that occur in the home or immediate surroundings, nor does it apply to assaults, which are covered in later chapters. This definition differs from the National Center for Health Statistics (NCHS) definition of home injuries in that it does not include drownings and playground injuries that occur in the home or on the home premises. These are discussed in Chapter 8. The term "residential injuries" is used to designate a broader range of environments in which injuries occur, including the home, but also nursing homes and day-care centers, the prevalence of which has grown dramatically over the past 30 years.

THE MAGNITUDE OF THE PROBLEM

In 1987, 20,500 deaths occurred in homes and on home premises, accounting for 22% of all fatal injuries in that year. The most common fatal residential injuries were falls, fires and burns, poisonings, suffocation/asphyxiation, and unintentional firearm injuries. Injuries are the leading cause of death among children. Twenty-two percent of children under the age of 6 suffer residential injuries each year. And among older persons, injuries rank sixth as a cause of death among 65- to 74-year-olds and seventh among those 75 years of age and older.[3] Eighteen percent of adults 75 years of age and over suffer injuries in the home.[4]

Information about nonfatal residential injuries is not readily available because of the poor quality of E code data on hospital discharge and ambulatory care visit records. In 1987 an estimated 3,100,000 persons suffered disabling injuries in the home; of these, 80,000 resulted in some permanent impairment. The costs of these injuries was estimated at $16.7 billion.[3]

RISK FACTORS

Residential injury deaths present a common bimodal distribution: the very young and the very old are at greatest risk. This distribution reflects the impact of exposure coupled with the developmental stages across the life span.

Infants and toddlers experience the environment through their senses and, as they develop, through physical movement. They learn based on their experiences, trial and error, repetition, imitation, and identification.[5] Preschool children enjoy physical activity, especially running and jumping. They explore their environment, experiment with cause-and-effect relationships, and engage in dramatic play. Children often overestimate their newly developed physical abilities and engage in activities that are unsafe or beyond their abilities.[5]

Childhood injuries show distinct patterns across age groups,[6,7] reflecting how and where children spend most of their time. Studies of injuries treated in the emergency room indicate that falls from furniture or nursery equipment are the major source of injury for infants and young children.[8,9] As children become more mobile and independent, spending more time outdoors and on their own, residential injuries become less common. Sports and recreational equipment or activities account for half of all nonfatal injuries to teenagers.[6]

Children possess certain advantages when traumatically injured. They have greater recuperative power than adults because they usually experience less associated or preexisting illness[10] and their wounds frequently heal faster.[11] A fracture of the femur may heal in 6 weeks in an infant but may take 6 months or more in an adult.[12] In addition, children's pliable bones often resist injury while elderly people's brittle bones localize energy and therefore fracture more readily.[13] Children also have certain disadvantages in dealing with trauma, ranging from the more critical nature of blood loss, abdominal and head injuries, and fractures to the more disruptive psychological impact on a child's emotional development.[14]

The elderly are also at high risk for residential injuries. They may perceive danger less accurately because of deteriorating sensory awareness and acuity.[15] Compensating for losses in physical coordination[16] or withstanding the physical stress of an injury becomes more difficult because of osteoporosis and other causes of bone fragility.[17] And they may respond more slowly to warning signals because of impaired intellectual or mental function.[18] Alcohol and drugs, including inappropriate or misused prescription medications, often contribute to these impairments. In addition, the fact that many elderly people live alone puts them at increased risk of injury. The major causes of injury death among elderly people include motor vehicle injuries, falls, suffocation, fires and burns, and poisonings.[3]

Males are at greater risk for injuries until the age of 75.[3] They experience more deaths from injuries than females for each cause of residential injury. The reasons for these differences are not clear, but risk-taking behavior among males appears to be a major determinant.[19]

Residential injuries also vary by income. This has been studied in depth among children. An analysis of childhood and adolescent deaths in Maine in 1976–80 found that low-income children were five times more likely to die as a result of fire than were higher-income children.[20] In a Baltimore study, the death rate from residential fires was more than three times higher in the lowest-income residential census tracts than in the highest.[21] Children in low-income families also have higher rates of repeat injuries.[22] In part, this may be explained by the greater number of environmental hazards found in the homes of low-income families (e.g., housing in poor repair, faulty wiring, use of space heaters).[23,24]

Certain home environments also increase the risk of injury. Mobile homes warrant special attention. Because of their relatively small floor area and the crowding of furniture, the risk of fire is increased. Ventilation is likely to be poor and exits limited. These conditions result in a more rapid spread of fire and faster accumulation of lethal gases.[25]

As new housing stock continues to grow rapidly, the potential for introducing well-designed, less hazardous environments is great. Architects, engineers, and builders are able to apply advances in electronics, fire-safe building materials, and redesigned furniture and household products to reduce injury risk in the home. The greatest risk, then, continues to be concentrated in the deteriorating housing of older urban neighborhoods.

Specialized environments, such as day-care centers, also must be considered. Day-care centers vary from formal licensed settings to family day care based in private homes. A number of studies have documented injury hazards in day-care environments.[26–29] In some instances, hazards have been reduced by inspecting the premises and educating day-care operators about safety.[27,28]

Nursing homes in which large numbers of frail elderly people live are another specialized environment. Among elderly persons in institutions, 10%–25% suffer a serious fall each year.[30] Persons living in long-term care institutions report the highest rates of falls compared to community-living or hospitalized elderly persons. This may be due in part either to the greater frailty of institutionalized elderly people or to reporting differences.[2] Further study of injuries in such settings is needed.

Alcohol and Residential Injuries

Alcohol is known to affect both the risk and the severity of injuries, and evidence suggests that this increased risk exists at every phase: before the injury, during the injury event, and into the postinjury phase.[31] As alcohol consumption increases, so do the risk and severity of motor vehicle injury,[32] but further study is needed to test whether the same holds for other injuries.[33]

Few researchers have examined the role of alcohol in residential injuries, but early studies suggest that alcohol is a major contributor. One study found that an alcoholic's risk of dying of a fall was 16 times greater than that of the general population.

The risk was also 10 times greater for burns and fires.[32] The same study found that alcohol was involved in seven of 10 fall fatalities occurring within 4 hours of injury. In another study, alcohol was present in about half of residential injury fatalities in persons 15 years of age and older.[34]

An impediment to studying the link between alcohol and residential injuries is the lack of systematic collection of information about blood alcohol levels. For example, in one study of hospitalized house fire victims, over 50% of the patients' charts showed no mention of either smoking or alcohol use despite the high association between these practices and house fires.[35] Recent alcohol abuse prevention efforts, such as reducing consumption through price increases (discussed in Chapter 6), are examples of the broad social policies toward alcohol control that need to be considered.[36]

A BRIEF HISTORY OF PREVENTION EFFORTS

The history of the home safety movement has been marked by surges and declines in activity involving the efforts of federal agencies, private groups, and dedicated individuals. The focus of these activities has been inspection to remove environmental hazards and education about consumer product safety, fire safety, and poison control.

Several events set the stage for the home safety movement. In 1951 the American Academy of Pediatrics (AAP) organized an accident prevention committee that called for "anticipatory guidance," a practice through which doctors would educate parents about how to control hazards in their child's environment.[37] In 1953 the University of Michigan School of Public Health and the W. K. Kellogg Foundation sponsored one of the first conferences on home accident prevention in cooperation with the National Safety Council, the U.S. Public Health Service, and the American Public Health Association (APHA). As a result, the Kellogg Foundation funded three community-based demonstration projects designed to advise parents about home hazards. Based in San Jose, California, Cambridge, Massachusetts, and Mansfield, Ohio, the projects involved physicians and nurses in home safety counseling and public health sanitarians in home safety inspections.[38] In 1956 the APHA Committee on Administrative Practice issued a statement on "Suggested Home Accident Prevention Activities for Health Departments." The statement followed a decade of groundwork in which the APHA encouraged state and local health departments to become involved in injury prevention. The statement presented an array of activities

for divisions within state and local public health agencies, including vital statistics, public health nursing, maternal and child health, environmental sanitation, and food and drug control. Home safety was seen as a "team" activity in which a variety of individuals within the agency had an important role to play.[39]

Early Federal Coordination Efforts

During the early 1960s federal injury control activities were spearheaded by the Public Health Service's Division of Accident Prevention. Several federally funded demonstration projects involved state and local public health practitioners in home hazard inspection activities.[40,41] Support for these programs diminished when the Division of Accident Prevention was dismantled in 1967 and its responsibilities were divided among the Food and Drug Administration, the Health Services and Mental Health Administration, and the Centers for Disease Control.

At the same time, then-Senators Warren Magnusson and H. L. Fountain appealed to President Johnson for support of injury prevention efforts. Fountain held a series of hearings to highlight the magnitude of injuries as a public health problem and the need for a continued federal role, and a law was passed to establish the National Commission for Product Safety. The final report of the commission made clear the need for an ongoing national effort to collect data about product safety. As a direct result, the Consumer Product Safety Commission (CPSC) was established in 1973, and the enforcement of toy, flammable fabrics, and labeling laws was placed under its jurisdiction (R. D. Verhalen, personal communication).

Fire and burn prevention efforts were advanced concurrently through several landmark community-based studies.[42-45] Even though these projects were inadequately designed, several of them advanced the knowledge base available to plan and implement community-based prevention programs.

The fourth thread interwoven into the history of home safety involved the poison control center movement, which was launched by AAP and the Illinois health department in the fall of 1953 in response to a need to update information on the toxic ingredients of common household agents. During the 1960s more than 600 local poison control centers were established; many of these were staffed by persons without training in clinical toxicology. It was not until 1978 that preliminary standards for regional poison control centers were published by the American Association of Poison Control

Centers.[46] Much progress has been made since then in establishing regional poison control centers nationwide. They now exist in all states.[47]

Growing Support for Home Safety

In the late 1970s the Centers for Disease Control (CDC) funded several extensive analyses of the causes and prevention of falls, burns, and recreational injuries. In addition, three pilot injury projects (in Virginia, California, and Massachusetts) funded by the Division of Maternal and Child Health in 1979 were instrumental in advancing the home safety movement. The California project focused on preventing burns, poisonings, and head injuries for children under the age of 4.[48] The program laid the foundation for ongoing work in San Diego County, including training and materials development and an annual statewide conference on childhood injury control.

The Massachusetts project, the Statewide Comprehensive Injury Prevention Program (SCIPP), began by planning, implementing, and evaluating a variety of prevention activities. It instituted a population-based surveillance system to collect data on injury-related deaths, hospital admissions, and emergency room visits for 3 years in target communities. The effectiveness of the program in reducing injury rates for burns, falls, poisonings, and motor vehicle occupant injuries was also evaluated.[49] Building on this work, SCIPP subsequently developed a series of modules targeted to various types of care providers.[50-53]

The leadership in home safety today can be found in foundations, professional groups, and local, state, and federal organizations. The Carnegie Corporation of New York, the Henry J. Kaiser Family Foundation, and others support demonstration projects addressing home injuries. APHA, AAP, and other professional groups are active in supporting research and practice in this field. And at the federal level, the Bureau of Maternal and Child Health and Resources Development now funds 20 projects in 15 states, many of which address home injury. CDC also actively supports home safety through its extramural grants program, funding research and state and local health department efforts to address home injuries.

RESIDENTIAL INJURY PREVENTION: STATE OF THE ART

The residential injury problem presents many opportunities for intervention. The field has evolved from a focus on education to change personal behavior to a recognition of the complex environment in which injuries occur. It is now accepted that ameliorating environmental hazards requires the use of a mixed strategy of legislation/enforcement, education/behavior change, and engineering/technology interventions. Reducing residential injuries also requires a multidisciplinary effort. The shared experience of engineers, architects, health care providers, elder care services providers, environmental health workers, day-care providers, long-term care providers, fire officials, poison control experts, and others is contributing to the growth of this field and the development of effective interventions.

Success stories are common: poison prevention packaging, flammable fabric laws, and the installation and enforcement of window guards are only a few examples. And yet we face two great needs today. Engineering/technology interventions are not available to address many types of residential injuries. And too often, programs that address residential injuries are not evaluated. Thus progress in reducing residential injuries is halting, at best. To continue to effect change in this field, "we must ensure that people find support for positive health behavior in their physical and social environments."[54]

The following discussion of interventions begins with comprehensive approaches to preventing residential injuries. It then focuses on the four major causes of residential injury death: falls, fire and burn injuries, poisonings, and suffocations. (Unintentional firearms injuries are discussed in Chapter 17.)

INTERVENTIONS

General Residential Injury Prevention Programs

Environmental hazard inspection programs. Few studies have examined exclusively the role of environmental hazards in residential injuries. Several studies of elderly patients seeking care after a fall found that 39%–52% of the falls were attributable to environmental causes.[55,56] These included slippery surfaces, loose rugs, objects on the floor, poor lighting, and broken stairs. In some cases, the hazards represented violations of building codes and thus required enforcement efforts. This was particularly true for inadequate lighting, poorly designed stairways, broken steps, and missing balusters or handrails.[57]

State sanitary codes often address such hazards in defining minimum standards for habitation.[58] The standards usually apply to all dwelling units and have the force of law, including criminal penalties for failure to comply. However, they vary greatly from state to state and are usually enforced locally by boards of health. The codes generally cover proper installation and maintenance of structural elements (stairways, floors, doors, windows, etc.), as well as kitchen, bath, heating, lighting, and electrical facilities.

A home injury prevention program for children, which studied the impact of the Massachusetts Sanitary Code, illustrates the potential impact of enforcement. While conducting home safety inspections, this study found an average of 11.1 code violations per housing unit. In one city, the study resulted in the first legal proceedings in 20 years for code violations. Follow-up visits revealed complete correction of these code violations, demonstrating the effectiveness of this approach, which combines engineering/technology with legislation/enforcement measures.[24]

Numerous home safety assessment forms are available for both the professional and layperson.[50,59] Using such assessments to perform audits in the homes of young children and the elderly, particularly in substandard housing areas, can be effective in reducing environmental hazards.[24]

One program now testing the effectiveness of home audits in reducing injuries to older people is the San Francisco Community and Home Injury Prevention Project for Seniors (CHIPPS). The program is conducting a study to determine if home audits, home repair, installation of safety devices, and personal medication counseling can reduce the risk of falls and other injuries among elderly people (M. T. Dowling, personal communication).

Recommendation: Environmental hazard inspection programs are promising interventions to reduce hazards in living environments. Establishing or enforcing existing building and sanitation codes is an integral part of these interventions. They should be used and monitored and their effects evaluated.

Day-care safety programs. Day-care centers present an important area for safety intervention that can be easily integrated into existing health department responsibilities. Because many state health agencies are responsible for licensing day-care centers, they can help to ensure that these environments have a minimum of injury hazards.

Two recent studies have shown reductions in injury hazards after inspecting day-care premises and educating operators about safety.[27,28] The Safe Day-care Module, developed by SCIPP, has been implemented by day-care licensers, early childhood educators, and day-care operators. It focuses on creating a safe environment, indoors and outdoors, uses a site safety checklist, and provides prevention and first aid information. Other states, including Vermont and Wisconsin, have tailored this module to fit their own licensing and code requirements. In addition, the Massachusetts Health Department's work in this area resulted in including safety as a major topic in a reference manual recently published by the National Association for the Education of Young Children.[62]

Recommendation: Day-care safety inspection programs are promising interventions to reduce hazards in day-care centers. They should be used and monitored and their effects evaluated.

Injury prevention education. Education/behavior change approaches to prevent childhood injuries have been tested, but evidence of their effectiveness is weak.[63–71] Most of these studies are discussed in the injury-specific sections of this chapter. The most effective programs appear to use simple, targeted messages in conjunction with legislation/enforcement and engineering/technology interventions.

AAP has recognized the importance of anticipatory guidance for more than 30 years. Its general injury prevention policy urges the use of approved child car restraints, smoke detectors that monitor the child's sleeping area, safe tap water temperatures, and window and stairway guards/gates to prevent falls and suggests stocking 1-ounce bottles of ipecac syrup. It also suggests that physicians caring for children should counsel parents in age-appropriate, season-appropriate, and locality-appropriate prevention strategies that reduce common serious injuries and reflect this counsel in the medical record.[72] It is meant to serve as a minimum standard of care in preventing childhood injuries. Several tools are available to help health care practitioners counsel parents on childhood injury prevention.[51,73] One Massachusetts package, Safe Child, uses a pictorial guide designed to minimize language barriers between practitioners and parents.[51]

APP operates The Injury Prevention Program, TIPP (now revised, TIPP II). The program targets the most common and severe injuries at each developmental stage from birth to 12 years of age. AAP is currently evaluating the effectiveness of the TIPP program. Many states encourage health care providers to use the TIPP materials in conjunction with well-child clinics. Schedules of recommended minimal safety counseling to be performed at specific visits are also commonly used. Handouts providing detailed information about specific injuries are available.

Another important avenue for promoting injury

prevention education is to integrate injury prevention with ongoing services at the federal, state, and local levels.[74] Programs that serve young children and the elderly, in particular, could most easily do so and could offer the most promise of impact. Such efforts can be very cost-effective, offering injury prevention education, and where possible, legislation/enforcement and engineering/technology interventions. For example, Gallagher et al. suggest that "Home-based program staff who interact with families have the opportunity to adopt some injury prevention strategies within their normal activities. These include visiting nurses, day-care licensers, lead paint programs, early intervention programs, and even rodent control programs."[24] Support for these activities must come from state and federal levels.

Recommendation: Injury prevention education programs administered by health professionals are promising interventions. They are most effective when they employ targeted messages and are used in conjunction with legislation/enforcement and engineering/technology interventions. Injury prevention measures should be integrated, wherever possible, in ongoing activities. They should be used and monitored and the results evaluated.

Falls

Falls are second only to traffic injuries as a cause of injury death across all ages, accounting for 11,300 deaths in 1987.[3] A recent examination of fatal falls noted that "in 1984, over 43 percent of the deaths (5,166) from falls occurred in the home. Almost 67 percent of children less than five who died from a fall died in the home. Of all persons 65 and over who died from falls, about 45 percent died from a fall in the home and about 15 percent from a fall in a residential institution."[75] Most at risk are the elderly, particularly those over the age of 75. More than half of all fatal falls involve persons 75 years of age or older, who comprise only 4% of the population.[76] "There are more deaths from falls among people 85 or older than there are motor vehicle occupant deaths among males, ages 18–19 (3,800 versus 3,000), and the death rate per 100,000 population is more than twice as high."[76] And among elderly people over 85, one of every five fatal falls occurs in a nursing home.[76]

Morbidity resulting from falls is extremely high among the elderly. Massachusetts, one of the few states that make these data available, found that 74.2% of all hospital discharges in 1983 for persons 65 years or older were due to falls.[77]

Children under the age of 5 are also at increased risk for falls, but these falls are not often fatal. The incidence of childhood falls is highest in infancy and declines throughout childhood.[6] In part, this signals the child's developmental process. Before the age of 1, falls often involve furniture such as changing tables[8] or baby walkers.[80] Among toddlers, fatal falls are more common from windows[66] or on stairs.[8] As the child gets older, falls from roofs[8] or playground equipment[82] are common.

The literature reveals little information about the role of socioeconomic status in falls, except among children in dense urban settings, where falls are disproportionately high.[19,80,81] Falls show little relationship to per capita income.[76] Whites consistently have the highest death rates from falls.

Environmental hazards, such as poor lighting, dangerous stairs, clutter, loose rugs, and slippery floor surfaces, play a major role in fall injuries. Among the elderly, environmental factors have been estimated to cause 18%–50% of falls in the home.[82–85] Many falls among older persons occur on a level surface,[56,86] but among young children falls from heights are more common.[66,80,81,87,88]

Drugs, including sedatives and hypnotics (e.g., barbiturates and long-acting benzodiazepines) are frequently associated with falls.[89,90] This is particularly true in nursing homes, where as many as 43% of the patients use such medications,[91] compared to 10% of elderly people living in the community.[92] If the medication cannot be discontinued, the dose might be decreased or an alternative drug substituted.[2] Alcohol, alone or in combination with other drugs, is also commonly associated with falls among the elderly. Although data are sorely lacking in this area, one study noted recent alcohol use in 13% of older persons with fall injuries.[34]

Safety equipment: window guards and safety gates. The most effective tool for preventing children's falls from windows is a window guard. Studies have documented the fact that window falls are more likely to occur in children living in the inner city and stress the importance of securing windows, balconies, and fire escapes.[19,80,81] The Children Can't Fly program of the New York City Health Department reduced window falls 50% by installing window guards in high-risk houses, passing legislation requiring the guards, educating the public about the hazard, and working with public health nurses to provide counseling and referral.[66] It is important to note that continued enforcement of the regulation was necessary to maintain its effectiveness.[93]

Recommendation: Window guards have been proven effective in preventing children from falling out of windows. They should be used and the results monitored.

Other fall prevention devices. Safety devices for the elderly include grab bars in the tub or shower or next to the toilet, nonskid mats in the tub or shower, easily grasped handrails on one or both sides of all stairs, and adequate lighting in all hallways and rooms.[2] These items have not been proven effective in preventing falls, but their use is considered prudent by geriatricians, given the physiological deterioration that accompanies aging.

In nursing homes, similar attention should be paid to environmental hazards and installation of safety devices. Bed height is particularly important, given the incidence of falls from bed or while getting into or out of bed in institutional settings.[2] Placing beds at the proper height (so that a patient's feet can touch the floor when the knees are bent at a 90° angle) could minimize such falls.

Having fallen, an elderly person is often unable to get up. One study showed a significant increase in fatalities and complications in persons who were rescued more than 1 hour after injury.[85] Providing alarms at floor level or devices that can be worn on the body could reduce this risk.[94]

Gates on stairways help to prevent young children's falls, which are frequent but usually not serious.[95] AAP recommends that safety gates be firmly attached and permanently mounted, when appropriate, with double closures that children cannot open. Accordion-type gates are dangerous and should not be used.[37]

Recommendation: Installing residential safety equipment is a promising intervention to prevent falls among young children and the elderly. It should be used and monitored and the results evaluated.

Education of elderly people about falls. Because falls are such a common problem among the elderly and are associated with many age-related changes, it is important to educate older people and their caregivers about prevention. As noted earlier, there is a correlation between falls and certain types of drugs,[89,90] so elderly people and their caregivers need to know the common side effects of drugs and the increased risk of falls they pose.

Elderly people can also benefit from learning about the physiological changes of aging and how to delay or minimize them. One of the most notable changes occurs as the skeletal system becomes more porous and less resistant to stress, a condition known as osteoporosis. Osteoporosis and other causes of bone fragility are not well understood, but the loss of bone mass is generally attributed to sev-

eral factors: hormonal changes, nutritional deficiency, and a decrease in physical activity.[96] Much research is now under way to identify the role of these factors in osteoporosis. Increasing the bone mineral content with fluoride,[97] calcium, and vitamin D supplements[98] and exercise[99,100] has been tested in older women. Results show that such increases may help to protect against loss of bone mass, but further study is needed to identify the precise nature of these prevention techniques.

Recommendation: Education programs targeted at the elderly, their caregivers, and the general public on dietary measures, drug management, and the importance of exercise in preventing falls are promising interventions. They should be used and monitored and the results evaluated.

Education of parents of young children about falls. Frequent falls in childhood and the lack of countermeasures suggest the need for parental education to prevent these injuries, but evidence about the effectiveness of such education is limited. One effort to counsel mothers about preventing falls during infancy showed a significant decrease in the percentage of falls among their children compared to other children.[63] The key appears to be delivering highly targeted messages integrated with engineering/technology and legislation/enforcement approaches. For example, education was an integral part of the Children Can't Fly program.[66]

Recommendation: Educating parents about their child's risk of falls is a promising intervention. Such efforts should use highly targeted messages that lead parents to recognize and remove fall hazards. The interventions should be implemented and monitored and the results evaluated.

Baby walkers. Between 55% and 86% of infants in this country are placed in walkers prior to walking.[101,102] The benefits of walkers in helping an infant learn to walk are negligible,[101] and the risk of injury is great, as documented by a number of studies.[102–107] AAP recommends that pediatricians counsel parents about the risks of injury from baby walkers as part of well-child care.[37] Others have called for a ban on baby walkers.[9,108,109]

Recommendation: Baby walkers are a significant cause of injury among infants. They should be banned, and health care providers should counsel parents of infants on their danger.

Fire and Burn Injuries

Burns are the fourth leading cause of injury death, causing 4,800 deaths in 1987.[19] Most at risk are children under the age of 5 and the elderly. Burn incidence rates are generally lower among the elderly than among the young, but the elderly suffer a higher fatality rate.[76]

There are six categories of burn injuries: flames (and smoke inhalation), scald, contact, and electrical, chemical, and ultraviolet radiation. The final two are not dealt with in this chapter, because chemical burns, commonly caused by the ingestion of lye or other caustics, can best be prevented using interventions recommended for poisoning prevention, and ultraviolet radiation burns are seldom severe enough to warrant medical treatment. Each of the other types of burns varies in severity and frequency and displays distinct patterns of risk by age and sex.

House fires cause 75% of all deaths from fires and burns, with young children and the elderly at greatest risk.[76] About half of all fires involve cigarettes, and in many cases the smoker has also been drinking.[110] Arson is a growing factor in many house fires, presenting a complex set of criminal, economic, and behavioral factors. An arson task force is a strategy through which resources can be directed to set policies, mobilize resources, and integrate the effects of law enforcement, the insurance industry, and others in addressing this problem.[111]

Fire-related deaths decreased 20% between 1978 and 1982 despite a relatively constant incidence of structural fires. The decrease has been attributed to increased use of smoke detectors as well as improved firefighting techniques and emergency medical systems.[112]

Among the elderly, fabric ignitions, one-third of which happen in the kitchen, are common. These burns are nearly always associated with cooking.[113] One study noted that 62% of non-house-fire burns among the elderly involved clothing ignition, compared with 30% for all younger persons.[114] Fabric ignitions among children have been dramatically reduced by the Flammable Fabrics Act of 1953 (amended in 1967). The Children's Sleepwear Standard of 1971 (amended in 1974) resulted in a marked decline in sleepwear-related burns.[115] Today such injuries are seldom seen among children.

Contact burns are most commonly sustained by children. They are usually small but deep burns, most often caused by heating devices such as wood stoves.[116] Scalds, also common among young children, are rarely fatal and commonly occur when hot liquid is tipped over in the kitchen.[117] Flammable liquids, such as gasoline, are among the most lethal causes of burn injuries among boys 6–16 years of age.[118] Three common scenarios are using gasoline in a basement or garage as a cleaning product, using lighter fluid to boost a charcoal grill, and, especially among the elderly, spilling lighter fluid while refilling a cigarette lighter, after which the spark ignites the clothing. Disposable cigarette lighters also pose dangers, especially for young children. CPSC is currently developing regulations that will require all cigarette lighters to be child resistant.

Electrical burns generally occur when children suck on extension and appliance cords, injuring their mouths or faces.[119] High-voltage electricity is responsible for relatively few but severe burn injuries. These commonly occur among boys between 7 and 16 years of age and sometimes require skin grafting and/or amputation.[120] Adolescent risk taking is involved in many of these incidents.

Smoke detectors. Smoke detectors are known to be a reliable, inexpensive means of providing an early warning of house fires. The one-time installation of a smoke detector and the need for only periodic maintenance (battery replacement in battery-operated models) makes it one of the most promising engineering/technology interventions available for preventing deaths from fires caused by burns and asphyxiation. Consideration also should be given to requiring installation of electric rather than battery-operated smoke detectors to eliminate the need for battery replacement. Battery-operated models should use long-life alkaline batteries.

Numerous studies have examined the efficacy of smoke detector giveaway campaigns or low-cost purchase opportunities with and without installation.[68,69,121] Results suggest that such programs should be carefully targeted at low-income neighborhoods known to have high proportions of children and/or elderly residents. However, it is essential to assess the need for smoke detectors in the target area. A program in New Hampshire planned to give away smoke detectors through well-child clinics but found that most homes in the target area already had a detector.[122] Similarly, other studies noted the increased usage of smoke detectors among low-income populations.[24,69]

Evaluation of the effectiveness of smoke detectors reveals that they reduce the potential of death in 86% of fires and the potential of severe injuries in 88%.[123] A summary of smoke detector legislation published in 1983 indicated that 29 states required smoke detectors in all new construction and 22 states required one or more classes of residential housing to be retrofitted with smoke detectors (usually at the point of sale).[124] The effectiveness of these laws has not been systematically evaluated, but a study of the effect of smoke detector legislation in Montgomery County, Maryland, revealed that 94% of owner-occupied, single-family homes in the county had a detector (although only 82% of the homes had a *working* detector). Only 70% of homes in a control site had a working detector.[125] Fire data suggest that as a county approaches complete detector coverage, the risk of residential fire deaths decreases significantly.[118]

Recommendation: Smoke detectors have proven effective in providing early warning of a fire. States and localities should enact and enforce legislation requiring smoke detectors in all new single-family and multifamily dwellings and should retrofit multifamily dwellings with smoke detectors. All such efforts should be monitored.

Sprinkler systems. Research has demonstrated that the effectiveness of smoke detectors is increased if a sprinkler system also is used, thereby dramatically reducing the spread of a fire. In 1,648 fires in New York City high-rise buildings from 1969 to 1978, sprinklers were rated over 98% effective in suppressing and extinguishing fires.[126] An insurance industry assessment suggests that residential sprinklers would reduce property loss fire claims by three-quarters.[123] Requiring smoke detectors in building codes and mandating their use through legislation have proved enormously effective. This approach should be vigorously pursued for residential sprinkler systems as well. Such legislative efforts should also consider retrofit requirements. Because of the considerable expense associated with retrofitting existing buildings, one expert suggests giving priority to high-risk buildings, such as old wooden apartment houses, and using new, lower-cost materials, such as plastics.[94]

Recommendation: Automatic sprinkler systems are effective in extinguishing fires. States and localities should require sprinkler systems in all new housing and make efforts to provide such systems in older multifamily housing units. These efforts should be monitored.

Enforcement of building codes. Enforcement of existing building codes is necessary to eliminate fire hazards in older, high-risk dwellings. These hazards include substandard wiring, improper storage of flammable products, and dangerous or unusable exits.

Research has suggested that hazardous heating devices are a factor in many house fires.[127] Improper or faulty electrical and heating equipment are also common causes. This is particularly true in substandard housing, which has been shown to be nine times more common in low-rental than in high-rental census tracts.[128]

To help bring about change in such neighborhoods, ensuring adequate housing services must be a priority. Several resources are available. The Model Standards project of the American Public Health Association and others provides guidelines for adequate housing services. The objectives focus on housing codes, substandard housing, surveillance, and follow-up care.[129] In addition, APHA and CDC jointly published a recommended housing maintenance and occupancy ordinance for state and local housing and health authorities.[130] CDC has also published a manual, *Basic Housing Inspection,* designed to be an integral part of comprehensive housing inspection training courses.[57]

Recommendation: Enforcement of existing building codes is a promising intervention to reduce the number of dwellings at risk for fires. Such efforts should be implemented and monitored and the results evaluated.

Fire-safe cigarettes. Cigarettes are estimated to cause 45% of all fires and 22%–56% of deaths from house fires.[128,131] Most cigarettes made in this country contain additives in both the paper and the tobacco that cause the cigarette to burn for as long as 28 minutes, even if left unattended. Despite its role in fatal fires, the cigarette is explicitly excluded from the jurisdiction of any federal regulatory agency.[132] The 1984 Cigarette Safety Act established a technical study group to determine the technical and economic feasibility of manufacturing fire-safe cigarettes.

The study confirmed that fire-safe cigarettes were possible.[133] Of 41 experimental cigarettes tested, one did not ignite mock-up or full-scale flammable furniture and six ignited 5%–35% of the mock-up furniture. The study also found that the benefits of producing fire-safe cigarettes outweigh the costs to the industry. If only cigarettes that were fire safe were smoked in this country, nearly 2,000 deaths and more than 6,000 burn injuries would be prevented annually. In recent years, legislation has been introduced regularly, but has not passed at either the state or federal level.

Although fire-safe cigarettes are a response to the problem of burn injuries, they do not address the massive public health hazards posed by smoking, including lung cancer, heart disease, and other major causes of premature death and disability. There is the danger that a "fire-safe" cigarette might be inappropriately marketed as a *safer* cigarette. We therefore encourage increased efforts to reduce smoking through public bans, education to prevent youth from beginning to smoke, and support services for those who desire to quit smoking. Further, we urge federal review of the current exemption of cigarettes from the oversight of FDA and other regulatory agencies.

Recommendation: Fire-safe cigarettes are effective in reducing the fire hazards associated with smoking. The federal government should enact legislation that requires cigarette manufacturers to make cigarettes self-extinguishing.

Fire safety education. Fire safety education is frequently directed to three groups: the elderly (e.g., the risks of fabric ignition, smoking materials), school-age children (e.g., stop, drop, and roll in the event of clothing ignition, crawl under smoke, what to do in case of a house fire), and the general public (e.g., installation and maintenance of smoke detectors and emergency preparedness in case of fire, including adequate escape routes, proper handling and storage of flammable materials, safe use of heating devices). The effectiveness of these efforts is unknown. Project Burn Prevention, a community-based public education program, was designed to increase awareness of burn hazards and reduce the incidence and severity of burn injuries. Carried out in the late 1970s, the study increased knowledge about burn injuries as a result of its school program but failed to reduce burn incidence or severity.[134] However, limitations of the study have been noted, such as the small size of the population and the brief exposure to its messages. Research about the effectiveness of pediatric counseling and subsequent use of smoke detectors is mixed.[64,67,68] The lesson drawn from these studies seems to be that single, focused educational messages are more likely to be effective than multiple messages.

The rapid public acceptance of smoke detectors—present in less than 5% of U.S. homes in 1970 but in 67% by 1982—is attributed in part to intensive media campaigns.[135] However, accompanying these educational efforts were reductions in price, educational programs by local fire departments, building code modifications, and legislation. It is within this multifaceted approach that fire safety education seems most effective.

Recommendation: Public education about fire safety and proper care for burn injuries is a promising intervention. Such education should be monitored and the results evaluated.

Reducing tap water temperature. In the United States, hot tap water causes more than 2,600 scald burns each year.[136] Approximately 25% of these burns require hospital admission. A 10-year review of such burns in Madison, Wisconsin, and vicinity revealed that 85% of the cases occurred to children, the elderly, and the physically or mentally disabled.[137] Child abuse was noted in about one-fourth of tap water scalds.[138]

Recent studies indicate that about 25% of the general public are unaware of the potential danger of hot tap water.[139,140] These studies indicate that receiving a free liquid crystal thermometer to use in testing hot water temperature was effective.[139,140] Written material rather than multimedia campaigns seemed to be an important contributing factor to this behavior.[139,140]

We applaud the action of several states that have passed laws requiring manufacturers to set new water heaters at a safe temperature and recent indications that water heater manufacturers may institute a standard requiring that hot water heaters be preset at 125°F. This will do much to make hot water heaters safer. However, the average life of a water heater is 10 years, and thus about 10 years are required to complete a water heater changeover.[141]

Recommendation: Testing tap water temperature and setting hot water heaters at 125°F are effective in preventing tap water scalds. Such efforts should be used and monitored and the results evaluated.

Flammability standards. Legislation regulating product flammability has proved effective in many instances. The Children's Sleepwear Standard has resulted in a marked decline in sleepwear-related injuries.[115] The 1973 Mattress Flammability Standard has also significantly increased consumer safety.[142]

Recommendation: Flammability standards have proven effective in reducing fire and burn injury risks. Such standards should be expanded to include furniture, bedding, clothing, and home-building materials. They should be enforced and monitored.

Poisonings

Deaths by poisoning by both solids and liquids (drugs, petroleum distillates, and cleaning agents) have decreased dramatically over the past 15 years, particularly among children under the age of 5. In 1970, 226 poisoning deaths of young children occurred; in 1985 only 55 occurred.[143] Indeed, aspirin poisoning, the single leading cause of childhood poisoning deaths (numbering 46 in 1972), had virtually disappeared by 1985. These declines are due in large part to the Poison Prevention Packaging Act of 1970, which requires childproof closures on many drugs, medicines, and household substances. Other factors contributing to this decline include dose limits per package[144] and improved emergency management of poisoning, particularly through regional poison control centers.[145]

Despite these changes, nonfatal poisoning remains a major cause of hospital admissions and emergency room care. For every poisoning death among children under the age of 5, 80,000–90,000 nonfatal cases are seen in emergency rooms and about 20,000 children are hospitalized.[143]

It is not known if nonfatal childhood poisonings have declined at the same rate as fatalities. A recent national study of hospitalizations resulting from childhood poisonings, however, offers insight about who is at greatest risk. From 1979 to 1983, 108,280 children under 10 years of age were hospitalized because of poisoning by a liquid or solid. Nonwhite children had a hospitalization rate for poisoning that was 2.4 times that of white children.[146]

Studies have shown that poisonings occur 10 times more often among children under the age of 5 than among elementary school children.[6,146] Children at greatest risk are 1- and 2-year-olds, who are 17 times more likely to be hospitalized for poisoning than any other age group under 20 years.[146] It is important to note that the environment away from home, particularly grandparents' homes, is a common site of childhood poisoning. One study also found that ipecac syrup, which can be used to induce vomiting, was less frequently available in these homes.[147]

Although data are limited, there is growing evidence that poisonings among teenagers are nearly as high as among children.[148] The elderly are also at increased risk.[149] Poison prevention efforts should be expanded to target teenagers and older persons as well as young children.

Unlike the dramatic reductions in poisoning by liquids and solids, poisoning by gases and vapors remains a significant problem. Carbon monoxide is a major cause of such poisoning, primarily through motor vehicle exhaust gases; in 1987, 600 deaths among all ages were due to motor vehicle exhaust gases.[3] There seem to be no differences among races for these poisonings, and they occur frequently in rural and low-income areas, a fact possibly related to the prevalence of older vehicles.[76] It is important to note that reporting practices involving suicides (i.e., reporting deaths as unintentional when in fact they are suicides) cloud the picture of who is most at risk for such poisonings.

Lead poisoning entails serious consequences, particularly for children. In the past two decades, knowledge of the effects of lead poisoning has increased substantially. When the Lead-Based Paint Poisoning Prevention Act of 1971 established child-

hood lead poisoning prevention programs, lead encephalopathy and other severe manifestations of lead poisoning were common.[150] Today these outcomes are rare, but childhood lead poisoning has not disappeared. A recent report documents three critical developments. First, long-term effects (particularly neurobehavioral, cognitive, and developmental) are increasingly observed in studies of children with lead levels much lower than were previously believed to be harmful. Second, the number of children exposed to lead at these new lower levels of concern is estimated to be several million. Third, the most common sources of lead in the environment (particularly lead paint in older housing and lead in dust and soil) are difficult and expensive to remove.[151]

Poison prevention packaging. The Poison Prevention Packaging Act of 1970 (PPPA) was intended to protect young children from hazardous substances commonly found in the home by requiring the use of special child-resistant containers. To date, 16 categories of household products have been regulated by PPPA. With certain exceptions, prescription drugs are included among these.[152]

Several studies have demonstrated the effectiveness of childproof packaging in reducing poisoning deaths and hospitalizations among the elderly and young children.[144,153,154] A study of pharmacist compliance with the law, however, found noncompliance 25%–30% of the time.[155] By improving pharmacists' compliance with the PPPA, poisoning should decline further. To effect this, public health agencies should review data for evidence of compliance and make the information available to the public, health professionals, and pharmacists. Information about legal interventions and regulatory actions should also be disseminated.

Using hospitalization as a measure of clinical toxicity, research has shown that overdoses of iron salts, tricyclics, and Lomotil are serious.[156] Even a very small amount of any of these is extremely toxic to children, yet dispensing these medications singly would be very inconvenient for consumers. A double-barrier approach has been suggested for these drugs, that is, dispensing them in opaque plastic strip cells or blister packs and then further enclosing them in childproof containers.[157] The plastic strip material could also be malflavored to further discourage toddlers from consuming it.[154] No new legislation would be required to carry this out. It could be accomplished if manufacturers, pharmacists, physicians, consumers, and regulatory agencies worked together to implement it.

Recommendation: The Poison Prevention Packaging Act has proved effective in reducing poisonings. The federal government should consider expanding the act to include double-barrier packaging for drug products that can lead to acute toxic conditions and strictly enforcing existing legislation.

Poison control centers. Poison control centers are an important community resource, offering information and guidance to the general public and health professionals about first aid and medical management of poisonings.[158] Certified poison control centers operate 24 hours a day, 7 days a week, with toll-free access within the service area. Trained, experienced staff are an essential feature of these centers. Several studies have demonstrated that regional poison centers provide better poison information than nonregional centers[145] and reduce excess emergency room visits for children's poisonings.[159,160] States should consider developing guidelines for the operation of poison control centers to ensure that high-quality services are accessible by telephone, toll free, for all residents.[161]

Recommendation: Regional poison control centers have proved effective in reducing emergency room visits for suspected poisonings. State agencies should designate such centers and monitor their activities.

Community-based prevention education programs. The literature suggests that education is an important component in preventing poisonings. At least two studies have demonstrated that intensive, community-centered education campaigns can be effective in changing parental poison-storage habits.[162,163] Another study showed that education by a pediatrician could improve parents' knowledge of first aid for poisoning.[64] However, another study was not successful in persuading parents to visit pharmacies to obtain ipecac syrup.[70]

The efficacy of mechanical devices, such as cabinet or other safety locks, in preventing poisonings is unknown. In general, education and environmental modification efforts, when incorporated in a multifaceted comprehensive approach to eliminating household hazards and educating parents about the dangers of poisonings, do not cause harm and may be effective. Such efforts should be undertaken only if they can be carried out at very low cost—for example, in connection with comprehensive home hazard inspections.

Recommendation: Educational efforts designed to change the poison storage habits of parents of young children are promising. Such programs should be used and monitored and the results evaluated.

Pictorial stickers to deter young children. Several self-adhesive labels are available and frequently distributed in connection with poison control activities. The stickers are meant to be placed on containers of hazardous materials in the home, and children are to be instructed not to touch anything with a sticker on it. There are several designs; one of the most widely used is "Mr. Yuk," developed by the National Poison Center Network, headquartered in Pittsburgh. Other stickers include the traditional skull and crossbones, now accompanied by a telephone, "Little NO-NO," and "Officer Ugg." Advocates of the stickers use them to raise public awareness about the dangers of poisons and the availability of poison centers.

At least two studies reveal that the stickers are not effective; they do not deter young children from touching, holding, mouthing, or attempting to open containers.[164,165] In fact, it appears children under the age of 3 are attracted to the stickers. One study postulated that the stickers foster a false sense of security among parents and lead them to be less careful about properly storing poisons and using products with safety closures.[165]

These studies may be limited to small sample sizes or out-of-home settings, but the results raise serious questions about the effectiveness of the stickers in deterring poisoning. Note, however, that the stickers should not be confused with stickers designed to be put on telephones giving the telephone number of the poison control center. The latter are an important source of information for parents in a poisoning emergency.

Recommendation: Pictorial stickers designed to deter young children from toxic substances have proven ineffective. Such stickers should not be used as a tool for poison prevention efforts.

First aid for poisonings. A number of studies have found that families with young children do not commonly have ipecac syrup (which causes vomiting) on hand or know or its appropriate use.[24,71,166] Several studies have shown that education about the use of ipecac syrup may be effective.[71,166]

In 1986 ipecac syrup was used in the management of more than 145,000 poison victims reported to the American Association of Poison Control Centers.[167] Studies suggest that the rapid initiation of treatment saves numerous lives and prevents morbidity in victims of unintentional poisoning and is most likely an essential factor in the decline in pediatric mortality over the last 15 years.[168] Ipecac syrup also helps lower health care costs by reducing emergency room care and hospitalization. For example, 1987 data from one poison center indicate that nearly 2,500 patients who contacted the regional poison center were safely treated at home with ipecac syrup.[169]

Because ipecac syrup should be administered only with guidance from a poison control center or a physician, public education about poison control center services and knowledge of the regional center's telephone number are essential complements to having ipecac syrup in the home. At least one study demonstrated that introducing these concepts and handing out free ipecac syrup is feasible in a brief intervention session.[71] AAP also recommends that such education be part of routine pediatric care.[170]

Recommendation: Programs that distribute ipecac syrup and poison control telephone number stickers to families of young children are promising interventions. The programs should also educate the public about correct emergency procedures for a suspected poisoning. The programs should be used and monitored and the results evaluated.

Lead poisoning. Childhood lead poisoning is one of the most common environmental diseases of children in the United States, yet in theory it is entirely preventable. Exposure to lead poisoning is particularly common in the Northeast, where a large percentage of homes were built before 1950. In the Northeast, from 1979 to 1983, lead poisoning accounted for 36% of all childhood poisoning hospitalizations.[146]

Lead poisoning results from complex social, political, and environmental factors and presents a serious health risk to children. It is clear that to ameliorate this problem, a multidisciplinary effort is needed to bring together public and private organizations that have the responsibility to act.

Recommendation: Screening programs to identify children who have lead poisoning and programs to remove sources of lead from the environment are promising interventions. They should be implemented and monitored and the results evaluated.

Suffocation

In 1987, 2,600 deaths were caused by suffocation/asphyxiation/strangulation. Two thousand of these were the result of the ingestion or inhalation of objects or food causing obstruction of respiratory passages; the remaining 600 were deaths by smothering, caused by bedclothes, plastic materials, and so forth, or suffocation by cave-ins, confinements in closed spaces (e.g., refrigerators), and mechanical strangulation.[3]

Children under the age of 4 and the elderly are at highest risk for such deaths. Children under the age of 4 account for 20% of suffocation deaths, the elderly for about 33%. The male/female ratio is about 4 to 1.[76] Death rates are higher in low-income areas. Death rates among blacks are similar to those among whites except for aspirations caused by nonfood materials.

Fatal choking in children typically involves ingestion or aspiration of small, round objects. One study found that four foods—hot dogs, candy, nuts, and grapes—together caused 40% of all specified food-related suffocations.[171] Several federal efforts have addressed children's risk of choking on nonfood items. The 1979 CPSC toy standards set requirements to prevent choking hazards in nonfood products targeted at children younger than 3 years of age.[172]

During 1968–79, death rates from suffocation increased among the elderly, especially from aspirating nonfood materials. About 200 deaths occur each year in residential institutions such as nursing homes and extended-care facilities.[76]

Alcohol and, less commonly, sedative or hypnotic drugs are important contributing factors to choking among the elderly.[76] More research is needed to identify effective choking prevention strategies.

Mechanical suffocation and strangulation deaths occur primarily among children. Suffocation is commonly caused by plastic bags, bedclothes, the plastic sides of playpens and cribs, entrapment in refrigerators and other appliances, or burial under falling earth and other materials.[173] The redesign of refrigerator interiors led to an 80% decrease in deaths in units sold in the period 1978–81 compared with those in 1966–68.[173]

Infant strangulation commonly occurs by hanging from pacifier cords, clothing, and high chair straps; wedging the head between crib slats, in accordion-style safety gates, or between the mattress and bed frame; and catching the head in electrically operated car windows.[94]

Education about choking risks. Education efforts to prevent choking injuries involve two major activities: counseling parents, the elderly, and caregivers about the choking hazards associated with certain objects and teaching the general public first aid management of a choking victim. A tool for health professionals called the "no-choke test tube" can assist in educating parents about the safe size of toys and other small objects for young children. This short cylinder has a diameter of 1.25 inches, the standard set by government regulation of small parts on toys. A toy that fits into the no-choke test tube is not safe for children under the age of 3.

A study of childhood asphyxiation by food commented on the importance of disseminating information on food choking risks, particularly regarding foods that are not packaged (such as grapes or carrots) and therefore cannot be labeled.[171] Product labeling, such as "Not intended to be given to children under four years of age. Fatal choking may result," and product modification (changing the shape or size of the food product where possible) are more likely to be effective in reducing fatal choking.

Much of the existing data on treatment of a choking victim has been anecdotal. AAP has revised its guidelines regarding first aid for the choking victim, and, in concurrence with the American Heart Association and the American Red Cross, recommends that the Heimlich maneuver be used except with infants under 1 year of age. For infants, back blows are the recommended treatment for choking.[174]

Recommendation: The guidelines of the American Academy of Pediatrics regarding management of the choking victim should be followed. To effect this, the Heimlich maneuver should be taught to health professionals and the general public. Parents, the elderly, and caregivers should also be counseled about choking hazards.

INTERVIEW SOURCES

Marie Bond, EdM, Health Educator, Childhood Injury Prevention Program, Boston City Hospital, Massachusetts, November 2, 1988.

M. Teresa Dowling, MPH, Director, Office of Senior Information, Referral and Health Promotion, City and County of San Francisco, California, December 27, 1988.

Robert D. Verhalen, DrPH, Associate Executive Director, Directorate for Epidemiology, Consumer Product Safety Commission, Washington, DC, September 9, 1988.

REFERENCES

1. Canellos PS. Boston Globe 1988 Nov 21:5.

2. Rubenstein LZ, Robbins AS, Schulman BL, Rosado J, Osterweil D, Josephson KR. Falls and instability in the elderly. J Am Geriatr Soc 1988;36:266–78.

3. Accident facts, 1988 edition. Chicago: National Safety Council, 1988.

4. Collins JG. Persons injured and disability days due to injuries: United States, 1980–81. Washington, DC: National Center for Health Statistics, 1985; DHHS publication no. (PHS)85-1577. (Vital and health statistics; series 10; no. 149.)

5. Stone LJ, Church J. Childhood and adolescence. 3rd ed. New York: Random House, 1975.

6. Gallagher SS, Fineson K, Guyer B, Goodenough SS. The incidence of injuries among 87,000 Massachusetts children and adolescents. Am J Public Health 1984;10: 1340–7.

7. Rivara FP. Epidemiology of childhood injuries. In: Bergman AB, ed. Preventing childhood injuries: report of the 12th Ross Roundtable on Critical Approaches to Common Pediatric Problems. Columbus, Ohio: Ross Labs, 1982:13–18.

8. Gallagher SS, Guyer M, Kotelchuck M, et al. A strategy for the reduction of childhood injuries in Massachusetts: SCIPP. N Engl J Med 1982;307:1015–19.

9. Rivara FP, Kamitsuka MD, Quan L. Injuries to children younger than 1 year of age. Pediatrics 1988;81:93–7.

10. Fallis JC. Initial assessment of the injured child. In: Care for the injured child: the hospital for sick children, Toronto, Canada. Baltimore: Williams & Wilkins, 1975: 6–7.

11. Salter RB. Musculoskeletal injuries—part II: Specific features in children. In: Care for the injured child: the hospital for sick children, Toronto, Canada. Baltimore: Williams & Wilkins, 1975:252–64.

12. Noer HR. Address delivered to the Society of Automotive Engineers, Children's Restraints Subcommittee meeting of the Society of Automotive Engineers Motor Vehicle Seat Belt Committee; 1966.

13. Waller J, Klein D. Society, energy and injury—inevitable triad? Research directions toward the reduction of injury in the young and the old [report of a conference]. Washington, DC: U.S. Department of Health, Education, and Welfare, 1971:1–37; HEW publication no. NIH 73-124.

14. Kane DN. Environmental hazards to young children. Phoenix, Arizona: Oryx Press, 1984.

15. Brocklehurst J, Robertson D, James-Groom P. Clinical correlates of sway in old age—sensory modalities. Age Ageing 1982;11:1–10.

16. Wolfson L, Whipple R, Amerman P, Kleinberg A. Stressing the postural response: a quantitative method for testing balance. J Am Geriatr Soc 1986;34:845–50.

17. Riffle KL. Falls: kinds, causes, and prevention. Geriatr Nurs 1982;3:165–9.

18. Rosen A, Campbell R, Villanueva J. Factors affecting falling by older psychiatric patients. Psychosomatics 1985;26:117–23.

19. Rivara FP, Bergman AB, LoGefgo JP, Weiss NS. Epidemiology of childhood injuries. II. Sex differences in injury rates. Am J Dis Child 1982;136:502–6.

20. Nersesian WS, Petit MR, Shaper R, Lemieux D, Naor E. Childhood death and poverty: a study of all childhood deaths in Maine, 1976 to 1980. Pediatrics 1985;75:41–50.

21. Mierley MC, Baker SP. Fatal house fires in an urban population. JAMA 1983;249:1466–8.

22. Westfelt JN. Environmental factors in childhood accidents. Acta Paediatr Scand (suppl) 1982;291:1–75.

23. Rivara FP. Epidemiology of childhood injuries. In: Matarazzo JD, Weiss SM, Herd JA, Miller NE, Weiss SM, eds. Behavioral health: a handbook of health enhancement and disease prevention. New York: Wiley, 1984:1003–20.

24. Gallagher SS, Hunter P, Guyer B. A home injury prevention program for children. Pediatr Clin N Am 1985; 32:95–112.

25. Neutra R, McFarland RA. Accident epidemiology and the design of the residential environment. Hum Factors 1972;14:405–20.

26. O'Connor MA, Boyle WE, O'Connor GT, Letellier R. Safety hazards and injury prevention practices in day care centers. Paper presented at the annual meeting of the American Public Health Association, New Orleans, Louisiana, Oct. 1987.

27. Wasserman RC, Dameron DO, Brozicevic MM, Aronson RA. On-site intervention reduces injury hazards in day care homes. Paper presented at the annual meeting of the Ambulatory Pediatric Association, Anaheim, California, 1987.

28. Davis WS, McCarthy PL. Safety in day care centers [Abstract]. Am J Dis Child 1988;142.

29. Wasserman RC, Dameron DO, Brozicevic MM, Aronson RA. Injury hazards in home day care. Paper presented at the annual meeting of the Ambulatory Pediatric Association, Anaheim, California, 1987.

30. Hogue CC. Injury in late life. I. Epidemiology. J Am Geriatr Soc 1982;30:183–90.

31. Waller J. Injury: conceptual shifts and preventive implications. Annu Rev Public Health 1987;8:21–49.

32. Eckardt MJ, Harford TC, Kaelber CT, et al. Health hazards associated with alcohol consumption. JAMA 1981;246:648–66.

33. Smith GS, Kraus JF. Alcohol and residential, recreational, and occupational injuries: a review of the epidemiologic evidence. Annu Rev Public Health 1988;9:99–121.

34. Waller JA. Nonhighway injury fatalities. I. The roles of alcohol and problem drinking, drugs, and medical impairment. J Chron Dis 1972;25:33–45.

35. Levine MS, Radford EP. Fire victims: Medical outcomes and demographic characteristics. Am J Public Health 1977;67:1077–80.

36. Room R. Alcohol control and public health. Annu Rev Public Health 1984;5:293–317.

37. Injury control for children and youth. Elk Grove Village, Illinois: American Academy of Pediatrics, 1987.

38. Proceedings of the first conference on home accident prevention, University of Michigan School of Public Health, W. K. Kellogg Foundation, in cooperation with the National Safety Council, U.S. Public Health Service, American Public Health Association, Jan 20–22, 1953. Ann Arbor, Michigan: University of Michigan Press, 1953.

39. American Public Health Association. Suggested home accident prevention activities for health departments. Am J Public Health 1956;46:625–30.

40. Rockland County, New York childhood injury prevention project, September 1959–April 1964. Washington, DC: U.S. Department of Health, Education, and Welfare, 1965.

41. The Philadelphia research demonstration project in accident control through small group discussion. Washington, DC: U.S. Department of Health and Human Services, 1964.

42. A project designed to demonstrate and test the community action approach as an educational method within the field of burn injury prevention in a six-county area in the Bootheel section of Missouri, November 1966–October 1969. Washington, DC: U.S. Department of Health, Education, and Welfare, 1972; publication no. (HSM)72-10008.

43. Barancik JI, Shapiro MA. Pittsburgh burn study. Pittsburgh: Environmental Health Program, Department of Public Health Practice, Graduate School of Public Health, University of Pittsburgh, 1972.

44. Feck G, Baptiste MS, Tate CL. Burn injuries: epidemiology and prevention. Accident Anal Prev 1979;11:129–36.

45. Fire in the United States: deaths, injuries, dollar loss and incidents at the national, state, and local level in 1981. 4th ed. Washington, DC: Federal Emergency Management Agency, 1980.

46. American Association of Poison Control Centers. Regionalization criteria. Vet Hum Toxicol 1978;20:117–8.

47. Harrington C, Gallagher S, et al. Injury prevention programs in state health departments: a national survey. Cambridge, Massachusetts: Childhood Injury Prevention Resource Center, Harvard School of Public Health, 1988.

48. Micik S, Miclette M. Injury prevention in the community: a systems approach. Pediatr Clin N Am 1985;32: 251–65.

49. Guyer B, Gallagher SS, Chang B, Azzara CV, Cupples LA, Colton T. Prevention of childhood injuries: evaluation of the Statewide Childhood Injury Prevention Program (SCIPP) (submitted for publication).

50. Statewide Comprehensive Injury Prevention Program. Safe home. Boston: Massachusetts Department of Public Health, 1986.

51. Statewide Comprehensive Injury Prevention Program. Safe child. Boston: Massachusetts Department of Public Health, 1986.

52. Statewide Comprehensive Injury Prevention Program. Safe daycare. Boston: Massachusetts Department of Public Health, 1986.

53. Statewide Comprehensive Injury Prevention Program. Safe school. Boston: Massachusetts Department of Public Health, 1987.

54. Norman R. Health behavior: the implications of research. Health Promotion 1986;25:2–5.

55. Morfitt JM. Falls in old people at home: intrinsic versus environmental factors in causation. Public Health 1983;97:115–20.

56. Waller J. Falls among the elderly: human and environmental factors. Accident Anal Prev 1978;10:21–33.

57. Centers for Disease Control. Basic housing inspection. Washington, DC: U.S. Department of Health, Education, and Welfare, 1976; DHEW publication no. (PHS)80-8315.

58. Commonwealth of Massachusetts. Minimum standards of fitness for human habitation. Chapter II, Massachusetts state sanitary code [105 CMR 410.000–419.000].

59. Pynoos J, Cohen E, Lucas C, et al. Home evaluation resource booklet for the elderly. Los Angeles: UCLA/USC Long Term Care Gerontology Center, 1986.

60. Tideiksaar R. Preventing falls: home hazard checklist to help older patients protect themselves. Geriatrics 1986;41:26.

61. Home safety checklist for older consumers. Washington, DC: U.S. Consumer Product Safety Commission, 1985.

62. Kendrick AS, Kaufman R, Messenger KP. Healthy young children: a manual for programs. Washington, DC: National Association for the Education of Young Children, 1988.

63. Kravitz H, Grove M. Prevention of accidental falls in infancy by counseling mothers. Ill Med J 1973;143:570–3.

64. Dershewitz RA, Williamson JW. Prevention of childhood household injuries: a controlled clinical trial. Am J Public Health 1977;67:1148–52.

65. Dershewitz RA. Will mothers use free safety supplies? Am J Dis Child 1979;133:61–4.

66. Speigel CN, Lindaman FC. Children Can't Fly: a program to prevent childhood morbidity and mortality from window falls. Am J Public Health 1977;67:1143–7.

67. Thomas KA, Hassanein RS, Christopherson ER. Evaluation of group well child care for improving burn prevention practices in the home. Pediatrics 1984;74:879–82.

68. Miller RE, Reisinger KS, Blatter MM, Wucher F. Pediatric counseling and subsequent use of smoke detectors. Am J Public Health 1982;72:392–3.

69. Shaw KN, McCormick MC, Kustra SL, Ruddy RM, Casey RD. Correlates of reported smoke detector usage in an inner-city population: participants in a smoke detector give-away program. Am J Public Health 1988;78:650–3.

70. Alpert J, Heagarty M. Evaluation of a program for distribution of ipecac syrup for the emergency home management of poison ingestions. J Pediatr 1966;69:142–6.

71. Woolf A, Lewander W, Filippone G, Lovejoy F. Prevention of childhood poisoning: efficiency of an educational program carried out in an emergency clinic. Pediatrics 1987;80:359–63.

72. Injury prevention. Elk Grove Village, Illinois: American Academy of Pediatrics, 1985.

73. TIPP: the injury prevention program. Elk Grove Village, Illinois: American Academy of Pediatrics, 1986.

74. Gallagher SS, Messenger KP, Guyer B. State and local responses to children's injuries: the Massachusetts Statewide Childhood Injury Prevention Program. J Soc Issues 1987;43:149–62.

75. Lambert DA, Sattin RW. Deaths from falls, 1978–1984. Morbid Mortal Weekly Rep 1988;37(suppl 1):21–6.

76. Baker SP, O'Neill B, Karpf RS. The injury fact book. Lexington, Massachusetts: Lexington Books, 1984.

77. Injuries in Massachusetts: a status report. Boston: Massachusetts Department of Public Health, 1987.

78. Wellman S, Paulson JA. Baby walker-related injuries. Clin Pediatr 1984;23:98–9.

79. Rivers RPA, Boyd RDS, Baderman H. Falls from equipment as cause of playground injury. Community Health 1978;8:178–9.

80. Bergner L, Mayer S, Harris D. Falls from heights: a childhood epidemic in an urban area. Am J Public Health 1971;61:90–6.

81. Garrettson LK, Gallagher SS. Falls in children and youth. Pediatric Clin N Am 1985;32:153–62.

82. Prudham D, Evans JG. Factors associated with falls in the elderly: a community study. Age Ageing 1981;10:141–6.

83. Droller H. Falls among elderly people living at home. Geriatrics 1955;10:239–44.

84. Perry BC. Falls among the elderly living in high-rise apartments. J Fam Pract 1982;14:1069–73.

85. Wild D, Nayak US, Isaacs B. How dangerous are falls in old people at home? Br Med J 1981;282:266–8.

86. Manning DP. Deaths and injuries caused by slipping, tripping, and falling. Ergonomics 1983;26(3):3–9.

87. Sieben RL, Leavitt JD, French JH. Falls as childhood accidents: an increasing urban risk. Pediatrics 1971;47:886–92.

88. Smith MD, Burrington JD, Woolf AD. Injuries in children sustained in free falls: an analysis of 66 cases. J Trauma 1975;15:987–91.

89. Tinetti ME, Speechley M, Ginter SF. Risk factors for falls among elderly persons living in the community. N Engl J Med 1988;319:1701–7.

90. Ray WA, Griffin MR, Schaffner W, Baugh DK, Melton LJ III. Psychotropic drug use and the risk of hip fracture. N Engl J Med 1987;316:363–9.

91. Ray W, Federspiel C, Schaffner W. A study of antipsychotic drug use in nursing homes: epidemiologic evidence suggesting misuse. Am J Public Health 1980;70:485–91.

92. Koch H. Utilization of psychotropic drugs in office-based ambulatory care. National Ambulatory Medical Care Survey, 1980 and 1981 (advance data from Vital and Health Statistics no. 90). Hyattsville, Maryland: US Public Health Service, 1982; DHHS publication no. (PHS)83-1250.

93. Safety in summer: window guards. New York Times 1984 Jul 14:48.

94. Smith GS, Falk H. Unintentional injuries. Am J Prev Med 1987;3(suppl 5):143–63.

95. Joffee M, Ludwig S. Stairway injuries in children. Pediatrics 1988;82:457–61.

96. Cohn SH, Vaswani A, Zanzi I, Ellis KJ. Effect of aging on bone mass in adult women. Am J Physiol 1976;230:143–8.

97. Goggin JE, Haddon W Jr., et al. Incidence of femoral fractures in postmenopausal women: before and after fluoridation. Public Health Rep 1965;80:1005–12.

98. Smith EL Jr., Reddan W, Smith PE. Physical activity and calcium modalities for bone mineral increase in aged women. Med Sci Sports Exerc 1981;13:60–4.

99. Aloia JF, Cohn SH, Ostuni JA, Cane R, Ellis K. Pre-

vention of involutional bone loss by exercise. Ann Intern Med 1978;89:356–8.

100. White MK, Martin RB, Yeater RA, et al. Effects of exercise on postmenopausal osteoporosis. Paper presented at the 28th annual meeting of the Orthopedic Research Society, Jan 19–21, 1982.

101. Kauffman I, Ridenour M. Influence of an infant walker on onset and quality of walking pattern of locomotion. Percept Mot Skills 1977;45:1323–9.

102. Rieder MJ, Schwartz C, Newman J. Patterns of walker use and walker injury. Pediatrics 1986;78:488–93.

103. Miller R, Colville J, Hughes NC. Burns to infants using walker aids. Injury 1975;7:8–10.

104. Fazen LE III, Felizberto PI. Baby walker injuries. Pediatrics 1982;70:106–9.

105. Kavanagh C, Banco L. The infant walker: a previously unrecognized health hazard. Am J Dis Child 1982;136:205–6.

106. Wellman S, Paulson JA. Baby walker-related injuries. Clin Pediatr 1984;23:98–9.

107. Stoffman J, Bass M, Fox AM. Head injuries related to the use of baby walkers. Can Med Assoc J 1984;131:573–5.

108. Berger L. Childhood injuries: recognition and prevention. Curr Probl Pediatr 1981;12:1–59.

109. Stewart AR. Head injuries and baby walkers. Can Med Assoc J 1984;131:1327.

110. McLoughlin E. The cigarette safety act. J Public Health Pol 1982;3:226–8.

111. Weisman HM. Arson resource directory. Washington, DC: Federal Emergency Management Agency, 1982.

112. Karter MJ Jr. Fire loss in the United States during 1982. Fire J 1983;77:44–60.

113. Beverly EV. Reducing fire and burn hazards among the elderly. Geriatrics 1976;31:106–10.

114. Rossignol AM, Locke JA, Boyle CM, Burke JF. Consumer products and hospitalized burn injuries among elderly Massachusetts residents. J Am Geriatr Soc 1985;33:768–72.

115. McLoughlin E, Clarke N, Stahl K, et al. One pediatric burn unit's experience with sleepwear-related injuries. Pediatrics 1977;60:405–9.

116. Yanofsky NN, Morain WD. Upper extremity burns from woodstoves. Pediatrics 1984;73:722–6.

117. Lewis PJ, Zuker RM. Childhood scald burns: an inquiry into severity. J Burn Care Rehabil 1982;3:95–7.

118. McLoughlin E, Crawford JD. Burns. Pediatr Clin N Am 1985;32:61–75.

119. Crikelair GF, Dhaliwal AS. The cause and prevention of electrical burns of the mouth in children. Plast Reconstr Surg 1976;58:206–9.

120. McLoughlin E, Joseph MA, Crawford JD. Epidemiology of high-tension electrical injuries in children. J Pediatr 1976;89:62–5.

121. Gorman RL, Charney E, Holtzman NA, Roberts KB.

A successful city-wide smoke detector giveaway program. Pediatrics 1985;75:14–18.

122. O'Connor MA. New Hampshire Home Injury Prevention Project: interim report. Submitted to the New England Network to Prevent Childhood Injuries; 1987 Sep 10.

123. An evaluation of residential smoke detector performance under actual field conditions: final report. Washington, DC: Federal Emergency Management Agency, 1980.

124. Senate Special Committee on Aging. Home fire deaths: a preventable tragedy. Hearing before the Special Committee on Aging. Washington, DC: US Government Printing Office, 1983 (98th Congress, 1st session).

125. McLoughlin E. Smoke detector legislation: its effect on owner-occupied homes [Dissertation]. Baltimore, Maryland: Johns Hopkins University, 1984:858–62.

126. An ounce of prevention. Washington, DC: Federal Emergency Management Agency, 1983.

127. Moyer C. The sociological aspects of trauma. Am J Surg 1954;87:421–30.

128. Mierley MC, Baker S. Fatal house fires in an urban population. JAMA 1983;249:1466–8.

129. Model standards: a guide for community preventive health services. Washington, DC: American Public Health Association, 1985.

130. Mood EW. Housing and health: APHA-CDC recommended minimum housing standards. Washington, DC: American Public Health Association, 1986.

131. Fire in the United States: deaths, injuries, dollar loss and incidents at the national, state, and local level in 1981. 4th ed. Washington, DC: Federal Emergency Management Agency, 1980.

132. O'Malley B. Cigarettes and sofas: how the tobacco lobby keeps house fires burning. Mother Jones 1979;56–63.

133. Maguire A. There's death on the block, there's hope in Congress. J Public Health Policy 1987;8:451–4.

134. McLoughlin E, Vince CJ, Lee AM, Crawford JD. Project Burn Prevention: outcome and implications. Am J Public Health 1982;72:241–7.

135. Residential smoke and fire detector coverage in the United States: results from a 1982 survey. Washington, DC: Federal Emergency Management Agency, 1983.

136. Nicholls CA. Accidents and injuries involving scald burns from tap water sources [contained in US Consumer Product Safety Commission memorandum of 1978 Oct 20 staff briefing packet on petition CP 78-15, tap water scalds]. Washington, DC: U.S. Consumer Product Safety Commission, 1978.

137. Katcher ML. Scald burns from hot tap water. JAMA 1981;246:1219–22.

138. Feldman KW, Schaller RT, Feldman JA, McMillon M. Tap water scald burns in children. Pediatrics 1978;62:1–7.

139. Katcher ML. Prevention of tap water scald burns: evaluation of a multi-media injury control program. Am J Public Health 1987;77:1195–7.

140. Katcher ML, Landry GL, Shapiro MM. Use of a liquid crystal thermometer in pediatric counseling about tap water burn prevention. Pediatrics (in press).

141. Walsh B. Technical background information for appliance efficiency, targets of water heaters (draft report). Washington, DC: Federal Emergency Administration, 1976.

142. Linneman P. The effects of consumer safety standards: The 1973 mattress flammability standard. J Law Econ 1980;23:461–79.

143. Centers for Disease Control. Update: childhood poisonings–United States. Morbid Mortal Weekly Rep 1985;34(9):26–7.

144. Clarke A, Walton WW. Effect of safety packaging on aspirin ingestion by children. Pediatrics 1979;63:687–93.

145. Thompson DF, Trammel HL, Robertson NJ, Routt Reigart J. Evaluation of regional and non-regional poison centers. N Engl J Med 1983;308:191–4.

146. Rodriguez JG, Sattin RW. Epidemiology of childhood poisonings leading to hospitalization in the United States, 1979–1983. Am J Prev Med 1987;3:164–70.

147. Polakoff JM, Lacouture PG, Lovejoy FH. The environment away from home as a source of potential poisoning. Am J Dis Child 1984;138:1014–7.

148. Trinkoff AM, Baker SP. Poisoning hospitalizations and deaths from solids and liquids among children and teenagers. Am J Public Health 1986;76:657–60.

149. Woolf A, Helsing K, Azzara C, Gallagher SS, Fish S, Lovejoy F. Poisoning hospitalizations in Massachusetts: a population based study of product type and injury severity. Paper presented at the annual meeting of the American Public Health Association, New Orleans, Louisiana, 1987.

150. Centers for Disease Control. Childhood lead poisoning—United States: report to the Congress by the agency for toxic substances and disease registry. Morbid Mortal Weekly Rep 1988;37(2):481–5.

151. Agency for Toxic Substances and Disease Registry. The nature and extent of lead poisoning in children in the United States: a report to Congress. Atlanta, Georgia: U.S. Department of Health and Human Services, 1988, DHHS document no. 99-2966.

152. Poison Prevention Packaging Act of 1970. Regulations, 16 CFR, 1700–1704.10, 1985.

153. Walton W. An evaluation of the Poison Prevention Packaging Act. Pediatrics 1982;69:363–70.

154. Done AK, et al. Evaluation of safety packaging for the protection of children. Pediatrics 1971;48:613–28.

155. Dole EJ, Czajka PA, Rivara FP. Evaluation of pharmacists' compliance with the Poison Prevention Act. Am J Public Health 1986;76:1335–6.

156. Palmisano PA. Hospitalization as a measure of clinical toxicity. Clin Toxicol 1980;16:377–80.

157. Palmisano PA. Targeted intervention in the control of accidental drug overdoses by children. Public Health Rep 1981;96:150–6.

158. Chafee-Bahamon C, Caplan DL, Lovejoy FH. Patterns of hospitals' use of a regional poison information center. Am J Public Health 1983;73:396–400.

159. Chafee-Bahamon C, Lovejoy FH. Effectiveness of a regional poison center in reducing excess emergency room visits for children's poisonings. Pediatrics 1983;72:164–9.

160. Marcus SM, Chafee-Bahamon C, Arnold VW, Lovejoy FH. A regional poison control system. Am J Dis Child 1984;138:1010–3.

161. Lovejoy FH, Caplan DL, Rowland T, Fazen L. A statewide plan for the poisoned patient: the Massachusetts poison control system. N Engl J Med 1979;300:363–5.

162. Maisel G, Langdoe B, Jenkins M, et al. Analysis of two surveys evaluating a project to reduce accidental poisoning among children. Public Health Rep 1967;82:555–60.

163. Fisher L, Van Buren J, Nitzkin J, et al. Highlight results of the Genesee regional poison prevention demonstration project. Vet Human Toxicol 1982;24(suppl):112–7.

164. Fergusson DM, Horwood LJ, Beautrais AL, Shannon FT. A controlled field trial of a poisoning prevention method. Pediatrics 1982;69:515–20.

165. Vernberg K, Culver-Dickinson P, Spyker DA. The deterrent effect of poison-warning stickers. Am J Dis Child 1984;138:1018–20.

166. Dershewitz R, Posner M, Paichel W. The effectiveness of health education on home use of ipecac. Clin Pediatr 1983;22:268–70.

167. Litovitz TL, Martin TG, Schmitz B. 1986 Annual report of the American Association of Poison Control Centers National Data Collection System. Am J Emerg Med 1987;5:405–45.

168. Centers for Disease Control. National poison prevention week: 25th anniversary observance. Morbid Mortal Weekly Rep 1986;35:149–52.

169. Litovitz T. In defense of retaining ipecac syrup as an over-the-counter drug. Pediatrics 1988;82:514–6.

170. Guidelines for health supervision. Elk Grove Village, Illinois: American Academy of Pediatrics, 1985:31.

171. Harris CS, Baker SP, Smith GA, Harris RM. Childhood asphyxiation by food. JAMA 1984;251:2231–5.

172. U.S. Consumer Product Safety Commission. Method for identifying toys and other articles intended for use by children under 3 years of age which present choking, aspiration, or ingestion hazards because of small parts. Washington, DC: U.S. General Services Administration, 1979.

173. Kraus JF. Effectiveness of measures to prevent unintentional deaths of infants and children from suffocation and strangulation. Public Health Rep 1985;100:231–40.

174. American Academy of Pediatrics. Revised first aid for the choking child. Pediatrics 1986;78:177–8.

Chapter 8: Recreational Injuries

Recreational activities such as swimming, hiking, tennis, and softball occupy an important place in the lives of many Americans. Children, especially, spend a great deal of time playing and participating in school- and community-sponsored sports and recreation programs. Because these pursuits are seen as healthy "fun," they are not thought of as dangerous. Yet they can be.

Many persons who are (or should be) concerned with recreational injuries have neither the epidemiological sophistication nor the programmatic expertise to recognize injury patterns or identify appropriate interventions. Public health practitioners can bring injury problems to the attention of such people and work with them to implement injury prevention and control programs. Potential collaborators include public school physical education teachers, persons in charge of municipal and county intramural athletic programs, people in parks and recreation departments, and legislators with an interest in public health and safety.

PROBLEMS WITH THE DATA

The data on recreational injuries are woefully inadequate. The United States Consumer Product Safety Commission's National Electronic Injury Surveillance System (NEISS) includes product-associated injuries affiliated with some of the more popular recreational activities, but the information it collects has limitations. One study concluded that NEISS is capable of identifying only 48% of all sports injuries.[1]

Other sources of data on recreational injuries also have substantial limitations. It has been estimated that only 33% of sports injuries can be identified by using E codes.[1,2] The National Safety Council's *Accident Facts* includes data on recreational injuries, but the council cautions that because its information "is not complete and the number of participants varies greatly, no inference should be made concerning the relative hazard of these sports or rank with respect to injury."[3]

There is little information about some recreational activities that would appear to be hazardous. A review of the data on skydiving, ballooning, gliding, and hang gliding found that "only a handful of reports could be located which document injury oc-

currences in persons who participate in those forms of recreational activities."[4] Others, such as skateboarding, have not been studied systematically enough to present a national picture.[5,6]

Exposure rates are very difficult to calculate, especially for unorganized or "informal" play, like "pickup" basketball games or surfing. Without exposure rates, it is difficult to compare the dangers of different activities.[1,2]

FOCUS OF THE CHAPTER

It is difficult to provide a systematic overview of recreational injuries because of the diversity of the activities. This chapter includes only recreational activities in which the number and/or severity of injuries, availability of interventions, and jurisdictional concern about safety make preventive efforts worthwhile. One of these categories, drownings, is an important cause of mortality. Others, such as playground and competitive sports injuries, result in large numbers of nonfatal injuries. Some activities, such as rock climbing and hang gliding, are not included because of the relatively small number of people who participate in these activities and the lack of data about them.

DROWNING

The Magnitude of the Problem

From 1978 to 1984, there were an average of 6,503 drownings annually in the United States.[7] Drowning, by definition, is fatal. In cases of "near-drownings," an individual is underwater long enough to suffer the consequences of oxygen deprivation, which can include brain damage. The medical, personal, and financial consequences of near-drownings can be devastating.[8-10] The number of potential drownings in which persons are rescued without serious medical consequences is unknown but believed to be substantial.[11,12]

Risk Groups and Injury Patterns

Males drown at a rate four times that of females.[7,13-17] Drowning rates are highest for children under 5 years of age and those between the ages of 15 and 24.[7,18] As with many injuries, drowning disproportionately affects those with low incomes.[19] Rates are highest among Native Americans. This disproportionate rate may be due to high rates of alcohol use, especially among adolescents.[20,21] It may also reflect the fact that substantial numbers of Native Americans live in Alaska, which has the highest drowning rate of any state. Alaska's high drowning rate has been attributed to the swiftness of its rivers, the frigid temperatures of its natural bodies of water, and the fact that many of its citizens are employed in jobs, such as fishing, that entail exposure to water.[19]

The rate for blacks is also disproportionate—twice that of whites.[7,13,14,18,22] However, white children between the ages of 1 and 4 have twice the drowning rate of blacks of this age, due largely to residential swimming pool drownings.[19] The other exception to the relationship between income and drowning is boat-related drownings.[19]

Forty to forty-five percent of drownings occur during swimming,[14,18,23] and 12%–29% are boat related.[7,14,18] As might be expected, more than half occur in June, July, and August, nearly half of these on weekends.[7,14,18,22] Outside Alaska, drowning rates are highest in the southern and western states.[19] Alcohol plays a substantial role.

Between one-half and three-quarters of drownings occur in lakes, ponds, rivers, and the ocean.[13,14,18,23] However, a very large percentage of drownings of young children occurs in residential swimming pools.[22,24,25] This rate varies by region. In Los Angeles, for example, half of all drownings (and 89% of the drownings of 2- to 3-year-olds) take place in residential pools. This rate is similar to that of other areas where pools are prevalent but many times the rate in areas where they are not.[16] The residential pool drowning rate for children under the age of 5 is increasing at roughly the same rate as the number of pools.[26] According to the U.S. Consumer Product Safety Commission (USCPSC): "The annual death rate to children under five years of age associated with this single consumer product is among the highest that has ever been considered by the commission."[25]

Children under the age of 5 do not understand the consequences of falling into deep water,[27] and they usually do not call out for help.[25,27] A majority of victims drown during lapses in adult supervision caused by chores, socializing, or phone calls.[17]

The danger of a child's drowning also increases with the number of children present because of the difficulty in supervising several children at once. A lapse in adult supervision does not have to be long. Forty-six percent of young pool drowning victims in

a USCPSC study were missing for 5 minutes or less. Nor do lapses of supervision always have to occur in the immediate area of the pool. The USCPSC found that almost half of the young children who drowned in backyard pools had last been observed in the house.[27]

One major study found that there was no physical barrier between the house and the pool in 72% of residential drownings.[17] In Sacramento County, California, 80% of young victims gained access to a pool directly from the yard or house. Sacramento's pool fencing law does not require that the pool itself be fenced, only the property on which the pool is located. Thus a child in the yard or in the house may have completely unimpeded access to the pool.[22] A study in Florida found that although 97% of the pools in which children drowned were fenced, 70% of the fences had gates that had been open at the time of the drowning.[28]

Residential spas, hot tubs, and Jacuzzis also can be hazardous. According to NEISS data, a yearly average of 1,788 injuries occurs in the residential use of these products.[29] Although the numbers of these injuries have been limited, it is likely that they will increase with the number of residential hot tubs and Jacuzzis. Because many of these tubs remain filled with water when they are not in use, they, like swimming pools, present a hazard to young children.[30]

Cases have been documented in which children were trapped by the vacuum created by a hot tub's drain. Entanglement of hair in drains also presents a drowning hazard.[31] Adults face different hazards. Because hot water can promote drowsiness, prolonged use or use in combination with alcohol or drugs (including some medications) can lead to hyperthermia-induced stupor and drowning.[30]

Interventions

Barriers for swimming pools. A USCPSC study of childhood pool drownings concluded that a barrier to keep the child out of a pool area would have prevented 70% of the incidents investigated.[25] Another study comparing a town that required that pools be separated from both the yard and the house by a fence with neighboring towns that had no such ordinance estimated that pool drownings and near-drownings could be cut in half by using such fences.[32] The efficacy of proper fencing has also been documented by a number of other studies.[24,33,34]

Research indicates that the most effective barriers are those that

• Restrict entry to the pool from yard and residence

• Use self-closing and self-latching gates with the latch placed at a height unreachable by small children

• Do not restrict the view of the pool (so that children within the fence are visible from outside and

• Are at least 5 feet high and have no vertical openings more than 4 in wide[27,35,36]

Effective barriers should supplement, not replace, adult supervision of children around swimming pools.

Recommendation: Ordinances requiring appropriate fencing have been demonstrated to be effective in reducing drowning rates associated with residential swimming pools. Municipalities and counties should enact, enforce, and monitor such ordinances.

Adult supervision of young children. Adult supervision of young children around water is the best way to prevent drownings.[35] Parents and other caregivers should be made aware of the importance of supervision, the characteristics of child drownings, basic rescue techniques, and cardiopulmonary resuscitation (CPR). The USCPSC warns that flotation devices and/or swimming lessons are not adequate substitutes for adult supervision.[36] Similarly, supervision is not a substitute for the use of effective swimming pool barriers but an adjunct to them.

Recommendation: Research has demonstrated that lapses in adult supervision contribute to the drowning of young children. Programs promoting the importance of such supervision should be used and monitored and their outcomes evaluated.

Life preservers, telephones, and resuscitation. Research indicates that the risk of drowning increases with the distance of the telephone from the pool. Leaving the immediate pool area to use the phone can result in leaving children unattended and increases the time necessary to obtain emergency medical assistance.[17] Very often no resuscitation attempts are made until emergency medical service personnel arrive.[22,37]

Recommendation: Research has indicated that the presence of life preservers, telephones, and immediate resuscitation can limit the medical consequences of immersion incidents. Programs that promote or require life preservers, ropes, and a telephone with the local emergency medical services number at swimming pools and controlled waterfront areas and the training of pool owners and waterfront personnel in CPR are a promising approach to the prevention and control of drownings. Such programs should be used and monitored and their outcomes evaluated.

Pool covers. Although the extent of the problem is not fully known, research indicates that children can become trapped under pool covers and drown and that very young children can drown in the puddles of rain water that form on the surface of these covers. It has been speculated that some small children attempt to walk on these covers, believing they are a solid surface, fall through, and drown.[17] More research is needed to determine the extent of these hazards and the types of design modifications that could affect them. Standards for swimming pool covers are being developed by the American Society for Testing Materials and should be available in 1989.[38] Until such design modifications and standards are available, these procedures should be followed when using pool covers.

• Pool covers should always be completely removed before swimming.

• Pools should never be left partially covered.

• Puddles of water should never be allowed to collect on pool covers.

Recommendation: Strategies to promote the proper use of pool covers should be used and monitored and their outcomes evaluated.

Drownproofing. Swimming lessons for very young children as well as lessons allegedly teaching infants and young children how to save themselves in cases of unintentional immersion (sometimes promoted as "drownproofing") have received a great deal of publicity in recent years. The effects remain controversial. Such training may actually increase a child's risk because of increased exposure to water, decreased fear, and the complacency of parents. Research indicates that such training can lead to other health problems for infants.[39] The American Academy of Pediatrics (AAP) maintains

Infants and toddlers usually are not coordinated enough to swim and breathe at the same time, so they cannot really be taught to swim effectively. Risks of attempting to teach children in this age group to swim include hypervolemic seizures, hypothermia, a false sense of security around water, and, of course, pool contamination from incontinence.[40]

The AAP counsels against organized swimming instruction for children under the age of 5 years and suggests that instruction not begin until age 4 or 5. It also warns that "even if young children know how to swim . . . they should not be considered water-safe. They may not retain their swimming skills in an emergency."[40] This position is supported by a number of other professional organizations and water safety specialists.[26,27,41]

Recommendation: "Drownproofing" has not been

demonstrated to be effective and may be hazardous. "Drownproofing" and swimming lessons for children under 4 years of age are not recommended.

Swimming lessons. Surprisingly little is known about the swimming capability of drowning victims or how they acquired their capability. Given the controversy surrounding the effects of other formal training designed to prevent injuries (such as driver education), it would be valuable to ascertain whether swimming skills are actually beneficial in the typical drowning incident and if formal swimming lessons by trained instructors have safety benefits over informal instruction by parents or peers.

Recommendation: Research should be conducted on the effects of both formal and informal swimming lessons on drowning.

Hot tubs, spas, and Jacuzzis. The Centers for Disease Control (CDC) and the National Pool and Spa Institute have issued guidelines for the safe design, construction, use, and maintenance of both commercial and residential hot tubs, spas, and Jacuzzis.[31,42-44]

Important considerations for the safe use of these products are as follows:
- Residential hot tubs, spas, and Jacuzzis should have barriers around them.
- Children in or around these products should be supervised by an adult.
- The maximum water temperature allowable is 104°F.
- Drains should be equipped with raised safety grates.
- People under the influence of alcohol, drugs, or certain medications should not use these products.
- Pregnant women should not use these products.

Recommendation: Municipalities and counties should enact and enforce minimum safety standard ordinances for both commercial and residential hot tubs, spas, and Jacuzzis based on CDC and National Pool and Spa Institute guidelines. The enforcement of such ordinances should be monitored and their outcomes evaluated.

AQUATIC SPINAL CORD INJURIES

The Magnitude of the Problem

Approximately 1,000 diving-related spinal cord injuries occur each year.[45] Diving accounts for approximately 10% of all spinal cord injuries and 60%–65% of all recreational spinal cord injuries.[27,46,47] Most result in permanent paralysis.[48]

Risk Groups and Injury Patterns

Most victims of diving injuries are adolescent and young adult males.[46,49] The typical aquatic spinal cord injury occurs when a diver's head strikes the bottom or side of a shallow body of water.[50] Many of these injuries take place in swimming pools,[24,49] most resulting from colliding either with the bottom of the deep end or with the section that slopes toward the shallow end of the pool, although some occur when people dive into the shallow end. Because diving is often learned informally from peers or parents, many divers are not taught to "steer up" to avoid collisions with the bottom.[49]

Diving injuries also occur in lakes, ponds, and rivers where murky water conceals the depth of the water, rocks, and other hazardous obstructions. Alcohol also plays a substantial role in aquatic spinal cord injuries. Although most estimates of the percentage of alcohol-related diving injuries are between 40% and 50%, some are as high as 80%.[24,46,49]

Little is known about the effects of warning signs or barricades in unsupervised areas where water depth or obstructions make diving hazardous. Any programs using these approaches should carefully evaluate their effects.

Interventions

Comprehensive spinal cord injury prevention programs. Following a Fourth of July weekend in which seven teenagers were rendered quadriplegic by diving injuries, the West Florida Regional Medical Center developed the "Feet First First Time" program, which encourages people to check water depth and bottom conditions before diving. This program is one component of Florida's Comprehensive Spinal Cord Injury Prevention Program, which, along with a spinal cord injury prevention program developed at the University of Missouri School of Medicine, became the basis for the National Head and Spinal Cord Injury Prevention Program of the American Association of Neurological Surgeons/Congress of Neurological Surgeons. This program has three components:
- Basic education for high school students on spinal cord injuries
- Student and media projects to reinforce this message in the community
- Promotion of legislative measures to prevent head and spinal cord injuries (such as mandatory safety belt and motorcycle helmet laws)

Both programs have specific content directed at

diving injuries. A guide for the implementation of such a program is available.[51] A comprehensive spinal cord injury prevention program including components for younger children is being developed by the Injury Prevention Research Center at the University of Alabama.

Recommendation: Comprehensive spinal cord injury prevention programs may be a promising approach for the prevention of diving injuries and other spinal cord injuries. Such programs should be used and monitored and their outcomes evaluated.

BOATING INJURIES

The Magnitude of the Problem

In 1987 there were 1,036 recreational boating fatalities and 3,501 reported recreational boating injuries. A reportable boating incident is one in which there is a loss of life, personal injury requiring medical attention beyond first aid, property damage exceeding $200, or the complete loss of a vessel. The U.S. Coast Guard estimates that only 10% of nonfatal incidents are reported.

Risk Groups and Injury Patterns

Coast Guard data indicate that most boating injuries occur in collisions between boats, but most fatalities result from a person falling overboard and drowning. As with swimming, a disproportionate number of recreational boating injuries occur on weekends.[52] One report estimated that between one-third and two-thirds of incidents resulting in a fatality may involve alcohol.[53]

Interventions

Personal flotation devices. Personal flotation devices (PFDs) are more commonly known as "life vests." Coast Guard research indicates that many drownings could be prevented if people wore PFDs while boating.[54–56] The Coast Guard has approved a number of types of PFDs based on their flotation ability.

• Type I: Will float people face up even if unconscious and in rough water. Bulky and cumbersome. Recommended for use in open water and ocean.

• Type II: May not always float unconscious people face up. Uncomfortable. Not suitable for rough water.

• Type III: Requires the active participation of the wearer to remain in an upright position. Comfortable. Recommended for use in inland water and activities such as small-boat sailing and water skiing.

• Type IV: "Life preservers" designed to be thrown, not worn.

• Type V: "Specials" designed for specific applications such as board sailing and commercial whitewater rafting.

• Type VI: "Hybrids" that are comfortable but need to be inflated for full flotation.

Recommendation: Appropriate use of personal flotation devices (PFDs) effectively reduces drownings associated with recreational boating. States should require their use. The enforcement of such efforts should be monitored and their outcomes evaluated.

Licensing. At present, few states require operating licenses for recreational boats. Maryland and Puerto Rico have recently begun to require a boating education course for those born after July 1, 1972, who wish to operate a motorboat. Several other states similarly require boating education for persons under 16.[57] Many states offer but do not require boating operating safety courses, usually taught by members of the Coast Guard auxiliary, marine police or volunteers.[58]

Recommendation: Researchers do not know how licensing would affect boating injuries, but it is clear that it has some safety benefits. States should mandate operator competency requirements for recreational boating based on boat size and engine power. Their enforcement should be monitored and their outcomes evaluated.

Alcohol. Most state "drunk boating" laws fail to specify a blood alcohol content (BAC) level. That boating lacks specific indicators of operating under the influence (such as weaving or crossing marked lanes) makes enforcement difficult. Specifying a BAC level would aid enforcement of these laws.

Recommendation: State governments should enact and strictly enforce regulations concerning operating recreational boats under the influence of alcohol and the presence of alcohol aboard recreational boats. A .05 BAC level should be taken as evidence of impaired operator capacity and lead to heavy fines and revocation of operating licenses. These regulations should be enforced and monitored and their outcomes evaluated.

PLAYGROUND INJURIES

The Magnitude of the Problem

NEISS reports that there are about 200,000 injuries involving playgrounds or playground equipment each year, about a quarter of which occur at schools and another quarter at other public playgrounds.[59] Forty-one percent involve residential playground equipment.[60] Approximately 61,000 preschool children are injured in this way every year. A third of these preschoolers are injured in day-care settings.[59] Most playground injuries are minor, but one study found that about one-quarter are severe, involving concussions, crush wounds, fractures, or multiple injuries.[61] The number of deaths was small, most of which were the result of head injuries.[62]

Risk Groups and Injury Patterns

Seventy-two percent of playground-related emergency room visits involved falls. Eighty-three percent of these were falls to the surface of the playground, usually from monkey bars and other products designed to be climbed or from seesaws and swings. Injuries sustained from falls to paved surfaces were double the rate of those to all other surfaces.[62]

Falling and striking playground equipment was the second most common cause of playground injuries.[62] Other causes include colliding with equipment while running and being struck by moving equipment such as swings and merry-go-rounds.[63] The number of injuries associated with each type of equipment reflects the proportion of each type of equipment on playgrounds.[62]

Interventions

Playground surfacing. Most playground injuries, especially the serious ones, involve falls to the surface of the playground. The medical consequences of the falls reflect both the height from which the child falls and the surface on which the child lands. The crucial factor in controlling these injuries is the surface material. The USCPSC maintains that surfaces such as asphalt and concrete "do not provide injury protection from accidental fall impact and are therefore unsuitable for use under public playground equipment." Studies by the USCPSC, among others, indicate that certain materials, including sand and some artificial materials, if properly installed and maintained, can reduce the severity of many playground injuries.[61,62,64] The USCPSC, the American Society of Testing and Materials (ASTM), and the American Association for Leisure and Recreation (AALR) have published comprehensive guidelines for playground surfacing.[63,65-67] The ASTM guidelines on playground surfaces will be available in early 1989.

Recommendation: Research has demonstrated that some surface materials produce more, and more se-

rious, injuries than others. States and municipalities should prohibit the use of concrete and asphalt surfaces in playgrounds and set standards for playground surfaces based on USCPSC and ASTM guidelines. Compliance with and enforcement of such standards should be monitored.

Playground equipment. Many elements should be considered in the design, construction, and maintenance of playgrounds and the selection, installation, and maintenance of residential playground equipment. These include selecting equipment without protruding bolts or sharp edges, locating equipment to minimize the potential of collisions during use, and proper maintenance. The USCPSC, ASTM, and AALR all offer recommendations for equipment construction, maintenance, and placement.[63,65-67]

Recommendation: Research indicates that equipment design, installation, and condition can affect the injury potential of a playground. USCPSC, ASTM, or AALR guidelines should be followed for the choice, installation, and maintenance of public and residential playground equipment. Efforts to promote or mandate compliance with these standards should be monitored and their outcomes evaluated.

Education. A pilot project in New York State conducted a series of workshops for individuals involved in the purchase, installation, maintenance, and supervision of public playgrounds. An evaluation of one of these workshops concluded that it had reduced playground equipment hazards by 42% and emergency room–treated playground injuries by 22.4%.[64]

Recommendation: Seminars on playground safety for individuals with authority over public and school playgrounds are a promising approach and should be used and monitored and their outcomes evaluated.

NONPOWDER FIREARM INJURIES

Nonpowder firearms use a spring mechanism, air pressure, or a gas cartridge to fire small steel balls (BBs) or lead pellets. These products can and do cause serious injuries. One report states that "readily available nonpowder firearms produce velocities well above those required to penetrate eyes, skin, and thin bone."[68]

The Magnitude of the Problem

More than 21,000 injuries associated with nonpowder firearms occurred each year from 1981 to

1986.[68] The typical serious injury involves the eye, although cases have been reported in which BBs have penetrated to the brain and heart.[68,69] Forty percent of BB gun eye injuries treated at one hospital caused legal blindness.[70] A study concluded that "nine out of ten children who suffer penetrating eye injuries caused by a BB will lose their eye."[70]

Seven percent of nonpowder firearm injuries require hospitalization, compared to an average of 4%–5% for all the consumer products about which NEISS collects information.[71] The morbidity rate for nonpowder firearms in the 5- to 14-year-old age group is 35.3/100,000, while the morbidity rate for powder firearms in the same age group is 11.21/100,000.[68] However, powder firearm injuries are usually more serious and more often fatal.

Risk Groups and Injury Patterns

A USCPSC study estimated that 90% of nonpowder firearm injuries were unintentional. More than 60% were to children in the 5- to 14-year-old age group, with the 15- to 24-year-old age group accounting for another 20%.[72]

Males are at far greater risk than females,[71] probably because boys play with these guns more than girls. Most of the injuries involve children unintentionally shooting a playmate or themselves while pumping up a gun or a ricochet from a target.[69]

Fifty-eight percent of the nonpowder guns associated with injuries are multiple-pump air guns.[71] USCPSC investigations found that parents do not think of nonpowder firearms as capable of killing.[69] However, contemporary multiple-pump nonpowder guns are much more powerful than the single-pump models common when today's parents were children.[69,73]

Interventions

Although a few states require permits to own nonpowder firearms and several restrict their sale to persons 18 years of age and over, most states show little interest in regulating these items.[68,74] There is no federal regulation of any of these products.[68]

A USCPSC study concluded that "Most nonpowder gun related injuries occurred when the gun was apparently performing in the way for which they were designed."[71] Product modifications may be possible that would, for instance, prevent the firing of the weapon while it is being pumped. However, there is no substantial literature on designing a "safer" nonpowder firearm. It has been suggested that the use of such guns be restricted to target ranges under adult supervision.[69] However,

use patterns imply that parents buy the products to provide the child with a gun that can be used without supervision.[71]

Prohibition. Nonpowder firearms are purely recreational products primarily used by children. They are not used for hunting, serious recreational target shooting, law enforcement, or defense of the home. The injury threat presented by these products, especially the threat to vision, is serious enough and the benefits of these products trivial enough to warrant their prohibition.

Recommendation: States should prohibit the sale, manufacture, and ownership of nonpowder firearms. Enforcement of such prohibitions should be monitored and their outcomes evaluated.

AMATEUR COMPETITIVE SPORTS

Competitive sports are diverse activities that range from automobile racing to backyard baseball. A discussion of the injury problem and safety and conditioning needs for every competitive sport would require a multivolume work. Even within a single competitive sport, the differences in levels of competition, player experience, supervision, and equipment affect the potential for both injuries and intervention.

Interest in injury control and reduction of competitive sports injuries is often predicated on jurisdiction rather than epidemiology. A high school football coach will be interested in football injuries not because these injuries affect his charges more seriously than traffic injuries do, but because he is responsible for the safety of the sports program.

The Magnitude of the Problem

A 1981 USCPSC study found reports of 105 deaths to those between the ages of 5 and 14 who participated in 15 of the most popular sports.[75] In this age group more fatalities are associated with baseball than with other sports. There were 51 baseball deaths in this age group from 1973–83. Thirty-five of these players were struck by a ball and 10 by a bat.[76] Although the head is most vulnerable to serious injury from a thrown or batted ball, an impact to the chest, especially in a child, can cause cardiac arrest and death.[75]

Football also has its share of fatalities. Football has resulted in 921 direct deaths and about 400 indirect deaths (primarily from heatstroke and overexertion) among all age groups between 1931 and 1987.[77]

The chances of being killed or disabled in amateur competitive sports are relatively small.[78–80] Although catastrophic and disabling injuries do occur in competitive sports, the most prevalent injuries are strains, concussions, and abrasions.[75] Only 2% of sports injuries result in hospitalization.[75] Less serious injuries are far more numerous. The USCPSC reported that there were 5 million medically treated injuries associated with 15 of the most popular sports during 1980. More than 1.7 million of these occurred to children between the ages of 5 and 14.[75] In 1987 the USCPSC estimated there were 58,957 track and field injuries, 35,585 wrestling injuries, 197,411 football injuries, 168,215 baseball injuries, and 219,361 basketball injuries.[72]

Football was found to be especially hazardous.[75,78,81,82] The National Athletic Trainers' Association found that 37% of all high school football players were injured at least once during 1987.[83] Other sports with high injury rates are baseball, basketball, wrestling,[75,78,84] and ice hockey.[85]

Serious eye injuries resulting from collisions with balls or a racquet striking a player's eye are increasing with the popularity of racquet sports. Estimates of these injuries run as high as 70,000–100,000 annually.[86-88] One report estimated that there is one serious eye injury to each 100 racquet sports players every year.[86]

Problems with Sports Injury Data

Data on exposure rates for competitive sports are difficult to calculate, especially for informal play.[78,89,90] It is estimated that more than 40% of competitive sports injuries occurred in informal play.[78,91] According to one commentator, "Epidemiological studies that review only organized sports programs will significantly underestimate overall injury rates . . . preventive and educational programs must be directed at the community as a whole and not just at organized programs."[91]

The calculation of exposure rates is further complicated for sports that include a number of different events, such as gymnastics and track and field, each of which entails different risks. Research has also shown that the risk in any sport may differ between practice and competitive play, that players with different roles within an individual sport are at risk for different types of injuries, and that injury potential may vary over the course of a game or a season. A study of a youth soccer tournament, for instance, found that injury rates were highest during the final rounds of competition, when, it is assumed, players are more aggressive.[92]

There is some evidence that a substantial percentage of children's sports injuries are associated with physical education classes.[78] Much more research about these injuries is needed.

It is unique to sports injuries that severity is often defined by the amount of time the injury prevents the player from participating in the activity. However, this varies among sports and among coaches and may even vary by incident, depending on how much a coach feels a particular player is needed to win a game. Among the studies cited in this section, a "major injury" is defined as causing the athlete to refrain from participation for periods ranging from 5 days to 3 weeks.[79,80,83,85,93]

Risk Groups and Injury Patterns

Contact sports have higher injury rates than noncontact sports. Although boys are injured at twice the rate of girls,[78,79] rates are similar if contact sports (in which boys participate more than girls) are eliminated.[79,84,93,94] The highest injury rate (and the most serious injuries) for secondary school girls occurs in basketball, the woman's sport with the most contact.[94]

Football is the sport with the highest rate of concussive injuries.[75] Contact between players causes more than 80% of football injuries to youths.[80] Certain practices, such as "head butting," play a major part in contact injuries.[85,95,96] Studies also have revealed a pattern of noncatastrophic spinal cord trauma among young football players.[96]

Athletes often return to play without proper medical attention or treatment after experiencing a concussion. This can lead to futher injury. In one study, 58% of players who experienced loss of awareness (a symptom of concussion) were not examined before resuming play. Many players did not realize that loss of awareness (or even loss of consciousness) was symptomatic of a concussion. In 60% of these cases, the decision as to whether a player should return to play was made by the player himself. Research indicates that the risk of experiencing a concussion was significantly greater for ice hockey and football players who had previously experienced such an injury.[85,95]

Research has also indicated that the majority of football players reporting neck or back pain returned to play the same day. As with concussion, the majority of decisions about returning to play were made by the athletes themselves.[96] And as with concussions, a return to play without medical evaluation or treatment may increase the chances of reinjury.[85]

Developmental considerations are also important. Children cannot tolerate the same levels of stress or exertion as adults; also, they cannot dissipate heat as effectively and thus are more susceptible to heatstroke.[75] Programs targeted at competitive sports also need to take into account the fact that although many injuries occur in competition, a large number also occur during practice.[76]

Interventions

There have been very successful interventions in sports injuries. Trampoline injuries tended to be especially severe, resulting in paralysis or death. An AAP warning and the refusal of insurance companies to provide coverage for these devices resulted in the elimination of trampolines from school gymnastic and cheerleading programs and a drastic decline in injuries.[75]

Prohibiting dangerous practices has been successful when the prohibitions have been enforced. A 1976 rule change prohibiting "head butting" in football led to a decline in fatalities.[77] However, many players still use illegal techniques, and coaches do not stop them.[95]

Evidence also exists for the effectiveness of personal protective devices. Ice hockey eye injuries declined when facial protective devices were made mandatory.[75] However, evidence suggests that such protection may make athletes feel safer and lead to more aggressive play and an increase in other types of injuries.[85]

Because of the sheer number and diversity of competitive sports and events, the injuries that occur, and the types of conditioning, safety equipment, and regulations that may affect these injuries, we cannot give specific interventions for every sport. Instead, we offer generic guidelines for all organized sports along with some specific interventions.

Comprehensive sports injury control programs. Specific safety guidelines exist for many sports and injuries. The AAP has issued recommendations for heat stress[97] and competitive athletics for elementary school children.[98] Various professional associations offer health and safety certification programs for coaches and athletic trainers, sport-specific safety guidelines, recommendations on when to return to play an athlete who has been injured,[99] and guidelines on preparticipation health evaluations.[100] Other sources for information include the American College of Sports Medicine and the journals *The Physician and Sportsmedicine* and the *Journal of Sports Medicine*. The following guidelines provide a framework for the creation of a comprehensive injury prevention and control program that can be used by schools, community groups, or others who oversee competitive sports activities.[101,102]

• Rules must be strictly enforced by officials.

• Dangerous and illegal techniques must be prohibited.

• The use of alcohol, tobacco, and other drugs by players must be prohibited.

• Head coaches and staff should be versed in first aid and prevention of injuries and illnesses.

• Personnel specifically trained in prevention and care of injuries and illnesses and in CPR should be available at all games and practice sessions. Protocols for the care of injured or ill persons should be developed.

• Appropriate criteria for conditioning programs should be established. Practice sessions and conditioning programs should be supervised by qualified personnel. The length of practice sessions should be limited. The number of double practice sessions each season should be limited to no more than 10.

• Games and practice sessions should be postponed or canceled during inclement weather.

• Equipment and fields should be inspected and maintained. The condition of field surfaces should be inspected regularly. Adequate distance between activity areas and hazards such as walls, fences, and benches should be provided. Barriers that cannot be removed should be padded with materials adequate to prevent injury in the event of a collision.

• Players should be outfitted with properly fitting equipment. Activity without the full complement of required equipment should be prohibited.

• Players should not return to play following injury without a physician's permission.

• Players must not return to play after experiencing loss of consciousness and/or spinal trauma symptoms. Only well-trained, qualified personnel should be allowed to move such persons. Injured players should be referred for medical assessment. Concussion symptoms, including loss of consciousness and/or awareness, should be assessed by qualified personnel. Frequent communication among team members should be maintained to detect subtle symptoms associated with concussion.

• The use of substances to reduce pain and swelling so players may be returned to activity should be avoided.

• Salt tablets and other medications should be provided only on medical order.

• Educational programs in prevention should be developed and include information on 1) risks associated with the athletic activity, 2) the necessity for conditioning programs, 3) general prevention of injuries and illnesses, 4) rules, 5) nutrition, 6) drug use/abuse, 7) proper fitting and wearing of equipment, 8) illegal and dangerous techniques, and 9) psychological and motivational factors. Players and parents should be included in the program, and players' learning should be evaluated regularly.

• Administrators should participate in efforts to

obtain qualified personnel and a budget adequate to furnish protective equipment and essential supplies.

• Criteria for physical examination, including screening for evidence of neurological and cardiovascular pathology, should be developed.

• Health history cards should be maintained for all players. These should include a record of past injuries and illnesses, neurological and/or cardiovascular symptoms, hospitalizations, surgery, medications, allergies, and a consent for treatment form signed by parents including the names of the players' physicians and health insurance companies. These cards should be kept available for use at all times to prevent treatment delays, to maintain a current health record, and to identify potential risks.

• An emergency plan incorporating ambulance service staffed by certified personnel and emergency hospitalization should be developed. All staff members should be aware of the plan.

• A comprehensive system for reporting and recording information on all athletes should be developed to identify the magnitude of the injury problem and to help identify and evaluate potential intervention efforts. Definitions of injury and illness that facilitate the detection of injuries and illnesses, incorporate the definitions of symptoms, and incorporate criteria on duration of symptom, level of discomfort, and possible activity limitation(s) should be developed or adapted.

Recommendation: Comprehensive injury prevention and control efforts in organized competitive sports programs are promising interventions. They should be used, monitored, and evaluated for their effect on injury rates and severity.

Maturity grouping. Suggestions have been made that child and adolescent sports programs should be structured by weight and physiological maturity rather than age. This may reduce injury rates by ensuring that children of vastly unequal weights are not pitted against one another. Not enough is known to evaluate the effectiveness of this intervention, but it is an idea that warrants further research.[104]

Recommendation: Further research should be conducted on the effects of maturity grouping in sports.

Eye protection. Although open (lensless) eyeguards offer some protection against injury in racquet sports, research has shown that balls can penetrate the openings of these eyeguards and cause serious ocular injury. Eyeguards with closed lenses offer substantially more protection.[87,88,104,105] Ap-

propriate eyeguard use can be promoted through league and court rules, municipal ordinances covering commercial facilities, and educational efforts directed at players and the operators of racquet sport facilities.

Recommendation: Research indicates that appropriate eye protection could lessen the rate and severity of eye injuries associated with racquet sports. Closed-lens eyeguards meeting the standards of either the Canadian Standards Association or the ASTM should be worn during racquet sports. Programs promoting the use of such eye protection should be used and monitored and their outcomes evaluated.

Boxing. Although research on boxing injuries is limited, one study of adult professional boxers found that 87% had sustained brain damage.[106] The Committee on Sports Medicine of the AAP has pointed out that "other sports offer the same conditioning opportunity with minimal or no risk of brain injury." They recommend that "children and young adults should be encouraged to participate in sports in which intentional head injury is not the primary objective of the sport."[107]

Recommendation: Children and young adults should be discouraged from participating in boxing. Schools, community groups, youth organizations, and other groups should be urged not to sponsor youth boxing programs. Efforts to discourage youth participation in boxing should be used and monitored and their outcomes evaluated.

Detachable bases. Baseball and softball bases designed to break away from their mooring posts have been developed in an attempt to reduce the number of injuries caused by players colliding with bases while sliding. Based on a study of the use of these bases in Michigan, the CDC estimated that 1.7 million baseball sliding injuries occur annually with a cost of over $2 billion and that "exclusive use of breakaway bases would reduce injuries to just over 70,000 (a 96% reduction) and medical costs to $24 million (a 99% reduction)."[108]

Recommendation: Detachable bases have been demonstrated to reduce the rate and severity of baseball and softball sliding injuries. Programs encouraging the use of detachable bases should be implemented and monitored and their outcomes evaluated.

ALL-TERRAIN VEHICLES (ATVs)

The Magnitude of the Problem

In 1984 NEISS data revealed a doubling of ATV-related fatalities as well as a dramatic increase in the number of nonfatal injuries from 8,000 in 1982 to 60,000 in 1984. In April 1985 the USCPSC decided to make further study a priority. Research demonstrated that ATVs represented a serious injury problem. In 1985 there were an estimated 85,900 ATV-related injuries treated in emergency rooms and 238 ATV-related deaths. Subsequent research revealed such vehicles continued to represent a substantial problem and that a considerable proportion of these injuries affected the head and brain. Head injuries were the major cause of death, accounting for 70% of the fatalities.[109–111]

Risk Groups and Injury Patterns

The USCPSC study found that 93% of ATV injuries in 1985 occurred during recreational use. Half were to children under the age of 16. Seventy-two percent of those injured were male.[111] Eighty-seven percent of the injuries involved three-wheeled ATVs. Sixty-eight percent of the injuries occurred when the vehicle hit a bump or another object and tipped or rolled over. Fifty-six percent of the injured were not wearing a helmet or any other protective equipment at the time of the injury.[112]

A follow-up study reinforced and refined the description of ATV injuries. It reported that three-wheeled ATVs, which were favored by younger drivers, were found to be less stable and thus more prone to tip and roll over than four-wheeled models. However, the additional weight and horsepower of the four-wheeled vehicles make them less controllable by young people, which counteracts any safety benefit gained from stability. Thirty percent of ATV injuries were alcohol related. Drivers with less than 1 month of experience had an injury risk 13 times that of the average driver. Eighty percent of users did not wear helmets. The risk of fatal or hospitalizable head injuries increased by a factor of three for those not wearing helmets, and the risk of injury increased by a factor of two.[111]

Interventions

In 1987 the USCPSC and the U.S. Department of Justice filed suit against ATV manufacturers to have this product removed from the market. An agreement reached in U.S. district court in April 1988 specified that manufacturers must carry out the following steps.
- Stop selling three-wheeled ATVs and buy back unsold units from retailers
- Offer safety courses to all new ATV purchasers and others who purchased vehicles in the preceding 12 months
- Place warnings on the risks associated with

ATVs on the vehicles, in owners' manuals, and at dealerships
- Participate in an advertising campaign demonstrating the dangers of ATV use
- Refrain from advertising that shows ATVs being used in patently dangerous ways
- Recommend appropriate age limitations for each model
- Not oppose state legislative initiatives for licensing and certification of ATV operators

The USCPSC can reopen this case in 1991 if injuries have not been significantly reduced.[113] Several bills aimed at reducing ATV injuries are currently before Congress.

Prohibition. Research indicates that three-wheeled ATVs are inherently unstable; this instability is compounded by their common use in driving over uneven and rugged terrain at relatively high speeds. The unreasonable risk of injury associated with these vehicles justifies a prohibition on their sale and import.

Recommendation: The federal government should prohibit the manufacture, import, and sale of all-terrain vehicles or parts which could be used to assemble them. Enforcement of such a prohibition should be monitored and its outcomes evaluated.

REFERENCES

1. Smith G, et al. Sports injury hospitalizations: incidence and coding problems. Paper presented at the annual meeting of the American Public Health Association, New Orleans, Louisiana. Oct 20, 1987.

2. Gerberich S. Sports injuries: implications for prevention. Public Health Rep 1985;100:670–1.

3. Accident facts, 1986 edition. Chicago; National Safety Council, 1986:75.

4. Kraus J. Alcohol and recreation-related injuries: a review of the epidemiologic evidence. Paper presented at the Conference on Research Issues in the Prevention of Alcohol-Related Injuries at the Prevention Research Center, Berkeley, California, Mar 3–4, 1986.

5. Maddox D. Skateboarding: the spill-and-skill sport. Physician Sportsmed 1978;6:108.

6. Jacobs P, Kellor E. Skateboard accidents. Pediatrics 1977;59:939.

7. Gulaid J, Sattin R. Drownings in the United States, 1978–1984. Morbid Mortal Weekly Rep 1988;37:27–33.

8. Allman F, et al. Outcome following cardiopulmonary resuscitation in severe pediatric near-drowning. Am J Dis Child 1986;140:571–5.

9. Fandel I, Bancalari M. Near-drowning in children: clinical aspects. Pediatrics 1976;58:573–9.

10. Brill D. The cost of drowning. In: Brill D, Micik S, Yuwiler J, eds. Childhood drownings: current issues and strategies for prevention. Orange, California: California

Drowning Prevention Network, North County Health Service, Orange County Trauma Society, pp 45–6 (undated).

11. Fergusson D, Horwood L, Shannon F. Domestic swimming pool accidents to pre-school children. NZ Med J 1983;96:725–7.

12. Geddis D. The exposure of pre-school children to water hazards and the incidence of potential drowning accidents. NZ Med J 1984;97:223–6.

13. Spyker D. Submersion injury: epidemiology, prevention, and management. Pediatr Clin N Am 1985;32:113–25.

14. Dietz P, Baker S. Drowning: epidemiology and prevention. Am J Public Health 1974;64:303–12.

15. Kraus J. Extent of the problem in California. In: Brill D, Micik S, Yuwiler J, eds. Childhood drownings: current issues and strategies for prevention. Orange, California: California Drowning Prevention Network, North County Health Service, Orange County Trauma Society, pp 31–6 (undated).

16. O'Carroll P, Alkon E, Weiss B. Drowning mortality in LA County, 1976–1984. JAMA 1988;100:380–3.

17. Present P. Child drowning study: a report on the epidemiology of drownings in residential pools to children under age five. Washington, DC: Division of Hazard Analysis, Directorate for Epidemiology, U.S. Consumer Product Safety Commission, 1987.

18. Smith J, et al. Drownings—Georgia, 1981–1983. Morbid Mortal Weekly Rep 1985;34:42–3.

19. Baker S, O'Neill B, Karpf R. Injury fact book. Lexington, Massachusetts: Heath, 1984.

20. Subcommittee on Homicide, Suicide, and Unintentional Injuries. Report of the Secretary's Task Force on Black and Minority Health. Vol 5. Washington, DC: U.S. Department of Health and Human Services, 1986:35.

21. Indian Health Service. Bridging the gap: report on the task force on parity of Indian health services. Washington, DC: Indian Health Service, 1986.

22. Wintemute G, et al. Drowning in childhood and adolescence: a population-based study. Am J Public Health 1987;77:830–2.

23. Centers for Disease Control. North Carolina drownings, 1980–1984. Morbid Mortal Weekly Rep 1986;35(40):635–8.

24. Final report of the data subcommittee of the National Swimming Pool Safety Committee. Washington, DC: U.S. Consumer Product Safety Commission, 1986.

25. Baxter L. 1986–1987 Consumer Product Safety Commission study of child drownings. In: Brill D, Micik S, Yuwiler J, eds. Childhood drownings: current issues and strategies for prevention. Orange, California: California Drowning Prevention Network, North County Health Service, Orange County Trauma Society, pp 37–44 (undated).

26. Baxter L. Report: infant drowning in swimming pools/spas. Washington, DC: Household Structural Products Program, U.S. Consumer Product Safety Commission, 1984.

27. Elder J. Human factors in drowning and near-drowning injuries. In: Brill D, Micik S, Yuwiler J, eds. Childhood drownings: current issues and strategies for prevention. Orange, California: California Drowning Prevention Network, North County Health Service, Orange County Trauma Society, pp 27–8 (undated).

28. Rowe M, Arango A, Allington G. Profile of pediatric drowning victims in a water-oriented society. J Trauma 1977;17:587–91.

29. National Electronic Injury Surveillance System product safety report. Washington, DC: U.S. Consumer Product Safety Commission, 1985, 1986, 1987.

30. Brown, V. Human factor analysis: spa associated hazards—an update and summary. Washington, DC: Division of Human Factors, Directorate for Hazard Identification and Analysis, U.S. Consumer Product Safety Commission, 1981.

31. Monroe B. Immersion accidents in hot tubs and whirlpool spas. Pediatrics 1982;69:805–7.

32. Milliner N, Pearn J, Guard R. Will fenced pools save lives? A 10-year study from Mulgrave Shire, Queensland. Med J Aust 1980;2:510–11.

33. Pearn J, et al. Drowning and near-drowning involving children: a five-year total population study from the city and county of Honolulu. Am J Public Health 1979;69:450–4.

34. Pearn J, Thompson J. Drowning and near-drowning in the Australian Capital Territory: a five-year total population study of immersion accidents. Med J Aust 1977;1:130–3.

35. Rennell D. Child drowning prevention through supervision and barriers: a report surveying current studies and recommendations. Alexandria, Virginia: National Pool and Spa Institute, 1987.

36. Children and pool safety checklist. Washington, DC: U.S. Consumer Product Safety Commission, 1988.

37. Gardiner S, et al. Accidental drownings in Auckland children. NZ Med J 1985;98:579–82.

38. Fitch T. Pool cover standards. Natl Drowning Prev Network 1988–89;2:4.

39. Kropp RM, Schwartz J. Water intoxication from swimming. J Pediatr 1982;101:616–8.

40. Committee on Accident and Poison Prevention. Injury control for children and youth. Elk Grove Village, Illinois: American Academy of Pediatrics, 1987.

41. Pearn J. Current controversies in child accident prevention: an analysis of some areas of dispute in the prevention of child trauma. Aust NZ Med J 1985;15:782–7.

42. Minimum standards for public spas. Alexandria, Virginia: National Pool and Spa Institute, 1985.

43. Standards for residential spas. Alexandria, Virginia: National Pool and Spa Institute, 1986.

44. Suggested health and safety guidelines for public spas and hot tubs. Atlanta, Georgia: Center for Environmental Health, Centers for Disease Control, 1985.

45. Maiman D, et al. Diving-associated spinal cord injuries during drought conditions—Wisconsin, 1988. Morbid Mortal Weekly Rep 1988;37:30.

46. Tator C, Edmonds V. Sports and recreation as a rising cause of spinal cord injury. Physician Sportsmed 1986; 14:157–67.

47. Kennedy EJ, et al., eds. Spinal cord injury: the facts and figures. Birmingham, Alabama: National Spinal Cord Injury Statistical Center, University of Alabama at Birmingham, 1986.

48. Gabrielson M. Diving's deadly toll. Paper presented at the National Environmental Health Association Conference, New Orleans, Louisiana, Jun 20, 1982.

49. Harris B. A profile of diving injury victims, accident locations, and prevention issues. Paper presented at the National Pool and Spa Safety Conference, Arlington, Virginia, May 14, 1985.

50. Centers for Disease Control. Aquatic deaths and injuries—United States. Morbid Mortal Weekly Rep 1982;31:31.

51. Cain K, Saxton C. Instruction guide for conducting the National Head and Spinal Cord Injury Prevention Project of the American Association of Neurological Surgeons and the Congress of Neurological Surgeons. Park Ridge, Illinois: National Head and Spinal Cord Injury Prevention Project, American Association of Neurological Surgeons/College of Neurological Surgeons, 1986.

52. U.S. Coast Guard. Boating statistics 1987. Washington, DC: U.S. Coast Guard, 1988.

53. Proceedings of the workshop on alcohol-related accidents in recreational boating. Washington, DC: Transportation Research Board, National Research Council, 1986.

54. Personal flotation devices research phase II (3 vols). Springfield, Virginia: Natl Tech Info Serv, 1978.

55. U.S. Coast Guard. Personal flotation device research reports sponsored by the Coast Guard. Washington, DC: U.S. Coast Guard, 1984.

56. Abstracts of personal flotation device research reports sponsored by the Coast Guard. Washington, DC: U.S. Coast Guard (undated).

57. National Association of State Boating Law Administrations. Mandatory boater education. Small Craft Advisory 1988;3(4):18–21.

58. National Association of State Boating Law Administrations. Boating safety education. Small Craft Advisory 1988;3:4–7.

59. Centers for Disease Control. Playground-related injuries in preschool-age children in the United States, 1983–1987. Morbid Mortal Weekly Rep 1988;37:6629–32.

60. Werner P. Playground injuries and voluntary product standards for home and public playgrounds. Pediatrics 1982;69:18–20.

61. Boyce W, et al. Playground equipment injuries in a large, urban school district. Am J Public Health 1984; 74:984–6.

62. Rutherford G. HIA hazard analysis: injuries associated with public playground equipment. Washington, DC: Directorate for Hazard Identification and Analysis-Epidemiology, U.S. Consumer Product Safety Commission, 1979.

63. A handbook for public playground safety: Vol 1. General guidelines for new and existing playgrounds. Washington, DC: U.S. Consumer Product Safety Commission, 1981.

64. Fisher L, et al. Assessment of a pilot child playground injury prevention project in New York State. Am J Public Health 1980;70:1000–2.

65. A handbook for public playground safety. Vol 2. Technical guidelines for equipment and surfacing. Washington, DC: U.S. Consumer Product Safety Commission, 1981.

66. Bruya LD, Langendorfer SJ. Where our children play: elementary school playground equipment. Reston, Virginia: American Association for Leisure and Recreation, 1988.

67. Bruya LD. Play spaces for children. Reston, Virginia: American Association for Leisure and Recreation, 1988.

68. Christoffel K, et al. Childhood injuries caused by nonpowder firearms. Am J Dis Child 1984;138:557–61.

69. Thackray J. The deadly toy: high-powered air guns. Sightsaving 1984;53:2.

70. Carter B. BB's: a major cause of permanent eye injuries. For Kid's Sake 1987;5:1.

71. Rutherford G. Special report: injuries associated with non-powder guns. Washington, DC: U.S. Consumer Product Safety Commission, 1981.

72. National Electronic Injury Surveillance System. Product safety report. Washington, DC: U.S. Consumer Product Safety Commission, 1988.

73. Blocker B, et al. Serious air rifle injuries in children. Pediatrics 1982;69:751–4.

74. Christoffel T, Christoffel K. Nonpowder firearms injuries: whose job is it to protect children? Am J Public Health 1987;77:735–8.

75. Rutherford G, Miles R. Overview of sports related injuries to persons 5–14 years of age. Washington, DC: U.S. Consumer Product Safety Commission, 1981.

76. Rutherford G, et al. Hazard analysis: baseball and softball related injuries to children 5–14 years of age. Washington, DC: Division of Hazard Analysis, U.S. Consumer Product Safety Commission, 1984.

77. Mueller F, Schindler R. Annual survey of football injury research 1931–1987. Orlando, Florida: American Football Coaches Association, 1988.

78. Zaricznyj B, et al. Sports-related injuries in school-aged children. Am J Sports Med 1980;8:318–24.

79. Garrick J, Requa R. Girls' sports injuries in high school athletics. JAMA 1978;239:2245–8.

80. Goldberg B, et al. Injuries in youth football. Pediatrics 1988;81:255–61.

81. Kraus J, Conroy C. Mortality and morbidity from injuries in sports and recreation. Am Rev Public Health. 1984;5:163–92.

82. Garrick J, Requa R. Injuries in high school sports. Pediatrics 1978;61:465–9.

83. Fact sheet. Greenville, North Carolina: National Athletic Trainers' Association, 1987.

84. Study shows 23% of high school basketball players injured this year. Greenville, North Carolina: National Athletic Trainers' Association, 1988.

85. Gerberich S, et al. An epidemiological study of high school ice hockey injuries. Childs Nerv Syst 1987;3:59–64.

86. Thackray J. How to score fewer racquet sport eye injuries. Sightsaving 1982;51:2–6.

87. Vinger P. Eye protection for racquet sports. J Phys Ed Recr Dance 1983:46–48.

88. Vinger P. Sports-related eye injury: a preventable problem. Survey Ophthalmol 1980;25:47–51.

89. Schor S. Relation of football injuries to exposure time. Am J Public Health 1984;74:1170–1.

90. Gerberich S, et al. Response from Dr. Gerberich, et al. Am J Public Health 1984;74:1170–1.

91. Goldberg B, et al. Children's sports injuries: are they avoidable? Physician Sportsmed 1979;7:93–101.

92. Maehlum S. Frequency of injuries in a youth soccer tournament. Physician Sportsmed 1986;14:73–9.

93. Shively R, et al. High school sports injuries. Physician Sports Inj 1981;9:46–50.

94. Chandy T. Grana W. Secondary school athletic injury in boys and girls: a three year comparison. Physician Sportsmed 1985;13:106–11.

95. Gerberich S, et al. Concussion incidence and severity in secondary school varsity football players. Am J Public Health 1983;73:1370–5.

96. Gerberich S, et al. Spinal cord trauma and symptoms in high school football players. Physician Sportsmed 1983;11:122–39.

97. Committee on Sports Medicine. Climatic heat stress and the exercising child. Pediatrics 1982;69:808–9.

98. Committee on Pediatric Aspects of Physical Fitness, Recreation, and Sports. Competitive athletics for children of elementary school age. Pediatrics 1981;67:927–8.

99. Cantu R. Guidelines for return to contact sports after a cerebral concussion. Physician Sportsmed 1986;14:75–83.

100. Strong W, Linder C. Preparticipation health evaluation for competitive sports. Pediatrics 1982;4:113–21.

101. Gerberich SG. Evaluation of injury/illness and health care provision for football participants in the secondary schools of Minnesota. Minneapolis: Minnesota State Planning Agency, 1981.

102. Gerberich SG. Analysis of high school football injuries and concomitant health care provision [Dissertation, 2 vols]. Minneapolis: University of Minnesota, 1980.

103. Whielson D. Maturity sorting: new balance for young athletes. Physician Sportsmed 1978:137–42.

104. Easterbrook M. Eye protection in racket sports: an update. Physician Sportsmed 1987;19:180–92.

105. Feigelman M, et al. Assessment of ocular protection for racquetball. JAMA 1983;250:3305–9.

106. Casson R, et al. Brain damage in modern boxers. JAMA 1984;251:2663–7.

107. Committee on Sports Medicine. Participation in boxing among children and young adults. Pediatrics 1984;74:311–2.

108. Janda D, et al. Softball sliding injuries—Michigan, 1986–1987. Morbid Mortal Weekly Rep 1988;37:169–70.

109. Marchica N. Memorandum: ATV deaths and injuries. Washington, DC: U.S. Consumer Product Safety Commission, 1988.

110. Kriel R, et al. Pediatric head injury resulting from all-terrain vehicle accidents. Pediatrics 1986;78:933–5.

111. Newman R. Hazard analysis: analysis of all terrain vehicle related injuries and deaths. Washington, DC: Division of Hazard Analysis, Directorate for Epidemiology, U.S. Consumer Product Safety Commission, 1987.

112. Newman R. Hazard analysis: survey of all terrain vehicle related injuries (1985 preliminary report). Washington, DC: Division of Hazard Analysis, Directorate for Epidemiology, U.S. Consumer Product Safety Commission, 1985.

113. Judge approves plan limiting sales of all-terrain vehicles. Boston Globe 1988 Apr 28:21.

Chapter 9: Occupational Injuries

Rapid industrialization at the turn of the century generated new dangers for American workers. Industries such as steel underwent rapid reorganization of manufacturing processes, with huge investments in new equipment and enormous increases in productivity. Work done by skilled craftsmen was often replaced by machine production. The pace of work was greatly increased, and 80- to 100-hour work weeks were not uncommon. With all these changes, little attention was given to worker safety. The U.S. Bureau of Labor reported in 1907 that the nation's 26 million male workers suffered 15,000 to 17,500 on-the-job deaths per year. Women and children were also employed in hazardous occupations. Conditions in the garment industry sweatshops, for example, received public attention when 145 young women died in New York's 1911 Triangle Shirtwaist fire. Children were also frequently injured while working.

These conditions gave rise to workers' compensation insurance systems, which, by 1920, were operating in all but six states. Under workers' compensation, injured workers exchanged their right to sue employers for a system that provided income benefits, medical payments, and rehabilitation benefits. Industry benefited from this system through more predictable costs and protection from lawsuits.

The first state-run industrial inspection programs began in the late 19th century. The first state agency to address occupational safety, the Massachusetts Department of Factory Inspection, was created in 1867. Massachusetts passed the first worker safety law in 1877, requiring guards on spinning machinery in textile mills. By 1930 most industrial states had passed occupational safety laws, although few were systematically enforced.

It was assumed that workers' compensation and state occupational safety laws would solve the problem of workplace death and injury, but the laws designed to prevent workplace injuries were weak and poorly enforced and excluded large segments of the work force. Employers regularly contested compensation claims, delaying benefits. Benefit levels often failed to keep pace with wage levels and the cost of living. During the 1930s and 1940s, faced with economic depression and world war, both federal and state governments reduced their occupational safety efforts, and "neglect of the job safety and health problem continued in the postwar era."[1]

THE OCCUPATIONAL SAFETY AND HEALTH ADMINISTRATION

Public attention to workplace safety reemerged in the late 1960s. The industrial injury rate had increased by 29% between 1961 and 1970. A series of disasters in the coal mining industry coupled with vigorous efforts by mine workers to obtain compensation for occupational lung disease led to passage of the federal Coal Mine Safety and Health Act of 1969. One year later, Congress passed the Occupational Safety and Health (OSH) Act of 1970, to "assure so far as possible every working man and woman in the Nation safe and healthful working conditions."

The act created the Occupational Safety and Health Administration (OSHA) and gave it the power to set standards, inspect workplaces, cite violations, impose penalties, and seek injunctions to shut down operations in cases of imminent danger. It also gave employees the right to call inspectors into the workplace, accompany OSHA personnel during inspections, and participate in review proceedings when employers contested citations. OSHA also was expected to provide training and education of employers and employees in the "recognition, avoidance and prevention of unsafe or unhealthful working conditions." Another provision of the act established the National Institute for Occupational Safety and Health (NIOSH) to carry out research, recommend standards, and conduct hazard evaluations.

OSHA's effect on injury rates has been the subject of controversy. Some studies have shown a slight decrease in injury rates associated with its activities, and others have found no effect at all.[2-4] OSHA's effectiveness may reflect the following limitations.

• Inadequate inspection and enforcement: There are 1,181 inspectors for over 4 million workplaces. Fewer than 2% of the nation's workplaces are inspected every year.

• Inadequate and outdated standards: Most of

the safety standards adopted by OSHA rely on those created by the American National Standards Institute, a voluntary standard-setting organization. Many of these standards have not been updated since they were originally developed in the 1960s, nor have standards been set for activities associated with many workplace injuries, such as the handling of heavy loads, repetitive motion, and vibration.

• Limited coverage: OSHA does not cover public employees, and significant exemptions are made for small businesses.

• Insufficient penalties: Penalties for violations of OSHA regulations are not severe enough to force employers to improve workplace conditions. The maximum fine that OSHA can levy has not changed since 1970 ($1,000 a day for a serious violation and up to $20,000 for a willful violation).

It is possible that OSHA's impact on occupational injuries may not be reflected in the data. Trends in injury rates often are influenced by factors unrelated to regulatory activity, and thus it is difficult to isolate the effect of regulatory programs on workplace injury rates. In addition, most studies evaluating OSHA's effectiveness have looked at total injury rates. This approach may mask the effect of regulatory activities on specific types of injuries or in specific industries.

THE MAGNITUDE OF THE PROBLEM

Estimates of the number of fatal occupational injuries in 1984 ranged from 3,740 to 11,500.[5,6] The reasons for this wide range include differences in the sources, methods, and case definitions of the various surveillance systems.[6] The Office of Technology Assessment's 1985 report *Preventing Illness and Injury in the Workplace* reviewed the data on occupational injuries and estimated that there are approximately 6,000 work-related deaths and from 2.5 to 11.3 million nonfatal occupational injuries each year.[7]

The social and economic consequences of workplace injuries are enormous. The costs include lost wages, medical expenses, insurance claims, production delays, lost time of coworkers, and equipment damage. The National Safety Council has estimated the total cost of workplace injuries in 1987 to be $42.4 billion. Back injuries alone account for $14 billion a year. Some 35 million work days were lost by workers injured on the job in 1987.[8]

The personal consequences of occupational injuries for injured workers and their families are more difficult to quantify. They include pain and suffering, stress and stress-related illnesses, economic insecurity, and a diminished quality of life, especially for permanently disabled workers.

Problems with the Data

The OSH Act of 1970 (public law 91-596) required that the secretary of labor "compile accurate statistics on work injuries and illnesses other than minor injuries requiring only first aid treatment and which do not involve medical treatment, loss of consciousness, restriction of work or motion, or transfer to another job." This task is performed by the U.S. Bureau of Labor Statistics (BLS), the statistical branch of the U.S. Department of Labor. The BLS also issues guidelines for the reporting of occupational illnesses and injuries by employers.[9] These records also are used to target OSHA inspection efforts at both industrial sector and individual plant levels.

Occupational safety advocates have consistently challenged the accuracy of both employer reports of injury and the BLS Annual Survey of Occupational Illnesses and Injuries, arguing that despite reporting and record-keeping regulations many employers underreport injuries in order to lower recorded rates of lost workdays caused by injuries. This concern was heightened when OSHA implemented a new enforcement policy in 1981, under which industries with a lost workday injury (LWI) rate less than the national average for manufacturing are exempted from regular inspections. In the industries that are not exempted, OSHA examines employer injury logs. If the LWI rate for a particular establishment is found to be below the national average for manufacturing, no workplace inspection will take place in the absence of a complaint. This system, it is argued, encourages employers to underreport injuries. In response to these and other concerns, Congress appropriated funds in 1984 for the BLS to study the quality of data generated by its annual survey. A panel of experts was convened by the National Research Council; its report, *Counting Injuries and Illnesses in the Workplace: Proposals for a Better System*, concluded that "the BLS data systems, in their current form, are inadequate for providing OSHA with the data it needs for maintaining an effective program for prevention of workplace injuries and illnesses."[10]

The BLS, in issuing its 1986 annual survey, expressed concern about "errors in company logs on which the survey is based." It reported on a study of 200 workplaces in Massachusetts and Missouri, where "about 90 percent of the establishments were in compliance with the requirement to maintain an OSHA log, but underrecording, especially of cases involving lost workdays, occurred."[8,11] The lack of accurate data on the extent and patterns of occupational injuries is a major impediment to understanding and attempting to control these injuries.

Risk Groups and Injury Patterns

Workers in agriculture (including forestry and fishing), mining, construction, transportation, and the public utilities industry are at greatest risk for occupational injuries.[11,12] Although the dangers of the mining, construction, transportation, and public utilities industries have long been recognized, the injury problem in agriculture only recently has come under scrutiny. Agriculture is now recognized as one of the nation's most dangerous industries.[13] The National Safety Council, using National Center for Health Statistics (NCHS) data, found that the agricultural, forestry, and fishing industries had an average of approximately 1,600 fatal injuries a year, for an injury mortality rate of 49 per 100,000 in 1987. These industries also had an average of about 160,000 nonfatal injuries each year.[8] Hazards faced by agricultural workers include tractor rollovers,[14] amputations by machinery,[15] electrocutions,[16] and pesticide poisoning.

Although there are many ways in which people are injured on the job, approximately one-third of all occupational injury fatalities are associated with over-the-road motor vehicles.[11] NIOSH estimates that an additional 11% of occupational deaths are associated with nonhighway industrial vehicles.[17] Workers who operate vehicles such as bulldozers, front-end loaders, cranes, forklifts, and tractors are at risk of injury and death from collisions and rollovers.[18,19] Even in manufacturing, more workers are killed by motor vehicles than by fixed machinery.[20] Falls also make up a large proportion of occupational injury fatalities, especially falls in construction work.[7,8]

NIOSH also has estimated that more than 11% of all traumatic occupational fatalities are the result of homicides.[5] Although data on work-related homicides are incomplete (and some surveillance systems do not include these injuries as occupational injuries), it is estimated that 1,600 such homicides occur every year.[21] Studies in Ohio and Maryland found that more than 70% of work-related murders involve the use of firearms.[20,22]

Studies in Ohio,[22] Texas,[23,24] and California[21] found that the rate of occupational homicide is highest for police and other law enforcement officers. The occupational homicide rate for police officers in California is ten times greater than the overall homicide rate for males. In Texas, the homicide rate for sheriffs, bailiffs, and other law enforcement officers is 44.4 per 100,000. These studies also indicated that occupational homicide rates are high for employees in the retail service sector, especially for cab drivers and those who work in establishments prone to armed robberies, such as gas stations and convenience stores. There is also some evidence that women working in convenience stores are at increased risk of rape.[25]

Although risk groups for occupational injuries are traditionally categorized by occupation, there is a growing awareness that other demographic groups are also at risk for particular injuries. This is usually a consequence of the types of jobs available to individuals in these groups, often jobs in industries that are not regulated or are under-regulated by OSHA. Research has indicated that it is precisely those firms that have an unskilled, low-paid work force that tend to use hazardous processes and substances in ways that present a danger to their employees.[26] These firms also tend to be smaller and nonunionized.

Workers in small businesses. Firms with ten or fewer employees are exempted from regular OSHA inspections and record-keeping requirements. Although there is evidence that workers in small firms have higher occupational injury rates than workers in larger companies,[27] this probably varies by type of business. A substantial percentage of small firms are concentrated in hazardous industries, such as construction, longshoring, logging, and meat cutting.[1] Small businesses are less likely than larger firms to have resources available for safety improvements.[1] In addition, small firms that are required to keep injury records are less likely to do so than larger firms.[28]

Public employees. A substantial percentage of the American work force is employed by federal, state, county, and municipal governments. The occupations are diverse, ranging from trash collection to health care and from education to police and fire protection.

It is difficult to ascertain the precise extent of the injury problem in the public sector because of the paucity of record keeping. However, many jobs in public employment have particular injury problems. Sanitation workers, for example, are at risk for lifting injuries, amputations, chronic joint conditions, and eye injuries. Firefighters commonly sustain burns, cuts, sprains and strains, broken bones, smoke inhalation, and eye injuries. Health care workers, the second largest public employee group, face safety hazards leading to back injuries, injuries from patient assaults, burns and cuts, and slips and falls. Needlestick injuries have taken on increased importance with the increasing prevalence of hepatitis B and AIDS. The Centers for Disease Control (CDC) has issued recommendations for the prevention of transmission of these diseases in health care settings, including transmission caused by needlesticks.[29]

Despite clear evidence that public employees face

significant health and safety problems on the job, state, county, and municipal employees are excluded from protection under the OSH Act of 1970. However, if a state chooses to assume responsibility for enforcing workplace safety standards under a state plan, its program must apply to public employees. Twenty states currently have state plans, and several states have "public employee only" plans. Such Public Employee Occupational Safety and Health (PEOSH) programs are discussed in more detail below.

Young workers. There is some evidence that young workers experience a disproportionate number of injuries at work. Some studies have shown that inexperienced workers have higher injury rates than workers with more job experience.[30-34] The occupational injury risk of the young has been attributed to the lack of the training and experience that help older, more experienced workers reduce the risk of injury.[35] It is also possible that the same developmental factors that contribute to the high levels of adolescent traffic and recreational injury rates may contribute to their risk for occupational injuries.

The Fair Labor Standards Act of 1938 (29 U.S.C. 201 and amendments), which includes federal provisions for child labor protection, generally applies to those under the age of 16. However, 14- and 15-year-olds can be employed in many retail and service jobs. Even younger children can legally work in agriculture. Children under the age of 12 can work on their parents' farms or on farms small enough to be excluded from the provisions of the act and in certain seasonal work. More than 25,000 children and adolescents are injured on farms each year. More than 300 of these injuries are fatal.[36] One study conducted in Indiana estimated that children 5–14 years of age are three times as likely to be injured doing farm work than are those in other age groups.[37] There is evidence that the farm environment itself, with a great potential for unsupervised play near hazards such as moving machinery, animals, and irrigation ditches, presents a substantial injury risk to children even if they are not working.[38] As with adults, a large percentage of farm injuries to children (and especially the more serious injuries) are associated with tractors and machinery.[36-38]

Women and minority workers. As is the case with many types of injuries, minorities are disproportionately affected by occupational injuries. One study showed that "the average black worker is found to be in an occupation 37 to 52 percent more likely to result in a serious injury or illness than the occupation of the average white worker, and this overrepresentation in hazardous jobs holds strong even after controlling for differences in education and on-the-job experience."[39] Some jobs traditionally held by women, though rarely considered hazardous, have high rates of some types of injuries. For example, hospital and nursing home personnel frequently sustain back injuries while moving or lifting patients. Ill-fitting personal protective equipment, which is often designed by and for men, has been cited as a problem for women in industry.[40] Light assembly operations and even secretarial work can produce cumulative trauma disorders.

Musculoskeletal injuries. Although most fatal occupational injuries involve crushing injuries, severe lacerations, burns, and electrocutions, less dramatic musculoskeletal injuries exact a huge toll in the workplace. The magnitude of this problem is not clearly defined because existing reporting systems on musculoskeletal injuries fail to distinguish between occupational and nonoccupational disorders. Grouped by parts of the body, back injuries are the largest category of occupational injuries.[41-43]

One analysis of workers' compensation data revealed that nursing aides and licensed practical nurses had the highest incidence of back strain injuries of all occupations. Other groups found to be at high risk were construction laborers, licensed practical nurses, garbage collectors, truck drivers, and registered nurses.[44]

There is evidence of some new problems in other sectors of the economy as well:

A projected increase in musculoskeletal injuries and disorders is already evident despite the move toward more sophisticated automation and the shift away from physical to mental work, for example, processing of information and service-related jobs. Remarkably, the introduction of modern office technology, such as computers, video display terminals, and optical scanners, designed to reduce physical labor, has generated claims of new, pervasive, and even more insidious sources of biomechanical stress to the musculoskeletal system. The majority of these claims are attributed to chronic repetitive motion and static and constrained postures. The magnitude of these problems is only beginning to emerge.[45]

Cumulative trauma disorders (CTDs) are caused by repeated strain to the hand, arm, wrist, or other parts of the body. Vibration, forceful and repetitive hand motions, awkward postures, and mechanical stress at the base of the palm are among the factors associated with CTDs. Over 23,000 occupationally related CTDs were reported in 1980.[46] One form of CTD, the prevalence and seriousness of which has only recently come to attention, is carpal tunnel syndrome, a disabling condition of the hand brought on by repeated flexing of the wrist or the application of arm-wrist-finger pressure. Although it was first studied in the garment industry, almost any repetitive motion, including typing, can create this condition.[7]

Other Experts in Occupational Safety

As with any injury problem, federal regulation is only one part of an overall strategy to prevent injuries in the workplace. There are other agencies, institutions, and groups that have an interest in the prevention and control of occupational injuries and are potential collaborators in state and local programs toward this end.

National Institute for Occupational Safety and Health

NIOSH is the federal agency responsible for education and research relevant to safety and health in the workplace. NIOSH studies and disseminates information about workplace hazards and preventive measures. This research includes the development of recommendations for limits of exposure to hazardous substances or conditions in the workplace—recommendations that OSHA uses in setting legal standards for exposure.[47] NIOSH also funds educational resource centers, which offer training and research training programs in occupational medicine, occupational health nursing, occupational injury prevention and safety, and industrial hygiene. In addition, these centers provide regional consultation services and continuing education programs relevant to these areas.

State OSHA Programs

Under the OSH Act of 1970, states may assume responsibility for enforcing workplace safety and health regulations if their regulations and enforcement are "at least as effective" as federal OSHA. Almost half of the states currently administer their own programs. New Jersey requires all local health departments to include an occupational health program as one of their core activities. The New Jersey Occupational Health Service (although it has regulatory authority only for public employees) provides the training for local staff members who conduct these activities. The training program includes an introduction to occupational health and safety and explanations of how to conduct a workplace safety inspection, how to assist employers and employees in making their workplace safer, and when and how to report violations to OSHA for federal action.

Public Employee Occupational Safety and Health Administrations

A number of states have instituted programs targeted at the health and safety of public employees: Public Employee Occupational Safety and Health (PEOSH) programs. New Jersey has had a PEOSH program since 1984. Its major activities include occupational safety and health promotion efforts, the promulgation of standards, an inspection program, and the application of some federal OSHA standards to public employment facilities within the state. The responsibility for enforcement was given to the state Departments of Labor, Health, and Community Affairs. Inspections are conducted in response to reports of injuries or fatalities or at the request of public employees who believe that a health or safety hazard exists in their place of employment. The Department of Health, under the PEOSH program, also conducts targeted investigations of state and local government facilities such as public schools, municipal utilities, and municipal garages.

The Departments of Health and Labor are also charged by the New Jersey PEOSH act to promulgate state standards for occupational safety where no federal standards exist or where federal standards are believed to be inadequate. New Jersey's Occupational Health Service has developed, or is developing, standards for firefighter protective equipment, carbon monoxide exposure in garages and vehicles, and other aspects relevant to public employment.

The New Jersey Department of Health also provides technical assistance to local governments and their employees on the provisions and requirements of PEOSH. It produces and disseminates information on occupational health and safety in the public sector, including hazardous substance fact sheets, information bulletins on occupational hazards and standards relevant to public employees, and information on the PEOSH program and the state's right-to-know law.[48,49] Legislation pending in Pennsylvania that mandates joint labor–management health and safety committees as mechanisms for promoting safety in public workplaces is also considered an innovative approach for public employee coverage at the state level.

State Workers' Compensation Agencies

A number of states have recognized the need for a preventive component in their workers' compensation systems and have established safety and health training and educational programs under their workers' compensation laws. New York, Michigan, Connecticut, and Massachusetts currently operate such programs, either through grants to outside organizations or through direct training and education by state agencies. Ohio has an occupational health and safety consultancy program funded by a 1% assessment on state workers' compensation premiums.

Unions

Safety issues often are included in union-negotiated contracts and pursued more informally through union grievance procedures. Unions commonly organize health and safety committees to inspect the workplace, educate members, and propose changes in work processes and practices. The presence of a union in a business has been shown to improve OSHA record-keeping requirements.[38]

Insurers

Many workers' compensation insurers provide loss-control services to the employers they insure. These services provide loss-control specialists who visit work sites to analyze conditions and provide recommendations to employers for improved workplace design, safer work practices, and employee training programs.

COSH Groups

There are approximately 25 committees on occupational safety and health (COSH groups) throughout the country. These nonprofit organizations bring together local labor unions and occupational health and safety professionals to provide safety and health assistance to workers and to provide education on workplace safety for many different audiences. COSH groups have been instrumental in legislation at the state and local levels on such issues as right-to-know laws and criminal prosecution for workplace deaths. Many draw on volunteer help from physicians, attorneys, industrial hygienists, and other professionals to provide technical assistance, training, and other services.

Universities

Occupational safety and injury control programs located in schools of engineering, public health and medicine, and business are important sources of information, training, and research on occupational safety issues. Many labor studies programs also offer occupational safety and health components.

Occupational Health Clinics

Physicians and other health professionals who specialize in work-related injuries and illnesses can be another source of information and assistance. They can be particularly valuable in the case of injuries whose occupational origin may not be readily apparent (e.g., some repetitive motion injuries).

U.S. Department of Agriculture Extension Safety Specialists

Some state branches of the Department of Agriculture's Extension Service maintain either full- or part-time safety specialists. These specialists engage in applied research and safety outreach programs. Through county extension services, they provide programs and workshops to chapters of farming organizations such as the Grange, the 4-H Club, the Farm Bureau, and the Future Farmers of America. They also engage in research relevant to agricultural occupational health and safety, usually in collaboration with the faculty and staff of a state college of agriculture (D. Baker, D. Murphy, personal communication).

PREVENTING OCCUPATIONAL INJURIES: THE STATE OF THE ART

As with all injuries, epidemiologic techniques can greatly enhance efforts to identify the agents of workplace injury and to design and validate strategies for their prevention. In addition, specialists in occupational safety and injury prevention have devised a specific framework for analyzing and intervening in problems in this area.

The Hierarchy of Controls

The model that occupational safety and health professionals commonly use is called the "hierarchy of controls." According to this paradigm, engineering controls, such as modifying machines or the work process or substituting a less hazardous substance or device, are the most reliable interventions and are preferable to other methods. These controls operate at the source of the hazard.

The second level of control attempts to prevent the transmission of the hazard to workers through environmental modifications such as ventilation systems or by separating the workers from hazardous machines. The third and least effective level includes personal protective equipment such as respirators or protective eyewear or administrative controls such as work rotation. These are considered the controls of last resort, to be used when engineering or administrative controls are not feasible.

The U.S. Office of Technology Assessment has identified the following advantages of engineering controls over personal protective equipment.

- They work day after day, with minimal human intervention.
- They provide the same level of protection to all workers and do not depend on a "good fit" with individual workers.
- They can control several routes of exposure simultaneously (e.g., through the skin and by inhalation). Personal protective equipment is generally limited to one exposure route.
- The effectiveness of many personal protective devices, especially under workplace conditions, has not been demonstrated.
- Primary reliance on engineering controls can promote the development of new technology to promote safety.
- When engineering solutions operate at the source of a hazard, other means can be used as back-up. There are no options for supplemental control when personal protective equipment is used as a first line of defense.
- Personal protective equipment is frequently burdensome to the worker and may itself create safety hazards (e.g., by impairing worker-to-worker communication).[7]

OSHA, since its inception in 1971, has required engineering controls and work-practice controls as the first line of defense against workplace hazards. However, occupational safety advocates are concerned that the agency is reconsidering this policy and moving toward performance-oriented standards that give employers flexibility in deciding how to comply with regulations.

The designers of programs for worker safety must take into account the social and economic variables that influence the dynamics of the workplace as well as the well-established principles of occupational injury control. Regulation and standard setting, education, worker participation, and economic incentives all must be considered in developing a comprehensive strategy for preventing injury and death in the workplace.

Ergonomics

Ergonomics has recently gained prominence in the United States in occupational hazard recognition and control. Ergonomics (literally "the laws of work") focuses on the interaction of the worker with machines, tools, work methods, and the work environment. The principle of ergonomic design is that "the machine should fit the worker, rather than forcing the worker to fit the machine."[7] Ergonomic studies in the laboratory and the field analyze the ways in which work practices, tool and machine design, and plant layout can conform to worker characteristics and capabilities. This model has been used for many years in European countries. It has proven useful in preventing cumulative trauma, such as repetitive motion disorders, as well as acute trauma in the workplace.

Worker Training and Education Programs

Worker training and education programs designed to prevent occupational injury are provided by a wide variety of organizations: employers, unions, state and local government agencies, hospitals, universities, and COSH groups. OSHA has included training requirements in several of its regulations. Worker training and education programs range from basic instruction on workplace hazards and the use of preventive measures (e.g., training in lifting techniques to prevent back injuries) to more complex educational programs that provide workers with the information and skills necessary to participate in hazard recognition and control.

The impact of these programs is difficult to evaluate. Some programs have been reported to be effective according to criteria such as the number of "safe behavior incidents" or other measures of worker performance.[50,51] However, most programs are evaluated using "process measures" such as the number of workers trained or the number of hours of training provided.[7]

In general, it is difficult to measure the success of worker training and education by the degree to which injury rates decrease for several reasons. Statistically, injuries resulting in lost work days are rare, and changes in the injury rate cannot easily be used to evaluate the effect of training (or other interventions) in a single workplace.[50] It is also very difficult to establish an adequate control group against which the trained work force can be compared. Few workplaces have similar work processes, machinery and products, environmental conditions, management styles, or work-force characteristics.

The reporting of injuries may actually increase as safety awareness increases through training and education programs. Increased reporting of injuries could be viewed as a positive effect of such a program. Finally, the goal of many training and education programs is not simply to improve the injury rate of a specific workplace or occupation. The goal may be to produce workers who act as safety specialists on the job, respond to new hazards as they arise, train their fellow workers in safety issues, and bring a labor perspective to occupational safety research, regulatory efforts, and programs. There is no simple way to measure the potential long-term impact on injury rates of this kind of worker education.

Current thinking about worker training and education is that, while rigorous evaluation of these programs is difficult, they are a necessary adjunct to the use of engineering controls, administrative controls, safe work practices, and personal protective equipment.[7] It has been suggested that worker training in occupational safety should begin before people enter the work force and are exposed to occupational hazards. Several high school–level occupational health and safety curricula are available. These programs provide students with a basic introduction to occupational safety, the recognition of hazards they may encounter in their work lives, and what they can do about such threats to their health.[52,53]

Interventions

The following interventions do not attempt to prescribe specific countermeasures for the thousands of hazards found in the nation's 4 million workplaces. Instead, they address a broad range of issues that are critical for the prevention of workplace injuries, such as stronger regulation, standard setting, and enforcement at the federal and state levels; the development of improved control technologies; and an expanded role for employees in creating safer workplaces.

Some of the interventions discussed in Chapter 6 also may be relevant to occupational traffic injuries. Little research has been done on the effects of preventive measures targeted specifically at intentional occupational injuries. Such measures include providing police officers with bulletproof vests and higher-powered weapons, limiting the amount of cash available in retail outlets, and placing bullet-resistant windows between cab drivers and their passengers. Because a large number of occupational homicides involve firearms, the interventions described in Chapter 17 may affect this type of interpersonal violence.

Several of the interventions discussed below, such as occupational health and safety consultancy programs and surveillance efforts, can be implemented by state and local health departments. Others, such as the criminal prosecution of negligent employers, require collaboration with other

(e.g., criminal justice) agencies. Those that require changes in the law, in the development and application of new standards, or in the allocation of public resources (e.g., restructuring workers' compensation programs or providing additional resources for OSHA enforcement) can be supported by public health practitioners in their roles as advocates for occupational safety and educators of the public and policy makers on the need for such changes. Finally, public health specialists have played important roles throughout the history of occupational safety and health efforts in this country. As researchers, practitioners, and advocates, they must continue to collaborate with other experts in the design, implementation, and evaluation of effective interventions.

Data Collection and Surveillance

The inadequacies of current statistics on occupational injuries and deaths have already been discussed. The National Research Council has made 24 detailed recommendations for the improvement of the BLS annual survey, OSHA reporting forms, the use of NCHS to obtain better estimates of the annual number of occupational fatalities, the evaluation of the record-keeping practices of individual establishments, and the dissemination of BLS occupational safety and health statistics.[10] A number of these measures could be applied at the state level to improve the quality of data available for surveillance and compliance activities. The American Association of Public Health passed a resolution at its 1988 annual meeting calling for a national uniform surveillance system for occupational illnesses and injuries.[54]

State health departments can play an important role in the surveillance of occupational injuries. The Colorado Department of Health produced a comprehensive picture of the state's occupational injury problem for the years 1982–84. Colorado's computerized surveillance system links information from state workers' compensation claim records, death certificates, the Fatal Traffic Accident Report system, Colorado OSHA, the U.S. Department of Labor's Mine Safety and Health Administration data base, and the U.S. Census. The Colorado Department of Health describes this system as including "data that are sufficiently accurate and complete to provide an understanding of Colorado's work-related injury and fatality problem."[27] The Department of Health is refining and expanding this system and plans to use the data it produces to develop and implement measures to reduce occupational injuries in Colorado.

The New Jersey State Department of Health is one of a number of NIOSH-funded Fatal Occupational Injury Surveillance Projects. This system links vital statistics records and medical examiner reports of fatal occupational injuries and forwards the results to OSHA for further action. The Occupational Health Service of the New Jersey Department of Health also conducts special studies in the area of occupational health, including injuries.[55]

A project in New Hampshire has shown that, despite certain limitations, workers' compensation records also can provide a valuable state-based method for collecting data on occupational injuries.[56]

Recommendation: States should improve the quality of data on occupational injuries and fatalities by linking workers' compensation files with OSHA inspection records and state vital statistics files, mandating the reporting of all fatal occupational injuries to the state agency responsible for workplace safety, and including information on occupation in hospital discharge data, state mortality files, and trauma registries.

Enforcement and Resource Allocation

OSHA enforcement. OSHA has never had adequate resources to accomplish the goals set for it by Congress. Recent cutbacks and new policies limiting workplace inspections have further weakened regulatory efforts. States that obtain federal funding to administer state OSHA programs have been adversely affected by national policies. The federal government determines staffing levels in these programs as well as the number of workplace inspections that can be conducted.[57] The results of research evaluating the effects of OSHA inspections, like the results of research analyzing the impact of government standards, are mixed. However, recent research has indicated that "OSHA inspections do appear to work in the short term . . . not by permanently correcting specific safety violations, but by temporarily instilling a greater consciousness of transient factors related to injuries."[4] This study also concluded that more frequent inspections may lead to further decreases in injury rates.

Recommendation: OSHA enforcement activities should be strengthened. The number of inspectors should be increased, larger penalties should be mandated and levied, and OSHA's "records inspections" policy should be eliminated. Enforcement activities should be monitored and their outcomes evaluated.

Standards. Federal standards for occupational safety are widely considered inadequate. Standards do not exist for some hazardous work situations. Most of the OSHA standards codified in 1971 were industry consensus standards that were meant to serve as interim standards until new ones were developed. Few new permanent standards have been issued, partly because of the time-consuming standard-setting process and because new or upgraded standards are frequently challenged by industry. As a result, the current standards are inadequate in the most hazardous industries (agriculture, mining, and construction); they still do not cover some well-recognized hazards, such as working in confined spaces; they do not take into account new technologies, such as robotics; and they do not incorporate the principles of ergonomics that have been found effective in preventing both acute and cumulative trauma. It is likely that the effect of standard setting has been diluted because standards have been inadequate and weakly enforced. Increased OSHA resources, streamlined standard-setting procedures, and strengthened enforcement could improve this situation significantly.

The experience of state regulation of occupational safety in California provides evidence of the effectiveness of standard setting for injury prevention. Before June 1987, California operated a state OSHA program for both public and private sector employees. The state promulgated approximately 2,400 safety standards, compared to 700 under federal OSHA. For certain occupational activities regulated by the state program but not by federal OSHA, dramatic reductions in injury rates were achieved. For example, a high-hazard construction permit system led to a 50% reduction in injuries and deaths associated with trenching.[58]

NIOSH has provided recommendations for preventive measures and limits of exposure for many workplace hazards. These recommendations can be used to formulate state or federal standards for protecting worker safety.[47]

State legislatures can also pass laws to protect workers from hazards where standards are inadequate. Additional or improved standards are especially needed to address falls from elevations, working in confined spaces, controlling hazardous equipment during maintenance and servicing, robotics, trenching and shoring during excavation, the use of precast concrete forms, and in building construction hazards related to the handling of heavy loads and repetitive motion.

Recommendation: OSHA and NIOSH should improve standards addressing specific workplace safety hazards. States that administer OSHA plans should move to adopt such standards in the absence of federal action. The use of such standards should be monitored and their outcomes evaluated.

Criminal prosecution for workplace deaths and injuries. In 1985 three managers of Film Recovery Systems, Inc., in Illinois received 25-year jail sentences for murder after a worker died of acute cyanide toxicity. The firm had deliberately withheld information from workers about the chemicals they used and had even removed labels to prevent workers from identifying dangerous substances. In the three years since that verdict, over 200 firms or executives in many states and industries have been charged with criminal violations when workers were harmed by hazards on the job.

Over half the states have criminal corporate liability laws, but some states have created programs specifically to address workplace deaths and injuries. For example, Los Angeles County has developed a program in which every workplace death is treated as a potential homicide, and an investigative team set up by the district attorney immediately goes to the site of every occupational fatality.

Although this trend toward local and state criminal prosecutions has not been evaluated to determine its impact on workplace injury and death rates, it may prove to be a strong deterrent to gross and negligent violations of the law by employers. Criminal prosecutions are not a substitute for effective regulatory enforcement programs, but they may be appropriate where existing penalties and enforcement mechanisms have failed to deter employers from knowingly creating a danger for employees. An official of the Los Angeles Environmental Crimes/OSHA Division expressed the view that "the regulatory process hasn't done the job. When a human life can be reduced to the 'cost of doing business,' something's got to give, and believe me, one company officer spending two days in jail is worth any fine [against the firm] you can impose."[59] However, state courts have overturned a number of these convictions on the grounds that the OSH Act supercedes such prosecutions at the state level. It is expected that one of these cases eventually will come before the U.S. Supreme Court for a final determination of the legality of the actions.[60]

Recommendation: Employers whose disregard of workplace safety results in serious injury or death should be prosecuted. Efforts to prosecute such employers should be monitored and their effects evaluated. However, recent questions about the legality of such prosecutions by states under the OSH Act

suggest that implementation of this strategy at the state and local levels be delayed until the Supreme Court has decided whether such actions are permissible under federal law.

U.S. Department of Agriculture extension safety specialists. Whether a state branch of the USDA Extension Services employs a safety specialist, and whether this specialist is employed full- or part-time, is at the discretion of the state land-grant college of agriculture. Such specialists usually are either faculty members or graduate students affiliated with these colleges. As of 1986, 32 states had such specialists, although the number fluctuates almost yearly (D. Baker, D. Murphy, personal communication). Even though no substantial evaluation of the activities of these specialists has been carried out, providing a permanent, full-time safety specialist position in every state extension service would be a first step toward upgrading the injury prevention and control activities of the USDA Extension Service.

Recommendation: Every state USDA Extension Service should include at least one full-time occupational health and safety specialist.

Legislative Reform

Public employees. State, county, and municipal employees face many of the same hazards and require the same protections as their counterparts in the private sector. Although a number of states include public employees in their OSH programs or have instituted PEOSH programs, more comprehensive coverage is desirable.

Recommendation: Federal and state laws protecting occupational safety and health should be expanded to include public employees. Such efforts should be monitored and their outcomes evaluated.

Workers' compensation. Although it is commonly believed that the cost to employers of workers' compensation is an economic incentive for workplace injury prevention, there is little empirical evidence to support this view. The U.S. Office of Technology Assessment has identified several factors that may limit the preventive aspects of compensation.

> First, all insurance schemes spread losses; therefore, the insurance function of workers' compensation means that employers who cause injuries do not bear their full costs unless they are self-insured or pay premiums that are directly tied to their injury and illness experience. Second, benefit levels represent less than the full social costs of injuries and illnesses.

Third, some injuries and most illnesses are not compensated because a claim is never filed, or they are inadequately compensated because the claim is denied or delayed. To the extent that these factors reduce the fraction of the costs of injuries and illnesses that are borne by employers, they reduce incentives for prevention.[7]

Reforms at the state level, including tying workers' compensation insurance premium rates to safety records, have the potential to increase the economic incentives for prevention in workers' compensation.

Recommendation: Workers' compensation systems should be structured so that employers with high rates of occupational injuries bear a greater proportion of the medical costs of these injuries, creating a financial incentive for lowering their firm's injury rate. The effects of such efforts should be evaluated.

Right to refuse hazardous work laws. Under OSHA law, workers have a limited right to refuse unsafe work. A worker is protected from punishment for refusing unsafe work if the worker has a reasonable belief that there is a real, imminent danger of death or serious injury; the worker has sought correction of the dangerous condition from his or her employer; the danger is so imminent that it cannot be eliminated quickly enough through normal OSHA procedures; or the worker has no reasonable alternative. Situations in which all these conditions are met are rare, and this provision is enforced exclusively at the discretion of the secretary of labor. Employees can also bring unfair labor practice charges under the National Labor Relations Act to secure protection against discrimination for refusing unsafe work. However, the worker must present "ascertainable, objective evidence" supporting his or her judgment that an "abnormally dangerous condition" exists.

It is difficult to estimate the number of injuries and deaths that would be avoided if employees had full legal protection in refusing unsafe work. In countries where this right is protected, it appears to provide an incentive for management and labor to reach agreement on safety issues, and the right is rarely, if ever, abused.[61]

Statutory protection for the health and safety "whistleblower" is inadequate. Extensive problems with OSHA's handling of worker complaints of employer retaliation for health and safety activities have been reported.[62] Recent legislative initiatives were designed to create stronger and more uniform protection in this area. Several bills currently before Congress provide protection for a worker who refuses hazardous work, discloses employer viola-

tion of a federal law, or participates in a federal agency proceeding regarding the unsafe activities of an employer.

Recommendation: States and the federal government should enact legislation to ensure the right of workers to refuse hazardous work. Such legislation must prohibit discrimination against employees who exercise this right and protect employees who report unsafe working conditions or practices.

Labor–management health and safety committees. Joint labor–management committees are an integral part of occupational safety efforts in a number of countries. In the Canadian province of Quebec, establishments of 20 or more workers are required to have joint committees. These committees have an equal number of labor and management representatives and have the authority to approve safety training plans, select protective equipment, investigate injuries, and resolve employee complaints. In Sweden, committees are mandated in all workplaces of more than 50 employees and have one more labor than management representative. These committees can veto plans for new machines, materials, or processes that present safety or health hazards, decide how to spend the firm's health and safety budget, and shut down unsafe operations.

Joint labor–management health and safety committees exist in many U.S. workplaces, as well, and have been encouraged by OSHA. A 1986 survey found that 49% of collective bargaining agreements called for joint committees.[63] The state of Washington also mandates joint committees in all workplaces of 11 or more employees. Although some of these committees have the responsibility to correct and prevent hazardous conditions in the workplace, many others have only advisory power.[63]

Evaluations to date of joint labor–management health and safety committees indicate that their effectiveness depends on whether they are invested with the power to make changes in the workplace and whether both management and labor are committed to them as a vehicle for improving workplace safety. A study by the Center for the Quality of Work Life indicated that labor–management committees increased management concern for safety and were effective in identifying hazardous conditions. Joint programs in construction are credited with a 40% reduction in injuries.[63] Another study concluded that, while some labor–management committees seem to improve workplace safety, as measured by OSHA complaint inspections and serious citations, they "may have little or no impact in situations where they are only formalities."[64] A

1988 report from the U.S. Department of Labor stated

Labor–management committees for occupational safety and health have an important role to play in the overall efforts to provide safe and healthful workplaces for all working men and women. As currently structured, however, their role will be most often limited to solving such problems as access to information, plant monitoring, medical surveillance, training, housekeeping, and use of personal protective equipment. Resolving problems that require reorganization of a worksite, redesign, or investment in capital requires a committee structure with significantly more authority than most in existence in the United States today.[63,64]

Recommendation: Joint labor–management health and safety committees with the authority to correct unsafe working conditions are a promising approach. Such committees should be mandated through federal law or state initiatives, or both. Their activities should be monitored and their outcomes evaluated.

Other State and Local Activities

Workers' compensation–funded state occupational health and safety consultancy programs. One promising approach for state involvement in occupational health and safety activities is that of Ohio's Division of Safety and Hygiene (DSH). This agency was organized in 1925 within the Ohio Industrial Commission, which oversees the state's workers' compensation program. The DSH is supported by an assessment of 1% of the premiums paid into Ohio workers' compensation program. There are seven regional offices within the state of Ohio.

DSH has no regulatory authority but engages in a wide variety of activities relating to occupational safety and health. It provides consultants to companies requesting technical assistance in making their workplaces safer. DSH staff members use a team approach and work with representatives from labor and management, training them in workplace hazard recognition and correction. Staff members specializing in several particularly hazardous industries (such as construction) are available. The DSH helps create an occupational safety and health program for participating companies. Chemical hazard and ergonomic inspections also are available upon request.

DSH publishes a monthly magazine, the *Ohio Monitor*, and produces and disseminates a wide assortment of educational materials, including pamphlets, posters, and warning stickers. It operates an occupational safety and health library that is open to the public and has a substantial collection of films and video tapes on specific aspects of occupational safety and health for loan to businesses and other groups. DSH provides speakers for presentations to unions, employee groups, trade associations, and community organizations. It provides technical assistance to trade associations and chambers of commerce that wish to become involved in occupational safety and health activities.[65,66]

The DSH Research and Statistics Section produces comprehensive annual reports on occupational injuries within Ohio as well as special reports on particular types of injuries.[67-70] It also engages in special research projects in collaboration with the Ohio Health Department, the University of Cincinnati, and NIOSH.[22,25,71,72]

Recommendation: State occupational health and safety consultancy programs funded by an assessment on workers' compensation premiums are a promising approach. Such programs should be used and monitored and their outcomes evaluated.

Small businesses. Programs that bring the experience of health and safety professionals to small firms could improve worker safety in this sector of the economy. It is difficult for many small businesses to bear the costs of occupational safety programs and services. States might provide organizational and financial assistance for the start-up of cooperative efforts through trade associations, workers' compensation carriers, or other mechanisms. Group cooperative occupational safety and health services have proved successful in a number of Western European countries.[7,73]

Recommendation: Agencies responsible for occupational safety should encourage the sharing of resources among small businesses for functions such as recordkeeping, industrial hygiene, safety engineering, medical surveillance, occupational health services, and worker training.

Safety contests and award programs. Some incentive programs have been at least temporarily effective in promoting positive health behaviors such as safety-belt use. Programs that award employees with certificates or pay raises for low injury rates sometimes are used in industry with the stated goals of fostering safety awareness, reducing injuries, and cutting compensation costs. Some firms have reported decreases in reported injuries after using such competitions.[51]

However, these programs provide a strong incentive for employees and supervisors not to report injuries, or to downplay their seriousness. Given the pervasive problem today of underreporting occupational injuries, these programs should be discouraged. Safety contests for workers also provide little incentive for employers to make improvements in equipment design or work practices that may cause or contribute to workplace injuries. A poll of industrial safety officers and National Safety Council staff members rated such promotions to be of least importance in industrial safety programs.[74]

Recommendation: Incentive awards and contests among employees should not be used to attempt to reduce injury rates.

INTERVIEW SOURCES

David Baker, Secretary-Treasurer, National Institute for Farm Safety, Washington, DC, January 19, 1989

William Field, PhD, Safety Extension Specialist, Purdue University, West Lafayette, Indiana, January 13, 1989

Dennis Murphy, PhD, Associate Professor of Agricultural Engineering, Pennsylvania State University, University Park, January 23, 1989

REFERENCES

1. Ashford N. Crisis in the workplace: occupational disease and injury. Cambridge, Massachusetts: MIT Press, 1970.

2. Assessment of OSHA and NIOSH activities. Preventing injury and illness in the workplace. Washington, DC: Office of Technology Assessment, 1985(Apr):257–72; report no. OTA-H-256.

3. Viscusi W. The impact of occupational safety and health regulation, 1971–1983. Rand J Econ 1986;87:567–80.

4. Robertson L, Keene J. Worker injuries: the effects of workers' compensation and OSHA inspections. J Health Polit Policy Law 1983;8(3):581–97.

5. National traumatic occupational fatalities, 1980–1984. Morgantown, West Virginia: National Institute for Occupational Safety and Health, 1987 Apr 27 (revised Jun 1987).

6. Stout-Wiegand N. Fatal occupational injuries in US industries, 1984: comparison of two national surveillance systems. Am J Public Health 1988;78(9):1215–7.

7. Preventing illness and injury in the workplace. Washington, DC: Office of Technology Assessment, 1985(Apr).

8. Accident facts: 1988 edition. Washington, DC: National Safety Council, 1988:34.

9. Bureau of Labor Statistics. Record-keeping guidelines for occupational injuries and illnesses. Washington, DC: US Government Printing Office, 1986.

10. Counting injuries and illnesses in the workplace: proposals for a better system. Washington, DC: National Research Council, 1987.

11. BLS reports on survey of occupational injuries and illnesses in 1986. BLS News 1987, Nov 12.

12. National traumatic occupational fatalities 1980–1984. Morgantown, West Virginia: National Institute for Occupational Safety and Health, 1987 Jun 11.

13. Wilkerson I. Farms, deadliest workplace, taking the lives of children. New York Times 1988 Sep 26:1.

14. Goodman RA, Sikes RK, Rogers DL, Mickey JL. Fatalities associated with farm tractor injuries: an epidemiologic study. Public Health Rep 1985;100:100–3, 329–33.

15. Beatty M, Zook E, Russell R, Kinkead L. Grain auger injuries: the replacement of the corn picker injury? Plast Reconstr Surg 1982(Jan):96–102.

16. Helgerson S, Micham S. Farm workers electrocuted when irrigation pipes contact powerlines. Public Health Rep 1988;100(3):325–8.

17. Centers for Disease Control. Prevention of leading work-related diseases and injuries. Morbid Mortal Weekly Rep 1984;33:16.

18. Smith J, et al. Farm-tractor associated deaths—Georgia. Morbid Mortal Weekly Rep 1987;36(37):481–2.

19. Karlson T, Noren T. Farm tractor fatalities: the failure of voluntary safety standards. Am J Public Health 1979;69(2):149.

20. Baker S, Samkoff J, Fisher R, van Buren C. Fatal occupational injuries. JAMA 1982;248(6):692–7.

21. Kraus J. Homicide while at work: persons, industries and occupations at high risk. Am J Public Health 1987;77(10):1285–8.

22. Hales T, Seligman PJ, Newman SC, Timbrook LL. Occupational injuries due to violence. J Occup Med 1988;30(6):483–7.

23. Davis H. Workplace homicides of Texas males. Am J Public Health 1987;77(10):1290–3.

24. Davis H, et al. Fatal occupational injuries of women, Texas, 1975–84. Am J Public Health 1987;77(12):1524–7.

25. Seligman PJ, Newman SC, Timbrook LL, Halperin WE. Sexual assault of women at work. Am J Industrial Med 1987;12:445–50.

26. Robinson J. Job hazards and job security. J Health Polit Policy Law 1986;11(1):1–17.

27. Colorado population-based occupational injury and fatality surveillance system report, 1982–1984. Denver: Health Statistics Section, Colorado Department of Public Health (undated).

28. Seligman P, Sieber W Jr, Pederson D, Sundin D, Frazier T. Compliance with OSHA record-keeping requirements. Am J Public Health 1988;78(9):1218–9.

29. Update: universal precautions for transmission of human immunodeficiency virus, hepatitis B virus, and other bloodborne pathogens in health-care settings. Morbid Mortal Weekly Rep 1988;37(24):377–88.

30. Cohen A, Smith M, Cohen H. Safety program practices in high vs. low accident rate companies: an interim report (questionnaire phase). Cincinnati, Ohio: National Institute for Occupational Safety and Health, 1975.

31. Davis RT, Stahl RW. Safety organization and activities of award-winning companies in the coal mining industry. Pittsburgh, Pennsylvania: US Department of the Interior, Bureau of Mines, 1964.

32. Shafai-Saharai Y. An inquiry into factors that might explain differences in occupational accident experience of similar size firms in the same industry. East Lansing, Michigan: Division of Research, Graduate School of Business Administration, Michigan State University, 1971.

33. Jenson R, Sinkule E. Press operator amputations: is risk associated with age and gender? J Safety Res 1988;19(3):125–33.

34. Jones I, Stein H. Effects of driver hours of service on tractor-trailer crash involvement. Washington, DC: Insurance Institute for Highway Safety, 1987.

35. Hagglund G. What causes accidents and injuries? Madison, Wisconsin: University of Wisconsin School for Workers, 1980.

36. Rivara F. Fatal and nonfatal farm injuries to children and adolescents in the United States. Pediatrics 1985; 76(4):567–73.

37. Field WE, Tormoehlen RL. Analysis of fatal and nonfatal farm accidents involving children. Paper presented at the winter meeting of the American Society of Agricultural Engineers, Chicago, Illinois, December 1987.

38. Cogbill T, Busch H, Stiers G. Farm accidents in children. Pediatrics 1985;76(4):562–6.

39. Robinson JC. Racial inequality and the probability of occupation-related injury or illness. Milbank Mem Fund Q/Health Soc 1984;62(4):568.

40. Boston Women's Health Book Collective. Our jobs, our health: a woman's guide to occupational health and safety. Boston: Massachusetts Coalition for Occupational Safety and Health, 1983:22.

41. Back injuries associated with lifting (work injury report). Washington, DC: Bureau of Labor Statistics, 1982(Aug).

42. Bond MD. Low back injuries in industry. Indust Med Surg 1970;39:204–8.

43. Levitt SS, Johnson TL, Beyer RD. The process of recovery: patterns in industrial back injury, part 1. Indust Med Surg 1971;40:7–14.

44. Klein BP, Jenson AC, Sanderson LM. Assessment of workers' compensation claims for back strains/sprains. J Occup Med 1984;26:443–8.

45. Proposed strategies for the prevention of leading work-related diseases and injuries. Washington, DC: Association of Schools of Public Health, 1986:20.

46. Occupational injuries and illnesses in the United States by industry, 1980. Washington, DC: Bureau of Labor Statistics, 1980.

47. NIOSH recommendations for occupational safety and health standards: 1988. Morbid Mortal Weekly Rep 1988;37(s-7).

48. Occupational health service: background and overview. Trenton, New Jersey: Occupational Health Service, New Jersey Department of Health, 1987(Sep).

49. Crowell K. PEOSHA: occupational health. New Jersey Municipalities 1987(Feb):16–17.

50. Cohen HH, Jensen R. Measuring the effectiveness of an industrial truck lift safety training program. J Safety Res 1984;15:125–35.

51. Cohen A, Smith M, Anger WK. Self-protective measures against workplace hazards. J Safety Res 1979; 272:287.

52. Job hazard recognition program. Anchorage, Alaska: Alaska Health Project, 1983.

53. Improving health and safety in the workplace. Newton, Massachusetts: Education Development Center, 1982.

54. APHA endorses some state health insurance efforts. Nation's Health 1988(Dec):9–10.

55. Surveillance of occupational illnesses, injuries, and hazards in New Jersey. Trenton, New Jersey: Division of Occupational and Environmental Health, Occupational Health Service, New Jersey Department of Health, 1988(Jul).

56. Use of workers' compensation claims of worker-related illnesses, New Hampshire, January 1986–March 1987. Morbid Mortal Weekly Rep 1987;36(43):703–20.

57. DeVito A. Testimony before the New York State Assembly committees on labor and governmental employees. New York: New York State Public Employee Federation, 1988 Mar 2.

58. Schaffer A. Cal OSHA and federal OSHA: significant differences. Sacramento, California: Senate Industrial Relations Committee, 1987(Dec).

59. Local officials set up new program as others urge caution over trend to use criminal prosecution for workplace hazards. Occup Safety Health Rep 1986(Apr 10):1132–6.

60. Glaberson W. States are toppling workplace-injury convictions. New York Times 1988 Sep 19:1.

61. Witt M, Early S. The worker as safety inspector. Working Papers 1980;7(5):21–9.

62. The deadly dilemma: when OSHA fails to protect the worker's right to a workplace free of health and safety hazards. Madison, Wisconsin: Wisconsin Committee on Occupational Safety and Health, 1987(Oct).

63. The role of labor–management committees in safeguarding worker safety and health. Washington, DC: US Department of Labor, 1988.

64. Boden L, Hall J, Levenstein C, Punnett L. The impact of health and safety committees. J Occup Med 1984; 26(11):834–44.

65. 1989 catalog. Columbus, Ohio: Division of Safety and Hygiene, Industrial Commission of Ohio, 1989.

66. Training catalog. Columbus, Ohio: Division of Safety and Hygiene, Industrial Commission of Ohio, 1988 (revised).

67. Ohio 1987 occupational injury and illness statistics. Columbus, Ohio: Division of Safety and Hygiene, Ohio Industrial Commission, 1987.

68. Ohio 1980–1985: average costs—occupational injuries and illnesses. Columbus, Ohio: Division of Safety and Hygiene, Ohio Industrial Commission (undated).

69. Highlights of 1985 industrial accidents in Ohio resulting in back injuries. Columbus, Ohio: Division of Safety and Hygiene, Ohio Industrial Commission (undated).

70. Highlights of 1985 industrial accidents in Ohio resulting in eye injuries. Columbus, Ohio: Division of Safety and Hygiene, Ohio Industrial Commission (undated).

71. Seligman PJ, Halperin WE, Mullan RJ, Frazier TM. Occupational lead poisoning in Ohio: surveillance using workers' compensation data. Am J Public Health 1986;76(11):1299–1302.

72. Sinks T, Mathias CGT, Halperin W, Timbrook C, Newman S. Surveillance of work-related cold injuries using workers' compensation claims. J Occup Med 1988;30(6):492–502.

73. Ashford N. The special problem of the small firm. In: Crisis in the workplace: occupational disease and injury. Cambridge, Massachusetts: MIT Press, 1970:366–73.

74. Planek T, Driessen G, Vilardo FJ. Evaluating the elements of an industrial safety program. Nat Safety News 1967(Aug):60–3.

Chapter 10: Violence and Injury

The Dallas police found the body of William Burdette Perrin, Jr., slumped over a chair in John Curtis's office with a single bullet wound. The two lawyers had been drinking vodka. There was no apparent motive, police said. In Atlanta, 2-year-old Larry Lamar Bass was rushed to a hospital with a gunshot wound. Two of his relatives had been arguing over a portable radio. At New York's Pace University, a 21-year-old student was robbed and raped at knifepoint in a campus bathroom. And in Milwaukee, under new guidelines, the police began arresting all domestic violence suspects, even if the victim declined to press charges. They made 39 arrests in 31 hours. These were among the many stories that appeared in the morning papers on Friday, May 2, 1986.

That Friday would be a landmark day, but not because of Perrin, Curtis, Bass, or the other victims of violence. On that day, 200 photographers around the country took nearly a quarter million pictures, a selection of which were published as *A Day in the Life of America*.[1] America in that book is slick and beautiful—and without violence. But certainly no day—or place—in America can be understood completely if violence is ignored.

That America has had a violent history has long been accepted, both in fact and folklore. That violence is among our most pressing health problems is a much more recent realization. "Identifying violence as a public health issue is a relatively new idea," wrote Surgeon General C. Everett Koop in 1985. "Over the years we've tacitly and, I believe, mistakenly agreed that violence was the exclusive province of the police, the courts, and the penal system. To be sure, those agents of public safety and justice have served us well. But when we ask them to concentrate more on the prevention of violence and to provide additional services for victims, we may begin to burden the criminal justice system beyond reason."[2]

Injury prevention practitioners have an important role to play in understanding, controlling, and preventing violence in their communities. That some are already playing such a role is apparent from the interventions discussed in Chapters 11 through 17. We look at definitions of violence and the interconnections among the concerns of public health, crim-

inal justice, and the behavioral sciences. But the impact on the nation's health is clear from a brief look at recent statistics.

In 1986 nearly 53,000 deaths resulted from violence, including 21,400 homicides and 31,470 suicides.[3] Suicide is the eighth leading cause of death, and homicide is the 11th. And when all deaths are calculated in terms of years of potential life lost (YPLL), homicide ranks fourth and suicide sixth.[3] The extent of the problem varies dramatically by population group; among black men aged 15–34, homicide is the leading cause of death.[4]

Fatal outcomes are recorded in death certificates, but information about nonfatal violence is less well reported. As a result, we know less about nonfatal violence and the injuries it produces. In 1986, according to the FBI, there were 834,322 serious assaults.[5] However, when an earlier study compared FBI assault data with hospital records for the same period, it was found that nearly four times as many assault injuries are recorded in hospital records than had been reported to the police.[6]

Why is violence a concern for public health agencies and practitioners? What contributions can they make to its understanding and prevention? What have we learned about violence through public health's approach? And how can this approach and growing expertise best contribute to a multidisciplinary effort uniting health care, criminal justice, social service, and mental health professionals, who have long been in the forefront of society's response to violence? These are among the questions we consider before surveying the state of the art in violence prevention. One barrier to cross-agency collaboration is the lack of a common vocabulary. It is important, therefore, to look at some of the terms used in this chapter.

DEFINITIONS

Violence is defined most broadly as the use of physical force with the intent to inflict injury or death upon oneself or another. Our understanding of violence gains from perspectives contributed by several different disciplines. For public health, the fact that violent acts may result in physical injuries is the primary motivation for involvement. The be-

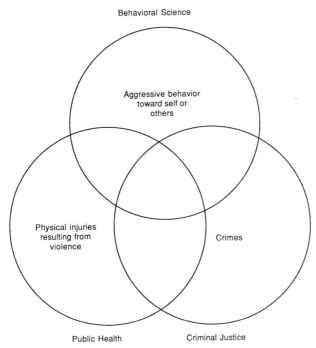

Figure 1. Overlapping perspectives of public health, criminal justice, and behavioral science.

havioral sciences see violence as forms of aggressive human behavior that can harm individuals and their families and communities. And criminal justice sees it as violations of the law, that is, criminal acts.

These perspectives overlap (Figure 1). Thus, a parent who severely beats his or her child would be of concern to each profession. Indeed, many of the injury-causing behaviors described in these chapters concern professionals from all three disciplines, but not all of the behaviors concern each discipline.

The perspectives of these disciplines do not overlap in several ways. Not all criminal acts are violent, nor do they all cause physical injury. Most burglaries lack violence and physical injury, for example; nor do all injuries resulting from violence involve criminal violations. A police officer who kills someone in the line of duty commits justifiable homicide, which is not a criminal act. And a parent's aggressive behavior toward a child may cause neither physical injury nor violate the law. Finally, domestic violence, long a source of injuries that were not treated criminally, only now appears on the agendas of all three disciplines.

The differences among disciplines are often reflected in their vocabularies. It is important, therefore, to clarify the terms that are used here and in subsequent chapters.

The term "violence" sometimes refers to specific activities or behaviors and sometimes to the outcomes (injuries) of those behaviors. The usage, in each case, will be apparent from the context. "Assault" is used generally to denote violent actions resulting in injury, whether fatal or nonfatal. When assault or "aggravated assault" is used to refer to specific criminal violations, this is noted in the text, and the term "sexual assault" is defined in Chapter 15. "Domestic violence" is generally used to encompass both spouse abuse (isolated acts of violence directed by one partner against the other) and "woman battering" (a syndrome of violent and controlling behavior that includes injury). As Chapter 13 makes clear, however, woman battering is the focus of interest in domestic violence in this book. The term "abuse," as in "child abuse" and "elder abuse," is used to denote violent acts occurring over time in intimate relationships.

The relationship between offender and victim is an important factor in interpersonal violence. Both homicides and nonfatal assaults are categorized, where possible, as occurring between family members, friends, acquaintances, or strangers.[5,16]

Injuries caused by interpersonal violence and suicide are sometimes characterized as "intentional" and contrasted with the "unintentional" (traffic, home, recreational, and occupational) injuries described in the preceding chapters.[7] However, the distinction is not made in this book for two reasons. First, although it may be convenient to make this distinction, some types of interpersonal violence—such as involuntary manslaughter—are not intentional.[8] At the same time, as was pointed out in Chapter 9, some types of occupational injuries may result from deliberate, knowing conduct. Second, writers in public health have used the term "intentional" in a very general sense to differentiate interpersonal violence and suicide from other forms of injury. But for criminal justice, intent is precisely defined, and establishing intent is often a critical element in proving guilt.[8]

In criminal justice, an action that causes injury may or may not be a crime, depending on the defendant's intent, or lack of intent, to cause harm. However, even this distinction lacks clarity; increasingly the criminal justice system has recognized, as does public health, that the use of alcohol and drugs and other social circumstances can play an important role in precipitating injuries. And, finally, it should be noted that the criminal justice system is involved in both intentional and unintentional injuries. The police and courts have long played an important role in the control of traffic injuries, for example.

Focusing on the injuries caused by interpersonal

violence and suicide rather than on the intentionality of the behavior does not mean that public health practitioners see the assessment of blame as unimportant. There is no suggestion that we give up the traditional, normative approach to crime on which society has relied. Criminal law plays an important role in educating members of every community about their obligations to one another.

There is no question that the assaults described in this and subsequent chapters are heinous. But by avoiding intentionality, by remaining free of the responsibility to determine who is at fault, public health practitioners are able to analyze violence from a larger perspective. They can focus on relationships, on circumstantial factors, and on the role of alcohol and firearms. They can complement the criminal justice system, which is preoccupied with the primary question of defining guilt.

VIOLENCE AND PUBLIC HEALTH

The morbidity and mortality of interpersonal and self-directed violence can be understood epidemiologically in terms applied to many other public health problems using factors of agent, host, and environment. Their patterns of occurrence can be studied, and public health approaches can be brought to bear on their prevention.

The public health approach to interpersonal violence and suicide is recent and still developing. It is a multidisciplinary process involving many other participants, some of whose efforts (e.g., the police) date from the last century. The beginning of an explicit concern within the public health community about violence can be traced to the mid-1970s. In part, this concern reflected attitudes of the wider society. During the preceding decade, through war, political assassination, and yearly increases in crime, Americans became obsessed with violence. Public opinion surveys, national commission reports, and the rhetoric of political campaigns all reflected a growing fear of violence.[9]

Since the first American police departments were established in Boston (1838) and New York (1845), the criminal justice system of police, courts, and corrections has served as society's bulwark against violence. Their primary interventions have been deterrence and incapacitation. Deterrence is the attempt to prevent crime through the threat of punishment. "Special" or "specific" deterrence seeks to deter an individual from commiting future crimes by imposing penalties for current criminal conduct. "General" deterrence relies upon the symbolic effect that the punishment of one individual may have on other potential criminals.[10] Incapacitation "involves removing an offender from general so-

ciety and thereby reducing crime by physically preventing the offender from committing crimes in that society."[11] Thus, the emphasis in police and criminal justice responses to violence has been on the apprehension of persons who have already committed the types of acts that public health would like to prevent.

During the 1970s, however, police administrators and criminologists began to recognize the limits of traditional crime-fighting techniques, the importance of prevention, and the contributions that other institutions could make. "Many of the problems coming to the attention of the police become their responsibility because no other means has been found to solve them. They are the residual problems of society. It follows that expecting the police (alone) to solve or eliminate them is expecting too much."[12] This is how one expert summarized the problem at the decade's end. It was in this societal and law enforcement context that public health's interest in injuries caused by violence grew, especially at the Centers for Disease Control (CDC).

"In 1977, when I became director of CDC," said Dr. William Foege, "we had a committee of outside experts advise us about priorities. They looked not only at standard mortality and morbidity rates, but also investigated the whole question of premature mortality, or productive years lost. While heart disease, cancer, and stroke led the list of causes of death, the leading causes of premature death were what we used to call 'accidents', homicide and suicide.

"We began looking at premature mortality, at what was causing the problem, and at what public health could do about it. There were many internal discussions. I traveled a great deal in those days, to speak to state health groups, and I always raised the issue of violence" (W. Foege, personal communication).

Simultaneously, CDC and other health agencies were helping to develop the surgeon general's first national agenda for health promotion and disease prevention. Published in 1979, *Healthy People* outlined 15 priority areas requiring national attention, including interpersonal violence.[13,14] The new CDC interest also encouraged health professionals from around the country to begin meeting to develop the health objectives for 1990 that were published in *Promoting Health/Preventing Disease: Objectives for the Nation*. Among the 220 specific health objectives were items addressing homicide, child abuse, and suicide.[15]

Work on violence continued in 1981 and 1982 in CDC's newly created Center for Health Promotion and Education. In 1983 the establishment of the Violence Epidemiology Branch provided new focus and visibility. In 1986 all of the agency's injury programs were reorganized into the Division of Injury Epidemiology and Control.

The year 1985 was significant in the developing public health role in addressing violence. In October, two important conferences reflected public health's growing interest in violence prevention. The New York Academy of Medicine's "Symposium on Homicide: The Public Health Perspective" was followed by Surgeon General Koop's "Workshop on Violence and Health." CDC and University of California at Los Angeles released *The Epidemiology of Homicide in the City of Los Angeles, 1970–1979*, a collaborative study of the demographic and situational patterns revealed in 4,950 homicide cases.[16]

The following year, the Secretary's Task Force on Black and Minority Health published its *Report of the Subcommittee on Homicide, Suicide, and Unintentional Injuries*. Building on the pioneering work of scholars such as Alvin Poussaint[17] and Ruth Dennis[18] and mandated to investigate minority population health problems marked by higher death rates than those found among the general population, the committee concluded that a great majority of these "excess" deaths resulted from interpersonal violence and suicide.[19]

Thus, within a relatively short period, it became clear that the toll of violence was a public health problem and, more important, that public health approaches could be brought to bear in helping to prevent violent death and injury. But what, precisely, could public health contribute?

PUBLIC HEALTH'S CONTRIBUTIONS TO VIOLENCE PREVENTION

Public health's first, critical contribution to the prevention of violence came with the very recognition that violence—by virtue of the enormous toll it takes in lives, health, and quality of life—is a health problem.[20] That identification opened the way for the application to violence of sophisticated epidemiologic techniques, the use of surveillance and other data collection systems, the identification of high-risk groups, and the development and implementation of preventive strategies.

The criminal justice system has focused on offenders, with less emphasis on victims and the victim-offender relationship. It has had difficulty "in attempting to approach the problems of crime from the point of view of the victim."[21,22] Thus, public health complements criminal justice's emphasis by looking at large numbers of cases and searching for underlying patterns. By identifying these patterns, epidemiologists seek to understand who is at high risk and what risk factors are associated with particular types of violence.

One important pattern first noted by criminologists and enlarged on by epidemiologists is that, contrary to popular belief, victims and perpetrators very often know each other. In 1977, for example, criminologists at New York City's Vera Institute of Justice asked why relatively few felony arrests resulted in felony convictions in the city's courts. To their surprise, they discovered that a high percentage of the cases that began as felonies were not prosecuted as such because "in every crime category from murder to burglary [these cases involved] victims with whom the suspect had had prior, or close, relations." There is an "obvious but often overlooked reality," they concluded. "Criminal conduct is often the explosive spillover from ruptured personal relations among neighbors, friends, and former spouses."[23] Many of these cases were not prosecuted because the victims refused to cooperate, or the cases were reduced in seriousness (to a nonfelony level) after the circumstances were understood.

As noted above, public health provides a model of injury resulting from interactions among the host, agent, and environment. This model highlights the importance of victim–offender relationships and establishes a framework through which these relationships can be understood and analyzed. This is one of the most important contributions epidemiologic analysis has made to violence prevention.

As in other areas of injury, the new knowledge revealed by epidemiologic research and data collection can become the basis for fashioning new preventive measures. In addition to improving the knowledge base for prevention, public health offers a new ally to those already working in violence prevention. This practical contribution is reflected in many of the interventions described in subsequent chapters.

Public health also addresses itself to the social norms and attitudes that accept violence and deter prevention. And against a long-standing public perception that violence is inevitable, public health defines it instead as "a concern to be addressed and remedied, not an inalterable fact of life."[24] Finally, defining violence as an important health problem brings public health and other health professionals into the growing constituency actively seeking to reduce the level of violence in America.

HOMICIDE: THE DIMENSIONS OF THE PROBLEM

Each of the behaviors described in Chapters 11 through 17—assaultive violence, child abuse and child sexual assault, domestic violence, elder abuse, and rape and sexual assault—can result in death as well as nonfatal injury. One group of researchers has argued persuasively that "homicide represents the final common outcome of assaultive behaviors that are both very diverse in their characteristics and many times more common than homicide."[25] Rather than view homicide prevention as a separate undertaking, we can therefore see it as depending on the prevention of behaviors that can culminate in homicide. For this reason, there is no separate chapter devoted to homicide; instead, each chapter, from 11 through 17, presents information about fatal outcomes.

Nevertheless, to understand the connection between violence and injury, it is useful to consider what we know about homicide. Although nonfatal outcomes of violence are far more numerous, we know more about homicide because we possess better information about the fatalities.

The definition of homicide varies not only from jurisdiction to jurisdiction but also from one reporting agency to another. The 21,400 homicides reported by the National Center for Health Statistics (NCHS) for 1986 include deaths resulting from "legal intervention." The FBI *Uniform Crime Reports* (UCR) for 1986 cites 20,613 instances of "murder and nonnegligent manslaughter" but does not count justifiable killings by law enforcement personnel.

NCHS homicide statistics are compiled from death certificates filed with local health authorities. The UCR is based on police reports filed with local law enforcement agencies. The system is voluntary, however, and not all police agencies choose to participate. A comparative analysis of NCHS and UCR homicide data for 1976–82 revealed that "the NCHS system reported an average of 9% more cases (each year) than did the FBI."[26] Part of this discrepancy results from the inclusion by NCHS of noncriminal ("legal intervention") homicides; part results from "incomplete voluntary reporting to the FBI by participating law enforcement agencies and lack of reporting by nonparticipating agencies."[26] The authors conclude, however, that "the existence of two national data sets on a cause of death such as homicide is a luxury seldom encountered in surveillance."[26] While NCHS reports are more complete, FBI data are more timely.

Using both data sets, epidemiologists have learned a great deal about homicide. Death certificates provide data to establish demographic, temporal, and spatial patterns. The use of firearms and the role of alcohol and other associated factors have been clarified. And because homicides frequently arise from many of the same circumstances that produce nonfatal outcomes, homicide data shed some light on nonfatal injuries as well.

In summary, research suggests that most homicide victims are young males who frequently are members of minority groups and who usually know their killers. Although the largest number of homicides occurs among whites, more lives are lost to homicide among blacks 15–34 years of age than to any other cause of death, including other injuries, suicide, heart disease, and cancer. Young black males are at especially high risk of death from homicide, with mortality rates 5 to 12 times greater than those of young white males.[27] (In fact, as noted below, it is socioeconomic status, rather than race, that accounts for these differences.)

The disparity between the races (and the sexes) can be shown by comparing years of potential life lost (YPLL) by sex and race. In 1985, YPLL related to homicide was 99.4 years per 100,000 white females, 273 years per 100,000 white males, 403.9 years per 100,000 black females, and a staggering 1,669.3 years per 100,000 black males.[28]

Homicides generally do not occur during the commission of a crime but arise from arguments and other noncriminal circumstances. Most homicides occur among members of the same race and usually involve firearms.[29] Alcohol and drug consumption are associated with homicides, except homicides of children.[30] The evidence for a link between drug use and homicide is less clear, although many homicides may be related to the business of buying, selling, or stealing drugs and to money earned in drug transactions.[31] Homicide rates are highest in large cities and, for both blacks and whites, in the West.[32] In any given year, the highest number of homicides is likely to occur in July, August, and December.[33]

VIOLENCE IN CONTEXT

Homicide is a form of injury in which multiple risk factors are often at play. In addition to the factors cited above, public health is concerned with the wider context in which violence occurs: the social norms, attitudes, and models that can encourage violent behavior. Some of these associations are less precise than the risk factors identified epidemiologically, but they are of interest nonetheless.

Culture and Attitudes

By the time a typical adolescent graduates from high school he or she will have been exposed to 18,000 television murders and 800 suicides.[34] And that does not count motion pictures, of which adolescents are major consumers.

The relation between televised acts of violence and individual behavior may be controversial, but it is nevertheless instructive to remember that television and motion pictures are the conduits of America's myths, of the stories that provide both models for and affirmation of everday behavior. There is evidence to suggest that the image of violence as an acceptable and effective tool for solving problems, whether across international borders, on the street, or around the home, may spill over into real behavior.

Other attitudes play a role as well. Child-rearing practices that convey to children the acceptability of violence can result in aggressive behavior both within and outside the home. The ideology of male dominance and acceptance of the need to "prove" such dominance when it is challenged are deeply ingrained in the culture. It is associated with killings of family members, friends, and strangers.[35] Homophobia and the general discrimination against gays have created a climate in which physical assaults against gays are not uncommon.[36] Racism and racial discrimination also stimulate violence.[37] This may be appreciated more easily in interracial violence, because racism motivates specific acts of violence (e.g., the home of a black family newly arrived in a white neighborhood becomes a target for arson; members of one race stranded in the neighborhood of another race are assaulted). And in homicides among persons of the same race, racism and racial discrimination contribute by helping to perpetuate structural conditions, including disparities in socioeconomic status, that are associated with violence.

Socioeconomic Status

The relations among race, socioeconomic status, and violence are complex and often confused. Racial minorities—most particularly black Americans—bear a disproportionately large burden of interpersonal violence and poverty. And poverty and interpersonal violence, including homicide, are intimately related.[38] The common practice of using racial but not socioeconomic classifications on death certificates, medical records, and police reports makes it hard to separate the contributions of race and socioeconomic status.

However, the research results are clear: When socioeconomic status is taken into consideration, the disparity between black Americans and the general population as both victims and perpetrators of violence becomes quite small.[39-42] Thus, although race

is associated with violence, socioeconomic status is the more significant risk factor. For example, a study of homicide rates in Chicago revealed that "the rates vary significantly according to the economic status of the community, with the highest rates of violent crime associated with the [poorest] communities. . . ."[43] Urban areas, especially inner-city areas, tend to have much higher rates of interpersonal violence than do suburban and rural areas.[44] The effects of both poverty and urban life on violence can be seen in the homicide statistics for the black community. In 1980, the homicide rate for black males between 15 and 24 years of age living outside of urban areas was 41 per 100,000. However, in urban areas, this rate approached 96 per 100,000.[45] And in some economically devastated urban black neighborhoods, the rate was as high as 142.6 per 100,000 (Boston City Hospital, unpublished observations).

Firearms

In 1986 more than 59% of homicides involved firearms.[5] They provide the mechanism for a similar percentage of suicides (57.3%).[46] In addition, they play an important role in nonfatal assaults and in both fatal and nonfatal occupational, recreational, and home injuries. Because of the number and broad range of types of injuries that involve firearms, Chapter 17 is devoted to a discussion of firearm injuries and proposed countermeasures.

Alcohol

The link between alcohol and violence has been demonstrated in study after study. One analysis of 588 homicides in Philadelphia, for example, indicated that alcohol had been used by the victim, the perpetrator, or both in nearly two-thirds of the cases.[47] A recent review of the literature cites 28 studies of alcohol and homicide, 14 of which found alcohol present in 60% of the cases and the great majority of which found alcohol present in at least one-third of the cases examined.[31] Studies of robbery, rape, assault, and homicide victims also reveal that if one of the parties (i.e., either victim or perpetrator) had been drinking the other party was likely to have been drinking as well. The association is particularly strong in homicides and rapes involving friends or acquaintances.[48]

Nevertheless, the authors of a major literature review conclude, "while research has shown that alcohol is involved in substantial proportions of . . . violent events, understanding of the ways in which alcohol consumption contributes to violence remains very limited."[49] Although alcohol may make an individual more impulsive (and less likely to measure the costs of a violent act) or more aggressive, it is nevertheless possible that alcohol abuse will lead to violence only if other conditions are present in the environment (e.g., a noisy barroom with patrons egging each other on). It is also possible that alcohol abuse and violence are each related to some other common cause such as an individual's personality traits. "Despite all the research that has been done on alcohol," two researchers concluded, "we do not yet know enough to choose with any certainty among" the possibilities.[31]

However, although we do not yet fully understand the relationship between alcohol and violence, we do know that some countermeasures can help to reduce alcohol abuse in general. Chapter 6 discusses these interventions.

Drugs

"The belief that there is a close and causal relationship between many forms of drug use and criminality and violent behavior probably forms the basis for many of our laws concerning drug use and drug users."[50] However, evidence to substantiate this belief has been hard to find. In the major epidemiologic study of homicide in Los Angeles, researchers were able to investigate alcohol use but found that "medical examiners screen for the presence of narcotics and other types of drugs . . . only when they suspect that one or more of these substances may be present."[29]

Some data are available that indicate that drug users commit violent offenses, including assaults and sexual assaults, at about the same rate as other offenders.[51] There have not been enough systematic studies of the effects of amphetamines, barbiturates, or hallucinogens on violence to establish any link. Some drugs, such as heroin, because of their known physical effects as narcotics, are unlikely to increase violence or aggressiveness by users.

But even if drugs do not make individuals physiologically more prone to violent behavior, they remain an important factor in homicides and nonfatal assaults. "A large number of assaults and murders can be attributed to the effort by one person to steal drugs or money from a drug dealer," concludes a recent review.[31] A 1975 study in Detroit, for example, indicated that a majority of homicide victims were either drug users or involved in drug dealing.[52] Increasingly open warfare among groups of drug dealers has killed participants and bystanders alike. The death toll of the "crack wars" chronicled by newspapers in Los Angeles, New York, and other major cities is an indication of the potent relationship between drugs and violence (see Chapter 11).

THE NEED FOR COLLABORATION

Public health offers a new approach to the prevention of violence and adds a new voice and a new set of tools and techniques. Yet while public health practitioners have much to contribute, they also have a great deal to learn from specialists in the fields of criminal justice, health care, social work, and mental health.

Collaboration at all levels, but especially among local injury-control practitioners and their counterparts, is critical both to break down barriers and to advance prevention and control activities. Differences in values, vocabulary, style, and institutional missions exist, as do long-standing misperceptions. "At worst," commented a criminal justice policy analyst, "public health sees criminal justice as solely concerned with locking up individuals, rather than with prevention. Criminal justice sees public health as a naive newcomer willing to subordinate all questions of civil rights or political decision making to its notion of health as the greatest public good. Certainly, it's not always that bad, but that does give you a sense of the tensions that must be overcome if they are to collaborate effectively" (M. H. Moore, personal communication).

Hospitals treat more violent injuries than ever are reported to the police. Health professionals play an important role in the diagnosis and treatment of nonfatal injuries resulting from violence. The criminal justice system (law enforcement, prosecution, the criminal trial courts, and corrections) is charged to protect lives, safety, and property; to preserve social and public order; to prevent crime; and to enforce the law by apprehending and prosecuting suspects, adjudicating the accused, and administering sanctions.[53] The social service and mental health systems include individuals and agencies that provide assistance to victims of interpersonal violence and suicide attempts. They also provide a broad range of preventive services and rehabilitation.

It is clear that the institutions that most collaborate in preventing violence view their missions quite differently. Indeed, even within one of these institutions there may be confusion about what constitutes preventable violence. For example, hospital emergency room (ER) personnel, confronted with a youth who has attempted suicide, will take immediate precautions if he says, "Why are you bothering, I'm only going to try again when I leave here?" When an adolescent being treated for a stab wound says, "I'm going to get the guy who did this and you'll be seeing him later," no such response is evoked among the ER crew (D. Prothrow-Stith, personal communication).

To begin to understand how the orientations, histories, and roles of these institutions can inhibit or enhance collaboration, we can look briefly at how they interact in the area of child abuse, highlighting some of the barriers to collaboration. (For more detailed information about the prevention of child abuse and child sexual assault, see Chapter 12.)

The Health Care System

Hospital personnel and medical professionals play important roles in the diagnosis, reporting, and treatment of child abuse. However, whereas diagnosis and treatment are the health care system's primary responsibility, the legal requirement that cases be reported brings medical personnel into the less familiar world of child protective agencies, the police, and the courts.

Although hospitals and health professionals are required to report cases of child abuse, they do not always do so.[54] Police are seldom called in to investigate nonfatal instances of child abuse treated in health care settings. "As a result timely evidence is not gathered which in the long run would serve both the social service systems and the courts to protect the child" (L. Jetmore, personal communication). Some health care professionals argue that abuse reports are not treated seriously, that reporting can render effective treatment impossible or make a situation worse, or that, "in order to help a family, a physician must, in effect, condemn the parents. . . ."[55]

The Criminal Justice System

The criminal justice system and its components play a variety of roles in child abuse cases, including case finding and reporting, assisting child protective agencies, investigating cases, placing children in protective custody, arresting perpetrators, and prosecuting offenders.[56] Nevertheless, one specialist concluded that "police agencies do not like to investigate most cases of child maltreatment" because they believe that child protective services are better suited to the task and can assure that services are provided more rapidly and because they believe that prosecution is often unsuccessful.[56] Indeed, police tend to underreport child abuse to the same extent as other mandated reporters.[56]

The size and decentralization of the criminal jus-

tice system affects its role. There are 20,000 law enforcement agencies in the United States, some with overlapping investigative responsibilities.[57] And this confusion does not end at the criminal courthouse door. Child abuse victims and their families are often involved in such related civil court actions as divorce proceedings, dependency hearings, and removal actions.[57]

The Social Service System

The development of child protective agencies in every state has been society's basic response to the problem of child abuse. These agencies are the primary vehicle through which social workers respond to child abuse cases.

Procedures and requirements vary from state to state, but social workers are generally expected to interview all children (and their parents) about whom reports of maltreatment have been received. Cases that are substantiated by ("screened into") the system are investigated further. A wide variety of services, from counseling, therapy, or support programs for the parents to removal of the child from the home and placement in foster care, may be provided by the agency or other social service or mental health providers.[58]

The passage of mandatory child abuse reporting laws and increasing awareness of the problem in some cases has resulted in a volume of reports so large that it has overwhelmed many child protective services. In 1986 some 2 million reports were filed; each required an investigation, and the provision of services was deemed necessary in more than half the cases.[59]

The police may seriously underreport child abuse, but it is also true that child protective services frequently fail to call upon police investigative capabilities. They "rarely involve the criminal justice system in efforts to protect abused and neglected children. Nationwide, fewer than 5% of substantiated cases result in criminal prosecution."[56] For this reason, a growing number of states have amended their child abuse statutes to require that child protective services notify law enforcement agencies of particularly serious cases, a category that varies from state to state.[56,58]

The Mental Health System

Mental health workers are more recent participants in child abuse treatment and prevention. This has been both because social service agencies have also provided mental health services in the past and because public and private mental health agencies have been slow to recognize the need for their expertise.[60]

Mental health agencies (particularly the network of community mental health centers), as a primary resource for individuals and families in emotional difficulty, inevitably become involved in the identification and treatment of child abuse cases. In addition, they are called upon by child protective services to assist in psychiatric assessments and in the development of treatment plans. Many of the same issues identified in the social service system apply to mental health professionals as well.

Consequences of the Lack of Collaboration

Even a brief look at the overlapping roles played by health professionals and representatives of the criminal justice, social service, and mental health systems in child abuse prevention and treatment highlights the need for collaboration and the tensions that inhibit it. The failure to collaborate can mean that critical information is not shared. Police who are called in to investigate a child homicide may learn that earlier nonfatal assaults by the same offender against the same child went unreported. Or a child protective service worker may learn in legal proceedings that an abusive parent has a history of violence reflected in records to which the agency had not been given access. The results can be tragic.

For example, on April 20, 1985, 10-month-old Dale-Lyn Crenshaw died in Manchester, Connecticut. Months earlier, after allegations that she was being physically abused by her father, Dale-Lyn had been placed in temporary foster care by the Connecticut Department of Children and Youth Services (DCYS). She was returned to her parents after they agreed to a program of counseling and other services. Several months later, she was dead. In August 1986, her father was convicted of her murder and sentenced to 25 years.

After studying the case, a *Hartford Courant* reporter concluded that the agencies responsible for protecting Dale-Lyn had "no reliable way of gaining the intimate contact with a family that seems necessary to ensure a child's safety. The single, simple, overriding impression is that neither the DCYS, nor the court, nor any of the private agencies on which they depended knew Dale-Lyn's family very well."[61]

Thus, although sharing information can be an important factor in protecting children and other victims of abuse, it is not the only factor in prevention. In Atlanta, the Centers for Disease Control (in collaboration with the Georgia Department of Human Resources, the Atlanta health department, and the Atlanta Department of Public Safety) is investigating the patterns of prior agency contacts and interagency contacts in cases of domestic assault or homicide among family members or other intimates. Personnel in a number of agencies have told investigators that simply getting more information will not guarantee that it will be processed or investigated (L. E. Saltzman, personal communication).

It may be that the way the information is managed—the protocol developed for using the data within and among agencies—will be as important as the fact that it is being shared. A case management system may be necessary for the shared information to have any practical value. The beginnings of such a system are embodied in some of the recently amended state child abuse statutes mentioned above. In Massachusetts, for example, the law mandates interdisciplinary communication and cooperation among agencies involved in the most severe cases of physical and sexual abuse. "Never before have professionals from law enforcement, mental health and social services had to work so closely together to clarify one another's role, and to coordinate their services both to the child victim and offender."[58]

THE ORGANIZATION OF CHAPTERS 11 THROUGH 17

The next seven chapters survey the state of the art in the prevention of interpersonal injuries and suicide. As we have seen, this is a broad and complex area and one that is essentially new to public health practitioners. Few practitioners will be called upon to deal with the full range of injuries described. To maximize the value of the chapters as references, each contains background material and interventions addressed to a specific type of violence.

The order of the chapters is in keeping with the idea that the relationship between victims and perpetrators is a central factor in understanding and preventing violence. Thus, they proceed along the following continuum: Chapter 11 is about assaultive violence among strangers and nonfamilial acquaintances; Chapter 12 concerns relations between children and their parents in terms of child abuse and child sexual assault; Chapter 13 focuses on domestic violence, particularly on woman battering by intimates or former intimates; and Chapter 14 explores violence directed at the elderly by their grown children.

The type of relationship is obviously not the only category for organizing such material. We have used it because it highlights an important element in our understanding of violence. But the organization is only useful to a point. Thus, Chapter 15, which concerns rape and sexual assault, includes all of the types of relationships described in the other chapters: those between strangers, acquaintances, and intimates. Chapter 16 describes what we know about the understanding and prevention of suicide. Finally, because firearms are ubiquitous in almost every category of violence, we devote Chapter 17 to the prevention of firearm injuries. Each chapter provides information on the magnitude of the problem; what is known about high-risk groups, risk factors, and social and cultural factors that influence the behavior; and a descriptive evaluation of available interventions.

Although the organization of these chapters may sometimes reflect the lack of clarity that still marks this new field, the central message of the chapters is clear: Violence is very much a public health concern. We have considered the contributions that public health can make to prevention and stressed the importance of collaboration with other concerned agencies and individuals. It is our hope that the following chapters will provide the tools that practitioners can use to begin the job.

Each chapter includes recommendations for the application of specific interventions. There is, however, a more general recommendation that applies to interpersonal violence and suicide as a whole.

Recommendation: Interpersonal violence and suicide are major threats to health as well as to peace and security. Decision makers and practitioners must recognize that these injuries are more than a criminal justice or mental health problem. They are major public health problems that can be understood and prevented with the same strategies and techniques used for other injuries.

Researchers and program developers should develop and test new interventions to address interpersonal violence and suicide. There are few models and much uncertainty about the effectiveness of many available interventions. Therefore, the greatest need is for interventions that are designed with specific, measurable objectives. Evaluations of these interventions should be widely disseminated. There is a critical need for useful and appropriate partnerships among criminal justice and public health professionals and other agencies with respect to the control and prevention of interpersonal violence and suicide.

INTERVIEW SOURCES

William Foege, MD, MPH, Executive Director, The Carter Center of Emory University, Atlanta, Georgia, August 2, 1988

Captain Larry Jetmore, Hartford, Connecticut Police Department, April 6, 1988

Mark H. Moore, PhD, Professor of Criminal Justice, J.F. Kennedy School of Government, Harvard University, Cambridge, Massachusetts, June 23, 1988

Deborah Prothrow-Stith, MD, Massachusetts Commissioner of Public Health, Boston, November 29, 1988

Linda E. Saltzman, PhD, Behavioral Scientist, Intentional Injuries Section, Centers for Disease Control, Atlanta, Georgia, September, 17, 1988

REFERENCES

1. Smolan R, Cohen D. A day in the life of America. New York: Collins, 1986.

2. Koop CE. Introduction. Source book and background papers prepared for the surgeon general's workshop on violence and public health, Leesburg, Virginia, October 1985:i–ii.

3. National Center for Health Statistics. Annual survey of births, marriages, divorces, and deaths, United States, 1986. Monthly Vital Statistics Rep 1988;35(13):1–48.

4. Centers for Disease Control. Premature mortality due to suicide and homicide—United States, 1984: perspectives in disease prevention and health promotion. Morbid Mortal Weekly Rep 1987;36(32):531–4.

5. Federal Bureau of Investigation. Crime in the United States: uniform crime reports, 1986. Washington, DC: US Department of Justice, 1987:21.

6. Barancik JI, Chatterjee BF, Greene YC, Michenzi EM, Fife D. Northeastern Ohio trauma study, vol 1: magnitude of the problem. Am J Public Health 1983;73(7):750.

7. Baker SP, O'Neill B, Karpf RS. The injury fact book. Lexington, Massachusetts: Lexington, 1984:17–37.

8. Chamelin NC, Evans KR. Criminal law for policemen, 2nd ed. Englewood Cliffs, New Jersey: Prentice Hall, 1976.

9. Weiner NA, Wolfgang ME. The extent and character of violent crime in America, 1969 to 1982. In: Curtis LA, ed. American violence and public policy. New Haven, Connecticut: Yale University Press, 1985:19.

10. Nagin D. General deterrence: a review of the empirical evidence. In: Blumstein A, Cohen J, Nagin D, eds. Deterrance and incapacitation: estimating the effects of criminal sanctions on crime rates. Washington, DC: National Academy of Sciences, 1978:95–139.

11. Blumstein A, Cohen J, Nagin D, eds. Deterrence and incapacitation: estimating the effects of criminal sanctions on crime rates. Washington, DC: National Academy of Sciences, 1978:64.

12. Goldstein H. Improving policing: a problem-oriented approach. Crime and Delinquency 1979(Apr):236–58.

13. Healthy people: the surgeon general's report on health promotion and disease prevention. Washington DC: US Government Printing Office, 1979; DHEW publication no. (PHS)79-55071.

14. Healthy people: the surgeon general's report on health promotion and disease prevention—background papers. Washington, DC: US Government Printing Office, 1979, DHEW publication no. (PHS)79-55071A.

15. Promoting health/preventing disease: objectives for the nation. Washington, DC: US Department of Health and Human Services, 1980.

16. Loya F, Mercy JA, Allen NH, Vargas LA, Smith JC, Goodman RA, Rosenberg ML. The epidemiology of homicide in the city of Los Angeles, 1970–79: a collaborative study by the University of California at Los Angeles and the Centers for Disease Control. Atlanta, Georgia: US Department of Health and Human Services, Public Health Service, Centers for Disease Control, Violence Epidemiology Branch; 1985(Aug).

17. Poussaint A. Why blacks kill blacks. Ebony 1970(Oct).

18. Dennis RE. Homicide among black males: social costs to families and communities. Public Health Rep 1980; 95(6):556.

19. Report of the secretary's task force on black and minority health; vol 5: homicide, suicide, and unintentional injuries. Washington, DC: US Department of Health and Human Services, 1986:1–2.

20. Rosenberg ML. Violence is a public health problem. In: Maultiz RC, ed. Unnatural causes: the three leading causes of mortality in America. Philadelphia: College of Physicians of Philadelphia, 1988:149 (Transactions and Studies of the College of Physicians of Philadelphia; series 5; vol 10).

21. Lamborn L. Toward victim orientation in criminal theory. Rutgers Law Rev 1968;22:733–68.

22. Karmen A. Crime victims: an introduction to victimology. Belmont, California: Brooks/Cole, 1984.

23. Felony arrests: their prosecution and disposition in New York City's courts. New York: Vera Institute of Justice, 1977:135.

24. Rosenberg ML, Mercy JA. Homicide: epidemiologic analysis at the national level. Bull NY Acad Med 1986; 62(5):376.

25. Rosenberg ML, Stark E, Zahn MA. Interpersonal violence: homicide and spouse abuse. In: Public health and preventive medicine. Norwalk, Connecticut: Appleton-Century-Crofts, 1986:1399–426.

26. Rokaw WM, Mercy JA, Smith JC. Comparability and utility of national homicide data from death certificates and police records. Washington, DC: Intentional Injury Section, Epidemiology Branch, Division of Injury Epidemiology and Control, Centers for Disease Control, US Department of Health and Human Services (undated).

27. O'Carroll PW. Homicides among black males 15–24 years of age, 1970–1980. In: Public health surveillance of 1990 injury control objectives for the nation. Morbid Mortal Weekly Rep 1988;37:53–60.

28. Biometrics Branch, Epidemiology Branch. Premature mortality due to homicides—United States, 1968–1985. Morbid Mortal Weekly Rep 1988;37(35):543–5.

29. University of California at Los Angeles, Centers for Disease Control. The epidemiology of homicide in Los Angeles, 1970–79. Washington, DC: US Department of Health and Human Services, 1985.

30. Gelles RJ. The violent home. Beverly Hills, California: Sage, 1974.

31. Wilson JQ, Herrnstein RJ. Crime and human nature. New York: Simon and Schuster, 1985.

32. Centers for Disease Control. Homicide surveillance, 1970–78. Washington DC: US Department of Health and Human Services, 1983.

33. Cheatwood D. Is there a season for homicide? Criminology 1988;26:287–306.

34. Coleman L. Suicide clusters. Winchester, Massachusetts: Faber and Faber, 1987:97–8.

35. Jackson T. Violence and the masculine ideal: some qualitative data. In: Stinmetz S, Straus M, eds. Violence in the family. New York: Harper and Row, 1974.

36. Mohr RD. Gays/justice: a study of ethics, society, and law. New York: Columbia University Press, 1988.

37. Silberman CE. Criminal violence, criminal justice. New York: Random House, 1980.

38. Williams KR. Economic sources of homicide: reestimating the effects of poverty and inequality. Am Sociol Rev 1984;49:283–9.

39. Centerwall BS. Race, socioeconomic status, and domestic violence, Atlanta 1971–72. Am J Public Health 1984;74:813–5.

40. Loftin C, Hill RH. Regional subcultures and homicide: an examination of the Gastil-Hackney thesis. Am Sociol Rev 1974;39:714–24.

41. Willin KR. Economic sources of homicides: reestimating the effects of poverty and inequality. Am Sociol Rev 1984;49:283–9.

42. Parker RN, Smith MD. Deterrence, poverty, and type of homicide. AJS 1979;85:614–24.

43. Wilson WJ. The truly disadvantaged: the inner city, the underclass, and public policy. Chicago: University of Chicago Press, 1987.

44. Mercy J, et al. Patterns of homicide victimization in the city of Los Angeles, 1970–1979. Bull NY Acad Med 1986;62(5):423.

45. Homicide among young black males—United States, 1970–1982. Morbid Mortal Weekly Rep 1985;34.

46. Centers for Disease Control. Suicide surveillance, 1970–80. Washington, DC: US Department of Health and Human Services, 1985:4.

47. Wolfgang M. Patterns in criminal homicide. Philadelphia: University of Pennsylvania Press, 1958.

48. Room R. Region and urbanization as factors in drinking practices and problems. In: Kissin B, Begleiter H, eds. The pathogenesis of alcoholism: psychosocial factors (The biology of alcoholism; vol 6). New York: Plenum, 1983:555–604.

49. Mercy JA, Davidson LE, Goodman, RA, Rosenburg ML. Alcohol and intentional violence: implications for research and public policy. Background paper for the National Institute on Alcohol Abuse and Alcoholism Conference on Research Issues in the Prevention of Alcohol-Related Injuries, Berkeley, California, March 1986.

50. Oakley R. Drugs, society, and human behavior. St. Louis: C.V. Mosby, 1983.

51. Moore M. Controlling criminogenic commodities: drugs, guns, and alcohol. In: Wilson JQ, ed. Crime and public policy. San Francisco: Institute for Contemporary Studies, 1983:125–44.

52. Monforte JR, Spitz WU. Narcotic abuse among homicides in Detroit. J Forensic Sci 20:186–90.

53. Holten NG, Jones ME. The system of criminal justice, 2nd ed. Boston: Little, Brown, 1982.

54. Hampton RL, Newberger EH. Child abuse incidence and reporting by hospitals: significance of severity, class, and race. Am J Public Health 1985;75(1):56–60.

55. Newberger EH. The helping hand strikes again: unintended consequences of child abuse reporting. J Clin Child Psychol 1983;12(3):307–11.

56. Besharov DJ. Child abuse: arrest and prosecution decision-making. Am Criminal Law Rev 1986;24(2):315–77.

57. Toth PA, Whalen MP. Investigation and prosecution of child abuse. Alexandria, Virginia: American Prosecutors Research Institute, National Center for the Prosecution of Child Abuse, 1987.

58. Calhoun G, Docherty G. The response of DSS to reports of child sexual abuse. In: The child abuse reporting law: The Middlesex County experience. Cambridge, Massachusetts: Middlesex County Child Abuse Project, 1986.

59. Daro D. Confronting child abuse. New York: Free Press, 1988.

60. Lauer JW, Lourie IS, Salus MK, Broadhurst DD. The role of the mental health professional in the prevention and treatment of child abuse and neglect. Washington, DC: US Department of Health, Education, and Welfare, 1979(Aug); DHEW publication no. (OHDS)79-30194:XIII.

61. Lang J. The short, unhappy life of Dale-Lyn Crenshaw. Northeast Magazine, Hartford Courant 1987 (Mar):12.

Chapter 11: Assaultive Injuries

Late on a May night, a young couple sat in a parked car on a residential street in Wheaton, Maryland. A man approached the open window, brandished a gun, and demanded money. The couple did not resist, but as the young man reached for his wallet, the gunman shot him in the face and then fled on foot. The woman immediately drove off with her injuried companion in search of help. The police later described his condition as "stable" and told reporters that they were still searching for the armed robber.[1]

An armed robbery on a dark street by a stranger motivated by the desire for economic gain is the stereotype of violent crime in America. In fact, many Americans are killed and injured in the course of muggings and armed robberies. However, 54% of all violent crimes are committed by people known, at least by sight, to their victims.[2]

Some of this violence takes place in the course of relationships that periodically explode into physical violence and injury. These relationships—between parents and children, intimate domestic partners or former partners, and grown children and their parents—are discussed in later chapters, as are rape and sexual assault, which cut across the spectrum of relationships.

Even the violence that takes place outside the relationships often involves people who know one another. According to the FBI, only 13% of the homicides occurring during 1986 were committed by persons known to have been strangers to their victims. Even if we assume that the victims and assailants were strangers in all cases in which the relationship is unknown, then at least 47% of the homicides in 1986 were committed by persons acquainted with or related to their victims.[3] Although information on nonfatal assaults is meager, a study of patients treated for assaultive injuries in a San Francisco emergency room found that 47% were acquainted with their assailants.[4]

The stereotype that violence is motivated by a desire for economic gain is also only partially true. Approximately 65% of homicides occur during or after an argument or other nonfelony circumstance.[5] The *Uniform Crime Report* for 1986 reports that only 19% of homicides involved robberies, arson, or other felonies, while 38% were precipitated by arguments.[3] A study in Los Angeles revealed a similar pattern.[6] Even apparently inconsequential arguments, when fueled by alcohol, machismo, or frustration, can escalate into serious violence and homicide.

DEFINITIONS

This chapter is concerned with assaultive injuries among strangers and acquaintances. An assault is the use of physical force by a person or persons to inflict injury or death on another. Most interventions to reduce assaultive injuries do so by attempting to reduce the immediate cause of those injuries, that is, assaults. Thus, much of this discussion is about violence—the use of physical force with the intent to inflict injury or death upon oneself or another—rather than injuries per se.

The criminal justice system categorizes assaultive behavior by its actual or potential consequences and thus treats homicide or attempted homicide more seriously than aggravated assault ("an unlawful attack by one person upon another for the purpose of inflicting severe or aggravated bodily injury"[3]) or nonaggravated assault. Because the circumstances preceding and the interventions for all assaults are similar, this discussion addresses them together. This discussion also refers to robbery, defined by the FBI as "the taking or attempting to take anything of value from the care, custody, or control of a person or persons by force or threat of force or violence and/or by putting the victim in fear."[3] Robberies also are always potentially injurious (unlike, for example, burglaries, which generally take place when the victims are not at home).

Street crime—muggings and robberies—also creates its share of injuries. This violence has consequences for the health of its victims, including their mental health, as a consequence of the fear it generates; people (especially the elderly and women) become afraid to move about in their own communities. Although reliable statistics are not available, there is speculation that a not insubstantial percentage of robberies, especially robberies in which drugs are involved, occur among people who have at least a passing acquaintance. Such incidents are not likely to be reported to the police unless a serious injury occurs.

Political violence, civil disorders, and serial and mass murders are neither as common nor as representative as armed robberies, barroom brawls, and the like, and are not included in this chapter. The discussion of "organized" crime is limited to the phenomenon of gang violence involving the "crack" trade. Firearm injuries are discussed in Chapter 17.

THE MAGNITUDE OF THE PROBLEM

There were over 20,000 homicides in 1986. As with many injuries, mortality represents only the tip of the assaultive injury pyramid. It has been estimated that "the ratio of actual assault to homicides is probably far greater than 100:1."[7] The FBI reported that there were 834,332 aggravated assaults in 1986, for a rate of 346.1 per 100,000. This represents an increase of 25% over 1982 and 56% over 1977.

Robberies have declined by 2% since 1982, but the 542,775 robberies reported in 1986 represented a 32% increase over 1977, for a rate of 225 per 100,000.[3] The number of nonaggravated assaults is not available but is probably substantial. Evidence is accumulating that law enforcement data substantially underestimate the number of nonfatal, injury-producing assaults that take place in this country.[8]

RISK FACTORS

The perpetrators and victims of assaultive injuries often resemble each other in educational background, psychological profile, and reliance on weapons.[9] Other research has found that more than a quarter of homicides are "victim-precipitated" (i.e., the victim was the first to strike a blow or show a deadly weapon).[10] In many cases both victim and perpetrator have been drinking.[11] Indeed, the distinction between perpetrator and victim sometimes may have more to do with who is injured or killed than with who initiated the violence.

Sex

Most violence is committed by men. Eighty-three percent of male homicide victims are killed by men. Ninety percent of female homicide victims also are killed by men.[3] Thirty percent of female homicide victims in 1986 were killed by their husbands or boyfriends, while only 6% of male homicide victims were killed by their wives or girlfriends.[3]

Socioeconomic Status

In the majority of homicides, killer and victim are of the same race.[12] The poor are disproportionately at risk for assaultive injuries, and racial minorities—particularly black Americans—bear an especially large share of this burden. Urban areas, and especially inner-city areas, tend to have much higher rates of interpersonal violence than do suburban and rural areas.

Age

The amount and lethality of contemporary violence perpetrated by young people is unprecedented in the history of the United States.[13] People under 25 years of age accounted for 48% of arrests for violent crime and 41% of arrests for murder during 1986.[3] More than 40,000 people under the age of 18 were arrested for violent crimes in that year.[3] Three to four children under 18 were arrested for murder every day.[14] However, arrest statistics tell only part of the story. Not every violent incident results in a police report, much less an arrest. Juveniles commit between eight and 11 serious crimes for every one that leads to an arrest.[15]

Thirty-six percent of all assaults and 40% of all robberies reported by people 12–19 years of age occurred in school. For the younger members of that population—12–15 years of age—the numbers are even more dramatic: 50% of assaults and 68% of robberies occurred in school.[16] In 1983, the Boston Commission on Safe Schools found that 28% of students (38% of boys and 18% of girls) in the school system were carrying potentially lethal weapons (D. Prothrow-Stith, H. Spivak, personal communication). This problem has become so serious in New York City that authorities have instituted an experimental program in which all students entering selected high schools are searched with metal detectors.[17]

Reported instances of interpersonal violence in the schools, like arrest statistics, reflect only a small part of the problem. The 1979 *Safe School Report to the Congress*[16] found that 80%–90% of incidents of school violence, including those involving weapons and those that resulted in injuries, were not reported to police. Most of this violence is committed by students against other students and does not involve intruders from the outside.[16]

Some reports have implicated television in the amount of violence among young people. "By the time an average child graduates from high school," one study concluded, "he or she will have spent 22,000 hours of accumulated viewing time before the television screen and only 11,000 hours of classroom time."[18] The National Institute for Mental Health found that an average of 80.3% of all television programs contain violent acts. A typical television program contains 5.21 violent acts.[19] By the age of 16, the average child will have witnessed approximately 16,000 television murders.[18]

The people at greatest risk for violence are the ones who watch the most television. Television viewing is inversely related to income, education, and employment.[20] Boys watch more violent television than do girls.[21] Research suggests that people who are exposed to a high degree of televised violence become more prone to commit violence themselves in resolving conflicts and to accept violent behavior from others.[20,22–25]

Adolescence itself may contribute to violence among the young. A major developmental stage of adolescence involves separation from the family and the creation of an individual identity. This process can include resistance to traditional values and authority. Adolescents can be deeply narcissistic and have an exaggerated sense of self. This narcissism, combined with the physiological changes of puberty and the development of a sexual identity, can create an extreme sensitivity to real or perceived insults about competence, physical appearance, sexual prowess, and other personal characteristics. Adolescent concerns with sexual identity can lead to behavior that young people believe is consistent with being male or female. For boys, a preoccupation with aggressiveness, bravery, and physical prowess can lead to attempts to prove themselves through risk taking and violence, especially when they are challenged by or in front of peers.[26]

Drugs and Alcohol

Both drugs and alcohol play large roles in violence in America. Studies show that at least half of all perpetrators and victims of homicides had been drinking and that a large percentage of violence occurs in places where alcohol is consumed.[6] This phenomenon is linked to alcohol's power to disinhibit. Some drugs may have a similar effect. PCP ("angel dust"), in particular, has been blamed for causing violent behavior.[27] Because blood alcohol content tests are far more commonly done on homicide victims than are tests for narcotics, marijuana, and other controlled substances, the role of alcohol in violence is far better documented than that of other drugs.

Drugs, however, play two very important and well-documented roles in interpersonal violence. Some, especially heroin and cocaine, are both addictive and expensive. Many users of these drugs commit burglaries, robberies, and other crimes to support their addiction, and the crimes often involve violence or the potential for violence. And because these substances are both illegal and valuable, their marketing involves considerable violence: violence among drug dealers over turf, violence between dealers and their customers over price and quality, violence between dealers and those who see them as robbery targets, and violence among drug users over drugs. Figures on the contribution of drugs to violence are not readily available. "Drug-related homicides" are usually defined as homicides that occur during drug sales.[11] However, such calculations overlook much of the violence related to "turf" issues. Whether a mugging in which the perpetrator escapes was motivated by a need for drugs is also impossible to determine.

The National Institute on Drug Abuse estimates, conservatively, that at least 10% of homicides and assaults in the United States are drug related.[28] One study found that in 1980 there were over 2,000 drug-related homicides and more than 460,000 drug-related assaults nationally.[29] Another study showed that more than one-third of the men killed in Manhattan in 1981 were victims of drug-related homicides. One-quarter of the homicide victims were killed in robberies or burglaries, at least some of which were probably drug related.[11] Other studies have found that drug abuse and violent crime tend to occur in the same neighborhoods and that users of certain types of illicit drugs tend to be arrested for serious crimes against persons at a far greater rate than do persons who do not use those drugs.[30] Sixty percent of those arrested in Washington, D.C., tested positive for cocaine and 40% for PCP.[31]

The rise in popularity of "crack," a smokable, extremely potent, and very addictive form of cocaine, is making a substantial contribution to violence in urban areas. Research indicates that as many as half the murders in New York City are now drug related.[32] Both Boston and Washington, D.C., are also experiencing sharp increases in murders, many of which are the result of drug-related violence.[31,33]

Youth Gangs

Several risk factors—socioeconomic status, youth, urban residence, and drug use—have converged in the violence of contemporary youth gangs. Research has demonstrated that increased homicide rates were associated with the establishment of networks for the marketing of alcohol in the late 1920s and early 1930s and cocaine and heroin in the late 1960s and early 1970s.[34] This pattern apparently is being repeated. The marketing network for crack is largely controlled by youth gangs who battle over turf using sophisticated and powerful weaponry.[35–38]

Competition over drug markets adds to traditional youth gang violence over turf and status. The California Task Force on Youth Gang Violence estimated that there are between 400 and 800 gangs operating in California and that there are more than 50,000 gang members in Los Angeles alone.[39] "Drive-by" shootings (often used as an initiation for new members) have become common.[40] The crack trade has been held responsible for a 25% annual increase in the Queens, New York, murder rate for the past two years.[41] Warring drug gangs in upper Manhattan are estimated to have killed at least 500 people over the last 5 years. Police say that the youths involved in these killings take pride in executing their rivals in broad daylight. "We sell drugs and we kill," is how one New York gang member described his group's life.[42] And the carnage is not restricted to gang members. Bystanders also are injured and killed.[43] Los Angeles gangs and gang violence are spreading to other cities, including Portland, Oregon, and Spokane, Tacoma, and Seattle, Washington, in search of new markets.[44]

Youth gangs involved in the drug trade are using more weapons and far more lethal weapons than ever before. The zip guns of the 1950s and Saturday night specials of the 1970s have been replaced by large-caliber, automatic and semiautomatic weapons and even machine guns. According to the *New York Times*, the weapon of choice among upper Manhattan gangs is a 9-mm semiautomatic machine pistol that retails for $1,000 on the black market.[42] The availability and use of these weapons, among even a small portion of urban youth, does not bode well for the future.

INTERVENTIONS

Many of the traditional interventions that attempt to reduce levels of violence and violent crime are strictly within the purview of the criminal justice system. These include the enforcement of laws against violent crime, imprisonment, mandatory sentences, increased police patrols, and drug interdiction. This discussion focuses on interventions in which public health practitioners can play an active role. A number of these interventions target young people. It is hoped that effective intervention for such violence will reduce not only the level of youth violence but also the violence these individuals engage in after they mature. Interventions directed at children not yet engaged in serious violence or directed at violence that has yet to harden into adult behavior patterns may reduce the need for rehabilitation efforts later.

Interventions Directed at Drug and Alcohol Abuse

There are many types of programs that attempt to break their clients' addiction to or abuse of controlled substances. The techniques used in such programs are varied, including group therapy, methadone maintenance, and peer support. Research indicates that effective substance abuse rehabilitation programs can reduce not only drug use but criminal activity.[45–49]

Although the public health community has usually thought of interventions designed to reduce alcohol abuse as a way of controlling traffic injuries, such interventions may also reduce the number of assaultive injuries in which alcohol is involved. A number of these interventions have been discussed in Chapter 6. These include the 21-year-old drinking age, higher alcohol excise taxes, the Model Alcoholic Beverage Licensee Liability Act of 1985, server training programs, and TEAM (Techniques for Effective Alcohol Management). Additional suggestions for interventions can be found in other documents.[50–53]

A successful substance abuse prevention program at an Albuquerque, New Mexico, middle school also reduced violent behavior among the students, including fights, classroom disruptions, and sexual threats and attacks.[54] Research has demonstrated that the use of alcohol and the use and distribution of other controlled substances have strong links to interpersonal violence.

Recommendation: Drug and alcohol prevention and rehabilitation programs are promising approaches to the prevention and control of assaultive injuries. They should be used and monitored and their outcomes evaluated.

Parenting Programs

Parent training programs seek to teach parents skills that can be used to enhance their children's social, intellectual, physical, and emotional development. These programs address issues such as authority and discipline (especially the use of nonviolent and constructive disciplinary techniques), and communication and problem-solving skills (both for the parent and the child). Although they are not aimed at reducing violence per se, these programs have been found to have a beneficial effect on the aggressive behaviors of both the parent and the child.[21,55]

A number of parenting programs are available through commercial and nonprofit sources as well as from federal agencies, such as the Administration for Children, Youth, and Families of the Department of Health and Human Services. These programs include "Parent Effectiveness Training," "Looking at Life," and "Systematic Training for Effective Parenting."[56] A project of the Center for

Studies of Minority Group Mental Health, National Institute of Mental Health (NIMH), adapted several successful parenting programs to the cultural backgrounds and educational levels of lower-income black and Mexican-American parents.[56]

Recommendation: Programs designed to improve parenting skills and reduce aggressive behavior among parents and their children may be able to decrease parents' violence against their children and subsequent violence by the children against others. Such programs should be used and monitored and their outcomes evaluated.

Interventions Directed at the Effects of Television Violence

Legislative interventions. A number of bills have been introduced in Congress to mandate or promote the reduction of the amount of violence on television. One recent bill, S-844, introduced by Senator Paul Simon (D-IL), would exempt from antitrust laws any actions by the broadcasting industry to alleviate the negative impact of televised violence. Legislation designed to affect media content can raise serious constitutional issues.

Educational interventions. Another approach that attempts to mitigate the effects of televised violence among younger viewers uses school-based educational programs to teach children that television violence is unrealistic and that violence in the real world has serious medical and social consequences. One experimental program, carried out among grade-school populations in the United States, Finland, and Poland, met with significant success in changing children's perceptions of television violence.[57,58] However, it is not known whether these programs had any long-term effect on the children's behavior.

Recommendation: Research on the effectiveness of school-based programs in directly addressing the effects of televised violence is inconclusive. Further research should be conducted.

School-Based Interventions

Schools are an important arena for violence prevention efforts. They contain a captive audience, many of whose members are at risk for interpersonal violence. Teachers, in addition to their role as educators, serve as important role models.

Administrative interventions. One approach to reducing the levels of violence in public schools involves the removal of violent or disruptive students to alternative schools. The alternative schools provide programs designed to affect both students' violent behavior and their educational achievement.

These efforts have not been without controversy. Some critics maintain that moving violent and disruptive students to separate programs is segregation, not a way of solving the problem, and "that these schools are little more than youth prisons which encourage class [discrimination] and alienation."[59] Advocates of the programs, however, insist that they provide the extra attention and special services the students need as well as a positive alternative to expulsion.

Alternative programs are limited by two factors: their high cost and their restricted target group. The characterization of some students as "violent" may be misleading. A student who becomes involved in a string of altercations with a fellow student at school may be labeled "violent." A student who becomes involved in a similar string of altercations outside school will escape this designation.

Because of differences in the design and population of the many alternative programs, each must be evaluated individually. However, studies have indicated that such programs can be effective.[59]

"School improvement programs" attempt to upgrade a school's educational outcomes through a combination of curricular, administrative, and teaching reforms, and activities that give students, staff, administration, and parents a greater sense of the school as a community. These programs usually promote features that research has shown to be characteristic of schools with successful educational outcomes[60-63] and good discipline.[64] They include daily full-day attendance, direct supervision of all activities, mandatory counseling for parents and students, a rigorous work load, clear and consistent goals, high standards and expectations, and a democratic climate.[59]

Research suggests that these programs not only improve educational outcomes but also can improve discipline in the school.[65] Some programs include components specifically designed to achieve the latter goal.[66] One such program is the Core Curriculum in Preventing and Reducing School Violence and Vandalism developed by the Center for Human Services in Washington, D.C. It is a 5-day, 34-hour program for teachers in planning and evaluating a school violence and vandalism prevention program. Topics include clarifying the school discipline policy, improving the school climate, interpersonal relations and conflict management, physical design strategies for reducing violence and vandalism in the school, and the community as a problem-solving resource.[67]

Curricula. Some recently developed educational programs seek to change attitudes toward violence and to teach interpersonal skills useful in resolving conflicts. These programs are based on the success of substance abuse education programs that focus on interpersonal and decision-making skills and information about the medical and social consequences of individual behaviors.[68-70]

One "conflict resolution" program was developed by the Institute on Promoting Interpersonal Development in School-Age Children at the Harvard University Graduate School of Education. This program shows educators how to teach the social and cognitive skills that enable children to resolve conflicts nonviolently.[71,72] Other programs, like the Violence Prevention Curriculum for Adolescents,[73] specifically address adolescents' behaviors and attitudes about violence. This 10-session high school curriculum teaches students about the extent to which they are at risk for violence, the elements usually involved in a violent or potentially violent situation, and positive ways to deal with anger and arguments.

A number of other conflict resolution and violence prevention curricula are available for various grade levels.[74-78] Although evaluative data are not conclusive, these curricula seem promising. At least one study has shown that they can be successful in increasing social problem-solving abilities and preventing attitudes that promote aggressive behavior.[79]

Recommendation: Administrative efforts such as general school improvement programs, alternative schools for disruptive students, and school-based curricula designed to change the understanding of and attitudes toward interpersonal violence as well as behavior are promising approaches to prevention of interpersonal violence. Such efforts should be used and monitored and their outcomes evaluated.

Community-Based Interventions

Violence-specific programs. The effects of school-based violence prevention curricula are limited by the influences and constraints young people face outside school. Many of the children at greatest risk for violence have dropped out of school. Thus, some programs attempt to address violence among youth outside the school environment. The Health Promotion for Urban Youth Violence Prevention Project based at Boston City Hospital, for example, is a comprehensive demonstration project aimed at young people in Roxbury and South Boston, two of Boston's poorest neighborhoods. It seeks to create a community environment in which the lessons of the Violence Prevention Curriculum for Adolescents[73] (described above) can be reinforced.

The staffs of four neighborhood health centers

offer violence prevention counseling to adolescents using these centers. Posters, pamphlets, a video presentation, and other educational materials are available in waiting areas. An instrument for identifying high-risk youth was developed and is being used to screen all adolescents treated for assaultive injuries in the emergency room of Boston City Hospital (the public hospital serving the populations of the two target neighborhoods). A special violence prevention clinic at the hospital offers services to youth found to be at high risk, including comprehensive assessment, educational interventions, counseling, and therapy. A referral network provides additional secondary services for youth at risk. A public education campaign reinforces violence prevention messages in the target neighborhoods.[80]

Another model for comprehensive youth violence intervention and prevention is Philadelphia's Crisis Intervention Network (CIN). In 1975, in response to increasing levels of gang-related violence, CIN began as a municipal agency fielding teams of negotiators to work among gangs. It became an independent nonprofit social service agency in 1979 and is now involved in a full range of activities to prevent violence and encourage young people to behave in ways that will decrease their risk for violence.

CIN's intervention teams are trained in negotiation strategies. They respond to rumors and tips of impending violence reported to a 24-hour hotline and monitor situations with a potential for gang violence, such as block parties and street festivals. One team specializes in racial and ethnic conflicts. In a cooperative arrangement with the probation department, CIN staff members are assigned to youths and young adults on probation or parole after gang-related convictions. Regular contact with gang members allows CIN to track gang activities and encourage the peaceful settlement of disputes.

CIN is also involved in programs that attempt to steer urban youth away from lives of violence. In conjunction with the family court, CIN runs a day treatment program that provides counseling, skills training, tutors, and recreational activities for adjudicated youth who otherwise would be incarcerated. CIN offers youth development workshops in schools, churches, and community centers to educate and motivate high-risk youth and assist them in developing self-esteem, leadership skills, positive values, and a career orientation. It also acts as a referral center for the Philadelphia Private Industry Council's employment training programs, sponsors job readiness training programs, and runs summer employment, athletics, and drama programs.

CIN reports that it has helped reduce gang-related homicides in Philadelphia from a high of 32 in

1974 to an average of 2.18 a year since 1975. Similar programs have been established in Los Angeles, Chicago, and Miami.[81]

Youth development programs. Community-based recreational and cultural programs such as those sponsored by police athletic leagues, Boys Clubs, Girls Clubs, and YMCAs are a traditional approach to deterring children from delinquent behavior. Their rationale is twofold. First, it is thought that children in structured situations with adult supervision are less likely to become involved in delinquent behavior than children who are unsupervised. Second, many believe that involvement in recreational programs nurtures attitudes and behaviors that prevent delinquency (e.g., increased self-esteem, a willingness to play by the rules, and a capacity to get along with others). One report suggests that increased funding for such programs might also provide an alternative to drug abuse, "particularly in urban areas where access to drugs is easy and organized alternatives are few."[27]

Some organizations, such as the Boys Club, target outreach programs to high-risk youth. Staff members are trained to respond to the special needs of these children and coordinate services with other social service agencies and organizations.

The Boys Club reports that, of the almost 9,000 children who have entered local clubs under the targeted outreach program, 38% have demonstrated a positive change in school performance.[82] In a survey performed for the Boys Club by the Louis Harris Organization, about 78% of Boys Club alumni (including those who joined through targeted outreach) said that "the Boys Club helped me to avoid trouble with the law."[83]

Other community delinquency prevention programs go beyond recreational activities and target a number of at-risk activities. The primary purpose of Positive Youth Development Initiative, Inc. (PYD), of Wisconsin is to help the staff of community organizations develop skills to prevent at-risk behavior among teenagers. This nonprofit organization and others like it around the country provide training, counseling, and facilitating services to community organizations concerned with youth. PYD provides workshops on community needs assessment, substance abuse, suicide prevention leadership training for youth, and school climate improvement. It also conducts workshops and training for community group volunteers and staff in stress management, AIDS, and sexual abuse.

Recommendation: Community-based violence prevention and youth development programs are promising approaches to the prevention and control of assaultive injuries. Such programs should target youth at high risk for violence. They should be used and monitored and their outcomes evaluated.

REFERENCES

1. Man shot in Maryland robbery attempt. Washington Post 1988 May 22.

2. BJS data report: 1986. Rockville, Maryland: Bureau of Justice Statistics, 1987(Sep).

3. Crime in the United States: uniform crime reports, 1986. Washington, DC: Federal Bureau of Investigation, 1987.

4. Sumner B. Interviewing persons hospitalized with interpersonal violence-related injuries: a pilot study. In: US Department of Health and Human Services. Report of the secretary's task force on black and minority health; vol 5: homicide, suicide, and unintentional injuries. Washington, DC: US Department of Health and Human Services; 1986(Jan).

5. Centers for Disease Control. Premature mortality due to suicide and homicide: U.S. 1984. Morbid Mortal Weekly Rep 1987;36(32).

6. University of California, Centers for Disease Control. The epidemiology of homicide in the City of Los Angeles: 1970–79. Washington, DC: US Department of Health and Human Services, 1985.

7. Rosenberg ML, Mercy J. Homicide: epidemiologic analysis at the national level. Bull NY Acad Med 1986; 62(5):376–94.

8. Barnacik I, Chatterjee BF, Greene YC, Michenzi EM, Fife D. Northwestern Ohio trauma study; vol 1: magnitude of the problem. Am J Public Health 1983;73(7):746–51.

9. Dennis RE. Homicide among black males: social costs to families and communities. Public Health Rep 1980; 95(6):556.

10. Loftin C. Assaultive violence as a contagious social process. Bull NY Acad Med 1986;62(5):551–2.

11. Tardiff K, Gross E. Homicide in New York City. Bull NY Acad Med 1986;62(5):413–26.

12. Report of the secretary's task force on black and minority health; vol 5: homicide, suicide, and unintentional injuries. Washington, DC: US Department of Health and Human Services, 1986.

13. Brown RM. Historical patterns of violence in America. In: Graham HD, Gurr R, eds. Violence in America: historical and comparative perspectives; vol 1: report to the National Commission on the Causes and Prevention of Violence, 1969;35–65.

14. Calhoun J. Violence, youth, and a way out: testimony before the Select Committee on Children, Youth, and Families, 9 March 1988. Washington, DC: National Crime Prevention Council, 1988.

15. Wolfgang M, Figlio R, Sellin T. Delinquency in a birth cohort. Chicago: University of Chicago Press, 1972.

16. Boesel D, et al. Violent schools, safe schools: the safe

school report to the Congress. Washington, DC: National Institute of Education, 1978;31–2.

17. Daley S. Five high schools to use metal detectors. New York Times 1988 May 5:B1.

18. Pearl D. Television and children: a report on major research. Paper presented to the American Medical Association, September 1983.

19. Signorielli N, Gross L, Morgan M. Violence in television programs: ten years later. In: Television and behavior: ten years of scientific progress and implications for the eighties; vol 2: technical reviews. Rockville, Maryland: National Institute of Mental Health, 1982:168.

20. Television and behavior: ten years of scientific progress and implications for the eighties; vol 1: summary report. Rockville, Maryland: National Institute of Mental Health, 1982.

21. Slaby RG, Roedell WC. The development and regulation of aggression in young children. In: Worell J, ed. Psychological development in the elementary years. New York: Academic, 1982.

22. Liebert RM, Neale JM, Davidson ES. The early window: effect of television on children and youth. New York: Pergamon, 1973.

23. Zuckerman D, Zuckerman B. Television's impact on children. Pediatrics 1985;75:233–40.

24. Slaby RG, Quarfoth GR. Effects of television on the developing child. In Camp BW, ed. Advances in behavioral pediatrics; vol 1. Greenwich, Connecticut: JAI Press, 1980.

25. Huesmann LR. Television violence and aggressive behavior. In: Television and behavior: ten years of scientific progress and implications for the eighties; vol 2: technical reviews. Rockville, Maryland: National Institute of Mental Health, 1982:132.

26. Spivak H, et al. Dying is no accident: adolescents, violence, and intentional injury. Ped Clin N America (in press).

27. Goldstein PJ, Hunt D, Des Jarlais DC, Deren S. Drug dependence and abuse. In: Amler RW, Dull HB, eds. Closing the gap: the burden of unnecessary illness. New York: Oxford University Press, 1987.

28. Harwood H, et al. Economic costs to society of alcohol and drug abuse and mental illness. Rockville, Maryland: National Institute on Drug Abuse, 1984.

29. Goldstein P, Hunt D. Health consequences of drug use. Atlanta, Georgia: Carter Center, 1984.

30. McBride D, et al. Drugs and homicide. Bull NY Acad Med 1986;62(5):497–508.

31. Molotsky I. Capital's homicide rate is at a record. New York Times 1988 Oct 30, Sec A, p. 20

32. Pitt D. Slayings in New York heading for record high. New York Times 1988 Oct 27. Sec B, p. 8.

33. Murphy S. Boston homicide rate up 36%. Boston Globe 1988 Dec 28. 1 & 14.

34. Zahn M. Homicide in the twentieth century United States. In: Inciardi J, Faupel C, eds. History and crime. Beverly Hills, California: Sage, 1980.

35. Stover D. A new breed of youth gang is on the prowl and a bigger threat than ever. Am School Board J 1986 (Aug):19–24.

36. US Senate, Committee on the Judiciary, Subcommittee on Juvenile Justice. Gang violence and control, hearings Feb 1983. Washington, DC: US Government Printing Office, 1983.

37. Feldman PA. Drug-peddling street gang holds neighborhood in fear. Los Angeles Times 1987 Nov 16; sect 30:1.

38. LAPD asserts that surge in youth gang violence in first half of 1986 is due to street drug sales and related battles over turf. Los Angeles Times 1986 Jul 17;sect 2:1.

39. Report from the task force on youth gang violence. Los Angeles Times 1986 Jan 7; sect 2:2.

40. Holgum R. Six killed, 4 hurt as truce drive among street gangs begin. Los Angeles Times 1987 Nov 30; sect 2:1.

41. James G. Murders in Queens rise by 25%: crack described as a major factor. New York Times 1988 Apr 20:1.

42. Raab S. Brutal drug gangs wage war of terror in upper Manhattan. New York Times 1988 Mar 15:B1,B5.

43. Exchange of insults and blows turn to tragedy. Los Angeles Times 1985 Dec 21; sect 2:1.

44. Girdner B. Portland, Oregon, struggles to cope as gangs step up violence. Boston Globe 1988 Nov 3:100.

45. Tobler N. Meta-analysis of 143 adolescent drug prevention programs: quantitative outcome results of program participants compared to a control or comparison group. J Drug Issues 1986;16(4):537–67.

46. Schaps E. A review of 127 drug abuse prevention program evaluations. Lafayette, California: Pacific Institute for Research and Evaluation, 1980(Feb).

47. Bell C. Prevention research: deterring drug abuse among children and adolescents (NIDA research monograph 63). Rockville, Maryland: National Institute on Drug Abuse, 1975.

48. Glynn T, et al., eds. Preventing adolescent drug abuse: intervention strategies (NIDA research monograph 47). Rockville, Maryland: National Institute on Drug Abuse, 1983.

49. Times F, Ludford J, eds. Drug abuse treatment evaluation: strategies, progress, and prospects (NIDA research monograph 51). Rockville, Maryland: National Institute on Drug Abuse, 1984.

50. Stoudemire A, et al. Alcohol dependence and abuse. In: Amler R, Dull H, eds. Closing the gap: the burden of unnecessary illness. Am J Prev Med 1987;4(suppl):9–18.

51. Moore MH, Gerstein DR. Alcohol and public policy: beyond the shadow of prohibition. Washington, DC: National Academy Press, 1981.

52. Gerstein DR, ed. Toward the prevention of alcohol problems: government, business, and community action. Washington, DC: National Academy Press, 1984.

53. Olson S, Gerstein DR. Alcohol in America: taking action to prevent abuse. Washington, DC: National Academy Press, 1985.

54. Lopez C, Pendley L. Substance abuse prevention program—Albuquerque, New Mexico. Morbid Mortal Weekly Rep 1987;36(44):729–30.

55. Parke RD, Slaby RG. The development of aggression. In: Mussen PH, ed. Handbook of child psychology; vol 4. 4th ed. New York: John Wiley, 1983:615.

56. Elvey KT, et al. Black parent training programs. Rockville, Maryland: National Institute of Mental Health, 1982.

57. Eron LD, Huesmann LR. Integrating field and laboratory investigations of televised violence and aggression. Paper presented at the annual meeting of the American Psychological Association, Montreal, Canada, 1980.

58. Huesmann LR, et al. A combined laboratory and field study of the reduction of aggressive behavior. Paper presented to the International Society for Research on Aggression, Haren, Netherlands, July 1980.

59. Alternative schools for disruptive youth: NSSC resource paper. Malibu, California: National School Safety Center, 1987:6.

60. Edmonds RR. Some schools work and more can. Soc Policy 1979:28–32.

61. Edmonds RR. Effective schools for the urban poor. Educational Leadership 1979:15–27.

62. Boyer EL. High school: a report on secondary education in America. New York: Harper and Row, 1983.

63. Lightfoot SL. The good high school. New York: Basic, 1983.

64. Lasley TJ, Wayson WW. Characteristics of schools with good discipline. Educational Leadership 1982 (Dec):28–31.

65. Friday PC, Elrod HP. Schools and delinquency: strategies and conflicts in implementing delinquency prevention through organizational restructuring. Paper presented at the annual meeting of the American Society of Criminology, Montreal, Canada, November 1986.

66. Brookover W, et al. Creating effective schools: module 7. Holmes Beach, Florida: Learning Publications, 1982.

67. Core curriculum in preventing and reducing school violence and vandalism. Washington, DC: Center for Human Services and National School Resource Network, 1980.

68. McCarthy WJ. The cognitive developmental model and other alternatives to the social deficit model of smoking onset. In: Bell CS, Battjes R, eds. Prevention research: deterring drug abuse among children and adolescents (NIDA research monograph 63). Rockville, Maryland: National Institute on Drug Abuse, 1985.

69. Moskowitz J. Preventing adolescent drug substance abuse through drug education. In: Glynn TJ, et al., eds. Preventing adolescent drug abuse: intervention strategies. Rockville, Maryland: National Institute on Drug Abuse, 1983.

70. DeJong W. Project DARE evaluation results. Newton, Massachusetts: Education Development Center, 1986.

71. Yeates KO, Selman RC. Social competence in the schools: toward an integrated developmental model for intervention. Boston: Judge Baker Children's Center, 1987.

72. Selman RL, Glidden M. Negotiation strategies for youth. School Safety 1987:18–21.

73. Prothrow-Stith D. Violence prevention curriculum for adolescents. Newton, Massachusetts: Education Development Center, 1987.

74. Bickmore K. Alternatives to violence. Cleveland, Ohio: Cleveland Friends Meeting (undated).

75. Bigda-Peyton T, Bigda-Peyton F. Conflict resolution curriculum package. Cambridge, Massachusetts: Boston Area Educators for Social Responsibility (undated).

76. Kreidler W. Creative conflict resolution: over 200 activities for keeping peace in the classroom. Glenview, Illinois: Scott, Foresman (undated).

77. Preventing family violence. Boston, Massachusetts: Resource Center for the Prevention of Family Violence, Massachusetts Department of Public Health (undated).

78. Guerra N, Panizzon A. Viewpoints: solving problems and making effective decisions. Santa Barbara, California: Center for Law-Related Education, 1986.

79. Guerra NG, Slaby RG. Cognitive mediators of aggression in adolescent offenders: intervention. Dev Psych (in press).

80. Prothrow-Stith D, et al. The violence prevention project: a public health approach. Sci Tech Human Values 1987;12(3–4):67–9.

81. Crisis Intervention Network report, 1975–1987: new approach to youth violence. Philadelphia: Crisis Intervention Network, 1987.

82. Briefing paper: current update on the state of targeted outreach. New York: Boys Clubs of America, 1988.

83. Testimony to Boys Clubs. New York: Boys Clubs of America, 1987.

Chapter 12: Child Abuse

Intimidated, perhaps, by the size of the courtroom or by all of the adults around her, the young girl delivered her evidence haltingly. "My father and mother are both dead," said 8-year-old Mary Ellen Wilson. "I don't know how old I am. I have no recollection of a time when I did not live with the Connollys. I call Mrs. Connolly mamma. . . . I have never been allowed to go out of the room where the Connollys were, except in the night time, and then only in the yard. . . . I am never allowed to play with any children, or have any company whatsoever. Mamma has been in the habit of whipping and beating me almost every day. . . . I have no recollection of ever having been kissed by anyone —have never been kissed by mamma. . . . I have never dared to speak to anybody because if I did I would get whipped."[1]

Mary Ellen was not without friends, however. She was represented by a well-known New York attorney, and an important charitable organization backed her case. The evidence was irrefutable. Mary Ellen was removed from the Connollys' custody. Convicted of assault and battery, Mrs. Connolly was sentenced to a year's imprisonment.

Mary Ellen Wilson was America's first recorded child abuse victim; it was April 1874. Her benefactors were Henry Bergh and Eldridge T. Gerry, founder and lawyer, respectively, of the American Society for the Prevention of Cruelty to Animals (ASPCA). Indeed, Gerry had had to convince the court that children were members of the animal kingdom entitled to the same protections afforded animals under the law. Within the year New York State had passed the country's first child abuse law and Gerry had founded the first Society for the Prevention of Cruelty to Children.[2]

Mary Ellen Wilson may have been the first recorded child abuse victim, but she was hardly the first. That distinction belongs to some child who lived and died before the beginning of recorded time. Indeed, child abuse has been endemic to most societies.

DEFINITIONS

"Child abuse" is a general term encompassing physical abuse, psychological or emotional abuse, sexual abuse or sexual exploitation, and neglect. "Maltreatment" and "child maltreatment" are less precise terms used as synonyms for child abuse in the general sense or, in context, as synonyms for the specific form of abuse under discussion. All these terms describe the failure of a caretaker to provide responsible care for a child. Indeed, most mandatory child protection reporting systems limit abuse to activities by family or household members or other caretakers. They do not include the actions of strangers or acquaintances.

Because there is no standard definition of child abuse, each state has created its own legal definition. Michigan, for example, defines child abuse as "harm or threatened harm to a child's health or welfare by a person responsible for the child's health or welfare which occurs through nonaccidental physical or mental injury, sexual abuse or maltreatment . . . or negligent treatment."[3]

Child sexual abuse is the term used most frequently to describe any sexual contact between a child and an adult (and sometimes between juveniles, if there is sufficient age difference, use of force, or some other special circumstance). "Sexual exploitation" and "sexual victimization" are also used, though less frequently. "Thus," wrote David Finkelhor, "a stranger who exposes himself in front of a child is committing child sexual abuse. So is a father who has sexual intercourse with his daughter. Most sexual abuse of children falls somewhere between these two extremes. The categories of child sexual abuse are incest, pedophilia, exhibitionism, molestation, statutory rape and rape, child prostitution, and child pornography."[4]

Even the precision implied in these categories is illusory, however. For example, "incest" is used to define any sexual activity between family members, but in anthropological terms, incest refers only to sexual activity between blood relatives. In many jurisdictions, "family" is defined in much broader terms to encompass the actual living arrangements of those involved and may include relatives who do not regularly live with the child.

"The inability to reach closure on the issue of defining abuse," wrote one long-term researcher, "reveals the lack of coherent pro-child ideology among Americans."[5] Further, according to another, "The broader the definition of abuse, the more clear is its relation to 'normal' caregivers and their behavior with children, and the more serious the 'indictment' against society and its institutions."[6]

From the public health perspective, the inability to arrive at a precise—or even generally accepted—definition of child abuse limits our ability to understand accurately the dimensions of the problem and therefore to plan effective prevention programs. We can also see the effects of that lack of understanding in trying to specify the incidence of these injuries.[7]

THE MAGNITUDE OF THE PROBLEM

Because child abuse has been a reportable crime in every state for many years, statistics about its incidence are readily available. Since 1974, the American Association for Protecting Children (AAPC) and the National Committee for Prevention of Child Abuse (NCPCA) have collected national data, based primarily upon state reports, to document the dimensions of child abuse and neglect.

According to the latest NCPCA annual 50-state survey, an estimated 2 million reports of child abuse and neglect were filed in 1986. That represented an increase of 6% over the previous year. Over the past 10 years, such reports have increased by 184%.[8] The majority of these reports—58.5% in 1985, according to the U.S. House Select Committee on Children, Youth, and Families—allege neglect.[9]

Child homicide is now among the five leading causes of death in childhood, accounting for one in every 20 deaths of people below the age of 18, with the majority of infant victims killed by parents and relatives and older children more frequently killed by strangers, acquaintances, or unidentified perpetrators.[10] In addition, according to the most recent NCPCA survey, there were an estimated 1,200 childhood deaths resulting from maltreatment in 1986. "Overall," the report indicates, "the number of deaths due to maltreatment rose 23 percent between 1985 and 1986, compared to an increase of 12 percent between 1984 and 1985."[8] However, according to a recent report, "Statistics on the number of children who die from maltreatment are still unreliable. Local authorities are often hesitant to report maltreatment as the cause of death; the role of abuse and neglect may be ignored in deaths from 'natural causes' such as pneumonia, malnutrition, drowning, and sudden infant death syndrome. One study of death certificates and clinical records suggested that as many as 5,000 children die each year from maltreatment."[11]

Finally, charges of sexual abuse now represent the fastest growing category of abuse reporting. In 1985 more than 150,000 such reports were filed, a more than 12-fold increase in a decade.[12] They now comprise approximately 14% of all maltreatment reports.[8]

Such statistics as we have are based on *reports* of child abuse and neglect. In theory, this might bias our understanding of the true incidence of maltreatment in either direction. The total number of reports, for example, is greater than the total number of children involved. A child trapped in a long-term abusive situation may become the subject of many reports. In addition, each report is not evidence of abuse or neglect. Some reports (though a relatively small number, by all accounts) are frivolous. Others, while probably genuine, cannot be substantiated.

Concern over possible misclassification, frivolous reports, and the possibility of "false positives" (a situation incorrectly identified as child abuse in which none exists) pale, however, in the face of the massive underreporting of child abuse and neglect. And this is particularly true of the underreporting of both fatalities and child sexual abuse, research on which is "difficult, scarce, and extremely variable in quality."[13]

Hospitals, physicians, and other medical personnel have played an important role in developing an understanding of child abuse. Dr. Henry Kempe, a Colorado pediatrician and educator who coined the phrase "battered child syndrome" in 1961, set the stage for the first sustained examination of the problem for professionals and the general public.

Medical personnel are often called on to diagnose and treat the consequences of abuse and neglect. They are also, in every state, bound by law to report these cases to child protection agencies. One would expect a high degree of compliance. However, according to the U.S. Department of Health and Human Services' National Study of the Incidence and Severity of Child Abuse and Neglect (1979–1980), hospitals failed to report almost half of the cases that met the study's criteria.

An analysis of Health and Human Services data from hospitals in a stratified random sample of 26 counties (distributed over 10 states) revealed several important distinctions between cases that were reported to protective agencies and those that were not. "Social class and race," the analysis found, "are the most important perpetrator characteristics that distinguish reported from unreported cases of abuse." Black and Latino families were more likely to be reported than white families. Families in the lowest income levels had the highest rates of reporting, but among families earning $25,000 or more a year, barely a third of the cases of abuse were reported. Cases of emotional abuse or neglect tended, in general, to be underreported. Another study suggests that "class and race, but not severity, define who is and who is not reported by hospital personnel to child protective services."[14]

"If the reporting of child abuse is as biased by class and race as these data suggest," the researchers concluded, "there is a clear need for a critical review of the *system* as well as the *process* of reporting. . . . In selectively ignoring the prevalence of child abuse in more affluent, majority homes, we may be perpetuating a myth that child abusers are out there, and that homes like ours are free of violence."[15]

In a 1985 survey, more than 62% of the parents questioned admitted to having used physical violence (on a continuum ranging from pushing and slapping to the use of knives or guns) against their children in the previous year. Nearly 11% of the parents reported using severe violence: hitting, kicking, beating, threatening, or using knives or guns.[16]

The 1985 data indicate how poorly the figures on cases reported reflect the actual incidence of abuse. The findings suggest that 1.5 million children from birth to 17 years of age in two-parent families were subjected to very severe physical violence in 1984. In contrast, the official reporting data indicate that between 149,000 and 219,000 children were injured by physical abuse in 1984. These data, although admittedly drawn from two different types of sources, suggest that even in the case of severe physical abuse only one physically injured child in seven was reported to protective services.[11]

When one considers the incidence of child sexual abuse, the figures are so unclear that even good estimates are hard to come by. Researchers who have surveyed adults retrospectively about experiences with childhood sexual abuse have produced results that vary from study to study but in each case suggest a problem enormously greater than official reports indicate.

In a series of five surveys conducted between 1940 and 1980, one-fifth to one-third of all women reported some sort of childhood sexual encounter with an adult male.[17] And a major *Los Angeles Times* poll in 1985 revealed that "at least 22 percent of Americans have been victims of child sexual abuse, although one-third of them told no one at the time and have lived with their secret well into adulthood." Twenty-seven percent of the women and 16% of the men reported past abuse. Extrapolating to the U.S. population as a whole, it was concluded that 38 million adult Americans were sexually abused as children.[12]

What is unknown, of course, is how current rates of sexual abuse among children relate to these retrospective accounts. A consistent finding of these retrospective studies—and one with important lessons for prevention—is that the majority of abusers were already known to their victims and that one-third to two-fifths of them were family members. In a recent Los Angeles study, for example, only 21% of the women and 28% of the men reporting childhood sexual abuse said that they had been assaulted by a stranger.[13,18]

As this discussion demonstrates, there is much that we do not know about the actual incidence of child abuse in America. This is an important problem, and one to which attention must be paid. As a physician who recently conducted a major review of child abuse prevention programs for the Congressional Office of Technology Assessment put it: "We know that we are facing a problem of enormous proportions. Despite the varying definitions of child abuse, there really is a remarkable consensus on the nature of the problem we seek to address. Nor should we underestimate what we already know about how to respond" (H. Dubowitz, personal communication).

In the face of such conflicting data, we have chosen to adopt as the best estimate the figures contained in the U.S. Department of Health and Human Services' 1988 study of the national incidence and prevalence of child abuse and neglect:

• In 1986 an estimated 25.2 children per 1,000, or a total of more than 1.5 million children nationwide, experienced abuse or neglect.

• The majority of cases (63%) involved neglect; 43% involved abuse.

• The most frequent type of abuse was physical, followed by emotional abuse and then by sexual abuse, with incidence rates of 5.7, 3.4, and 2.5 children per 1,000, respectively.[19]

THE COSTS OF CHILD ABUSE

Child abuse carries high costs, both for individuals and society. Seriously abused and neglected children suffer permanent neurological, physical, and developmental damage. Sexually transmitted diseases are frequent sequelae of child sexual abuse, and unwanted pregnancies are not uncommon. Even less severely abused children may be cognitively, linguistically, and physically impaired. Lasting emotional and psychological damage often results. And the damage is not necessarily limited to the victim; the siblings of an abused child, particularly a sexually abused child, may suffer feelings of fear, anger, or helplessness or be affected in other ways.

Society sees the consequences of child abuse and neglect compounded from year to year and pays accordingly. Retrospective studies of institutionalized adults reveal a significant number of childhood abuse cases. As many as 80%–90% of juveniles arrested for delinquent acts report a history of abuse and neglect. Adults who mistreat their children and spouses frequently have histories of child maltreatment themselves.[11]

It is estimated that the child homicides committed in 1980 accounted for 93,000 years of potential life and productivity lost.[20] Another study conservatively estimates the immediate costs of placement and medical and therapeutic services for victims at $500 million a year. Another $600 million, it suggests, may eventually be required for the foster care of each year's victims or the juvenile detention of those abused children who themselves commit violent or criminal acts.[21] But even if the dollar figures are far from exact, the need to understand the risk factors for childhood maltreatment as an aid to prevention is very clear.

RISK FACTORS

The search for risk factors in child abuse has been hampered by the overall problems with child abuse data. However, some progress has been made.

Physical Abuse and Neglect

Childhood health and developmental problems such as prematurity, disability, congenital defects, and early childhood illness increase the risk for maltreatment. Abusive parents are more frequently marked by their own childhood maltreatment, low self-esteem, very young parenthood, single parenthood, and inadequate knowledge of or preparation for caretaking. Stress resulting from unemployment, overcrowded living conditions, and isolation from social support also can be a risk factor.[11] And more than one authority has pointed to the risk posed when "the family lives in a culture in which corporal punishment is sanctioned or encouraged."[2] It is likely that there is no single risk factor but rather a combination of many for abuse and neglect. No risk factors have been identified that distinguish cases of child abuse from those of neglect.

Child Homicide

According to the NCPCA, approximately half of the 1,200 children who died as a result of maltreatment in 1986 died as the result of physical abuse—either a single violent episode or the cumulative result of repeated severe beatings. The other half died as a result of severe neglect. The NCPCA also report

studies in which the majority of the children who died, nearly 75%, were 1 year of age or younger. Children who die, these studies indicate, are more likely to come from families with two caregivers and in which the caregivers are 16 to 20 years of age.[8]

In 1980, 501 child homicides were committed in this country. A study of those deaths found that children under 3 years of age and over 12 years of age are at greatest risk. Children under 2 years of age accounted for nearly one-fourth of all deaths. Black and other nonwhite male children were at highest risk (6.8 per 100,000), as compared with black and other nonwhite females (3.3 per 100,000) and white males (2.3 per 100,000).[20] It is important to note that in studies in which child homicide is correlated with socioeconomic status race ceases to be a risk factor and is replaced by such factors as the degree of household crowding.[22,23]

A Centers for Disease Control (CDC) analysis of FBI child homicide data from 1976 to 1979 indicated that 29% of the murders were committed by parents and another 35% by acquaintances. Ten percent were perpetrated by strangers; the perpetrators of another 26% could not be identified.[24]

The manner of death varies with age as well. A very high percentage of homicides in early childhood are caused by beatings. The percentage caused by firearms increases steadily with age. By 15 years of age the proportion of firearm homicides reflects adult patterns[10] (also G. Lapidus, personal communication).

Child Sexual Abuse

Most authorities agree that there are hypotheses about the risk factors for child sexual abuse but very little evidence to prove their effect. Research has failed to find differences in rates by social class or race. However, several factors have been consistently associated with a higher risk for sexual abuse and particularly for incest:
- A child lives with one of the biological parents.
- The mother is unavailable to the child because of employment outside the home, disability, or illness. (Unavailability of the mother, particularly sexual unavailability to the male parent figure, has also been noted as an associated factor.)
- The child reports that the parents' marriage is full of conflict.
- The child reports an extremely poor relationship with his or her parents, including child abuse.
- The child reports having a stepfather.[13]

"Although few studies have examined why these factors increase risk, poor supervision, emotional turmoil, neglect, and rejection may make a child vulnerable to the ploys of child molesters. In other words, as the result of conflicts and emotional deficits, the children are easier to manipulate with offers of affection, attention, and rewards in exchange for sex and secrecy."[13]

In addition, although the use of the term "child molester" above may conjure up the image of the stranger with candy against whom parents warn children, reality is different. Most studies indicate that children are sexually abused by adults they already know and trust, who often are members of their families.

Our culture has a distinctly ambivalent attitude toward children's sexuality. On one hand, there is an explicit belief that children are not sexual beings and that their sexual activity is to be postponed as long as possible. Any recognition of childhood sexuality is discouraged; the panicked responses to proposals for sex education classes for young children are evidence of this attitude. At the same time, there is the implicit message that children really are little adults, coy and seductive, a canvas on which the passions of their elders are projected. Vladimir Nabokov's fictional 12-year-old nymphet, Lolita, is merely a reflection of 10-year-old, nude Brooke Shields in *Playboy* ("the first prepubescent sex symbol in the world," according to the photographer who took the pictures).[12]

SOCIAL AND CULTURAL FACTORS

Dr. Henry Kempe once explained the general social unwillingness to address the problem of child abuse in terms of "three almost sacred sayings of our culture: 'Spare the rod and spoil the child,' 'Be it ever so humble, there's no place like home,' and 'A man's home is his castle.' "[2] More than one commentator has referred to the general acceptance of corporal punishment in society as a precondition for the existing level of child abuse. Despite the degree to which poor and nonwhite parents find themselves overly represented in the reporting process, there remains a general societal and bureaucratic reticence to interfere in the family. The final maxim highlights the ways in which traditional patterns of authority intertwine with abuse.

Perhaps the truly startling thing about severe child abuse is that, given the general level of violence directed at children, the most serious cases are—underreporting aside—relatively uncommon. Anyone visiting a large supermarket or mall on a busy Saturday will see children slapped, pushed, hit, and verbally demeaned in astonishing numbers. And all of this takes place in public. Although the state may stake its claim to a child's welfare under extraordinary circumstances (e.g., overriding a parent's religious objection to blood transfusion), the general belief that children "belong" to their parents is remarkably strong.

The presence of firearms and the abuse of drugs and alcohol are other sociocultural factors often associated with interpersonal injuries. The role of firearms in child homicides has already been mentioned. Alcohol and drug consumption, associated with all other types of homicide, generally are not associated with child homicide. And though alcohol is frequently linked with family violence, there is no evidence that it causes child abuse per se.[20] It may, however, raise a parent's general level of stress while lowering his or her inhibitions against harming the child.

Alcohol certainly does not cause adults who are not otherwise sexually attracted to children to become so. But here again it may lower inhibitions and act as a facilitator. As many as 60%–70% of incest victims in some surveys report that their fathers had been drinking at the time the first attack occurred.[12]

A BRIEF HISTORY OF PREVENTION

Nearly 100 years elapsed after Mary Ellen Wilson's rescue before the medical profession had a name for the problem it had been so slow to recognize: the battered child syndrome. "Coined for its shock value," according to one of Dr. Kempe's long-time collaborators, the term was first introduced at an October 1961 American Academy of Pediatrics symposium; the following year it was the title of a landmark article by Kempe and four colleagues in the *Journal of the American Medical Association*.[25,26]

Professional journal papers rarely have a profound and immediate effect on society, but Kempe's article opened a floodgate. Newspaper reports and magazine articles, both popular and professional, followed at a dizzying pace. Soon what had been a "dirty little secret" was accepted—albeit uncomfortably and with much lingering denial—as part of contemporary culture. In 1963 the popular television medical drama "Ben Casey" introduced the subject to an audience of 200 ABC network affiliates, and "Dr. Kildare" was not far behind.[2]

Even before publication of the paper, Kempe and his colleagues had met with representatives of the U.S. Children's Bureau to discuss the need for action. A subsequent meeting, also attended by lawyers and police officials, led to a model child abuse reporting law. By the late 1960s, all states had such laws on their books.

The federal Child Abuse Prevention and Treatment Act of 1974 required states to go beyond reporting physical abuse (the only requirement in many of the earlier laws) to report all maltreatment, including neglect and sexual abuse. This landmark legislation also established the National Center on Child Abuse and Neglect and provided funds for both federal and state activities. (The $15 million allocation in 1974 was more than doubled by 1987.[8]) Child protection services were strengthened, and a strong emphasis was placed on case finding.

Despite that emphasis, substantial numbers of abused and neglected children are never identified or, if identified, never reported. Even so, the cases that are captured by the system are more than enough to overwhelm it.

"Despite increased reports of child abuse, states [are] unable to provide needed services," reported the Select Committee on Children, Youth, and Families in 1987. "A majority of states," the committee contended, "report staff shortages, inadequate training, high personnel turnover, and a lack of resources for staffing as the principal barriers to improved child protection and child welfare services."[9] The results can be appalling. Child abuse victims can be removed from their parents and placed in foster homes where they subsequently die of maltreatment; there were 18 such cases in New York City in 1979, for example.[7]

A large number of discrete agencies and institutions play a role in identifying and resolving cases of child abuse and neglect. Health institutions, schools, the police and courts, and child protection agencies are the most obvious examples. As a result, Eli Newberger points out:

Although child abuse may indeed be defined by many people as a crime, most social policy in the U.S. has inclined toward a human service model for most victims. An awkward tension now prevails between the protagonists of the criminal process, who are skeptical of the utility of helping approaches and believe in the social deterrent functions of the criminal system, and the advocates of professional service to children, who are mindful of the value of good clinical work and are concerned about the unpredictable nature of the criminal system which may itself victimize children.[7]

The U.S. surgeon general, calling for improved responses to child abuse and child sexual abuse, has asserted that laws making this violence a crime will be ineffective until there is "informed cooperation and close collaboration of health, social service, law enforcement, and criminal justice professionals."[27]

In the last several years, a number of states have taken action to resolve this problem. In Massachusetts, for example, the 1983 Child Abuse Reporting Law amended the state's earlier legislation so that serious cases of child abuse are no longer the responsibility of any single state agency or department. Instead, the law mandates a multidisciplinary response involving both law enforcement and social service agencies. Each case referred to the district attorney is managed by a caseworker from the Department of Social Services (DSS), a representative of the district attorney's office, and a third party selected either for criminal justice experience or expertise in working with children. The team is responsible for evaluating the DSS service plan and for recommending whether the case should be prosecuted.[28]

"In implementing [Massachusetts] Chapter 288," wrote one participant, "we find that, as professionals, we have been brought into the child abuse arena together. We all come from different systems, each with its own set of priorities and its own difficulties. Each of us must work . . . to manipulate our systems to meet the needs of the child."[29]

PREVENTING CHILD ABUSE: THE STATE OF THE ART

After a year-long survey of child abuse prevention programs, researcher Jane Simmons concluded, "The kinds of programs currently available are only a beginning in primary prevention. If we wish to promote healthy parent-child interaction and to prevent children from becoming victims, we must provide opportunities for intervention at multiple stages of the life cycle."[30] Simmons discovered that many programs focusing on preschoolers actually target parents rather than children, yet in the early school years parents are virtually ignored. And although most adolescent programs focus on issues such as pregnant and parenting teens, parent-teenager interactions are overlooked.

What works, for whom, and under what circumstances is not known. "The reality is that children and families who are hurt by maltreatment cannot wait until research provides the answers. In the meantime, the best available information can be used to guide the development of policies and programs."[31]

General Recommendations

• As a basic step to improve reporting practices, states should use and monitor their mandatory reporting laws. The improved training of professionals who are required to file reports will reduce both underreporting and incorrect reports. Special attention must be paid, however, to the unintended effects of the criminalization of family problems and the causes of erroneous reports.[14]

• Police and social service agencies should build stronger collaborative relationships to strengthen their investigative capabilities and their potential for effective intervention. Many states publish analyses of their data in a central registry of abuse cases. For public health practitioners, these agencies can be potentially helpful collaborators as well as a valuable source of information.

• Coordinated, multidisciplinary data collection can provide states with an accurate picture of the child abuse problem they face. States that establish multidisciplinary child death review processes can learn the circumstances of death and apply those findings to prevention. However, they must persist in attempting to find better ways to share data and link the reporting of critical events before deaths occur.

The specific interventions discussed below are grouped by the sites in which they take place: the health care system, the community, the workplace, social service agencies, and the schools.

INTERVENTIONS

The Health Care System

In-hospital rooming-in arrangements. In an experimental program at a major hospital, mothers were given extended contact with their babies in the days after birth. In addition, the infant's father was, with the mother's permission, able to visit whenever the infant was rooming in. Assignment to the experimental program or the hospital's normal routine was random. The researchers' follow-up evaluation indicated that significantly fewer of the babies in the experimental program were later abused or neglected.[32] It may be that the mother's receptivity to such a program is an important consideration. According to one state health official, "Forcing a first-time mother who is scared about caring for a baby may be counter-productive" (W. A. Myers, personal communication).

Home health visitor programs. In one program recently evaluated in New York, nurses visited the homes of first-time mothers during the prenatal and infancy periods. These mothers were either teenagers, unmarried, or of low economic status. During postnatal visits, the nurses provided parent education about infant development, encouraged the involvement of family members in child care and support of the mother, and linked the family to other health and human services. During a two-year period, those mothers at highest risk had fewer instances of certified child abuse or neglect. They developed better parenting skills, and their children were seen less frequently by physicians for injuries.[33] Programs that use lay visitors, often under the supervision of public health nurses, also have been favorably evaluated.

Recommendation: Child abuse prevention efforts directed at enhancing the parenting skills of young and at-risk parents, including both in-hospital and home-visiting programs, are a promising intervention. They should be used and monitored and their outcomes evaluated.

The Community

Self-help groups. Parents Anonymous is a self-help, mutual-aid group that was established in 1970. There are now some 1,500 chapters around the world. Based on the experience of groups such as Alcoholics Anonymous, the parents' group provides a forum to help parents develop self-esteem, decrease social isolation, and improve their knowledge of child development and child rearing. These groups can provide services for parents who would otherwise avoid contacts with social service agencies.[11]

Child care. The growth in the number of children requiring day care has been exponential in recent years. According to a 1984 study, approximately 60% of women with children 3–5 years of age and 50% of those with children under 3 years of age are employed outside the home.[34] Some studies project that as many as 10.4 million children will require day care by 1990.

"The availability of quality, safe child care," according to one recent review of programs, "is a critical component of healthy functioning for many families. For some, it may be an important component of child abuse prevention."[11] It is also clear that the provision of low-cost, quality, safe child care is an objective that has not been met. Indeed, several instances in which child care facilities have become the setting for abuse, particularly sexual abuse, only emphasize the importance of this requirement.

Recommendation: Community-based interventions, such as self-help groups, coupled with the provision of safe, affordable day care, are promising interventions. They should be used and monitored and their outcomes evaluated.

The Workplace

Workplace education and support programs. The Family and Community Project at an Illinois electronics factory (supported by the Ounce of Prevention Fund) provides a wide range of services to employees, including speakers on issues affecting working mothers, counseling and referral to social services, and programs to reduce family isolation. Another program, the School-Aged Child Care Project in Massachusetts, created a workplace seminar curriculum for parents of school-aged children. The curriculum provides education on a broad range of issues affecting child development and parenting.[11]

Employer-supported child care. According to a recent study by the U.S. Department of Labor, barely 10% of the nation's employers provide specific benefits or services to help their workers arrange for child care.[35] In some cases, this amounts to helping employees locate community-based programs, although in other instances child care is provided at the worksite. Boston's Stride Rite Corp., for example, operates two child care centers that provide services to 90 children, including some from the community.

Recommendation: Workplace-based parent education and support programs as well as the extension of employer-supported child care are promising interventions. They should be used and monitored and their outcomes evaluated. The possible effects of other worksite interventions on reducing child abuse, such as flexible time and parental leave policies, should be studied.

Social Service Agencies

Parent education. In an effort to address primary prevention, social service agencies have developed a number of strategies to overcome deficits in parenting capabilities. Inadequate parenting skills, a poor understanding of the normal needs and development of a child, and a history of maltreatment in a parent's own childhood are often associated with abuse and neglect. In Minneapolis, for example, the Minnesota Early Learning Design (MELD): Blending Information and Support for New Parents program offers both education and peer support for new parents. The two-year-long program, which uses volunteers (often parents) as facilitators, covers a broad range of issues in child development, health, family management, and parenting skills.[36]

Parent aide programs. To address the problems of family stress and isolation, communities have developed parent aide programs, in which trained, professionally supervised persons, some volunteers, some paid, help parents whose children are at risk of maltreatment. Services may be concrete

and specific, such as providing transportation or home management education, or they may be oriented toward reducing isolation and stress. The evaluation of such programs has been limited, but a recent review indicates that they may play a significant role in preventing child abuse.[11]

Crisis and emergency services. Although the exact composition of services may vary from program to program, typical community-based crisis and emergency services include crisis nurseries for short-term care, parental stress hotlines, and emergency caretaker and homemaker services. Nashville's Comprehensive Emergency Services Project, for example, includes these services as well as an emergency family shelter and 24-hour emergency intake for abuse and neglect cases. The Nashville program resulted in significant reductions in the number of neglect and dependency petitions filed, the number of children removed from their homes, and the number of reported children subsequently abused or neglected.[37]

Recommendation: Social service agencies' parent education and parent aide programs show promise as complements to crisis-intervention and emergency services. They should be used and monitored and their outcomes evaluated.

The Schools

Sexual abuse prevention education programs. School-based programs to prevent sexual abuse have grown rapidly during the 1980s. They include the use of films, theater, puppets, skits, coloring books, songs, role playing, and other techniques. Some are brief, lasting barely an hour or two; others last several days. But all of them are intended to accomplish several aims. First, they teach the concept of sexual abuse (sometimes called "bad touching" or "touching in private places"). They teach children to refuse and run away from such overtures, no matter who the perpetrator. Finally, they encourage children to report such incidents to parents, teachers, or other adults.

The Child Assault Prevention Project (CAP), based in Columbus, Ohio, is among the best known and best evaluated of these programs. Originally developed by a small group of Women Against Rape members in 1978, more than 100 CAP projects are now under way in 18 states, Canada, and Great Britain. In all of its various forms, CAP provides teacher in-service training, a parents' program, and a children's workshop.

A recent review of more than two dozen evaluations (many involving multiple programs) led to several conclusions about this enormously popular effort at prevention.[38-40] "The overwhelming and irrefutable message of the evaluation studies is that children do indeed learn the concepts they are being taught," commented the reviewers. It is also true that older children learn the material better than younger children. However, there is serious decay in the retention of information over time. In addition, while some increase in fear and anxiety was noted in some children, most were not so affected. Another important positive indication was that these programs do prompt some children to disclose past abuse. The implications of these disclosures are serious, requiring school personnel to report the abuse and refer the child for appropriate treatment.

Even in such cases, according to one study, "Sexual abuse prevention programs are likely to be effective primarily with younger children who are being abused by someone outside their families. Far less clear is the value of such programs in preventing the most prevalent kinds of abuse, mainly incest and voluntary relationships between older children and pedophiles."[12]

However, the central question for these programs—the extent to which they prevent child sexual abuse—remains unanswered. There is no evidence that they do, nor has an adequate research design been developed that would address the issue.

Recommendation: School-based programs to prevent sexual abuse are a promising intervention. They should be used and monitored and their outcomes evaluated.

Proposals for Additional Interventions

Besides the interventions described above, we emphasize the need for the development of programs in several other areas. In addition to the hospital-based programs discussed above, hospitals should be used more often to provide basic primary prevention programs, including programs to educate caregivers.

Apart from the general need for child care discussed above, programs to provide a basic level of training and certification for caregivers must be developed. Ensuring adequate compensation for caregivers is equally important. In addition to day-care, other programs are necessary to reduce the possibility that latch-key children, who must look after themselves for some portion of the day, will be abused.

In the workplace, more can be done with flexible time and parental leave policies to support parents.

Community-based or social service programs for pregnant and parenting teenagers might be a useful addition to prevention in these areas.

Finally, both because of the physical damage it does and because of the damaging social attitudes it reinforces, corporal punishment should be banned in all schools and institutions. (According to the Department of Education, more than 1 million children were subjected to in-school physical punishment during the 1980–1981 school year.[11]) At the same time, information about and training in creative parenting should be made available to provide alternative methods for the teaching and disciplining of children.

INTERVIEW SOURCES

Howard Dubowitz, MD, Department of Pediatrics, School of Medicine, University of Maryland, College Park, June 9, 1988

Garry Lapidus, MS, The Connecticut Coalition to Prevent Childhood Injuries, Hartford, November 4, 1988

Woodrow A. Myers, MD, State Health Commissioner of Indiana, Indianapolis, April 28, 1988

REFERENCES

1. The New York Times 1874 Apr 10.

2. Heins M. The "battered child" revisited. JAMA 1984;251(24):3295–300.

3. Michigan child protection law, cited in Janice Humphreys. Child abuse. In: Campbell J, Humphreys J, eds. Nursing care for victims of family violence. Reston, Virginia: Reston, 1984:124.

4. Finkelhor D. Implications for theory, research, and practice. In: Nelson M, Clark K, eds. The educator's guide to preventing child sexual abuse. Santa Cruz, California: Network, 1986:179.

5. Gil DG. Violence against children. Cambridge, Massachusetts: Harvard University Press, 1970.

6. Garbarino J. The human ecology of child maltreatment. J Marriage Fam 1977;39:722.

7. Newberger EH. Child abuse. Surgeon General's workshop on violence and public health source book. Atlanta: Centers for Disease Control, 1985(Oct).

8. Daro D, Mitchel L. Deaths due to maltreatment soar: the results of the 1986 annual fifty state survey. Chicago: National Center on Child Abuse Prevention Research, National Committee for Prevention of Child Abuse, 1987.

9. US House of Representatives, Select Committee on Children, Youth, and Families. Abused children in America: victims of official neglect. Washington, DC: US Government Printing Office, 1987.

10. Christoffel KK. Homicide in childhood: a public health problem in need of attention. Am J Public Health 1984;74(1):68.

11. Meyers M, Bernier J. Preventing child abuse: a resource for policymakers and advocates. Boston: Massachusetts Committee for Children and Youth, 1987(Nov).

12. Crewdson J. By silence betrayed: sexual abuse of children in America. Boston: Little, Brown, 1988.

13. Finkelhor D. Child sexual abuse. In: Surgeon General's workshop on violence and public health source book. Atlanta: Centers for Disease Control, 1987 Oct 27–29.

14. Newberger EH. The helping hand strikes again: unintended consequences of child abuse reporting. J Clin Child Psychol 1983;12:309–15.

15. Hampton RL, Newberger EH. Child abuse incidence and reporting by hospitals: significance of severity, class, and race. Am J Public Health 1985;75(1):56–60.

16. Straus M, Gelles RJ. Societal change and change in family violence from 1975 to 1985 as revealed by two national surveys. J Marriage Fam 1986;48:465–79.

17. Herman J. Father–daughter incest. Cambridge, Massachusetts: Harvard University Press, 1981.

18. Siegel JM, Sorenson SB, Golding JM, Brunam MA, Stein JA. The prevalence of childhood sexual assault: the Los Angeles epidemiologic catchment area project. Am J Epidemiol 126(6):1141–53.

19. Study findings: study of national incidence and prevalence of child abuse and neglect. Washington, DC: US Department of Health and Human Services, 1988.

20. Rosenberg ML, Gelles RJ, Holinger PC, et al. Violence: homicide, assault, and suicide. In: Amler RW, Dull R, eds. Closing the gap. Am J Prev Med 1987;4 (suppl): 164–78.

21. Daro D. Confronting child abuse: research for effective program design. New York: Free Press, 1988.

22. Jason J, Andereck ND. Fatal child abuse in Georgia: the epidemiology of severe physical child abuse. Child Abuse Neglect 1983;7:1–9.

23. Centerwall BS. Race, socioeconomic status, and domestic homicide: Atlanta, 1971–72. Am J Public Health 1984;74(8):813–5.

24. Centers for Disease Control. Child homicide—United States. Morbid Mortal Weekly Rep 1982;31(22): 292–3.

25. Silverman FN. Child abuse: the conflict of underdetection and overreporting. Pediatrics 1987;80(3):441–3.

26. Kempe CH, Silverman FN, Steele BF, Droegemueller W, Silver HK. The battered child syndrome. JAMA 1962;181:17–24.

27. Koop CE. The surgeon general's letter on child sexual abuse. Washington DC: US Department of Health and Human Services, 1988.

28. Kepler K. Overview of the child abuse reporting law. In: Swagerty E, Marcus B, eds. The child abuse reporting law: the Middlesex County experience. Cambridge, Massachusetts: Middlesex County Child Abuse Project, 1986:5–6.

29. McNamara P. The role of the victim witness advocate. In: Swagerty E, Marcus B, eds. The child abuse reporting law: the Middlesex County experience. Cambridge, Massachusetts: Middlesex County Child Abuse Project, 1986.

30. Simmons JT. Programs that work: evidence of primary prevention of child abuse. Houston: Greater Houston Committee for Prevention of Child Abuse, 1986.

31. Dubowitz H. Child maltreatment in the United States: etiology, impact, and prevention. Background paper prepared for the Congress of the United States. Washington, DC: US Congress, Office of Technology Assessment, 1987(May):73; contract no. 633-3705.

32. O'Connor S, Vietze P, Sherrod K, Sandler H, Altemeier W. Reduced incidence of parenting inadequacy following rooming-in. Pediatrics 1980;66(2):176–82.

33. Olds DL, Henderson CR, Chamberlain R, Tatelbaum R. Preventing child abuse and neglect: a randomizing trial of home nurse visitation. Pediatrics 1986;78(1):65–77.

34. Newberger CM, Meinicoe LH, Newberger EH. The American family in crisis: implications for children. Curr Probl Pediatr 1986;16(12):669–739.

35. Employers offer aid on child care. New York Times 1988 Jan 17. Sec A, p. 25.

36. Reineke R, Benson P. MELD final evaluation report. Minneapolis: Minnesota Early Learning Design, 1981.

37. Burt M. Final results of the Nashville comprehensive emergency services project. Child Welfare 19;55:661–4.

38. Finkelhor D, Strapko N. Sexual abuse prevention education: a review of evaluation studies. In: Willis DJ, Holder EW, Rosenberg M, eds. Child abuse prevention. New York: Wiley (in press).

39. Binder RL, McNiel DE. Evaluation of a school-based sexual abuse prevention program: cognitive and emotional effects. Child Abuse Neglect 1987;11:497–506.

40. Kenning M, Gallmeier T, Jackson T, Plemons S. Evaluation of child sexual abuse prevention programs: a summary of two studies. Paper presented at the national conference on family violence, Durham, New Hampshire, July 1987.

Chapter 13: Domestic Violence

No, I don't think I wanted to hurt her. I think I wanted her to be afraid. Not like I walk in the door and she's afraid. But to be afraid to, like if we was in an argument and it was getting heated, you know, and a little screaming, yelling going on, to be afraid, you know. "Leave it alone. Don't get [John] upset." I think that's what I was doing.

—Batterer's account[1]

The most common feature of domestic violence among adults is that, to an overwhelming extent, it is perpetrated by men against women. Though woman battering is an ancient practice, it has only recently been recognized as a health problem. Variously excused and permitted over the centuries, battering was brought to public attention by the women's movement of the 1960s and 1970s, which examined rape and battering in a political context.

Nationally, about 1 million women use emergency medical services each year for treatment of injuries related to battering.[2] In addition, "battered women comprise a significant percentage of rape victims, suicide attempts, psychiatric patients, mothers of abused children, alcoholics, and women who miscarry and abort."[3] Clearly, this makes battering a major health concern. Yet until recently it was an unexamined phenomenon. The redefinition of domestic violence begun in the 1970s asserts that the cultural, legal, and religious structures of a male-dominated world have distorted the nature and extent of the problem, retarding progress toward solutions.

This chapter proposes that prevention and intervention strategies be comprehensive and integrated. It further urges that these strategies be implemented in a manner that holds the perpetrator, not the victim, accountable for the violence.

DEFINITIONS

"Domestic violence" is the most commonly used term for the incidents described in this chapter, and it is retained as the title. However, one of the chapter's themes is the need to redefine the field. The chapter therefore focuses on battering as a continuum of violent and controlling behaviors that includes injury. The sex of the victim is the most highly consistent factor in adult battering: It appears that at least 95% of serious partner assaults are against women.[4] Although women may hit men during conflicts, there is no evidence of a "battered man" syndrome comparable to the syndrome among women.

The classification of this violence by victim allows us to examine the cultural dynamics that permit the widespread battering of women by intimates, a focus that is diluted by sex-independent terms such as "spouse abuse" and "domestic violence." Furthermore, the term "wife battering" excludes single, divorced, or lesbian women who are battered. As with the elderly, children, and others at risk for violence owing to their legal, economic, and cultural status in society, there is a need to examine women's social position as a component in the phenomenon of battering.

Besides physical violence, battering often includes forced isolation, belittling verbal abuse, threats, intimidation, and the restriction of access to money, transportation, and other resources. Physical violence, the aspect of battering most frequently measured and examined, may include slapping, punching, kicking, choking, and attacks with weapons. Women also report being shot, stabbed, and bludgeoned. Injuries tend to be on the head, neck, chest, breast, abdomen, and perineum rather than the extremities. Rape and traumatic injuries during pregnancy are also indicators of possible abuse. Past traumas such as strokes, miscarriages, or suicide attempts and a history of headaches, sleep disorders, or vague, unspecified complaints can also be symptoms of battering.

Sexual assault is a common element of battering. Between 30% and 50% of all rape victims seen in emergency rooms are battered women.[5] Sexual assault includes not only nonconsensual sexual activity but the insertion of foreign objects in the woman's orifices and violence to genital and breast areas.

Batterers often isolate their victims, tracking them and timing their movements, blocking their contact with family, friends, and other support networks (including health care providers), and restricting their access to transportation. This isolation sometimes leads to imprisonment of the woman in the house, one room of the house, the garage, or a closet. The batterer often justifies this behavior with jealous accusations of infidelity.

Threats against a woman's life and physical safety and the safety of her children and other family members are another tool of control. Batterers may threaten to kill themselves, to initiate the removal of children from the home, and to withhold needed money for rent or food.

It may be difficult for those outside the family to detect the spectrum of behaviors involved in battering. Abusive men often "project an image to others as friendly, warm, outgoing, and helpful. Their inflexibility and controlling behavior tends to be restricted to their partners and (but not always) their children."[6]

THE MAGNITUDE OF THE PROBLEM

Between 2 and 4 million women are physically battered each year by husbands, former husbands, boyfriends, or lovers.[3,4] Approximately half of these women are single, separated, or divorced.[7] Between 21% and 30% of all women in this country have been beaten by a partner at least once.[8,9] The physical violence associated with battering is recurrent, and it escalates over time. Woman battering has "an incremental/developmental sequence, which, if unchecked, will result in increased physical, psychological, and social morbidity of the victims."[10] A Texas survey showed that 47% of batterers physically beat their partners at least three times a year.[2] One national study showed that battered women are assaulted by their partners at a median frequency of 2.4 times a year.[2] In a study of 109 battered women, the violence in 25% of the cases was reported to last from 45 minutes to more than 5 hours.[11]

Abuse may be the single most common source of serious injury to women, accounting for almost three times as many medical visits as traffic injuries.[3] Studies indicate that approximately 20% of women using emergency rooms are battered.[12] Furthermore, 14% of battered women, compared to 1% of nonbattered women, return to emergency rooms with 10 or more injuries.[13] However, battered women who seek help from health facilities are often not identified as battered. Recent studies show that only 4%–5% of domestic violence episodes are accurately identified as such by emergency room personnel, even though women's presenting injuries follow an identifiable pattern.[14]

The National Coalition Against Domestic Violence reported that, in 1986, 115,000 women accompanied by 165,000 children made use of shelters for battered women. Many more made use of other services; 330,000 women were turned away from shelters owing to lack of space (C. Allan, personal communication).

Studies show that women in this country are more at risk for assault, rape, and murder by a male partner than by a stranger.[15–18] A recent Massachusetts Department of Public Health survey of homicides in the state found that every 22 days a woman in the state is murdered by her husband or boyfriend.[19]

Although these statistics are alarming, the reality may be worse. The Bureau of Justice Statistics found that, in a 12-month period, only 52% of battered women called the police.[16] The inadequate system of reporting and data collection combined with the need for strict confidentiality make it impossible to document the true extent of the problem and to trace the effects of intervention attempts.

THE COSTS OF BATTERING

The burden of battering is enormous in both social and financial terms, with costs extending well beyond those related to the immediate injury. One epidemiologic review of violence among intimates found that "an estimated 40 percent of all female injury episodes are presented by battered women," and for only 4.4% of these is "the current episode the first at-risk injury."[13] Hospitalizations for abusive injuries appear to be no less common than for nonabusive injuries.[12]

Recently studies in health care settings and shelters for battered women have revealed sequelae of battering that are both costly and, until now, have remained largely unrecognized. In examining the records of battered women, researchers have begun to discern a pattern of additional physical and psychiatric complaints that appears to result from battering.[12] These include problems related to old injuries, including headaches, pain in the back and ribs, and unspecified pain. Sleep disorders, anxiety, depression, unspecified fear, "vague complaints," speech impairments, hyperventilation, and other stress-related problems may occur. Unfortunately, instead of being recognized as indicators of possible abuse, these complaints have earned women labels such as "hysterical" and "hypochondriac." If these conclusions are recorded in a medical file, the result may be less responsive behavior on the part of medical personnel.

When women's experiences are minimized in this way, battered women are less likely to seek out health care or protection. Of battered women treated in emergency rooms, 10% attempt suicide at least once, and 5% make multiple attempts. Twenty-six percent of all suicide attempts by women are preceded by abuse.[3]

One in five pregnant women are battered. Battered women are pregnant more often but carry to term less often.[20] Other gynecological problems may result from abuse, including persistent abdominal pain and painful intercourse.[3]

Another problem linked to woman battering is child abuse. Studies indicate that 45% of mothers of abused children are themselves battered.[3] In these instances, the abuse of the woman begins before the abuse of the children.[3] Children of battered women may also suffer from sleep disorders, eating disorders, depression, nightmares, respiratory distress, and a range of physical complaints.[21] In addition, when a mother and her children flee a violent domestic situation, this may result in drastically worsened economic circumstances.

Other costs of battering include those incurred by victims as they attempt to end the violence, often making use of police, courts, emergency shelters, advocates, and other resources. If the woman decides to leave her home, the emotional and financial costs of flight can be enormous. It is not unusual for a batterer to pursue a woman from community to community and even from state to state, making her and her children domestic refugees.

For the battered woman, the costs of staying in the relationship are clearly quite steep, but so are the costs of leaving, particularly as batterers often continue and intensify their attacks even after a separation or divorce.[3] The 1984 National Crime Survey found that 19% of domestic violence was perpetrated by an ex-spouse. Another 40% was perpetrated by a spouse, although there is no indication of how many of these couples were separated.[22] The attitudes that prompt people to ask why women stay deny the predicament in which many battered women find themselves.

RISK FACTORS

A significant amount of research has been done to determine the factors in a woman's background, circumstances, or psychological makeup that put her at higher risk of abuse. Yet, despite attempts to identify "risk markers," recent studies have found that there is little that distinguishes this group of women from others.[14] However, there are some broad indicators. Possibly the most consistent finding of risk factor studies is that battered women tend to be young; this fact alone, however, is not a useful risk marker for practitioners. Black women appear to be at greater risk than white women, although it is unclear whether this represents genuinely increased risk or greater willingness to report violence. For example, one survey noted that black women are twice as likely to report abuse as white women.[8]

In addition, "the risk of battering is greatest where a woman has a higher occupational and educational status than her partner or where the man is unemployed or consistently underemployed."[3] Other researchers have noted the importance of "isolating factors" (e.g., rural residence, alienation from family and friends, language differences, physical handicaps); after battering has occurred, these isolating factors become an important risk factor for future abuse.[4]

Although research on men's violence toward women raises a number of complicated issues, it is sometimes forgotten that men's violence is men's behavior. Therefore, it is not surprising that the effort to explain male behavior by examining the characteristics of women has been largely unsuccessful.[23]

Battering and Homicide

An all too common end point of battering is homicide, committed by either the batterer or the victim. Usually, there are numerous opportunities for interventions before a homicide occurs. In a Chicago study of 132 women imprisoned for the murder of abusive husbands, all reported having called the police at least five times before the final incident.[24] Similarly, reviews of police records in Kansas City, Missouri (1976), and Detroit (1984) show that in 90% of domestic homicides, police had been called to the address at least once for a domestic disturbance. In 54% of cases police had been called at least five times.[25,26] These studies show only those opportunities for police intervention. It may be assumed that a portion of these women also had contact with health care and other institutions.

Men commit 92% of all homicides in this country. Of all male-perpetrated homicides, 14% are against their partners. Women commit 8% of all homicides in this country, and, of these, 51% are against partners. Thus, women kill much less often than men, but when they do kill it is more likely that the victim will be a partner.[27]

Women also tend to kill differently and for different reasons. In a study of 83 homicide incidents between intimate (men and women) partners, most women who killed men were acting in self-defense, but most men who killed women did so in response to the victim's attempt to leave the relationship. In the same study, men were found to be "more violent in the homicides," while women tended to inflict just one wound, "suggesting still more support for a lesser tendency for females to initiate violence because of rage or a desire to inflict punishment, and a greater likelihood of their engaging in self-defense related homicide."[28] Other studies indicate that women who kill partners do so after a history of abuse and in self-defense.[29-32]

One revealing study shows that female-perpetrated homicides began to decline in 1979 immediately following the establishment of alternatives for battered women, such as emergency shelters, legal remedies, and heightened community awareness. During the same period, however, male-perpetrated homicides in at least half the states increased.[27] This study calls into question the assumption that simply leaving an assaultive mate will end the violence.

Alcohol and Battering

Statistics linking alcohol abuse and battering vary widely, but it is safe to say that alcohol is present as a factor in a significant number of cases reported to police and battered women's shelters. Alcohol, and drugs to a lesser extent, are highly associated with all homicides, with the exception of child homicides.[3] It appears likely that battering situations that involve alcohol tend to be more serious, thus making them more likely to be reported.[33] However, it is generally conceded by experts in the field that alcohol use is not to be considered a causal factor in battering.[3,34]

The use of alcohol by a battered woman greatly compromises her ability to receive help. Police and the courts may view a woman's alcohol abuse as a provocation of violence and, as a result, excuse the battering. Shelters often do not accept women who are actively drinking, requiring either a period of sobriety or strict sobriety in the shelter. The latter condition may seem impossible for an alcoholic woman, especially in crisis, without specifically structured supports. Alcoholism also tends to isolate women more than it does men. This isolation, in conjunction with the isolation battering enforces, serves to create further barriers to intervention.[35] The problem of substance-abusing battered women requires special attention, both through the design of shelter and advocacy programs and through heightened sensitivities of health care and intervention practitioners.

SOCIAL AND CULTURAL FACTORS

Laws have been enacted recently to protect women and hold batterers accountable for their actions; unfortunately, the implementation of these laws lags, creating a contradiction between law and practice. Criminal justice agencies have been reluctant to treat battering as a crime rather than an interpersonal problem. Battering does differ in one key way

from assaults by strangers: The perpetrator has continuing access to and extended control over the victim. But instead of treating such assaults with greater seriousness, police and the courts, until recently, have chosen to minimize their involvement. Even the inevitable murders resulting from abuse have been excused in many instances as "crimes of passion."

The case of Clarence Burns is illustrative. In June 1983 in Denver, Colorado, Mr. Burns was awaiting sentencing after pleading guilty to second-degree murder. He had shot his 38-year-old wife, Patricia Ann, five times in the face after she said she was leaving him. Before sentencing the defendant, the judge said of him, "his mental and emotional condition, combined with the sudden heat of passion caused by a series of highly provoking acts on the part of the victim affected the defendant sufficiently so that it excited an irresistible passion." The judge then sentenced Mr. Burns to serve a two-year, work-release sentence in the Denver County jail so that he could support his teenage son.[1] Routine justification for violence and murder based on the perpetrator's "loss of control" and the victim's "provocation" excuses batterers from taking responsibility for their violence.

Consistent with this culturally supported neglect, some modern theories about battering have emerged that place the blame for the violence on the victim. Some professionals to whom women turn for help have seen the women as "troublemakers" who provoke the violence, "masochists" who enjoy it, or "professional victims" who martyr themselves to violence. Such stereotypes do not acknowledge the difficulty of finding help and alternatives, the number of times the woman asked for help, or the number of times she tried to leave. The depiction of a woman's threat to leave as a "trigger" and the claim that battered women share a "need" with the abuser to remain together only serve to excuse the violence.

Similarly, research and theories of battering that focus on the psychological makeup of the individual batterer underestimate the impact of male socialization and the social acceptance of violent male behavior. Richard Gelles in 1973 criticized researchers in child abuse for a parallel trend, suggesting that a focus on internal personality factors was too narrow to provide for sound research or the development of effective intervention strategies.[36] In other words, if battering is the result of individual pathology and not supported by broad cultural attitudes, why have sanctions against batterers not been harsher and more consistently applied?

David Adams of Emerge, a Boston-based program that counsels batterers, states that "violence is a controlling act. When a man is violent towards his wife, he quite often gets what he wants, whether that be more space, his dinner, or the last word in an argument."[6] For a man to be a batterer, "his belief system must include the following: (1) the belief that violence is a legitimate way of solving problems, and (2) a belief that it is okay for a man to control women."[37] These beliefs are fairly common within our society.

In addition, continuing unequal status in this country renders women vulnerable in other ways. A fear of destitution keeps some women in abusive relationships. Single women with children, for example, now make up a great majority of those living below the poverty line. According to the U.S. Census Bureau in 1984, the poverty rate for female-headed households was five times the rate for families with two adult partners. To make matters worse, the actual number of poor people had increased by 4.5 million and the economic status of families headed by women had declined.[38] Victim blaming, inconsistent responses, and denying or minimizing the violence all serve to perpetuate the problem; continued, institutionalized social and financial inequality act in consort with these to place women in continuing danger of abuse and death at the hands of intimate partners.

PREVENTING DOMESTIC VIOLENCE: THE STATE OF THE ART

The programs for battered women that emerged in the 1970s stemmed from the women's movement of the 1960s. Although these programs "articulated a multiplicity of philosophies, they shared one common belief: battered women faced a brutality from their husbands and an indifference from social institutions that compelled redress."[4] This movement, unique in its community roots and the involvement of battered women, has spawned interventions that emphasize protecting victims, stopping violence, expanding resources and options for victims and assailants, and early identification, referral, and public education.[13]

A variety of interventions have been introduced at different levels to address the problem of woman battering. These interventions include emergency shelter; civil and criminal justice responses; hospital, health care and psychiatric service protocols; primary prevention; and community education. Allied with these efforts are counseling programs for men who batter.

The rating of services by battered women themselves is a measure of the effectiveness of various responses. A survey of 1,000 battered women conducted by *Woman's Day* magazine indicated that more traditional agencies, though widely used, are considered less effective than battered women's shelters and women's groups. Medical professionals were rated least effective.[39]

Interventions for woman battering have historically targeted the abused woman and aimed to eliminate future violence. Primary prevention programs, the principal concern of this chapter, must be informed by these programs' understanding of battered women and those who beat them.

INTERVENTIONS

Grassroots Programs

According to the National Coalition Against Domestic Violence, more than 1,200 sheltering organizations for battered women exist in the United States. Their services include 24-hour hotlines; emergency shelter; advocacy with legal, social service, and other public and private institutions; child advocacy; peer counseling; and community education. Once established, these services are used at a rate that often exceeds their capacity. For example, in Massachusetts, where 31 shelters for battered women exist, 3,862 families were turned away during 1986 for lack of space (C. Allen, personal communication).

The major source of funds for many programs is the state government. The level of support offered varies widely from state to state, and no consistent source of federal funds exists. Although some battered women's programs recently have gained access to some federal monies, the funding is generally of short duration, limited, and not specific to battering.

The design and coordination of a well-integrated, systematic response to battering must involve advoacacy groups as policy makers, trainers, and monitors. This will ensure that efforts are responsive to the needs of battered women as they define them. A model for community intervention is the Domestic Abuse Intervention Project (DAIP) of Duluth, Minnesota, involving community, police, and court responses. After a man is arrested for battering a woman, a network puts the woman in touch with a local shelter, offers counseling to the man when he is released from jail, and introduces the woman to the court process. Additionally, the court process is streamlined and is designed to protect endangered women.

The Duluth project is being carefully monitored for quality, equitable application of the law, and responsiveness by a policy committee of eight battered women, half of whom are women of color.

In the Duluth project, the courts, police, battered women's programs, and mental health systems work together to provide a model for comprehensive, well-integrated action. This project is notable not only in its consistency but in its involvement of community advocacy groups as policy designers, trainers, implementers, and monitors.

All 50 states now have coalitions or associations that represent the state's battered women's programs. These groups and individual, community-based women's programs may be willing to cooperate toward a more effective and comprehensive system of interventions.

Recommendation: Comprehensive community-based programs promise interventions that should continue to be used and monitored and their outcomes evaluated. Such programs must involve advocacy groups as policy makers, trainers, and monitors.

Education Programs

Public education campaigns. Many excellent models exist for public education campaigns on woman battering. The material may focus on one or more of the following topics.

• Defining the problem. Placing battering within a social context, identifying common elements, and defining violence as the responsibility of the batterer helps women identify themselves as battered.

• Identifying resources. Directed at battered women, this effort outlines the relief available through the civil and criminal justice systems, from shelter and advocacy groups, and from other public and private agencies.

• Deterrence. Aimed at batterers, this effort outlines the criminal and civil penalties for battering. This is effective only when it is coordinated with a consistent criminal justice response. Even then, it takes time to convince batterers that the system is prepared to treat violations seriously.

Public education campaigns include radio and television public service announcements; television news specials, features, or dramatizations; newspaper ads, feature stories, or editorials; and ads on billboards, in public transportation terminals and vehicles, and at businesses, social service offices, and other public gathering spots.

Development of school curricula. Over the past decade, curricula on sex roles, sexual abuse, sexual assault, and woman battering have been developed. Although excellent materials are now available to help children and adolescents examine, interpret, and prevent violent and controlling behavior and to understand the way sex stereotypes function, there is often resistance to introducing these subjects into school systems.

Corporate programs. Corporations around the country have developed programs to treat family violence as a problem that undermines the health and performance of their employees. Numerous companies offer seminars, management training, and counseling and referral services.[40]

Recommendations:

• Public education campaigns on battering show promise for increasing awareness of the problem and available services. Such programs should be implemented and monitored and their outcomes evaluated, but they must also be supplemented by other programs in the criminal justice, community, and health care sectors.

• Curricula for elementary and secondary schools are promising interventions. We recommend that they be used and monitored and that their outcomes be evaluated.

• Education and referral services at corporations are promising interventions. They should be used and monitored and their outcomes evaluated. However, we caution against programs that use inappropriate mental health interventions or in any way attempt to hold the victim responsible for battering.

Health Care Systems

The health care system is often the first place to which a battered woman turns for help, placing health professionals in a central role.[41] If battered women are not identified, they remain at risk. Protocols for identifying and working with battered women in emergency rooms, primary care clinics and offices, and obstetrics offices are being used by practitioners around the country, though probably in a small proportion of such facilities.

The effectiveness of this strategy is supported by the experiences of a growing number of health care professionals.[42] One study in Los Angeles showed that before standardized protocols were used, 5.6% of female trauma patients were identified as being battered; after protocols were instituted, 30% were so identified.[43]

Protocols must be coordinated with consistently implemented criminal justice and community intervention techniques to be effective. There are many good models that may be used in developing a local program. The Domestic Violence Training Project, a program for health care professionals in New Haven, Connecticut, has produced training programs and protocols for emergency room personnel.[44] Other materials may be recommended by the Nursing Network on Violence Against Women, which has produced a critical listing of protocols[45]; or they may be available from the National Coalition Against Domestic Violence and from the National March of Dimes, which has produced a videotape and accompanying protocol entitled "Crimes Against the Future," designed for health care practitioners working with pregnant women.

A good protocol model includes (1) the identification of suggestive injury and complaint patterns, (2) a review of medical history, (3) a confidential interview with the woman when battering is suspected, (4) a lethality assessment with battered women, (5) validation of the battered woman's experience, (6) the presentation of options for safety and referrals to other services, and (7) the avoidance of victim blaming.

The identification of child abuse yields an excellent opportunity to provide options to battered women. Many women who suffer battering for years finally seek help only when the violence touches their children. Forty-five percent of mothers of abused children are themselves battered; in the large majority of cases the man is the batterer of both mother and child.[46] Child protective workers often blame the mother for not effectively protecting the child, and her own abuse often is not recognized or taken into account. The identification of child abuse provides an excellent opportunity to offer services to battered women. Interventions must be closely coordinated with community and criminal justice agencies, and the woman's confidentiality must be carefully guarded. Children's Hospital in Boston has designed Awake, the only project of its kind, to identify battered women through the identification of abused children. This project works closely with the battered women's programs in its area.

Mental health systems. Protocols for mental health practitioners have been produced by the National Coalition Against Domestic Violence.[47] Such protocols enable mental health practitioners to idenfity the extent of violence, measure the potential for lethality, validate the woman's experience, provide options for her safety, avoid blaming the victim, and respect her right to self-determination.

Counseling for batterers is an intervention that may be used as part of an alternative sentencing strategy. It is important, however, that such a sentencing structure include an immediate and serious response to repeated violence, the violation of court orders, or absence from the counseling group. A system of probation with counseling as one condition provides a milieu for this.

There are a number of independent counseling programs for batterers. One of these, Emerge, has developed a widely respected program. "Because battering and control are seen as the fundamental issues, therapeutic interventions are provided which directly challenge the abusive man's attempt to control his partner through the use of physical force, verbal and nonverbal intimidation and psychological abuse."[48] Emerge provides training for other independent, court-mandated projects nationwide. Emerge staff have produced much of the more comprehensive and data-rich research on batterers and techniques for working with them. They coordinate their counseling work with that of area shelters and their training work with the national battered women's movement.

Counseling that attempts to address the problems of the batterer and the battered woman as a unit is inadvisable and, in many cases, dangerous. Protocols for mental health practitioners released by the National Coalition Against Domestic Violence state that couples therapy allows the abuser to blame the victim. By seeing the couple together, the therapist erroneously suggests that the partner shares responsibility for the abuser's behavior.[47] The primary treatment goal of any mental health work related to domestic violence should be cessation of the violence, not the salvaging of the family unit.

Recommendation: Protocols for identifying and working with battered women in health care and mental health facilities are a promising intervention that should continue to be used, monitored, and evaluated.

Civil and Criminal Justice Systems

Police. A call to the police can be a first step for battered women seeking help. Thus the quality of the police response is critical. It should include (1) training for officers on the proper, effective handling of domestic calls; (2) clear and consistently applied guidelines requiring arrest, at least where probable cause of a felony exists and on violations of court orders; (3) a system that makes protective orders accessible to officers; (4) communication of the victim's rights at the scene; (5) referral of the victim to resource groups for support, advocacy, legal aid, and other services; (6) transportation to a medical facility, if necessary; and (7) reporting and data collection.

The manner in which police discharge these duties has a significant impact. In an experiment conducted in Minneapolis, it was found that arrest was a more effective response than advising or separation, interventions traditionally used by police. In addition, the study found that there is a significant reduction in repeat violence if police not only arrest the batterer but listen to the victim.[26] Indeed, a 1977 court decision found that an experimental program in which New York City's police were trained in and encouraged to use mediation techniques when responding to domestic disturbances failed to protect battered wives.[49]

A training curriculum for law enforcement officers, developed by the Family Violence Project of the San Francisco District Attorney's office, has been used since 1982 throughout the state.[50] Its authors report that since 1982 they have seen dramatic increases in police documentation of domestic violence cases, an increase in arrests, and an increase in prosecution.

Courts. Civil and criminal courts are being used increasingly by women seeking protective orders and the prosecution of batterers. Each state now has its own legislation providing for injunctive orders against abusive family members.

Most experts agree that the judicial response to battered women should include the following:

• Emergency protective orders, with no contact between the victim and abuser. The children should be released to the victim. Temporary support should be established, and the victim should have exclusive access to the common household. These orders should be available on a 24-hour basis and should be communicated immediately to those police departments responsible for enforcement.

• A graduated system of punishments in response to violations of protective orders, with the violation of protective orders carrying criminal sanctions.

• Consistently applied policies and guidelines for judges and district attorneys in the handling of familial attacks.

• Compensation for victims of family violence.

• Referrals to groups able to provide shelter, advocacy, legal aid, and other services aimed at ending the violence.

Unless it is properly conducted, this intervention strategy may impair the effectiveness and credibility of all other strategies. Although monitoring by advocacy groups cannot replace evaluation, the monitoring of the court response is vital, as is its coordination with other intervention strategies.

The attitudes of court representatives and district attorneys have often undercut the intent of abuse laws. The well-publicized case of Pamela Nigro Dunn may serve to illustrate the point. This 24-year-old Massachusetts woman appeared in court four times before a judge who belittled her and her fears, reprimanding her for taking up the court's time "when it has more serious matters to contend [with]."[51] Though the judge granted protective orders, he refused to provide a police escort to accompany Dunn to secure her belongings. To protect herself, she asked her mother to meet her at a bus stop after work. It was at the bus stop, as her mother watched, that Pamela Dunn, pregnant, was kidnapped and shot by her husband. She was later found dead at a dump site.

Court-ordered mediation has been an effective tool in certain types of disputes, but it is not appropriate for use in battering cases for a number of reasons.[52] The use of court-ordered or court-referred mediation serves to decriminalize battering in jurisdictions where criminal sanctions have only recently (if at all) been used consistently. Even as an adjunct to court orders and other sentences, a mediation order sends a message of shared responsibility for the violence.

Because the point of separation appears to be the most potentially fatal time for many battered women,[27,28] forcing contact through mediation is unreasonable and dangerous. Although some battered women may choose to participate in mediation at some point during the relationship, this must remain a voluntary decision.

Subpoenas that force the victim's testimony have not been shown to be effective. One study on the role of criminal court judges in battering cases states that judges report an increased willingness to cooperate on the part of women "when they receive information, emotional support, community referrals and trial preparation from victim advocates who are assigned to each case, and when criminal justice officials are attuned to victim needs and concerns."[51] However, in most localities these components are provided by independent community groups that cannot directly guarantee the woman's safety vis-a-vis either the batterer or the court system.

A comprehensive, consistent, and credible response system may be the most profound incentive to the battered woman's cooperation in prosecution.

Recommendations:

• Improved police training and response practices are promising interventions that should be used, monitored, and evaluated. There is much disagreement about arrest policies in misdemeanor cases. Consequently, arrest procedures should be more closely evaluated. Police responses should be monitored and evaluated and police department policies should be consistent and well coordinated with courts and other sources of intervention.

Chapter 14

The story of aging
posing themes. On
icans are living lc
housing, recreation
income security tha
the nation's history
is another. Increasii
lation are reports (
especially those wh
In the 1960s, we le
syndrome." In the
lence of spouse ab
decade, yet anothe
elder abuse—has b

• A 74-year-old \
fall is invited to mc
find herself confine
to come upstairs,
forces her back do
arm in the process.

• Eighty-four ye
family, Uncle Hora
dren's toys, underf
A parent aide repc
inals and left on
There is no medica

• Unable to walk
live at home only v
curses and yells at
mity. She restricts
prisoner in her owi

The proportion
over is growing m
group. About 13%
years from now it
group is increasing
the next century,
have increased sev

As the populatic
leave the elderly
About 85% of the
least one chronic i
chologically, and 1
5% of the elderly r
time.[6] Thus, the r
being met by famil
or in the elder's ho
the same places.[7]

• The continued enforcement of existing legislation and the implementation and evaluation of improved court response practices are promising interventions, especially in coordination with other strategies. These practices should be used and monitored and their outcomes evaluated.

• Because there is no evidence that court-ordered or court-referred mediation or the use of subpoenas to force testimony are effective, these practices should be curtailed, pending further research.

Possibilities for the Future

There are untried but promising steps that should be taken toward the prevention of domestic violence: promoting legislative change to increase penalties for battering and to decrease inequalities for women, educating and training judges and other judicial personnel, designing programs for families at risk of violence, and developing interventions that address women of all races, capabilities, and ages. In light of the recent findings concerning partner homicides, every intervention should include lethality assessments with battered women to help them plan for their safety. (See materials developed by the Pennsylvania Coalition Against Domestic Violence, Harrisburg, Pennsylvania.)

A national system for monitoring the incidence and severity of adult domestic violence must be established. In addition, agencies responsible for responding to the victims of domestic violence must design ongoing surveillance systems to identify batterers. "Without such systems and procedures, many victims of repeat domestic violence go undetected while many other victims fail to receive timely assistance and protection because their circumstances are improperly documented."[53]

INTERVIEW SOURCES

Charlene Allen, Organizer, Massachusetts Coalition of Battered Women's Service Groups, Boston, April 25, 1987

REFERENCES

1. Ptacek J. Wifebeaters' accounts of their violence: loss of control as excuse and as subjective experience. Durham, New Hampshire: University of New Hampshire, 1985:75 (master's thesis).

2. Straus M, Gelles RJ, Steinmetz SK. Behind closed doors: a survey of family violence in America. New York: Doubleday, 1980.

3. Rosenberg M, Stark E, Zahn M. Interpersonal violence: homicide and spouse abuse. In Last JM, ed. Public health and preventive medicine. 12th ed. Norwalk, Connecticut: Appleton-Century-Crofts, 1986.

4. Schechter S. Women and male violence. Boston: South End, 1982.

5. Roper M, Flitcraft A, Frazier W. Rape and battering: a pilot study. New Haven, Connecticut: Department of Surgery, Yale Medical School, 1979.

6. Adams D. Stages of anti-sexist awareness and change for men who batter. Paper presented at the 92nd annual convention of the American Psychological Association, Toronto, Canada, 1974.

7. Preventing domestic violence against women. Washington, DC: Bureau of Justice Statistics, 1986:3.

8. Shulman MA. Survey of spousal violence against women in Kentucky. Lexington: Kentucky Commission on Women, 1979; study no. 70927001.

9. Teske RHC, Parker ML. Spouse abuse in Texas: a study of women's attitudes and experiences. Huntsville, Texas: Criminal Justice Center, Sam Houston State University, 1983.

10. Report of the surgeon general's workshop on violence and public health. Washington, DC: Health Resources and Services Administration, 1986.

11. Dobash RE, Dobash R. Violence against wives: a case against the patriarchy. New York: Free, 1979.

12. Stark E, Flitcraft A, Frazier W. Medicine and patriarchal violence: the social construction of a private event. 1979;9(3):491–4.

13. Stark E, Flitcraft A. An epidemiological review. In: Hasselt VN, et al., eds. Handbook of family violence. New York: Plenum, 1985.

14. Goldberg WG, Tomlanovich M. Domestic violence victims in the emergency department. JAMA 1984;251 (24):3263.

15. Finkelhor D, Yllo K. Rape in marriage: a sociological view. In: Finkelhor D, Gelles R, Hotaling G, Strauss M, eds. The dark side of families. Beverly Hills, California: Sage, 1983.

16. Langan PA, Innes CA. Preventing domestic violence against women. Washington DC: US Department of Justice, Bureau of Justice Statistics, 1986.

17. Lentzner HR, DeBerry MM. Intimate victims: a study of violence among friends and relatives. Washington, DC: US Department of Justice, Bureau of Justice Statistics, 1986.

18. Russell DEH. Rape in marriage. New York: Macmillan, 1982.

19. Barber C. Violence in Massachusetts: the epidemiology of homicide in Massachusetts, 1977–1983. Boston: Massachusetts Department of Public Health, 1987:15.

20. Helton AS, McFarlane J, Anderson E. Battered and pregnant: a prevalence study. Am J Public Health 1987; 77(10):1337–9.

21. Rosenbaum A, O'Leary KD. Marital violence: characteristics of abusive couples. J Consult Clin Psychol 1981;49:63–71.

22. Bureau of Justice St
criminal victimization in
US Department of Justice

23. Hotaling G, Sugar
markers in husband to w
knowledge. Violence and

24. Lindsey K. When
murder or self defense?

25. Domestic violence a
and Kansas City. Wash
1976.

26. Sherman LW, Berk
of arrest for domestic ass
72.

27. Browne A, William
women at risk: its relati
trated partner homicide.
Criminology 1987.

28. Cazenave N, Zahn
domination: police repoi
cago and Philadelphia.
meeting of the American
Georgia, 1986.

29. Totman J. The mur
criminal homicide. San
ciates, 1978.

30. Daniel AE, Harris PV
ferred for pre-trial psyc
study. Bull Acad Psychia

31. Wilbanks W. The fe
County, Florida. Crimin

32. Chimbos PD. Mai
spousal homicide. San
ciates, 1978.

33. Kantor GK, Strauss
tered women, police u
Paper presented at the
Meetings, Montreal, 198

34. Cantrell LA. Alcoho
project for the Division c
ment of Health and Socia
Department of Health ar

35. Friedman C. Docui
Jeopardy project. Sacram
1987(Mar):2–6.

36. Gelles RJ. Child abu
logical critique and refor
1973;43:611–21.

37. Adams D, McCormi
group approach. In: Ro

of an elderly person's money or property; and 4) neglect is the failure to provide an elder with life's necessities—food, shelter, clothing, medical care. (It is sometimes categorized as either "passive" neglect, resulting from ignorance or stress, or "active" neglect, which is intentional or malicious.)

Our understanding of elder abuse has evolved considerably in the last decade. Much of the early publicity depicted nursing home residents tormented by impatient, overworked employees. Like the accounts of child abuse in day-care centers, these cases received vivid media attention. But, as with child maltreatment, the study of elder abuse is increasingly focusing on the home, where these vulnerable family members receive most of their care.[10] Early inquiries into elder abuse relied heavily on analogies to child and spouse abuse. However, experts in the field today emphasize that the dynamics of elder abuse are quite different.

"It is necessary to consider the uniqueness of being old in relationship to patterns of family violence. The validity of the extrapolation of findings from younger abused groups to elderly abused people has yet to be substantiated," wrote Jordan I. Kosberg, a pioneer in the field.[24] Although the suggestion that elder abuse springs from the same cycle found in child and spouse abuse[25] (those who were abused as youngsters go on to become victims or perpetrators of abuse in their adult lives) has not been documented, many researchers see a connection. But is is a very different kind of cycle, combining imitation with retaliation. Rather than simply passing the abuse suffered in childhood on to a new generation, as in the model of abused children growing up to become child abusers, this cycle may involve abused children retaliating directly against their now-elderly abusers.[1,26–28]

THE MAGNITUDE OF THE PROBLEM

There are no major studies of the national prevalence of elder abuse. Our understanding of the problem comes from small samples of the elderly population, surveys of professionals who work with the elderly, and case–control studies of abused and nonabused elderly. Although critics have pointed out flaws in these studies (such as a reliance on interviews with professionals rather than with victims and their families, the failure to use comparison groups, and imprecise definitions of elder abuse),[29] they have been used to assess national prevalence rates, resulting in estimates that as many as 10% and as few as 1% of the nation's elderly have been abused. The most recent report of the House Select Committee on Aging estimates that each year more than one million (4%) of older

Americans are physically, financially, and emotionally abused by their relatives.[30]

According to another report, 10% of Americans over 65 years of age may have suffered some kind of maltreatment,[31] and 4% may be victims of moderate to severe abuse.[30] A 1981 Illinois study estimated that 4% of that state's elderly had suffered abuse.[32] Two years earlier, an Ohio study reported that 9.6% of all the elderly seen by state agencies over 1 year had been abused.[13] Finally, a 1988 survey of 2,000 Boston-area elderly disclosed a regional abuse rate that, if extrapolated to the nation as a whole, would indicate that between 700,000 and 1.1 million persons—3% of the elderly population—have been abused.[11]

The types of abuse also vary in frequency. One study finds that physical abuse is most commonly reported, with 45% of the physically abused elderly reporting that something had been thrown at them; 63% had been pushed, grabbed, or shoved; 42% had been slapped; and 10% had been hit with a fist, bitten, or kicked.[11] Yet other studies indicate that passive neglect is the most common form of elder abuse.[33] Still others find that psychological abuse prevails.[14,15]

The conflicts among these numbers are heightened further because elder abuse, like other kinds of interpersonal violence, is underreported. Indeed, one study suggests that only 1 in 14 cases is reported;[11] earlier investigations indicated that only 1 in 6 came to public attention.[34] Underreporting can result from the elderly's fears of reprisal or removal from the home and their resistance to outside involvement in family affairs. One-third of the elderly who were judged to be abused in an Ohio study denied it.[13] There is further confusion about the number of individual elderly who are abused because such abuse is believed to be a recurring event in up to 80% of the cases.[35]

Sometimes it is difficult to differentiate between the natural decline during aging and symptoms that result from abuse or neglect.[36] Many of the symptoms of abuse—broken bones, bruises, malnutrition, and lethargy—could also be unintentional injuries common in the normal aging process. The skin of the elderly is more tender, falls are common, and drug side effects are often present. In addition, the mental status and memory of the elderly can be impaired, leading to some false or unverifiable accounts of abuse.

The Need for Data

"We believe that the most critical need in the area of domestic violence against the elderly at present is for information; without improved data, plans for intervention into the problem of elder abuse are, at best, educated guesses; and, at worst, opportunistic political compromises," concluded the Surgeon General's Workshop on Violence and Public Health in 1985.[28]

The workshop identified a number of underused data sources, including the FBI's *Uniform Crime Reports*, which record the relationship between the assailant and the victim in homicide cases. The *National Crime Survey*, developed from interviews with about 62,000 American households, is another valuable resource on victimization. The workshop suggested that other criminal justice documents (e.g., police, court, and prison records) be investigated for information bearing on elder abuse. Also, the records of state agencies that respond to cases of elder abuse were cited as a potentially valuable source of data.[28]

Although there is widespread disagreement over the number of the elderly being abused, there is greater consensus that the number will increase. Many researchers predict that greater awareness of the problem and the growing number of the elderly in our population will fuel the increase in reported cases.

RISK FACTORS

The risk factors most frequently cited for elder abuse are

• impairment, which increases dependence on and vulnerability to others

• pathology of the abuser, including mental illness, retardation, and alcoholism

• poor internal family dynamics

• external stress, such as unemployment and low income

• demographic and social changes resulting in more of the elderly placing increased demands on their families.[37]

No single theoretical model explains elder abuse, nor is the relation among these risk factors clear. It is possible, however, to combine what is known about risk and describe a composite high-risk elderly person as well as the person likely to abuse him or her. These portraits, however, should be viewed with caution. Their characteristics often reflect biases of the research: Studies that reveal abusers to be white and middle class, for example, are probably reflecting the characteristics of the population that was surveyed.

The High-Risk Elder

The composite abused elderly person is a white widow who lives with relatives. She is very old and physically or mentally impaired. Her annual income is less than $10,000, and she is a source of stress in the family. Physically abused elderly persons live with the abuser in 75% of the cases. Recent studies, however, suggest that men may be at greater risk because they are more likely to live with others as they age and thus become more dependent.[7,11] Yet these same studies show that women are at risk for more severe abuse.[11]

Several other risk factors have been identified for the elderly victim. Problem drinking or drug abuse reduces the person's ability to care for himself or herself; in addition, elderly persons with alcohol problems may live with an alcoholic spouse or child. Those who internalize the blame for their own condition may make themselves more vulnerable to abuse. Alternatively, an elderly person who is excessively loyal to a caregiver may fail to report ongoing abuse and thus allow it to continue. A fundamentally stoic personality may yield the same result. Social isolation of the elderly may prevent identification of the abuse by outsiders. Finally, past abuse may presage abuse in the future.[24]

High-Risk Abusers

The typical abuser is between 40 and 60 years of age and is a relative of the victim. The abuser is middle class, white, under stress, and somewhat financially dependent on the elderly person. The abuser is more likely to be male and half the time is a child of the victim. A 1982 study found that 43% have substance abuse problems.[38] The average abuser has cared for the victim for 9.5 years; 10% have provided care for 20 years or more.[39] Other risk factors for the abuser include conditions that would leave him too impaired to render appropriate care (such as mental, physical, or emotional illness), caregiving inexperience, economic troubles, abuse as a child, a lack of activities outside the house, a blaming or unsympathetic personality, unrealistic expectations, and a hypercritical nature.[24]

"Dependency seems to play a critical role in elder abuse, but it is not clear who is depending on whom in these abusive relationships."[28] Dependency, usually cited as a risk factor for the abused, may also be a factor for the abuser. Some research indicates that a cause of abuse may be the continued dependency of the abuser on the abused, usually for housing and finances.[28,38,40] In a recent case–control study of 42 abused elderly persons and the same number of controls in Massachusetts, New York, and Rhode Island, researchers found that the abusers were more likely to be dependent on their elderly victims than were the control-group relatives.[26]

Similarly, generationally inverted families (where the elderly person depends on a child for care) are characterized by abuse and neglect as methods of last resort for both the elder and his or her caregiver. One study indicated that 15% of the elderly slapped, hit with an object, or threw something at their adult child.[41]

In addition elder abuse is not unknown within a marriage. One-quarter of the cases in one study[42] and 58% in another[11] involved husbands and wives. Sometimes the abuse had been a long-standing feature of the marriage; in other cases, researchers refer to "late-onset spouse abuse." "The underlying dynamic is that an elder is most likely to be abused by the person with whom he or she lives. Many more elders live with their spouses than with their children." When abuse among the elderly living with spouses is compared to those living with children, the incidence rates for spouse abuse are slightly lower.[11]

SOCIAL AND CULTURAL FACTORS

Society's perception of elder abuse encompasses forms of violence (e.g., slapping or yelling) that are rarely considered part of child or spouse abuse. This is an indicator that, as a society, we generally believe that any sort of physical violence toward an elder is unacceptable.

At the same time, however, factors such as discrimination against the elderly and a tolerance for violence can encourage maltreatment. There has been no systematic research on the relation between elder abuse and age discrimination, but its importance has been widely suggested. Because "old age has become synonymous with loss of personal power, disability, and lack of control over one's life,"[4] mistreatment of the elderly may be perceived as somehow less egregious than abuse of a younger person. And the old may view abuse and neglect as deserved or unavoidable because they internalize society's negative stereotypes.[4]

Because a disproportionate number of elderly abuse victims are women, many believe that sex discrimination may be a contributing factor. "With advancing age, sexism and ageism link and work together to create a poisonous climate for many women," wrote two experts.[4] Further, being old and female is frequently linked with being poor and thus dependent.

Other social norms, such as the sanctity of family privacy, keep elder abuse under wraps. In fact, one of the public's first encounters with elder abuse came at a Congressional hearing whose 1979 report was subtitled, "the hidden problem."[34] Almost a decade later, a researcher could still write of elder abuse as an "invisible problem."[24] Despite nearly a decade of research, elder abuse remains hidden because of the physical isolation of the elderly, the reluctance to admit abuse, and continuing ignorance of the problem among professionals.

A BRIEF HISTORY OF PREVENTION

Until the late 1970s, elder abuse was not regarded as a discrete, serious problem in our society. It surfaced first in Great Britain in 1975 when G. R. Burston described cases of "granny-bashing."[43] Three years passed before Americans were introduced to the "battered parent" problem in the journal Society.[44] Congressional hearings followed in 1979. Repeated efforts since 1980 by Congressman Claude Pepper (D-FL) and Congresswoman Mary Rose Oakar (D-OH) to establish a national center on elder abuse have been unsuccessful, but the federal government has taken action. In 1984, elder abuse was addressed in legislation that established a National Clearinghouse on Family Violence. In addition, the final report of the attorney general's Task Force on Family Violence, issued in September 1984, included several recommendations bearing on elder abuse.[45]

All 50 states have adopted elder abuse legislation. The laws vary from state to state, with some mandating reports of abuse and others relying on voluntary reporting. Forty-three states (including jurisdictions such as the District of Columbia, Guam, and Puerto Rico) operate mandatory, statewide reporting systems. In some states, certain professional groups are required to report abuse, and in others anyone with knowledge or suspicion of abuse must report it. Responsibility for these reports usually rests with a state human service agency, law enforcement agency, or local human service agency.[46]

RISK ASSESSMENT INSTRUMENTS

Health care providers are often the only professionals who see abuse victims, because the abused often do not seek help unless they are severely injured. "Excellent detection skills and careful screening can make the difference between successful intervention or repeated incidents of abuse or neglect," wrote Terry Fulmer, a specialist in elder abuse treatment.[47]

Several researchers are developing tools and surveys to assist in the diagnosis and prediction of abuse. The Beth Israel Hospital elder assessment team was created in Boston in 1981 to develop staff skills in the assessment and detection of elder abuse, to develop an instrument to aid in that process, and to validate the instrument.[36] Prompted by mandatory reporting laws, the team surveyed all patients over 70 years old.

A 1984 evaluation of the Beth Israel experience gave the assessment instrument high marks. Emergency department staff members "have become more discriminating regarding possible elder abuse victims and are more likely to question physical or mental decline as a possible result of poor care instead of assuming it to be the inevitable result of the aging process. The staff members report a greater sensitivity toward the elderly and believe this program makes a difference in the care of our aged patients."[36]

Two Detroit-based gerontologists have found nine surveyable items that are 94% accurate in predicting elder abuse or neglect. The survey asks the elderly person to describe both himself or herself and the caretaker. The nine items include the removal of money or property from the elder, threats, whether the elder's needs are being met by others, the presence of illness, and whether the elder is a source of stress. Questions about the caregiver ask whether he or she tries to get the elder to act against his or her best interests, whether he or she demonstrates an inappropriate awareness of the elder's condition, whether he or she is dependent on the elder for support, and whether he or she is a persistent liar.

These indicators accurately distinguish between cases involving abuse and control cases that were similar in terms of age, sex, race, income, and physical and mental impairment. "We feel that these indicators have an advantage beyond the traditional demographic indicators for predicting, preventing, and ameliorating abuse and neglect of the elderly."[48] As in other areas of interpersonal violence and suicide, the development of increasingly accurate, easily usable risk assessment tools for elder abuse is a high priority for the future.

ELDER ABUSE PREVENTION: THE STATE OF THE ART

A 1985 study by the House Select Committee on Aging, Subcommittee on Health and Long-Term Care, revealed that while states spend an average of $22 a child for protective services, only $2.90 is spent for each elderly person.[30] Until very recently, little has been done to prevent elder abuse before it occurs, and some cases will always defy the most sophisticated strategies for prevention.[1] Although many primary prevention programs are under way, few have been evaluated.

A review of primary prevention efforts reveals that much of what qualifies as elder abuse preven-

tion is not so labeled. Instead, programs focus on awareness of aging and its problems.

None of the interventions discussed below has been evaluated; they are presented as examples of programs that show promise. Targeted at both the abused and the abuser, the interventions are organized by the place in which responsibility for the effort lies: the family, the health care system, social services, public educators, community groups, workplaces, school systems, and individuals. Because injury prevention programs are most likely to deal with the physiological effects of injury, interventions specific to financial and material abuse are not discussed.

INTERVENTIONS

Public Education Campaigns

Several states and communities have launched public awareness campaigns that target both the victim and the abuser. Designed to promote awareness of elder abuse as a community concern, these programs take many forms.

North Carolina's Department of Human Resources has developed a public education package that includes a brochure, poster, booklet, public service announcements, related materials, and guidelines for their use. In Wisconsin, several projects aim to increase awareness of elder abuse. The Great Lakes Intertribal Council contracted with the Wisconsin Bureau on Aging in 1986 to develop culturally appropriate methods of publicizing elder abuse following the discovery in 1985 that a general public awareness campaign had not reached the Native American population. Community Care Organization of Milwaukee County, Inc., has developed a publicity campaign including billboards, bumper stickers, and publications. Texas and Pennsylvania are also currently developing publicity programs. The Texas Adult Protection Services Program has created a slide show in conjunction with the Federation of Women's Clubs. They plan to survey nine professional groups to assess changes in their level of understanding. In Pennsylvania, a statewide public education campaign is nearing completion. The Yakima Indian Nation in Washington has developed a model training, education, and law reform effort. CANE, the federally funded Clearinghouse on the Abuse and Neglect of Elders, in operation since February 1987 at the University of Delaware contains 2,000 references (books, articles, audiovisual materials, pamphlets, etc.) in computerized archives.

Recommendation: Public education to increase the awareness and reporting of elder abuse is an intervention that shows promise. Such programs should be used and monitored and their outcomes evaluated.

School Programs

School programs for adolescents are designed to dispel stereotypes and myths about aging, raise awareness of elder abuse, and encourage positive interactions with elders. Tenth-, 11th-, and 12th-graders in a large midwestern high school were exposed to a curriculum that included interviewing older people about their lives and a discussion of aging. An evaluation revealed that "the curriculum had a positive effect on the experimental students' psychological development . . . [and that their] attitudes toward older people were significantly more positive." Many students continued their involvement with the elders after the school project concluded.[49]

A secondary school–based health education program in Austin, Texas, also demonstrated that "students enjoyed the elder abuse awareness sessions, learned new and important information, had attitudes changed, and were able to identify specific behaviors that they might engage in to positively interact with elderly people in their lives."[50] The pilot five-day program was integrated into health and psychology classes.

CANE is currently field-testing a middle-school curriculum that will be incorporated into health and education and civics courses. "Its goal is to provide basic knowledge about aging with an eye toward what happens if you're not alert to those processes," said Suzanne Steinmetz, a sociologist at the University of Delaware College of Human Resources and one of the pioneers in research on elder abuse. The curriculum emphasizes the similarities between the two age groups (e.g., that both are experiencing dramatic physical changes, that braces and dentures can cause similar eating restrictions). The program also describes how adolescents can help elders, either through career choices or volunteerism.

Recommendation: School-based programs that aim to educate the young about the aging process and the problem of elder abuse are promising interventions that should be used and monitored and their outcomes evaluated.

Community Projects

A variety of groups reach out to their communities to help families meet the needs of their elder members.

Religious institutions. A 12-hour church-based program for caregivers of noninstitutionalized elders was conducted by the Institute of Gerontology at the University of the District of Columbia. Two hundred eighty-two caregivers graduated from the course, which was given at eight churches. Designed as mutual help groups, the programs emphasized education and resource sharing. Graduates reported more efficiency at caregiving but did not show evidence of making better use of formal service agencies. One year after the program, however, three of the mutual help groups continued to meet.[51]

Workplaces. International Business Machines Corporation (IBM) recently inaugurated a nationwide elder-care consultation and referral service for its employees and will soon offer lunchtime seminars on caring for an elderly family member. The Elder Care Referral Service is intended to make the search for care easier and faster, thereby reducing stress and concern. The program also has the goal of enhancing existing community efforts to help the elderly.[52]

Other community projects. The San Francisco Consortium for Elder Abuse Prevention has developed a comprehensive program that draws from the existing services of about 50 public and private agencies. The program, coordinated by Mount Zion Hospital and Medical Center, offers training, case consultation, multidisciplinary case review, advocacy, community education, and outreach. "The San Francisco Consortium has met with varying degrees of success in accomplishing its goals and objectives . . . the complexity of the model has led to some confusion among Consortium members, about the roles and functions of the model's diverse components."[53] The consortium is currently working with four other communities to determine the model's replicability.

The Jewish Family and Children's Service of Greater Boston and the Boston University School of Social Work designed Family Centered Community Care for the Elderly, a project that establishes a formal relationship between service agencies and families. The partnership is a complementary relationship in which the agency provides counseling and the family is trained to assume as much responsibility as possible for case management tasks (transportation, social opportunities, the selection of day care or respite care, in-home services, etc.). The researchers' evaluation of this intervention is pending.[54]

A study of community involvement in combating abuse in New York nursing homes was conducted in 1983. Through an affiliation with Friends and Relatives of the Institutionalized Aged, researchers found that community involvement strategies (e.g., advocacy, ombudsman programs, reporting, community receivership) are effective for case finding and mediation. Community strategies were weak on investigative authority and enforcement powers. The researchers recommend "strategies that combine the special strengths of community involvement with the special powers of government that government cannot responsibly delegate."[55]

Other promising programs include the Senior Advocacy Volunteer Program in Madison, Wisconsin, which matches adult volunteers with elderly persons who have experienced or are believed to be in danger of abuse. Its goals are to improve the elders' lives, lessen the hidden nature of abuse, and increase elders' awareness of helping agencies.

The Duke University Medical Center's curriculum ("Prevention of Elder Abuse and Neglect: Coping with Family Conflict") is designed for community groups that are concerned with aging and the elderly. The curriculum includes a presentation script, videocassette, and group discussion materials. The program also includes evaluation forms for the participants.[56]

The University Center on Aging at the University of Massachusetts Medical Center rents and sells a series of films on elder abuse and neglect in the family. The series, which was the first-prize winner of the Retirement Research Foundation's 1987 National Media Awards, includes three 20- to 30-minute tapes.

Adult day-care. Adult day-care is a community-based service that helps the elderly maintain or improve their ability to remain in the community and offers relief to their primary caregivers. Through social events, family assistance, meals, transportation, health care, exercise, counseling and referral services, and therapies, these programs are often linked with other agencies and services in the community. The National Institute on Adult Daycare, part of the National Council on the Aging, Inc., collects, prepares and disseminates information on all aspects of day-care. Examples of thriving adult day-care programs can be found all over the country.

Texas established its first elder day-care program 13 years ago in Austin. Since then, North and South Austin Adult Day Care has grown to include a total of three facilities managed by Lutheran Social Services of Texas and funded by the state (through Medicaid) and the United Way. Some 60 elderly persons, most between 70 and 80 years of age, attend the day-care programs each day. "They need to be with their peers as well as other people and their caregivers need a break," commented staff member Marlene Schneider. In addition to the day-care program, a 24-hour residential pilot facility was established to care for elders identified by the adult protective service as being in danger of abuse. Individuals at high risk for abuse may remain at the 10-bed shelter until new residential arrangements are made for them (M. Schneider, T. Myers, personal communication).

The Annandale Elderly Daycare Center in Annandale, Virginia, has dual goals: to provide an alternative in the health care system that allows the elderly to remain in their homes and communities and to help families who need care for their elderly relatives during the day. The center, run by the county health services department and funded by the county general tax fund and participant revenue, offers health care and supervision, activities, socialization, meals, and many specialized services. Links to other social services, mental health centers, hospitals, home health agencies, etc., are maintained through established channels of communication and through a bimonthly county interagency meeting.[57]

The Respite Care Program in Wisconsin offers short-term hospitalization for the frail elderly to relieve the caregivers. Designed to enable an older person to remain in the community and preclude the need for institutionalization, the project also offers the elderly a program of therapy and activities.[58]

Sitters who act as "family friends" while relieving relatives of caregiving duties appear to be a successful service in England. The sitters are hired and assigned by a collaborative medical care/social service agency. An evaluation revealed that the sitters are well received and in increasing demand.[59]

Recommendation: Community-based programs that offer services to families with elders are promising interventions. They should be used and monitored and their outcomes evaluated.

Self-Help Measures

Alabama's statewide elder abuse prevention program targets abusers and others at risk of becoming abusers. Therapists conduct group meetings that instruct participants in adult development and aging, problem-solving and management skills, and self-control training. Participants are evaluated at the beginning of the program, after eight weeks, and after one year and are compared to a control group of successful caregivers.[60]

Many agencies offer programs that encourage the elderly to help themselves. "We have to remember that mistreatment is largely associated with personal limitations or isolation, and the people most likely to be mistreated are also the least likely to initiate self-protective measures." With that acknowledgment, the American Association of Retired Persons suggests that the elderly be taught to pursue social habits, business transactions, and family relations in ways that will protect them, much as homeowners are advised to protect themselves against burglary.[1]

Recommendation: Programs to help the elderly help themselves are promising interventions. They should be used and monitored and their outcomes evaluated.

Programs for Caregivers

The Natural Supports Program of the Community Service Society of New York is designed to assist family members in caring for their elders. Activities include assistance in problem solving, emotional support, and mediation of family problems originating in caregiving. The program has demonstrated some success.[61]

Education Development Center, Inc., in Newton, Massachusetts, has developed, in conjunction with the Benjamin Rose Institute in Cleveland, Ohio, educational materials ("Where Do We Go from Here?") for those who work with the elderly and their families. These materials are designed for groups and include a 9-minute film and a leader's guide. They emphasize the universality of problems related to caring for the elderly, help families recognize the common emotions of guilt, anger, frustration, powerlessness, and anxiety, and present problem-solving techniques.[62]

In the same vein, CANE has developed a film ("In Their Best Interest") with accompanying educational materials. Through the use of vignettes, the film portrays the complex task of caregiving and describes strategies of communication and problem solving.

Recommendation: Education and training to understand the caregiving process are promising interventions. They should be used and monitored and their outcomes evaluated.

Additional Recommendations

Because a large percentage of abuse cases involve adult children with histories of alcoholism, mental illness, or a lack of coping skills, the committee recommends that primary prevention be directed at adult children who live with their elderly parents and who are financially or emotionally dependent on their parents. Such programs might include counseling, housing and employment programs,

and mental health and alcoholism services designed for at-risk families.

INTERVIEW SOURCES

Marlene Schneider, North and South Austin Day Care, Texas, April 15, 1988

Therese Myers, North and South Austin Day Care, Texas, April 18, 1988

REFERENCES

1. Douglass RL. Domestic mistreatment of the elderly: towards prevention. Washington, DC: American Association of Retired Persons, 1987.

2. US House of Representatives, Select Committee on Aging. Elder abuse: the hidden problem. Washington, DC: US Government Printing Office, 1980.

3. Wolf RS, Pillemer KA. Working with abused elders: assessment, advocacy, and intervention. Worcester, Massachusetts: University of Massachusetts Medical Center, University Center on Aging, 1984.

4. Quinn MJ, Tomita S. Elder abuse and neglect: causes, diagnosis and intervention strategies. New York: Springer, 1986.

5. Aging America: trends and projections. 1985–1986. Washington, DC: Federal Council on the Aging, 1987.

6. Hudson MF. Elder mistreatment: current research. In: Pillemer KA, Wolf RS, eds. Elder abuse: conflict in the family. Dover, Massachusetts: Auburn House, 1986:125–66.

7. Douglass RL. Domestic neglect and abuse of the elderly: implications for research and service. Fam Rel 1983;32:395–402.

8. Ansberry C. Abuse of the elderly by their own children increases in America. Wall Street Journal 1988 Feb 3:1.

9. Reinharz S. Loving and hating one's elders: twin themes in legend and literature. In: Elder abuse: conflict in the family. Dover, Massachusetts: Auburn House, 1986:25–48.

10. Yin P. Victimization and the aged. Springfield, Illinois: Charles C. Thomas, 1985:106.

11. Pillemer K, Finkelhor D. The prevalence of elder abuse: a random sample survey. Gerontologist 1988;28(1):51–7.

12. Johnson T. Critical issues in the definition of elder mistreatment. In: Pillemer KA, Wolf RS, eds. Elder abuse: conflict in the family. Dover, Massachusetts: Auburn House, 1986:167–96.

13. Lau EE, Kosberg JI. Abuse of the elderly by informal care providers. Aging 1979(Sep–Oct):10–5.

14. Block MR, Sinott JD. The battered elder syndrome: an exploratory study. College Park, Maryland: University of Maryland Center on Aging, 1979.

15. Douglass RL, Hickey T, Noel C. A study of maltreat-

ment of the elderly and other vulnerable adults. Ann Arbor, Michigan: University of Michigan, Institute of Gerontology, 1980.

16. Sengstock MC, Liang J. Identifying and characterizing elder abuse: final report submitted to NRTA–AARP Andrus Foundation. Detroit, Michigan: Wayne State University, Institute of Gerontology, 1982(Feb).

17. Fulmer T, Ashley J. Neglect: what part of abuse? J Long-Term Care 1986;5(4):18–24.

18. Callahan JJ. Elder abuse programming: will it help the elderly? Urban Soc Change Rev 1982;15:15–19.

19. Crystal S. Social policy and elder abuse. In: Pillemer K, Wolf R, eds. Elder abuse: conflict in the family. Dover, Massachusetts: Auburn House, 1986:331–40.

20. Elder abuse and neglect: recommendations from the research conference on elder abuse and neglect; 1986. Durham, New Hampshire: University of New Hampshire, Family Research Laboratory, 1986.

21. Salend E, Kane RA, Satz M, Pynoos J. Elder abuse reporting: limitations of statutes. Gerontologist 1984;24:61–9.

22. Katz KD. Elder abuse. J Fam Law 1979–80;18(4):695–722.

23. Pedrick-Cornell C, Gelles RJ. Elder abuse: the status of current knowledge. Fam Rel 1982;31:457–65.

24. Kosberg JI. Preventing elder abuse: identification of high risk factors prior to placement decisions. Gerontologist 1988;28(1):43–50.

25. Straus MA, Gelles RJ, Steinmetz S. Behind closed doors: violence in the American family. New York: Dodd, Mead, 1980.

26. Pillemer KA. Risk factors in elder abuse: results from a case–control study. In: Pillemer KA, Wolf RS, eds. Elder abuse: conflict in the family. Dover, Massachusetts: Auburn House, 1986:239–63.

27. Council on Scientific Affairs. Elder abuse and neglect. JAMA 1987;257(7):966–71.

28. Pillemer KA. Violence against the elderly. In: Surgeon general's workshop on violence and public health source book. Washington, DC: Health Resources and Services Administration, 1985(Oct):EA3.

29. Pillemer KA, Wolf R. Introduction. In: Pillemer KA, Wolf R, eds. Elder abuse: conflict in the family. Dover, Massachusetts: Auburn House, 1986:xvii.

30. US House of Representatives, Select Committee on Aging, Subcommittee on Health and Long-Term Care. Elder abuse: a national disgrace. Washington, DC: US Government Printing Office, 1985, committee publication no. 99-502.

31. Clark CB. Geriatric abuse: out of the closet. J Tenn Med Assoc 1984;77:470–1.

32. Crouse JS, Cobb DC, Harris BB, Kopecky FJ, Poertner J. Abuse and neglect of the elderly in Illinois: incidence and characteristics, legislation and policy recommendations. Chicago: Illinois Department of Aging, 1981(Oct).

33. Hickey T, Douglass RL. Mistreatment of the elderly in the domestic setting: exploratory study. Am J Public Health 1981;71:500–7.

34. U.S. Congress, House Select Committee on Aging. Elder abuse: the hidden problem. Washington, DC: US Government Printing Office, 1980.

35. O'Malley TA, Everitt DE, O'Malley HC, et al. Identifying and preventing family mediated abuse and neglect of elderly persons. Ann Intern Med 1983;98:998–1005.

36. Fulmer T, Street S, Carr K. Abuse of the elderly: screening and detection. J Emerg Nurs 1984;10(3):131–40.

37. Elder abuse and neglect: a guide for practitioners and policy makers. San Francisco: National Paralegal Institute, 1981.

38. Wolf R, Strugnell C, Godkin M. Preliminary findings from three model projects on elderly abuse. Worcester, Massachusetts: University of Massachusetts Medical Center, Center on Aging, 1982(Dec).

39. Streib G. Older families and their troubles: familial and responses. Fam Coord 1972;21:5–19.

40. Hwalek MA, Sengstock MC. Assessing the probability of abuse of the elderly. Presented to the Gerontological Society of America, 1984.

41. Steinmetz S. Dependency, stress, and violence between middle-aged caregivers and their elder parents. In: Kosbert JL, ed. Abuse and maltreatment of the elderly. Littleton, Massachusetts: John Wright-PSG, 1983.

42. Wolf R, Pillemer K, Godkin M. Elder abuse and neglect: report from three model projects. Worcester, Massachusetts: University of Massachusetts Medical Center, Center on Aging, 1984.

43. Burston GR. Granny battering. Br Med J 1975(Sept):592.

44. Steinmetz SK. Battered parents. Society 1978(Jul–Aug):54–5.

45. Final report. Washington, DC: US Attorney General's Task Force on Family Violence, 1984.

46. American Public Welfare Association, Elder Abuse Project. A comprehensive analysis of state policies and practices related to elder abuse. Report no. 1: a focus on legislation, reporting requirements, funding appropriations, incidence data, special studies; Washington, DC: National Association of State Units on Aging, 1986(Jul).

47. Fulmer T. Elder abuse assessment tool. Dimens Crit Care Nurs 1984;3(4):216–20.

48. Hwalek MA, Sengstock MC. Assessing the probability of abuse of the elderly: toward development of a clinical screening instrument. J Appl Gerontol 1986;5(2):153–73.

49. Hatfield T. Alternative generation education: a developmental curriculum for secondary schools. Int J Aging Hum Dev 1984;19(3):223–34.

50. Peterson FL, Clifford PR, Covar K. The elder abuse prevention project: promoting awareness in schools. Paper presented to the American Public Health Association, New Orleans, Louisiana, 1987.

51. Haber D. Church-based programs for black care-

givers of noninstitutionalized elders. J Gerontol Soc Work 1984;7(1/2):43–55.

52. Elder care referral service. Armonk, New York: International Business Machines Corporation, 1988.

53. Nerenberg L, Garbino S. San Francisco consortium for elder abuse prevention: a history and description. San Francisco: San Francisco Consortium for Elder Abuse Prevention (undated).

54. Simmons KH, Ivry J, Seltzer MM. Agency–family collaboration. Gerontologist 1985;25(4):343–6.

55. Doty P, Sullivan EW. Community involvement in combatting abuse, neglect, and mistreatment in nursing homes. Milbank Med Fund Q/Health Society 1983; 61(2):222–51.

56. Gwyther L, Gold D. Prevention of elder abuse and neglect: coping with family conflict. Durham, North Carolina: Duke University Medical Center, Center for the Study of Aging and Human Development, 1987.

57. National Institute on Adult Day Care. Developing adult day care, an approach to maintaining independence for impaired older persons. Washington, DC: National Council on the Aging, 1983:82.

58. Hasselkus BR, Brown M. Respite care for community elderly. Am J Occup Ther 1983(Feb);37(2).

59. A sitting at home with grandma service. Health Soc Serv J 1984(Jul 5):800–1.

60. Baumhover LA, Pieroni R, Scogin F. Breaking the cycle of violence against elders: the Alabama experience. J Elder Abuse Neglect (in press).

61. Getzel GS. Social work with family caregivers to the aged. Social casework. J Contemp Soc Work 1981 (Apr):201–9.

62. Vince C. Where do we go from here? Newton, Massachusetts: Education Development Center, 1983.

Chapter 15: Rape and Sexual Assault

When I'm alone in the house at night, every night just before I fall asleep, I think I hear someone. I try to tell myself, "No, no, it's just the house making noise, it's just the cats." What am I afraid of? I'm not afraid of burglary. What I live in fear of is not that the noise I hear is someone who's going to take my TV but someone who's going to attack me.[1]

In a recent 10-year period (1973–1982), 1.5 million women—or 100,000 a year—were victims of rape or attempted rape.[2] The trauma of rape and sexual assault is often compounded by further harassment, shame, or blame from a society with vast misconceptions about the crime.

Without an accurate understanding of rape and sexual assault, why they occur, and what works for prevention, sexual victimization will continue to be a costly social and public health problem. The price is high and takes many forms. Rape is a trauma that disrupts the physical, psychological, social, and sexual aspects of the victim's life, creating an enormous demand for medical, legal, and other supportive services.

DEFINITIONS

The definitions of rape and sexual assault vary widely. There are no broadly accepted definitions. Differences occur in what constitutes an act of rape, whether the relationship between victim and assailant is pertinent, and the role of force. How rape is defined is important because it determines our perception of the scope of the problem, how we treat its victims, and how we think it can be prevented.

The FBI defines rape as the carnal knowledge (vaginal penetration) of a woman by force and against her will by a man who is not her husband. Attempts to commit rape by force or threat of force are included, but other sexual assaults—sexual victimizations defined as less extreme than rape, such as oral copulation, penetration by an object or device, or nonconsensual sexual activity between husband and wife or known relations—are not included. Most state legal codes define forcible rape as the FBI defines it, but each modifies the definition.

On the other hand, the National Crime Survey (NCS), instituted in the early 1970s by the Bureau of Justice Statistics (BJS), uses a broader definition of rape. In its survey, the BJS allows each victim to define rape for herself. "If she reports that she has been the victim of rape or attempted rape, she is not asked to explain what happened . . . no one in the survey is ever asked directly if she has been raped."[2] This survey does pick up more reports of rape than do FBI records, but it still is believed to underestimate the magnitude of the problem. The problem of underestimation will be discussed below.

Discrepancies in definitions are also found in the research literature. Some authors define rape narrowly as vaginal penetration by the male sex organ. Others' definitions include "any sexual intimacy forced upon one person by another."[3]

One author distinguished between "aggravated" rape (committed by a stranger, by force, or with a weapon) and "simple" rape (forced sex without consent by only one man who is known to the victim and who does not beat her or attack her with a gun). If only aggravated cases are considered, the author concluded, then rape is a relatively rare occurrence. However, "If the simple cases are considered . . . then rape emerges as a far more common, vastly underreported, and dramatically ignored problem."[4]

Many cases that fit a legal definition are still not considered rape if the assailant is an acquaintance or relative. Yet rape does occur among persons who are related. The intersection of rape and relationship can be described in the following categories.

• *Acquaintance rape* involves individuals who know each other prior to the rape, including relatives, neighbors, or friends.

• *Date rape* is intercourse within a relationship but without the consent of the woman and when harm or the threat of harm is used by the man.

• *Marital rape* occurs when the victim and offender are spouses.

• *Stranger rape* occurs when victim and offender have no relation to each other.[5]

As we learn more about how and with whom a rape or sexual assault occurs, we recognize the need for a more inclusive definition of rape that focuses "upon the intent of the acts and their impact upon the victim, rather than only the specific acts performed."[6]

It is important to distinguish between victimization as crime and victimization as injury. The law may define forcible genital rape as rape and forcible oral rape as sexual assault, but that does not convey the degree of injury sustained by the victim.

In a recent Massachusetts court case, a young woman testified that, as the result of a sexual attack, "My sleep has become seriously disturbed . . . the little sleep I get is often riddled with nightmares of sexual violence. My anxiety and tension are increasing. I cry easily, anger at others even more easily, and am overwhelmed by the uncleanliness that I feel. My relationship with the man I love is being severely taxed" (F. W. Lindsay, personal communication). This case involved a sexual assault and unwanted genital touching by a stranger, but the words and feelings are similar to those recorded by victims of forcible rape.

This chapter is concerned, therefore, with the prevention of sexual victimization. Except insofar as studies of the magnitude of the problem described in the next section distinguish explicitly between rape and sexual assault, the term "rape" will refer to all forms of sexual victimization, including forcible rape, attempted rape, and other acts of unwanted sexual aggression.

THE MAGNITUDE OF THE PROBLEM

The FBI considers rape the second most serious crime, following murder. From 1977 to 1984, the rate of rape increased 21%, the largest increase among major crimes.[7]

According to the FBI, whose *Uniform Crime Reports* are based on state and local police reports of forcible rapes, 90,434 forcible rapes and attempted rapes were reported in 1986; 73 of every 100,000 females were victims. This is an increase of 2% over 1985 rates, and an increase of 10% over 1982. Eighty percent of the cases involved forcible rape; the remainder were assaults or attempts to commit forcible rape.[7]

NCS, which estimates the large number of rapes never reported to the police, reported in 1983 that 154,000 rapes and attempted rapes occurred, or approximately 1 for every 600 females 12 years of age and over.[8] In 1986, NCS estimated that approximately 120 of every 100,000 women were victims of rape or attempted rape, a considerable increase over the FBI's 1986 *Uniform Crime Report* (M. J. Rand, personal communication). NCS estimated that over the 10-year period 1973–82, 1.5 million rapes or attempted rapes occurred.[8] (BJS collects information on rape for both sexes. Between 1973 and 1982, about 84 cases with male victims were reported, producing a national estimate of 123,000 rapes or attempted rapes, or 0.15 for every 1,000 males.)[2]

Some private researchers conclude that the magnitude of the problem is considerably greater. In a San Francisco study, 24% of the women reported at least one completed rape, and 41% reported at least one completed or attempted rape. The figure rose to 44% when marital rape was included.[9] By estimating the number of rapes that occurred in the 12 months prior to the survey, the researcher could compare the findings to government statistics. She found a rate 13 times higher than the total incidence reported by the *Uniform Crime Reports* and seven times higher than statistics gathered by NCS.[9] The study also concluded that there is a 26% probability that "a [San Francisco area] woman will be the victim of completed rape at some time in her life, and a 46 percent probability that she will become a victim of rape or attempted rape."[10] Although it is speculative to compare prevalence and probability rates between San Francisco and other cities nationwide, these findings on rape and attempted rape "provide concrete evidence of the inadequacy of current methods used to measure the magnitude of these crimes."[9]

Although a few studies have reported men raped by women, most of these incidents occurred in prisons or in juvenile institutions.[11] The prevalence rate for male rape is unknown, since rape is even less likely to be reported in these cases than among female rape victims. Although there are some indications, as in the BJS statistics above, that the rape of men may be more common than previously thought, it is quite apparent that the risk of becoming a victim is much higher for women than men.

Underreporting

According to one estimate, only one in five rapes is reported; another says it is only one in 20.[12] Still other sources estimate that the rate of reporting is between one in 10 and four in 10.[13] The Law Enforcement Assistance Administration estimated that for every reported rape, three to 10 rapes are not reported.[14] BJS reports that only about half of the victims of rape or attempted rape surveyed from 1973 to 1982 stated that they had reported the crime to police. One factor in underreporting is the FBI's "unfounding" process, which allows the police to dismiss a case if a victim's report of forcible rape or attempted rape is not considered believable.

The responsibility for underreporting extends beyond statistics-gathering practices. The victim herself may hesitate to report rape. "The social stigma traditionally attached to rape makes the experience difficult for many victims to discuss."[2] Shame, guilt, a fear of retaliation, and a reluctance to expose herself to a "second rape" by an insensitive medical or law enforcement establishment are reasons a woman may not report rape.[15] A rape by a friend, date, or spouse, sometimes called "hidden rape,"[16] may not be reported by a victim because she fears disbelief from law enforcement officials or because she herself may question whether a "real" rape occurred.

Practitioners report that women who do not know the rapist are more likely to report the incident to the police. The Columbus Rape Crisis Center statistics for 1981 indicate that more than 54% of rapes are committed by someone familiar to the victim.[17] Indeed, a recent study in Seattle noted that a former relationship was the most relevant factor in underreporting to police.[4] In Massachusetts, nearly two-thirds of women who sought the services of rape crisis centers knew their assailants, and most did not report the incident to the police.[4] A study of women who were assaulted primarily by acquaintances revealed that in "assaults that deviated from stereotypes of 'real' rape but that met legal definitions . . . of the women only 27 percent believed their experience qualified as rape."[5]

The justice system itself can contribute to a victim's unwillingness to report an attack. "The laws and the rules of evidence unique to rape are at least partially responsible for the unwillingness of victims to report rapes."[18] A woman's actions are more likely to be the focus of judicial scrutiny than are a man's. "Women's respectability and actions—the evidence of consent and force—not men's behavior, are the object of judicial inquiry."[19]

Further discouraging women from reporting rape is the requirement of many states that evidence of "adequate resistance" be submitted to prove that they were really raped. Despite the fact that most states have enacted rape reform laws that change the consent standard or eliminate the requirement that a victim's testimony be corroborated, many state statutes still allow a woman's sexual history or evaluations of the victim's status or character as admissible evidence.

THE COSTS OF RAPE

Rape is a traumatic crisis that disrupts the physical, psychological, social, and sexual aspects of the victim's life. The most common injuries, besides the rape itself, are bruises, black eyes, and cuts. The total cost of medical expenses for victims from 1973 to 1982 was almost $73 million.[2]

Clinical researchers have also identified a rape trauma syndrome—a progression of emotional responses to being raped.[20] The acute phase can last days or months and is characterized by general stress symptoms such as fear, humiliation, disbelief, rage, disturbed eating and sleeping patterns, and gastric and genitourinary complications. Victims often experience flashbacks, loss of concentration, acute anxiety, and nightmares. As a protective behavior, victims can also become numb in reaction to the rape and deny or suppress the trauma. If denial is strong, a victim may discontinue prosecution in the desire to forget about the rape.

In the second long-term or reorganization phase, the victim begins to rebuild her life (although some researchers believe it is extremely difficult to return to a pre-assault state).[5] In this period, as in the initial months, each rape victim will cope differently depending on prior stresses such as economic insecurity, mental health problems, or lack of support from family or friends. Regardless of other variables that affect the rate of recovery, many women experience recurring distress or fear as well as disruptions in their sexual lives. Following emergency care, a victim may require either short-term crisis counseling or long-term mental health treatment.

RISK FACTORS

Women of all ages, racial groups, economic backgrounds, and educational levels are vulnerable to rape, but not all are equally at risk. At highest risk are young, single, black women from poorer urban communities. Their attackers are also young and black.

Age

A review of 11 empirical studies reveals that the vast majority of rape victims are between the ages of 13 and 25. Although researchers cluster age groups differently, the peak age of rape is within the 13- to 19-year-old age group, with 14 and 15 being the most frequent ages.[3] As one research team points out, "It is difficult to say for certain what the high risk age group actually is, because so many cases are not reported. It may be that mature adults conceal their victimizations more than adolescents and young adults."[3]

Like the victim, the rapist is generally 15–24 years of age, with a median age of 23 years.[9] In most states, a rapist under 15 years of age is not legally responsible and is protected from criminal prosecution.[3]

Race and Socioeconomic Status

According to FBI's *Uniform Crime Reports,* NCS victimization studies, and reseachers who examine the association of rape and race, black females are disproportionately victimized by rape and attempted rape.[3] In a 1971 study in Philadelphia, rape was estimated to be 12 times more frequent for blacks than for whites, even though blacks accounted for only 10% of the population.[21] The high rate of rape for black women is consistent for every age group.

BJS data indicate that most rapes involve victims and offenders from the same racial group. In assaults involving one offender, victim and offender were the same race 70% of the time for white victims and 89% of the time for black victims. When there were two or more offenders, victims and offenders were the same race 52% of the time for white victims and 75% of the time for black victims.[2]

The 1971 Philadelphia study also disputes the widespread cultural myth that black men rape primarily white women. It found that in 93% of rapes, victims and offenders were of the same race.[21] The percentage of same-race rape in subsequent studies varies from 57% to 90%.[21]

Almost all reported cases of interracial rape involve black men raping white women. However, many researchers note that black women may be reluctant to report being raped by a white man because of "the tradition of white power/black powerlessness."[22]

Data on the socioeconomic status of victims are sparse; the FBI does not systematically collect this information. However, BJS reports that rape victims are usually members of low-income families. From 1973 to 1982, about half of all victims reported a family income of less than $10,000. (This figure has not been adjusted for inflation over the 10-year period and is therefore understated in current dollars.)[2] In addition, one federal government commission in the late 1960s concluded that a disproportionate number of arrested rapists are black and from the poorest areas.[23]

Other Factors

Does a victim's behavior put her at risk for rape? Theories that so claim have been criticized by feminists and many other researchers who believe it devalues the trauma of rape by making the victim share responsibility for her own rape and by rationalizing male sexual aggressive behavior. The numerous studies that have been conducted to determine what distinguishes a victim of rape from a woman who has not been raped are all inconclusive.[5]

Researchers have attempted to understand the behavior and motivation of rapists by studying their psychological makeup, attitudes supportive of rape, and hostility toward women. Most researchers in this area have based their studies on clinical samples derived from incarcerated offenders, a methodological weakness that is frequently criticized because most rapists are never convicted.[9]

Victim and Offender Alcohol Consumption

Alcohol is widely associated with rape, and many researchers feel that some causal relationship exists because alcohol is considered a disinhibitor of aggression.[24] In one study, more than half of the convicted rapists were drinking at the time of their offense.[25] For some rapists, "the desire to rape is only present when they are drinking."[9] Although researchers are cautious about asserting that alcohol causes rape, many do believe it is used as an excuse for deviant behavior by the rapist.[9,26]

There has been relatively little research to determine whether a victim's consumption of alcohol puts her at higher risk for rape. A two-year study in Denver (1970–72) found that about one-quarter of rape victims had used alcohol at the time of the offense.[27] Another study found that 111 of 1,223 rape victims, or 9%, were under the influence of alcohol.[28]

SOCIAL AND CULTURAL FACTORS

Although feminists, violence experts, and victims have worked steadily over the past two decades to change society's view of and response to rape, erroneous perceptions continue to thrive: Women ask for rape by their clothing, behavior, and actions; any woman can resist a rapist if she really tries; women can only be raped by someone they do not know; rapists are mentally ill or sexually perverted. These beliefs are held by many in the general public and by professionals, police, assailants, and victims themselves. Clinicians and researchers call them "rape-supportive belief systems."[29]

Recent studies have documented the extent of social and cultural conditions that lend support to rape. One study tested whether rape-supportive attitudes of men in general were a predictor of rape. "Twenty-five percent of the college men . . . interviewed said that they would be willing to use force to get sex, and 51 percent said they would be willing to attempt rape if they were sure they could get away with it. [The study] concluded that these male college students held an attitude of entitlement to use force to coerce sex in a dating situation."[30]

A study of college men in 1981 found that a man's rape-supportive attitudes rather than psychological characteristics were what determined his level of sexual aggression. Sexually aggressive men considered sexual aggression normal, had conservative beliefs about female sexuality, readily accepted rape myths, viewed women as primarily responsible for preventing rape, and maintained traditional beliefs about women's roles.[31] Thus, the more a man accepts culturally sanctioned sexual stereotypes, the greater the likelihood that he will maintain rape-supportive beliefs. It should be noted that both of these studies examined attitudes, not behavior.

It is suggested that rapists justify their aggressive behavior by charging that the victim precipitated the rape.[32] These men "fail to perceive accurately the degree of force and coerciveness that was involved in a particular sexual encounter or to interpret correctly a woman's consent or resistance."[16]

The mass media play a prominent role in perpetuating the images and values that sanction violent aggression against women. Studies in recent years have documented media exposure that either subtly or explicitly condones male aggressive behavior, "objectifies" women, and portrays female sexuality as a commodity, creating an atmosphere of tolerance and rationalization to commit rape.[32]

Pornography is viewed by many feminists as another powerful element in a system that allows violence against women.[33] In 1970, when aggressive pornographic depictions were less common than today, the Presidential Commission on Obscenity and Pornography concluded that there was no direct causal link between pornography and sexual crimes. Sixteen years later, however, it asserted that "the vast majority of depictions of violence in a sexually explicit manner are likely to increase the incidence of sexual violence in this country.[34]

Researchers approach the study of rape with an increasing awareness of its social and cultural underpinnings. A consensus is emerging on the following assumptions.

• Rape is an act of power and control. A rapist asserts his power by using physical force or threats of bodily harm or by forcing a woman to submit to sexual acts against her will. It can occur between strangers, acquaintances, or intimate partners.[6,35]

• Rape is the result of the power differences between men and women in a male-dominated culture. These power differentials are reinforced through cultural beliefs, attitudes, and institutions like the family, religion, education, and politics.[9,35]

• Men who rape are not primarily psychotic; instead they express violence and rage through assault and injurious sexual attacks to resolve conflict. Sexual violence is culturally supported through the mass media, pornography, and by behavior learned through family violence.[6,17,32,36]

• Rape can happen to all women regardless of age, race, social class, educational level, or occupation.[9,17]

RAPE PREVENTION: THE STATE OF THE ART

Services for rape victims have increased in quantity and improved in quality since the early 1970s, but relatively little work has been done to measure the impact of prevention strategies. Some social science researchers describe the difficulty of "devising instruments, beyond self-reporting, that would assess actual behavior changes as well as attitude changes."[32] For programs that aim to eliminate or reduce the risk of rape, there is no clear link among theory, program, and impact.

When evaluations are done, they are generally short-term and, by using pre- and posttraining tests, typically measure educational programs to assess attitude change. The reasons for the inadequacy of evaluations in the rape prevention field are similar to those in other injury prevention areas: Scarce resources are used to maintain direct services, and a lack of funds all but eliminates the opportunity to determine systematically whether prevention efforts reduce the incidence of rape and sexual assault and the threat of these assaults.

Since the organized antirape movement began in the 1970s, rape crisis centers have been and continue to be in the forefront of prevention efforts. Their overriding aim is to eliminate the social conditions that perpetuate rape and sexual assault; they work primarily through the promotion of social action and community change. Their strategies include community education and rape avoidance strategies as well as service reforms and legal advocacy.

INTERVENTIONS

The activities that address rape prevention fall into five categories: community education, skills training, legal reform, health care, and school-based programs. Information about specific rape crisis programs came from a research study by the

National Center for the Prevention and Control of Rape (NCPCR).[37] Although several community centers have posttraining evaluations built into their programs, long-range outcomes have not been measured.

Rape Prevention and Community Education

Community education efforts can challenge attitudes and behaviors about rape and help break down cultural support of rape. The two programs described below show the possible components of such efforts. Interagency collaboration, skilled leadership, open staff communication, and programming that combine victim services with social activism are critical to the operation of rape and sexual assault prevention programs.[29] Both programs have had a significant impact in educating their communities about rape and sexual assault. Without the aid of a formal, long-range evaluation, however, it is impossible to determine whether the incidence of rape and sexual assault dropped in these communities.

Community Action Strategies to Stop Rape (CASSR) was founded nearly 10 years ago by Women Against Rape of Columbus, Ohio.[38] CASSR attributed the prevalence of rape to three social conditions: misconceptions about rape, women's subordinate status and sex roles, and women's isolation from each other.[5] It conducted a 4-year research–demonstration project to evaluate four community prevention programs: a series of women's rape prevention workshops, a whistle alert program, a shelter program, and a women's rape prevention network. The project's goal was to determine which of the programs decreased women's vulnerability to rape and increased community concern and responsiveness to the problem.

The workshop participants were pretested, posttested when the workshop ended, and tested again 2 months later. Women in the prevention network received pretests when they joined and posttests 3 to 6 months later. The whistle program and shelters were monitored along with media coverage of rape and rape prevention. Using random samples, the community was questioned about the problem of rape once a year over a 3-year period.

When the demonstration period ended, the results were mixed but positive overall. Workshop participants' knowledge and attitudes about rape were more accurate, and they were more confident of their ability to confront others and defend themselves. On the whole, the general public demonstrated an increased awareness of the problem of rape and a belief in the feasibility of various prevention tactics, although many still believed in a number of rape myths.

Rape prevention through community education can also create community support for local prevention effort so that further changes within the system are possible. The Lexington, Kentucky, Rape Crisis Center (LRCC) has initiated an exemplary education program that has gathered local support, mobilized disparate agency resources, and influenced social change in the community through collaboration and the work itself. Established in 1974, the primary goals of the center are to end rape by raising the awareness level of the community, eliminating misconceptions about rape, educating varied community groups about rape, and working with law enforcement personnel to increase reporting and successful prosecution.[29]

Interagency cooperation has been a significant factor in LRCC's ability to raise the consciousness of the community, described as politically conservative. The LRCC staff provides rape awareness education to elementary and secondary schools, colleges, parent groups, and law enforcement agencies. It also offers in-service training to social workers, school counselors, and hospital personnel. The center fortifies its interactions with other organizations by publishing a quarterly newsletter on activities at the center and other general information about rape and sexual assault and by appearing on radio and television programs. Each year, one week in August is designated "Take Back the Night" week. The center also engages in lobbying efforts to support the passage of bills that affect rape and provides expert testimony for courtroom proceedings.

Recommendation: Community education programs are promising interventions. They should be used and monitored and their outcomes evaluated.

Skills Training

Programs that teach women how to defend themselves in threatening situations have been offered in many communities. Self-defense classes are a common means by which women are taught to increase their self-reliance and self-confidence with the hope that this will decrease the chance of rape. This approach allows women to take some control of the solutions to sexual assault and makes women feel more responsible for their own self-defense.[39]

Model Mugging, for example, is a self-defense and empowerment course that teaches women fighting techniques through the simulation of rape situations.[40] Studies of police department records have shown that rapists try to intimidate women and incapacitate them, so the 21-hour course uses martial arts techniques to teach women to knock out assailants. By learning powerful fighting techniques, women can deliver effective counterattacks. According to the program's own assessment, of the 6,000 women who have taken the course, 28 have subsequently reported being physically attacked. The 20 women who chose to fight knocked out the assailant within 5 seconds of the attack.

Alternatives to Fear (ATF) is a Seattle-based program that concentrates solely on preventive and educational interventions.[29] The goals of ATF are to train vulnerable individuals to use self-defense methods and other preventive strategies to avoid potential violence and to offer educational services that reduce victimization and influence social change.

In conjunction with self-defense courses adapted to a range of skill levels (including teens, the disabled, and the elderly), ATF offers rape prevention workshops and provides written materials and support services to public, private, and independent sexual assault prevention programs nationwide. Since 1981, nearly 2,000 women participated in their free workshops, which include risk information and self-defense demonstrations.

Recommendation: Rape prevention and skills training programs for women are promising interventions. They should be used and monitored and their outcomes evaluated.

Legal Reform

Rape law reform is intended to make victims more willing to report rape and to make court requirements and practices more sensitive to the victim. The last decade has produced widespread reforms in rape laws. The reforms have covered three general areas: the definition of rape, penalties for rape, and rules of evidence, such as corroboration and past sexual history of the victim.[41]

One study of rape reform legislation in six cities focused on municipalities that have enacted either a strong, moderate, or weak rape reform law. Strong reform, for example, may include changes such as eliminating standards for the degree of resistance, restricting evidence of past sexual conduct (called the rape shield law), and dropping the requirement that a victim's testimony be corroborated. The effectiveness of these reforms is unknown.

The researchers could draw no definitive conclusions but speculated that increased reporting, improved treatment of victims by judges, prosecutors, and defense attorneys, and an increase in arrests and convictions "might be reflections of changes in public attitudes" as a result of the women's movement rather than a direct result of the reform laws.[18] Nonetheless, along with shifts in public perception, changes in rape laws can be a catalyst for further attitudinal change. "The criminal law serves not only a general deterrent function. It also has a 'moral and sociopedagogic' purpose to reflect and shape moral values and beliefs in society. . . . [The new rape law] symbolizes and reinforces newly emerging conceptions about the status of women and the right of self-determination in sexual conduct."[42]

Recommendation: The impact of legal reform measures is unknown. Further research should be conducted on this intervention.

Health Care for Victims

Standards of care in hospital and health facilities can be designed to be comprehensive, interdisciplinary, and compatible with a victim's needs during the immediate crisis period. The Miami Rape Treatment Center (RTC) is one of the few rape crisis centers that operates in a hospital setting. The center views rape as an emergency and provides victims with immediate, comprehensive, high-quality care.[29] It provides direct care services through a physician-directed and professionally trained hospital staff (including a hotline, police advocacy, and counseling services), and functions as an educational resource center for the community and hospital personnel. The RTC administers the Interdisciplinary Rape Science Training Institute, which offers professional training to providers in social work, medicine, psychiatry, and criminal justice. Through the work of the institute, the RTC has cultivated interagency ties with local police, family service agencies, community mental health centers, and other public and private groups.

Setting formal standards for the evaluation and treatment of rape and sexual assault victims is another critical step. Massachusetts is one of the only states in the country that has established a uniform system for collecting and storing evidence of rape. Its standardized kit, developed by the Massachusetts Department of Public Safety Crime Laboratory, is used in hospitals throughout the state and is expected to help prosecutors win more convictions in rape cases. In the past, the variety of kits was confusing to hospital personnel who moved from one hospital to another. The result was a loss of evidence and improper use of the kits. By securing stronger evidence, there is a greater chance of successfully prosecuting a rape case, thereby sending the message that rape is a serious crime that women should report.

Recommendation: Health care protocols for victims of rape are promising interventions that should be used and monitored and their results evaluated.

School-Based Programs

School-based education and training programs that aim to change social patterns and sex role behavior may focus on acquaintance rape, sex role stereotyping, communication skills, peer violence and pressure, and assertiveness behavior and are usually taught to public school children and adolescents of both sexes. The goal of these programs is to help students identify and reject harmful behaviors and attitudes that prevent them from perceiving each other as individuals. The educational programs must be of some duration and intensity to produce attitudinal change.

As important as these programs may be, a 1977 study illustrated the difficulty of this approach.[43] Working with 5-, 10-, and 14-year-old children in three Boston schools, the researcher introduced special curricular materials and included peer group involvement during a 6-week intervention. She discovered that sex role stereotypes were independent of class, ethnicity, and occupational status of the mother and contended that television was a more powerful influence than the family in determining cultural stereotypes. The older the child, the more rigidly he or she adhered to sex role stereotypes. If educational interventions are to counteract sex role conditioning and play a role in rape prevention, the intervention must be introduced in the earlier grades and be consistent over time.

Recommendation: School-based rape prevention programs designed to change student behavior and attitudes about sex roles and aggression are promising interventions. They should be used and monitored and their outcomes evaluated.

INTERVIEW SOURCES

F. W. Lindsay, Cambridge, Massachusetts, August 16, 1988

Michael J. Rand, Bureau of Justice Statistics statistician, Washington, DC, March 14, 1988

REFERENCES

1. Schmich MT. In the "age of liberation," who is free? Boston Globe 1985 Jan 20:A25.

2. Bureau of Justice Statistics. The crime of rape. Washington, DC: US Department of Justice, 1985 (Mar).

3. Katz S, Mazur MA. Understanding the rape victim: a synthesis of research findings. New York: John Wiley, 1979.

4. Estrich S. Real rape. Cambridge, Massachusetts: Harvard University Press, 1987.

5. Koss M, Harvey MR. The rape victim: clinical and community approaches to treatment. Lexington, Massachusetts: Stephen Greene, 1987.

6. Burgess AW, Hartman CR. Rape and sexual assault. In: Surgeon general's workshop on violence and public health source book. Washington, DC: Health Resources and Services Administration, 1985(Oct):3.

7. US Department of Justice, Federal Bureau of Investigation. Crime in the United States: uniform crime reports. Washington, DC: US Government Printing Office, 1982:14.

8. Bureau of Justice Statistics. National crime survey: criminal victimization in the US. Washington, DC: US Department of Justice, 1983; NCJ-90541.

9. Russell DEH. Sexual exploitation. Beverly Hills, California: Sage, 1984.

10. Russell DEH, Howell N. The prevalence of rape in the United States revisited. Signs 1983;8:688–95.

11. Schultz LG. The child sex victim: social, psychological, and legal perspectives. Child Welfare 1973;52(3):147–57.

12. Brownmiller S. Against our will: men, women, and rape. New York: Simon and Schuster, 1975.

13. Clark L, Lewis D. Rape: the price of coercive sexuality. Toronto: Canada Women's Educational Press, 1977.

14. Law Enforcement Assistance Administration. Criminal victimization surveys in 13 American cities. Washington, DC: US Government Printing Office, 1975.

15. Rodabaugh B, Austin M. Sexual assault: a guide for community action. New York: Garland, 1981.

16. Koss M. Hidden rape: sexual aggression and victimization in a national sample in higher education. In: Burgess A, ed. Rape and sexual assault II. New York: Garland, 1988.

17. Intrepid Clearinghouse. Myths about rape. Columbus, Ohio: Community Action Strategies to Stop Rape, 1978.

18. Horney J, Spohn C. The impact of rape reform legislation. Lincoln, Nebraska: University of Nebraska, Department of Criminal Justice, 1987(Nov).

19. Stanko B. All too real. Women's Rev Books 1987(May):14. Available from Wellesley College Center for Research on Women, Wellesley, MA 02181.

20. Burgess A, Holstrom L. Rape trauma syndrome and post traumatic stress response. In: Burgess A, ed. Rape and sexual assault. New York: Garland, 1985.

21. Amir M. Patterns in forcible rape. Chicago: University of Chicago Press, 1971.

22. Williams JE, Holmes KA. The second assault: rape

and public attitudes. Westport, Connecticut: Greenwood, 1981.

23. Mulvihill D, Tumin M, Curtis L. Crimes of violence: a staff report submitted to the National Commission on the Causes and Prevention of Violence, vols 11–13. Washington, DC: US Government Printing Office, 1969.

24. Rada RT. Alcoholism and forcible rape. Am J Psychol 1975;132:444–6.

25. Rada RT. Clinical aspects of the rapist. New York: Grune and Stratton, 1978.

26. Scully D, Marolla J. Rape and vocabularies of motive: alternative perspectives. In Burgess A, ed. Rape and sexual assault. New York: Garland, 1985.

27. Selkin J. Rape: when to fight back. Psychol Today 1975:71–6.

28. Hayman CR, Lanza C, Fuentes R, Algor K. Rape in the District of Columbia. Am J Obstet Gynecol 1972; 113(1):91–7.

29. Harvey M. Exemplary rape crisis programs. Rockville, Maryland: National Institute of Mental Health, 1985.

30. Herman J. Sexual violence. Wellesley, Massachusetts: Wellesley College Stone Center for Development Services and Studies, 1984.

31. Koss MP. Hidden rape on a university campus. Rockville, Maryland: National Institute of Mental Health, 1981.

32. Community action strategies to stop rape. Proceedings of the National Conference on Rape Prevention Theory, Strategies, and Research. Columbus, Ohio: National Institute of Mental Health, 1978.

33. Wheeler H. Pornography and rape: a feminist perspective. In Burgess A, ed. Rape and sexual assault. New York: Garland, 1985.

34. US Attorney General's Commission on Pornography. Final report. Washington, DC: US Government Printing Office, 1986.

35. Graff S, et al. Freeing our lives: a feminist analysis of rape prevention. Rockville, Maryland: National Institute of Mental Health, 1978.

36. Scully D, Marolla J. Rape and vocabularies of motive: alternate perspectives. In: Burgess A, ed. Rape and sexual assault. New York: Garland, 1985.

37. National Center for the Prevention and Control of Rape. Exemplary rape crisis programs: a cross-site analysis and case studies. Washington, DC: US Department of Health and Human Services, Public Health Service, Alcohol, Drug Abuse, and Mental Health Administration, 1985.

38. Sparks C. Community action strategies to stop rape: final report. Rockville, Maryland: National Center for the Prevention and Control of Rape, National Institute of Mental Health, 1982; no. R18 MH29049.

39. Morgan M. Conflict and confusion: what rape prevention experts are telling women. Sexual Coercion Assault 1985(1):160–8.

40. Soalt M. Model Mugging, Inc. Women's self defense and empowerment (training program). Concord, Massachusetts.

41. Swift CF. The prevention of rape. In: Burgess A, ed. Rape and sexual assault. New York: Garland, 1985:417.

42. Loh W. What has reform of rape legislation wrought? J Soc Issues 1981(37):5.

43. Guttentag M. Prevention of sexism. In Abee G, Joffee J, eds. Primary prevention of psychopathology. Hanover, New Hampshire: University Press of New England, 1977.

Chapter 16: Suicide

Knowledge about suicide is elusive. We try to count suicides yet know that the numbers are wrong. We launch preventive programs without regard to whether they will work and then fail to evaluate them. It is time, therefore, to reassess what we think we know, to evaluate what has been done in the name of prevention, and to provide a firmer foundation for the development of new programs.

Each year, the self-inflicted deaths of 25,000 to 30,000 Americans are recorded by the National Center for Health Statistics (NCHS). In 1986 (the most recent fully analyzed year), the NCHS reported a total of 30,904 suicides.[1] These deaths accounted for 939,104 years of potential life lost (YPLL).[2] There is general agreement that suicides are consistently undercounted but less agreement as to the extent to which this is true (P. W. O'Carroll, personal communication). One federal working group estimated that suicides may be underreported by as much as 25%–50%,[3] but this estimate has not been confirmed by data. Coroners and medical examiners use different definitions of suicide, require different evidence (some only certify suicide if a note is found), and may yield to family or community pressures, either direct or unspoken. According to a recent analysis, "At best, we can estimate . . . that the sensitivity with which coroners and medical examiners certify true suicides varies from approximately 55 percent to 99 percent" (P. W. O'Carroll, personal communication).

The working group estimated 25%–50% underreporting of suicides. Based on the more conservative 25% figure, it is possible that the actual number of suicides in 1986 was closer to 37,000. The difference could be very important. As the working group's report pointed out, "These data in turn affect the course of health care research, the flow of resources, and, ultimately, public health policy."[3]

But whether suicide claimed nearly 30,000 lives in 1985 or more than 36,000, we do know that there are segments of society more seriously affected than others. Suicide in the United States is increasingly a problem of adolescents and the very old.[4] And we know that the picture of suicide in the United States has changed dramatically during the last two decades.

THE MAGNITUDE OF THE PROBLEM

During the 5 years from 1981 to 1985, there were 142,872 suicides recorded in the United States. The suicide rate during this period remained relatively stable, fluctuating between 12.2 and 12.4 per 100,000. This stability represented a high plateau, however. In 1970, for example, the rate was 11.6 per 100,000. It rose to 13.1 per 100,000 in 1977 and fell to 11.9 in 1980. By comparison, the rate for 1986 was 13.1. Suicide was the fifth leading cause of years of potential life lost, with approximately 900,000 years lost annually.[5,6]

Throughout this period, the age-adjusted suicide rate for males (18.0) was more than three times that for females (5.4), and the age-adjusted suicide rate for whites (12.1) was almost twice that for blacks and other races (6.7). The relations among these rates have remained relatively stable over time. What this stability masks, however, is the epidemic increase in suicide among the young, discussed later in this chapter.[5,7]

In 1970, slightly more than half of all suicides—among both men and women—used firearms. By decade's end, 57.3% of all suicides used firearms. The next most common method is hanging (14%). Poisoning by solids or liquids, the most common method used by females in 1970, was replaced in frequency by firearms (although poisoning remains the most common method by which females attempt but do not complete suicide). Jumping from high places, carbon monoxide poisoning by auto exhaust, drowning, and cutting oneself with a sharp instrument are less common methods.[8]

The frequency of suicide peaks during spring months, rises again in the fall, and is lowest in December.[9] Geographically, suicide rates in 1980 ranged from a low of 8 per 100,000 in New Jersey to a high of 22 per 100,000 in Nevada. Suicide rates are lowest in the Northeast and highest in the West. This may possibly reflect the greater availability of firearms in the western states,[10] but it may also suggest that some qualities of rural life, particularly loneliness, play a role.

In 1932, at the height of the Great Depression, the general suicide rate was approximately 17 per 100,000. It dipped briefly to 10 per 100,000 during the war years and then began the increase that has continued, with occasional annual variation, for 40 years. "Virtually all of the increase since 1963," one source notes, "has been in suicide by firearms."[6]

A Note on Attempted Suicide

Despite the popular notion that attempted suicides are merely unsuccessful suicides, indications exist that they represent a different, although overlapping, class of phenomena. Suicide attempts, unlike suicides, are not legally reportable events. Information on attempts, therefore, is even less accurate than the data on suicides. However, it is estimated that suicide attempts are eight times more common than completed suicides. In addition, it is known that while white males complete suicide three times more often than females, females are reported to attempt suicide three times more often than males. It is not known whether males are more reluctant to seek help and therefore are more reluctant to report their attempts. And it is not known whether the lower rate of suicide among females results from females choosing to use less lethal methods despite an equal desire to die. Changes in pharmaceutical dispensing practices and improvements in emergency medical care have substantially reduced the lethality of drug overdoses. The majority of female suicide completers, like males, use firearms. It is estimated that the cost of providing health care for people who have attempted suicide averages approximately $116.4 million a year.[11]

RISK FACTORS

Youth

Suicide is now the second leading cause of death among persons 15 to 24 years of age, following motor vehicle crashes. In 1950, people in this age group were responsible for less than 6% of all suicides. By 1980, they were responsible for 20%. The suicide rate among 15- to 24-year-olds increased from 4.5 per 100,000 to more than 12 per 100,000, an increase of more than 300% in three decades. And most of this increase was among white males.[7,12]

From 1950 to 1970, the firearm suicide rate for this age group increased at the same pace as suicides using other methods. In 1970, however, the use of firearms began a sharp and sustained increase, climbing three times faster than other methods among 15- to 19-year-olds and 10 times faster for 20- to 24-year-olds. The vast increase in youth suicide, therefore, is associated with a vast increase in the use of firearms as a method of suicide. And that increase parallels a proliferation of civilian firearms in the 1960s and 1970s, both in the number of households possessing guns and the number of guns owned.[13]

Youth suicide also raises special problems for prevention. "Suicide," wrote researchers at the Centers for Disease Control (CDC), "has traditionally been considered a mental health problem, and our approach to prevention was based on a portrait of the typical suicidal individual as an older, depressed, white male. The basic prevention strategy involved detection and treatment of depression. If, as some recent research suggests, most young persons at high risk for suicide are not depressed, the whole approach to prevention must be reexamined."[12]

The question is currently being reexamined. "Most youth suicides," write researchers at the New York Psychiatric Institute, "are *precipitated* by some stress—getting into trouble, breaking up with a boy or girl friend, problems at school, an argument with parents, etc. However, these are the common stresses of adolescence affecting countless teenagers every day who *do not* respond with suicidal behavior. In order to explain suicide, we have to look beyond the stressor to some feature of the individual's personality or to a co-existing mental illness."[14]

According to the project's preliminary findings, the risk groups for youth suicide (in descending order of seriousness) are these:

1. Boys who have made a suicide attempt serious enough to warrant inpatient psychiatric admission

2. Girls who have made a similar attempt

3. Boys who took an overdose but were not considered sufficiently disturbed to be admitted

4. Boys who have a major depressive disorder (including eating and sleep disorders and sexual dysfunction)

5. Girls who have a major depressive disorder (including eating and sleep disorders and sexual dysfunction)

6. Boys with a history of antisocial behavior

7. Girls who took an overdose but were not considered sufficiently disturbed to be admitted[14]

It is notable that uncomplicated depression (extreme sadness, melancholy, or dejection unaccompanied by severe eating disorders, sleep disorders, sexual dysfunction, etc.), long a staple among the "warning signs," is missing from this list of risk factors. The completion of research on risk factors will provide a sounder basis for efforts to prevent youth suicide.

Youth Suicide Clusters

A particularly disturbing aspect of youth suicide in recent years is suicide clusters: groups of deaths closely related in time and space, involving at least three or more completed suicides. As early as 1845, Aramiah Vrigham, founder of the *American Journal of Insanity*, wrote, "No fact is better established in science than that suicide is often committed from imitation." Although, as a contemporary researcher points out, "this fact has not been *conclusively* established with scientific research" (J. A. Mercy, personal communication).

For researchers and practitioners in the field, the question of imitative behavior is at the heart of suicide clusters.[15] Our understanding of clusters is incomplete. Many believe clusters occur through a process of "contagion," but this hypothesis has not been formally tested. "Nevertheless, a great deal of anecdotal evidence suggests that, in any given suicide cluster, suicides occurring later in the cluster often appear to have been influenced by suicides occurring earlier in the cluster."[16]

Although cluster suicides can be traced back to Greece in the fourth century B.C., serious study of the phenomenon began in this country only in the early 1980s. In Fairfax County, Virginia, for example, 20 teen suicides occurred during the 1980–1981 academic year. Attempts have been made to prevent cluster suicides by organizing community-based programs with the assistance of the CDC, but much research remains to be done.[17]

Suicide seen on television may increase the risk of suicide for certain susceptible individuals, although this has not been demonstrated conclusively.[17] One researcher believes imitation is a grave enough possibility that news of suicides should be kept from the media and that recent suicides should not be mentioned in schools.[17] The potential role of the media in stimulating imitative behavior among adolescents is particularly important because, although adolescents involved in cluster suicides may not have known each other (although some have), they have known of each other. It is not uncommon to find newspaper accounts of the deaths of other teens among the personal effects of a subsequent suicide.

According to the most recent research, the media's role is mixed. On the one hand, television news stories about suicide seem to trigger additional suicides. Recent studies suggest that dramatic, fictional television presentations about suicide do not have the same effect, although this remains an area of controversy among researchers.[18-21] Research will continue, but it is already clear that newspaper or television accounts of a teenager's suicide that glamorize or romanticize the act can endanger other adolescents.

The Elderly

In 1985, the last year for which suicide rates by age group are available, more than 5,700 persons 65 years of age or older committed suicide. Thus, the suicide rates for the elderly place them among those most at risk. The rate for persons 65 to 74 years of age was 18.5 per 100,000; for those 75 to 84 years of age it was 24.1; for those 85 years of age or older, the rate was 19.1 per 100,000.[22] Elderly suicides are disproportionately male, disproportionately white, and often recently widowed.[4] The elderly attempt suicide less frequently than other groups but complete the act more frequently. Older people also tend to communicate their suicidal intentions less and to use lethal weapons, often firearms, more often. Other characteristics of elderly suicides include a lack of employment, alcohol abuse, and solitary living arrangements, especially in deteriorating urban areas.

SOCIAL AND CULTURAL FACTORS

A recent study from Alaska provides clear insight into the relationship between suicide and two of the most ubiquitous elements in our culture, alcohol and firearms. A review of the records of 195 suicides in 1983–84 revealed that 76% of the deaths were caused by gunshot wounds. Further, 79% of the Native American suicides and 48% of the white Alaskan suicides had detectable levels of blood alcohol.[23]

There are several possible explanations for why alcohol is a risk factor for suicide. Alcohol or substance abuse, or both, may serve to facilitate the act. By reducing inhibitions and impairing the judgment of a person contemplating suicide, alcohol or drugs render the act more likely. "Very often alcohol is also associated with attempted suicide," noted one authority. "Alcohol is consumed during the act in approximately 20 percent of overdoses, and in more of those taken by men than by women. In addition, alcohol is often consumed shortly before an attempt."[24] Alcohol consumption may also be associated with other suicide risk factors, such as depression or other mental illnesses. Thus, the role of alcohol in American culture, described in Chapter 6, profoundly affects the suicide problem.

So, too, does the pervasive availability of firearms. "The choice of suicide methods," wrote one clinical psychologist, "seems to be affected both by the physical availability and the sociocultural acceptability of the various potential methods available."[25] The proliferation of firearms in the United States and particularly the growing number of handguns during the last two decades make them both physically available and culturally acceptable. Firearms are far more likely to be lethal than other methods. Thus, the vast majority of attempted firearm-related suicides are completed. Individuals contemplating suicide generally do not purchase guns for that purpose. They find them on closet shelves, in night-table drawers, or in other easily accessible places, ready for use.[26]

A Sacramento County, California, study revealed that from 1983 to 1985, handguns were used in 69% of firearm suicides, 65% for males and 88% for females.[27] In Minnesota, where hunting is more common, 64% of the firearm suicides in the 12- to 24-year-old age group (1980–81) used rifles or shotguns; 36% used handguns.[28]

SUICIDE PREVENTION: A BRIEF HISTORY

In 1970, the Center for the Study of Suicide Prevention of the National Institute of Mental Health sponsored a special conference on suicide in the 1970s. Leaders from around the country came together in Phoenix, Arizona, to provide an overview of the field and to develop directions for the future.

The conference report, *Suicide in the 70s*, published in 1973, contained more than 20 goals for research and prevention. Yet 6 years later, few of the goals had been met.[29] Important academic and clinical research was conducted during these years, but the major focus on prevention at the community level was the creation of crisis hotlines and suicide prevention centers, the efficacy of which has yet to be established.

In 1980, the U.S. Public Health Service made suicide prevention one of the key goals of its efforts to achieve measurable improvements in the health of the American people by 1990. *Healthy People: The Surgeon General's Report on Health Promotion and Disease Prevention* set broad goals, including a 20% reduction in the adolescent and young adult death rate. A companion volume, *Promoting Health, Preventing Disease: Objectives for the Nation*, set forth more than 200 measurable objectives to be achieved by 1990. Included among these is a reduction in the rate of youth suicide to 11 per 100,000 from a baseline of 12.4 per 100,000 in 1978. (It should be pointed out that at the midpoint of this effort—1985—the youth suicide rate had increased to 12.9.)

To achieve these goals, federal agencies have moved to coordinate their work with state and local community agencies in developing suicide prevention programs. The New Jersey and California suicide prevention education programs, discussed later in this chapter, benefitted from this cooperation.

More recently, in June 1985 the Secretary of Health and Human Services convened a task force on youth suicide. Chaired by the director of the National Institute of Mental Health, with extensive support from the CDC and its newly formed Division of Injury Epidemiology and Control, the task force was charged with coordinating activities among various federal agencies, Congress, state and local governments, private agencies, and professional organizations. Its major functions were to assess and disseminate information and to recommend and initiate activities. A priority was the more accurate reporting of suicides.[8]

SUICIDE PREVENTION: THE STATE OF THE ART

The 10 previous chapters showed that relatively few injury prevention interventions, including many that have long been in use, have been carefully evaluated. There are many reasons for this, including a lack of understanding of the importance of building evaluations into intervention projects, a focus on the provision of services rather than evaluation, and inadequate funds for evaluation. As a result, programs use scarce resources to replicate interventions that evaluations might have shown to be ineffective or even counterproductive.

Inadequate or nonexistent evaluation is a frequent problem in reviewing interventions directed at interpersonal violence and suicide. As noted below, interventions such as suicide hotlines, widely imitated because of their "self-evident" efficacy and ease of operation, turn out on careful evaluation to have little effect on the suicide rate.

Despite its role as a major cause of death, suicide is a rare event. Except in the case of suicide clusters, no community or individual school is likely to experience many suicides. Therefore, because these institutions are the common locus for preventive efforts, the evaluation of programs becomes statistically very difficult.

For example, a college of 10,000 students might experience no more than one or two suicides a year. Evaluating an intervention at that level is difficult. And the problem does not resolve itself on the larger community or state levels. Suppose that a particular intervention is intended to reduce the suicide rate among young males from 30 per 100,000 to 25 per 100,000. To detect statistically significant change (a probability of .05 or less) and ensure that no real effect is being missed (a confidence interval of 95%), an experimental sample of 2.9 million men and a control group of the same size would be required.[30]

As explained in Chapter 4, it is still possible, despite such difficulties, to design meaningful evaluations of injury prevention interventions. Doing so, however, puts a premium on the development of accurate, timely, and valid data on suicide and suicide attempts. One federal task force recommended training programs for coroners, medical examiners, and other officials whose judgment can affect the classification of a death by suicide, the creation of a uniform definition of suicide, the development of community-based surveillance systems for suicide attempts, and other measures (M. Rosenberg, et al., unpublished observations). The new data generated by a more accurate suicide reporting system could provide a sounder basis for evaluating the interventions described below and for creating the next generation of injury prevention programs.

General Recommendation: Because of the uncertainty about the effectiveness of suicide prevention interventions, the greatest need is for the evaluation of specific countermeasures. New interventions should be designed with specific measurable objectives and should be thoroughly evaluated.

INTERVENTIONS

Suicide Prevention and the General Population

Reduce the availability of the means of suicide. The reduction of the availability of the means of suicide, such as guns and pharmaceuticals, is an intervention that does not require individual action. During the 1960s, primarily for economic reasons, the cooking and heating gas available in Britain changed from coal-based gas to petroleum-based gas and finally to North Sea natural gas. One side effect of this change, 70% complete by 1971 and fully realized in 1975, was a drastic reduction in the carbon monoxide content of the product—from 10%–20% carbon monoxide to virtually none. As a result, gas is no longer a means of suicide in Britain, where it had long been the most common method used. In 1960, half of all British suicides died by gas.[31] What is most instructive is that the unavailability of lethal cooking gas was not offset by an increase in the use of other means. In 1975, the overall suicide rate in Britain was 35% lower than it was before the gas supply was detoxified.[32,33]

Suicide is a complex phenomenon in which elements of certainty and ambivalence, planning, and spontaneity seem to play significant and perhaps conflicting roles. The important lesson of the British experience is that, when a common and culturally acceptable means of suicide became unavailable, many would-be suicides did not simply choose an alternative method.

Because the most common and culturally acceptable agent of suicide in the United States is the handgun, applying the British lesson becomes a political question. Because handguns are implicated in many other areas of interpersonal violence, interventions directed at the availability of firearms are discussed in Chapter 17.

Handguns are not the only means of suicide whose availability might be reduced. It has been suggested, for example, that the following measures be explored: antisuicide barriers and nets at well-known jump sites, tighter prescription requirements for highly toxic psychotropic drugs, automatic ignition on gas fires and ovens, and devices to shut off a car's engine when dangerous levels of carbon monoxide build up inside the car.[31]

Recommendation: The British coal-gas experience demonstrates that reducing the availability of a means of suicide can be an effective intervention. Measures to reduce the means of suicide should be used and monitored in this country, and their outcomes should be evaluated.

Targeted hotlines. Many communities have suicide prevention hotlines. These have proven difficult to evaluate, and their effectiveness has not been demonstrated.[34] Some communities have attempted to target their hotlines by installing emergency telephones linked to a 24-hour crisis intervention center on bridges or other areas that are common sites for suicides.

The Mid-Hudson Bridge, located in Dutchess County, New York, averages some five suicides a year. Near Poughkeepsie, the Mid-Hudson is one of three major bridges connecting Dutchess with neighboring counties. After a rash of suicides in 1982, the Dutchess County Department of Mental Health and the New York State Bridge Authority began to consider an emergency telephone system. Installed in late August 1984, Helpline consists of call boxes on either end of the bridge. The call boxes, designed for simplicity, reliability, and the ability to withstand tampering, are connected to a 24-hour psychiatric emergency service, where trained personnel have immediate access to the police.

Between August 1984 and October 1986, the bridge phones were used 30 times. Several calls were unrelated to suicide (e.g., one reported a fire); however, 23 of the callers were apprehended and evaluated by the psychiatric emergency service. Six individuals (five of whom had not used the call boxes) jumped from the bridge during this period, one of whom survived the attempt. "The data suggest that most would-be jumpers are ambivalent enough about dying that they will reach out for help/contact if the opportunity exists, and, as a consequence, can be saved," wrote Kenneth Glatt of the Dutchess County Department of Mental Health.[35]

Recommendation: Targeted suicide emergency phone lines are a promising intervention that should be used and monitored and their outcomes evaluated.

Treatment of attempted suicides. Identifying those who attempt suicide who have treatable mental health problems and ensuring that they receive appropriate treatment are the aims of many hospital and clinic programs. Treatment, ranging from counseling to psychiatric hospitalization, is the primary intervention made available to those who attempt suicide. Because attempters remain at the highest risk to complete a suicide later, treatment can be both primary and secondary prevention; yet treatment as an intervention is generally beyond the mandate of a health department or injury prevention program.

Nonetheless, because treatment is often an intervention, injury prevention practitioners should understand how these interventions compare with the others discussed in this section.

Problems inherent in these programs include the assessment of the mental health of those who attempt suicide, the identification of adequate mental health services, appropriate referrals, and follow-up procedures.

Unfortunately, evaluations of treatment interventions have been, "simply descriptions of services or faintly disguised testimonials," according to a major review of the methodologic failures in these evaluations. The review suggests that, in the future, patients should be assigned randomly to treatment groups, at least 80% of those beginning treatment should be followed up, both statistical and clinical significance should be reported, all clinically relevant outcomes should be reported, all patients should be adequately described, and the intervention should be easily replicable in another setting.[36]

Recommendation: The identification and treatment of those who attempt suicide who have treatable mental health problems are promising interventions. Further research is necessary to determine which modes of treatment are most effective with specific patient populations. Treatment programs

should be used and monitored and their outcomes evaluated.

Crisis intervention services. Between the mid-1960s and the early 1970s, suicide prevention centers proliferated in the United States. These facilities were designed to provide a broad range of services, from telephone hotlines to individual and family counseling. Building in part on the experience of the Samaritan movement in Great Britain (a private, national movement begun in 1953 to provide confidential listening and befriending services to troubled people), these facilities spread rapidly, and by 1973 nearly every metropolitan center had at least one. As has often been the case in injury prevention, this commitment of money and resources was carried out without evidence that the centers were effective.

However, an analysis of the period of their greatest growth (1968–1973), which compared suicide rates in matched pairs of counties, did reveal one positive correlation. For young white females, the suicide rates decreased by 1.75 per 100,000 in the counties with prevention centers. Young white females constitute the major client group for the centers.

Suicide prevention centers may perform a variety of important roles in society, from reducing the general level of loneliness to providing an entry point into the mental health system. In terms of reducing the suicide rate, however, their impact remains to be demonstrated.[37–41]

In one community, nurses received additional training to augment insufficient mental health services. Under the sponsorship of the Center for Studies for Suicide Prevention of the National Institute of Mental Health (NIMH), a program was set up to train selected nurses already functioning as health counselors in rural Vermont communities. Thirty-five nurses, ranging from 23 to 68 years of age, participated in a 26-week course focusing on personality development, psychopathology, and suicide prevention. A descriptive follow-up study of three counties indicated that 25% of the nurse's clients were potentially suicidal and that their ability to identify and assist these individuals was an important addition to the creation of a triage system for mental health.[42]

Recommendation: Multipurpose crisis intervention and suicide prevention centers have had only a limited effect on the suicide rate. Further research is necessary to determine how best to target such programs. Existing programs should be rigorously monitored and their outcomes evaluated. In addition, nursing services that incorporate suicide counseling show promise. They should be used and monitored and their outcomes evaluated.

Proposals for untried interventions. Suicide prevention programs should be supported and evaluated in hospital emergency rooms, health maintenance organizations, and other health care facilities. These programs should be coordinated with schools, social service agencies, and religious organizations. Protocols should be developed for taking standardized clinical histories from those who attempt suicide and patients who belong to high-risk groups. To reduce the availability of the means of suicide, the amount of potentially lethal medications that can be obtained on a single pharmacy visit should be restricted. Finally, employee assistance and health insurance programs should be encouraged to provide services to employees if they or family members are at increased risk of suicide.

Preventing Youth Suicide

School-based programs. School-based educational programs are designed to educate administrators, faculty, students, and often parents about the warning signs of suicide and about available resources for help. They received their impetus from concern about the increasing levels of youth suicide in the late 1970s and early 1980s. This was further fueled by the spate of cluster suicides that began in 1983. Initially designed to help school personnel cope in the aftermath of a suicide, the programs soon incorporated a primary preventive focus and included students among their target audiences. There are approximately 120 such programs in use throughout the country (A. Garland, personal communication). Until very recently none of these programs had been evaluated professionally. And none of the small handful of evaluations has yet been able to demonstrate any correlation between the programs and a reduction in youth suicide (D. Shaffer, A. Garland, B. Whittle, unpublished observations).[43]

The most complete available evaluation, that of three New Jersey programs, reveals both the strengths and potential weaknesses of such programs. "Didactic programs of the sort we have studied," the authors reported, "have some effect on improving knowledge about resources and school policies, and willingness to use hotlines and mental health centers, but do very little to alter damaging and dangerous attitudes" among the students (D. Shaffer, A. Garland, B. Whittle, unpublished observations).

Useful as screening devices, the programs still provided only marginal help for the highest-risk students (i.e., those with a history of suicide attempts). Further, the researchers caution that great care should be taken to ensure that the most vulnerable students are not adversely affected by the program. And they urge long-term evaluative research to determine whether school-based programs can help reduce the suicide rate.

Recommendation: School-based suicide prevention programs are promising interventions that should be used and monitored and their outcomes evaluated.

Hospital-based programs. Hospitals can provide important services for some suicidal youths. As incicated above in the discussion of treatment services for those who attempt suicide, it is important to determine what types of services are appropriate and beneficial to individual patients as well as to evaluate the overall effect of treatment programs in reducing suicide.

The difficulties inherent in evaluating the effectiveness of hospital-based programs can be seen through the example of an intervention carried out from October 1981 to October 1983 at Boston City Hospital, with Brockton Hospital in Brockton, Massachusetts, serving as the control site. The program combined an educational curriculum for human service workers, health care providers, police, court personnel, and high school student peer leaders with direct services to at-risk adolescents. Support services, provided by a trained community outreach social worker, included setting up clinic appointments (and ensuring that they were kept), functioning as liaison with service providers, and acting as an advocate with the adolescents' family, school, the courts, and other community agencies.

The evaluation indicated that the program had increased the subjects' compliance with the medical regimen and had, to a lesser degree, facilitated early help-seeking by the adolescents. However, despite these benefits, the incidence of repeat suicidal or self-destructive behavior did not diminish.[44]

Recommendation: Hospital-based programs are promising interventions that can provide important services to some suicidal youths. Further research is necessary to determine which modes of treatment are most effective with specific patient populations. Treatment programs should be used and monitored and their outcomes evaluated.

Community crisis planning. With CDC's assistance, communities have begun developing plans to cope with cluster suicides. On November 16 and 17, 1987, CDC's Division of Injury Epidemiology and Control convened a meeting of suicide researchers, state health officials, federal agency representatives, and representatives of various communities to explore responses to suicide clusters. Held in Newark and cosponsored by the New Jersey Department of Health, the meeting followed by several months an apparent suicide cluster in which four Bergenfield youths had died.

Since the problem of cluster suicides was recognized earlier in the decade, CDC has been working with affected communities to develop model guidelines for response.[16] Several communities have put into practice plans whose elements reflect the guidelines.

The Bergenfield Mental Health Emergency Response Plan, for example, is broken down into precrisis programming, crisis operations, and postcrisis programs. Among its critical elements are the identification of at-risk populations prior to a crisis, the identification and assessment of community resources, a coordinated effort to locate those at risk after a crisis has begun, and the careful dissemination of public information in ways that do not glamorize those who commit suicide.[45-47]

Programs in college communities to prepare staff members to implement outreach efforts following a suicide have also been tried. Based on a literature review and a survey of 50 colleges and universities near Tarrytown, New York, Webb urges that all colleges develop written guidelines to be used in the event of a suicide and that training sessions be held yearly for all residence and support staff members. Following a campus suicide, the guidelines recommend setting up a crisis counseling service and contacting the deceased's friends and acquaintances as soon as possible. It encourages commemoration of the death, but cautions against glamorization or romanticization of the deceased. Among other suggestions, the guidelines recognize that the aftermath of a suicide may be felt for a long time.[48]

Recommendation: These are promising interventions that should be used and monitored and their outcomes evaluated.

Proposals for untried interventions. Programs should be developed for youth with multiple risk factors who fall outside the range of traditional programs (e.g., youth who are homeless, runaways, physically or sexually abused, prostitutes, substance abusers, or incarcerated).

Suicide Prevention and the Elderly

In no area of interpersonal violence and suicide is the lack of interventions—evaluated or not—so striking as in elderly suicide. Despite the clear presence of elderly white males among the highest risk groups for suicide, not a single intervention directed at this population has been identified that has been systematically implemented and evaluated.

Although the literature about elderly suicide is sparse, some suggestions for possible interventions exist: identifying depression in the elderly and ensuring better treatment, allowing individuals to work for as long as they are able, and making retirement a gradual process that involves counseling. Free annual medical examinations for those 60 years of age and older would increase the frequency of their contact with health professionals. Telephone service should be provided to any older person living alone. Other proposed steps range from the need to conduct further research to the need to restructure pervasive and detrimental societal attitudes toward old age.[4]

INTERVIEW SOURCES

James A. Mercy, PhD, Chief, Intentional Injuries Section, Epidemiology Branch, Division of Injury Epidemiology and Control, CEHIC, Centers for Disease Control, Atlanta, Georgia, May 16, 1988

Ann Garland, MA, Research Associate, New York State Psychiatric Institute, Columbia University, New York, New York, March 9, 1988

REFERENCES

1. National Center for Health Statistics. Advance report of final mortality statistics, 1986. Monthly Vital Statistics Rep 1986;37(6)(suppl).

2. Davidson LE, Berman AL, Murray D, et al. Operational criteria for determining suicide. Morbid Mortal Weekly Rep 1988;37(50):773–80.

3. Rosenberg ML, Davidson LE, Smith JC, et al. Operational criteria for the determination of suicide. J Forensic Sci 1988;33:1445–56.

4. Miller M. Suicide after sixty: the final alternative. New York: Springer, 1979.

5. US Department of Health and Human Services, Public Health Service, Centers for Disease Control. Suicide surveillance report: United States, 1970–1980. Washington, DC: US Government Printing Office, 1985(Mar):1–3.

6. Baker SP, O'Neill B, Karpf RS. The injury fact book. Lexington, Massachusetts: Lexington, 1984.

7. U.S. Department of Health and Human Services, Public Health Service, Centers for Disease Control. Youth suicide in the United States, 1970–1980. Washington, DC: US Government Printing Office, 1986(Nov).

8. McGinnis JM. Suicide in America—moving up the public health agenda. Suicide Life-Threat Behav 1987;17(1):18–32.

9. MacMahon K. Short-term temporal cycles in the frequency of suicide in the United States, 1972–1978. Am J Epidemiol 1983;117(6):744–50.

10. Markush RE, Bartolucci AA. Firearms and suicides in the United States. Am J Public Health 1984;74:123–127.

11. Rosenberg ML, et al. Violence, homicide, assault, and suicide. In: Amler RW, Dull HR, eds. Closing the gap. Am J Prev Med 1987;3(suppl):164–78.

12. Rosenberg ML, Smith JC, Davidson LE, Conn JM. The emergence of youth suicide: an epidemiological analysis and public health perspective. Annu Rev Public Health 1987;8:417–40.

13. Boyd JH, Moscicki EK. Firearms and youth suicide. Am J Public Health 1986;76(10):1240–2.

14. Shaffer D, et al. Review of youth suicide prevention programs. Albany, New York: Governor's Task Force on Youth Suicide Prevention, 1987(Jan).

15. Motto JA. Suicide and suggestibility: the role of the press. Am J Psychiatry 1967;124(2):252–6.

16. Centers for Disease Control. CDC recommendations for a community plan for the prevention and containment of suicide clusters. Morbid Mortal Weekly Rep 1988;37(S-6):1.

17. Coleman L. Suicide clusters. Boston: Faber and Faber, 1987.

18. Phillips DP, Carstensen LL. Clustering of teenage suicides after television news stories about suicide. N Engl J Med 1986;325:685–9.

19. Gould MS, Shaffer D. The impact of suicide in television movies. N Engl J Med 1986;315:690–4.

20. Phillips DP, Paight DJ. The impact of televised movies about suicide. N Engl J Med 1987;317:809–11.

21. Tanner MA, Murray JA, Phillips DP, Paight DJ, Gould MS, Shaffer D. The impact of televised movies about suicide. N Engl J Med 1988;318:707–8 (correspondence).

22. National Center for Health Statistics. Advance report of final mortality statistics, 1985. Monthly Vital Statistics Rep 1987;36(5)(suppl).

23. Hlady WG, Middaugh JP. Suicides in Alaska: firearms and alcohol. Am J Public Health 1988;78(2):179–80.

24. Hawton K, Catalan J. Attempted suicide: a practical guide to its nature and management. New York: Oxford University Press, 1982.

25. Boor M. Methods of suicide and implications for suicide prevention. J Clin Psychol 1981;37(1):70–5.

26. Browning CH. Epidemiology of suicide: firearms. Compr Psychiatry 1974;15:549–53.

27. Wintemute GJ, Teret SP, Kraus JF, Wright MW. The choice of weapons in firearm suicides. Am J Public Health 1988;78(7):824–6.

28. Gerberich SG, Hays M, Mandel JS, Gibson RW, Van der Herde CJ. Analysis of suicides in adolescent and young adults: implications for prevention. In: Laaser U, Senault R, Viefhues H, eds. Primary health care in the making. Berlin: Springer-Verlag, 1985.

29. Comstock BS. Suicide in the 1970s: a second look. Suicide Life-Threat Behav 1979;9(1).

30. Eisenberg L. Adolescent suicide: on taking arms against a sea of troubles. Pediatrics 1980:315–20.

31. Clarke RV. Politically palatable prevention of suicide. Paper presented at the annual meeting of the American Society of Criminology, November 1987.

32. Kreitman N. The coal gas story: United Kingdom suicide rates, 1960–1971. Br J Prev Soc Med 1976;30:89–93.

33. Brown JH. Suicide in Britain. Arch Gen Psychiatry 1979;36:1119–24.

34. Weiner IW. The effectiveness of a suicide prevention program. Ment Hygiene 1969;53(3):357–63.

35. Glatt KM. Helpline: suicide prevention at a suicide site. Suicide Life-Threat Behav 1987;17(4):299–309.

36. Streiner DL, Adam KS. Evaluation of the effectiveness of suicide prevention programs: a methodological perspective. Suicide Life-Threat Behav 1987;17(2):93–106.

37. Miller H, et al. An analysis of the effects of suicide prevention facilities on suicide rates in the United States. Am J Public Health 1984;74(4):340–3.

38. Lester D. Effect of suicide prevention centers on suicide rates in the United States. Health Serv Rep 1974;89(1).

39. Jennings C, Barraclough BM, Moss JR. Have the Samaritans lowered the suicide rate? A controlled study. Psychol Med 1978;8:413–22.

40. Holding TA. The BBC "befrienders" series and its effects. Br J Psychol 1974;124:470–2.

41. Bagley C. The evaluation of a suicide prevention scheme by an ecological method. Soc Sci Med 1968;2:1–14.

42. Marshall MCD. The indigenous nurse: a community crisis intervener. Semin Psychiatry 1971;3(2):264–70.

43. Suicide prevention program for California public schools. Sacramento: California State Department of Education, 1987.

44. Deykin E, et al. Adolescent suicidal and self-destructive behavior. J Adolesc Health Care 1986;7(2):88–95.

45. Bergenfield mental health emergency response plan. Bergenfield, New Jersey: Borough of Bergenfield (undated).

46. Division of Injury Epidemiology and Control, Center for Environmental Health. Guidelines for community-based approach to youth suicide prevention. Atlanta, Georgia: Centers for Disease Control, 1987(Mar) (draft document).

47. Whitlow J. Group focuses on the community in mapping action against clusters. Newark Star-Ledger 1987 Nov 17:38.

48. Webb NB. Before and after suicide: a preventive outreach program for colleges. Suicide Life-Threatening Behav 1986;16(4):469–80.

Chapter 17: Firearm Injuries

A bullet from John Hinckley's $29 handgun permanently disabled Presidential Press Secretary James Brady during the March 1981 assassination attempt against Ronald Reagan. Soon thereafter, his wife, Sarah Brady, became a fervent convert to handgun control. Now, as vice chairwoman of Handgun Control, Inc., she lobbies before Congress, in state capitols, and through the media for mandatory waiting periods, background checks, and other measures to restrict the sale of handguns. Sarah Brady wants to make it difficult for convicted felons, minors, drug addicts, fugitives from justice, and persons adjudicated mentally incompetent (categories established by the 1968 federal Gun Control Act) to purchase handguns.

No one would argue that Gail Buchalter fits any of these categories. Moreover, Buchalter would agree with Sarah Brady that such people should be denied the right to purchase handguns. Nevertheless, although Brady is part of a movement to restrict the ownership of handguns, Buchalter is a part of a handgun ownership explosion—one of the more than 12 million women who own handguns, many purchased since 1983. And, as an advocate of ownership, the young Los Angeles writer's photograph was on the cover of *Parade* magazine's February 21, 1988, issue, which contained her article, "Why I Bought a Gun."

"I am typical of the women gun manufacturers have been aiming at as potential buyers," Buchalter wrote. "I decided I would defend myself, even if it meant killing another person. I realized that the one-sided pacifism I once so strongly had advocated could backfire on me and, worse, on my son. Reluctantly I concluded that I had to insure the best option for our survival. My choices: to count on a cop or to own a pistol."[1]

THE MAGNITUDE OF THE PROBLEM

Accident Facts estimates that the use of firearms resulted in almost 31,000 fatalities in the United States during 1985. This included 17,363 suicides, 11,836 homicides, and 1,649 unintentional deaths.[2] It is difficult to find authoritative figures on firearm morbidity. One researcher estimated that there are at least five nonfatal firearm injuries for every fatality.[3] In 1975, the only year for which numerical estimates of morbidity are available, there were 896,778 nonfatal firearm injuries, including 703,778 assaultive injuries, 183,000 unintentional injuries, and 10,000 suicide attempts.[4] The Bureau of Justice Statistics estimates that each year there are at least 83,000 assaults, rapes, and robberies in which a gun is involved, resulting in at least 26,000 nonfatal injuries. Victims of such injuries spend an average of 16.3 days in the hospital, which is double the length of the average stay of persons injured with other weapons.[5] One researcher calculated that $500 million a year is spent on hospital care for handgun injuries.[6]

There are also many incidents in which a firearm is discharged or used to intimidate someone without producing an injury or as the prelude to other injury-producing violence, such as a beating or a rape. Firearms were involved in 34% of all reported robberies committed during 1986 as well as in 21% of all reported aggravated assaults.[7] There are probably a great number of incidents involving firearms in which injuries might have resulted had circumstances been slightly different (e.g., if someone had been a better shot, or more intoxicated, or more frightened).

Much information critical to understanding the role of firearms in both crime and injury is unavailable. As a summary of the available research reports, "At present, we do not know the total number of privately owned firearms in the United States except to the nearest few tens of millions, and we have even less knowledge of the kinds of firearms in private hands . . . how they are used, why they are owned, or how long they last."[4] The authors of this study estimated that as of 1978 there were at least 100 million firearms in the United States, although there could have been as many as 140 million.[4] Twenty-five percent to 30% of these are handguns.

The number of guns in America has been increasing at the rate of about 10 million a decade, although there was an especially marked increase (estimated at 40 million) between 1968 and 1978.[8] One-third to one-half of this increase, which was fairly consistent proportionally across all sociodemographic groups,[4] was probably a result of the postwar population bulge.

One of the reasons it is difficult to document the

number and distribution of firearms in the United States is that many of the guns used for criminal purposes are, at some point, stolen from others. Some 275,000 handguns reach criminal hands through theft each year.[8] One study found that one-tenth of Florida's gun owners had had a handgun stolen from them. That study also revealed that "inmates who had been convicted of breaking and entering [maintained] that handguns were often the most desirable merchandise to steal other than cash, since a handgun is easier to carry unnoticed from the premises, easier to dispose of for cash, and harder to trace than most other merchandise."[4]

RISK FACTORS

Firearms play an important role in many types of injuries. These include
• interpersonal violence, including occupational injuries, especially to law enforcement personnel and persons working in retail businesses targeted by armed robbers
• suicide
• unintentional injuries, including recreational injuries, primarily hunting mishaps, and residential injuries, especially to children and adolescents playing with their parents' guns.

Interpersonal Violence

Over 59% of the homicides committed in 1986 involved firearms. Handguns were used in 44% of the homicides committed that year.[7] The group most at risk for interpersonal firearm injuries is young black males. Handguns were involved in 54.1% and long guns (rifles and shotguns) in 16.9% of homicides of young black men.[9]

Interpersonal violence is also a leading cause of occupational injuries in the United States.[10] Although the overall rate of occupational homicides may seem low (1.5 per 100,000 workers), certain occupations have a very high risk of homicide. These include law enforcement, private security, and employment in businesses frequently victimized by armed robbers (e.g., taxis, eating or drinking places, convenience stores, and liquor stores). According to one study:

> Firearms accounted for 77 percent of all work-related homicides. . . . Over 92 percent of [murdered] supervisors or proprietors (mostly of food or dairy stores, convenience stores, liquor stores, or eating and drinking places) died from gunshot wounds, as did three-fourths of police or security guards,

waiters, waitresses, or bartenders, janitors or maids, or truck/bus drivers. Sixty-one percent of murdered taxi drivers were shot.[11]

In areas of the country where gun ownership is especially high, the homicide rates for certain occupations are staggering. In Texas, for example, the homicide rate for male police officers and detectives is 25.7 per 100,000. For sheriffs and bailiffs, the rate is 44.4 per 100,000. For male cab drivers, the rate is 78.3 per 100,000. Women working in gas stations have an occupational homicide rate of 13.3 per 100,000. For women working in food, bakery, and dairy stores, this rate is 3.6 per 100,000, and for those employed in restaurants and bars, 2.6 per 100,000. Firearms accounted for 81% of these occupational homicides.[10,12] Firearms pose a particular danger to police officers; most police officers killed in the line of duty are killed with guns.[13]

Suicide

Firearms are used in more than half of all suicides. The firearm suicide rate for young people is especially alarming. According to the National Institute of Mental Health:

> Suicide by firearms is the leading method of suicide for persons aged 10 to 24 throughout 1933 to 1982, and accounts for the majority of the increase in the overall rate for this age group. The firearm suicide rate increased 139 percent, from 2.3 per 100,000 in 1933 to 5.5 per 100,000 in 1982. The suicide rate for this age group by all other methods increased only 32 percent, from 2.5 in 1933 to 3.3 in 1982.[14]

Some researchers have speculated that the presence of a firearm in the house may be a risk factor for adolescent suicide.[15]

Unintentional Injuries

According to the National Center for Health Statistics, there were 1,756 unintentional firearm injuries in 1982.[16] Research indicates that the two largest categories of unintentional firearm injuries are hunting injuries and injuries to children or adolescents playing with firearms in their homes.[4,17] Groups with low incomes have the highest unintentional firearm mortality rates. Rates are also high in rural areas.[18]

A North Carolina study found that 16% of unintentional firearm fatalities involved children playing with guns. Children under 14 years of age comprised 31% of unintentional fatalities, and adolescents suffered the highest mortality rates.[19] Other studies have revealed similar age patterns. In California, the risk of being involved in an unintentional firearm fatality as either a shooter or victim peaks in the 10–14-year-old age group.[20] Research has suggested that children involved as the shooters in unintentional shootings "are almost certainly at increased risk for acute and chronic emotional and behavioral disturbances."[20]

A BRIEF HISTORY OF PREVENTION

Many interventions aimed at reducing firearm injuries are much more controversial than interventions for other types of injuries. Although there is continuing disagreement over mandatory safety belt and motorcycle helmet laws, for example, these issues do not generate nearly as much passion as does the debate over gun control. Few facts in the debate are accepted as such by partisans on either side.

The political feasibility of many firearm interventions is also the subject of debate. Some argue that the difficulty of passing gun control measures reflects a public consensus against such measures. Others maintain that it stems from a one-issue minority represented by a sophisticated and well-funded lobby, the National Rifle Association, whose power far outweighs the number of people it actually represents. Many interventions that attempt to decrease the availability of firearms or types of firearms even to limited numbers of people are perceived as a first step toward broader gun control measures and are vehemently opposed by people seeking to protect their own right to bear arms.

There are also powerful financial interests concerned with the outcome of this debate. According to American Firearms Industry, a Florida trade group, total nonmilitary sales of firearms and ammunition in 1985 were $8.9 billion, of which handgun sales accounted for $2.5 billion (R. Campbell, personal communication).

The debate about gun control often focuses on crime. Advocates of gun control measures bolster their arguments with statistics on crimes committed by individuals using guns as well as the injuries that result from these crimes. Opponents argue that it is imperative for the American public to bear arms to defend themselves against these criminals.

Opponents of gun control also maintain that gun ownership is protected by the Second Amendment to the Constitution and that it offers protection against the imposition of a totalitarian government. They say that firearms have important legitimate

recreational uses, primarily hunting and target shooting. Gun control advocates argue that, historically, weapon ownership has had no bearing on the potential for totalitarianism and that the health consequences of firearms as well as their illicit use in crimes far outweigh the benefits of their recreational use. Federal courts, including the U.S. Supreme Court, have consistently rejected the claim that the Second Amendment guarantees any blanket protection for civilian gun ownership.[21]

Two types of interventions usually are considered in attempts to reduce firearm injuries. Educational interventions are directed primarily at unintentional firearm injuries. The other strategy, limiting the number and types of people eligible to own firearms or the types of firearms that can be owned and carried, is generally referred to as "gun control."

The history of gun control efforts is complex, encompassing federal, state, and municipal laws of various kinds that are enforced to varying degrees. As early as the 17th century, Massachusetts prohibited the carrying of certain firearms in public places. A number of states prohibited the carrying of concealed weapons at different times during the 19th century. These laws were often vague and ill-enforced.[22] Public concern over criminal violence in the 1920s and early 1930s led to the passage of the National Firearms Act of 1934. This measure was directed at the weapons favored by organized crime: machine guns, sawed-off shotguns, and silencers. Additional legislation passed in 1938 mandated the licensing of dealers and manufacturers involved in interstate firearm transactions. It required dealers to keep records and prohibited firearm sales to people convicted of certain crimes. Beginning in 1958, serial numbers were required on all guns except .22-caliber rifles.

An increase in crime and civil disorders and the assassinations of John Kennedy, Martin Luther King, Jr., and Robert Kennedy led to the federal Gun Control Act of 1968. It banned interstate firearm shipments to people not federally licensed to have them and added several categories of people, including convicted felons and people with a history of substance abuse, to the group prohibited from purchasing handguns.

The provisions of state and local firearm laws as well as their enforcement vary from jurisdiction to jurisdiction. Some laws require a permit to purchase any firearm. In other areas and in some cities a permit is required to purchase a handgun but not a long gun (i.e., a rifle or a shotgun). Twenty-three states require a waiting period and police background check before a permit is issued, although the background check is often cursory. Some states and cities require that guns (sometimes just handguns) be registered with a state agency or the local police department. Some jurisdictions prohibit the carrying of concealed weapons, while others issue permits to do so on an "as needed" basis. The guidelines for such permits often allow great latitude to the licensing authority, which is usually the local police. At least 28 states have some sort of sentence enhancement for crimes committed using guns (i.e., an additional sentence beyond what would have been imposed for the commission of the same crime without a gun).[23]

The literature on the effects of interventions designed to reduce the availability or misuse of firearms is controversial and inconclusive. The debate, both public and academic, has centered more on crime control than on injury prevention. The number of handguns, especially unregistered handguns, already in the hands of the public and the fact that these guns are both portable and concealable make the implementation of most local or state gun control measures difficult.

Any consideration of interventions that seek to limit, regulate, or even document the possession of firearms will inevitably raise legal, ethical, and political questions. A program that seeks to implement any of these interventions should consider the intensity of this political controversy and its potential impact.

The involvement of firearms in injuries has not gone unnoticed by the public health community. *Promoting Health, Preventing Disease: Objectives for the Nation* called for a 25% reduction in the number of privately owned handguns by 1990.[24] The surgeon general's *Workshop on Violence and Public Health* in 1985 recommended "a complete and universal federal ban on the manufacture, importation, sale, and possession of handguns (except for authorized police and military personnel). . . ."[25] The same year, an editorial in the *American Journal of Public Health* called for an "attack" on the problem of handgun injuries.[26] The American Academy of Pediatrics, in the 1987 edition of *Injury Control for Children and Youth*, stated that, "as leaders in preventive medicine, pediatricians should recognize that gun violence is a critical public health issue as well as a criminal justice problem."[27] A recent editorial in the *New England Journal of Medicine* stated that "injury from firearms is a public health problem whose toll is unacceptable. The time has come for us to address this problem in the manner in which we have addressed and dealt successfully with other threats to public health."[28]

INTERVENTIONS

Research

The need for information on which to base firearms policy is critical. As one reviewer pointed out:

> The published literature is more noteworthy for what it does *not* show than for what it does. There is, it appears, scarcely a single finding in the literature that could be said to have been indisputably established. In part, this reflects the highly politicized nature of research in this area, but perhaps more importantly, it results from a near-total absence of sound and nationally generalizable data from which reliable information about weapons, crime, and violence might be extracted.[8]

Priorities for research in the area of firearm injuries include the following topics:
- the magnitude, characteristics, and costs of the morbidity and disability caused by firearms and the types of firearms that inflict these injuries
- the number, type, and distribution of firearms in the United States
- epidemiologic investigations focusing on quantifying the risks of injuries associated with the possession of firearms and factors that may modify the risks
- evaluations of regulations and other interventions that affect the risk of firearm injury.[28]

Recommendation: Research should be conducted to clarify the relationship among various types of firearms and injuries, and the impact of various interventions on firearm injuries should be evaluated.

Prohibiting the Purchase or Possession of Handguns

Restrictive licensing. Restrictive licensing prohibits the possession of handguns by any but those with a clearly demonstrated need. Under restrictive licensing, an individual wishing to purchase a firearm must apply for a license, which is granted only after an investigation by a law enforcement agency to ascertain whether the person should be prohibited from possessing a firearm by virtue of a previous felony conviction, an adjudication of mental incompetence, or a history of substance abuse. The applicant also must demonstrate a need to own such a weapon. For long guns, this need includes target shooting and hunting. For handguns, this need includes employment by a law enforcement or security organization or the routine carrying of large quantities of cash or valuables for business purposes. The applicant also must be certified in competence in the safe handling and use of firearms by a designated government agency or private organization.

A recent comparison of homicide rates and gun laws in two very similar cities—Vancouver, British Columbia, and Seattle, Washington—concluded that Vancouver's significantly lower homicide rate may be a consequence of the city's restrictive licensing policy.[29] Another study concluded that the implementation of a restrictive handgun law may have contributed to a reduction of homicides among acquaintances in Washington, D.C.[30]

Permissive licensing with waiting periods and background checks. Most state and federal laws that attempt to keep firearms out of the hands of individuals who may use them in injurious ways are permissive laws under which licenses for handguns are routinely issued unless a reason not to do so is found.[31] Federal law prohibits the owning of firearms by persons with prior felony convictions, a prior involuntary commitment to a mental institution, or a history of substance abuse or who are currently under indictment for a felony or are fugitives from justice. Federal law also prohibits the sale of handguns to those under 21 years of age and of any firearm to those under 18 years of age.

It is difficult to ascertain who has been involuntarily committed to a mental institution or who has a history of subtance abuse although data on felony indictments, convictions, and fugitives from justice are available from both federal and state criminal justice system files. A waiting period before any firearm could be purchased would allow the local police to check these sources. Because recidivism is high among criminals and homicides are sometimes preceded by a record of assaultive behavior by the perpetrator, background checks may prove effective at keeping firearms out of the hands of those most likely to use them for criminal purposes.[32]

Research on a relatively permissive licensing law in South Carolina revealed that in the three years after the law took effect there was a 28% decrease in homicides while nationally during the same period there was only a 10.2% decrease. Handgun homicides accounted for 94% of South Carolina's reduction.[6] Other studies also indicate that permissive licensing systems reduce the rate of handgun injuries, although the extent of the reduction varies.[33]

Permissive licensing laws appear to be politically acceptable. Sixty-six percent of the nation's population is currently subject to background checks before or after purchasing a handgun.[4] Many jurisdictions have more restrictive laws.[34]

Because state criminal justice files vary in the percentage of felony convictions and indictments included, improving these data sources would be an important component in improving the accuracy of firearm checks. The FBI's capability of handling requests from local police departments for such checks would also have to be expanded if their files were to be used for such background checks.[33]

Recommendation: States should consider firearm licensing alternatives such as restrictive licensing and permissive licensing with waiting periods and background checks. All such efforts should be monitored and their outcomes carefully evaluated.

Selective prohibition on the carrying of firearms. Prohibiting the carrying of firearms in high-risk areas might reduce the injuries resulting from disputes, robberies, and other incidents. Such a prohibition would give the police and courts greater power over people who carry guns for criminal purposes and would limit the likelihood that these weapons will be used in crimes of opportunity. The number of existing laws of this nature indicates that their implementation should not present insurmountable political problems.[34]

Recommendation: Municipalities and states should consider prohibiting the carrying of firearms except by law enforcement and security personnel in areas where people are at high risk for firearm injuries. All such efforts should be monitored and their outcomes carefully evaluated.

Enforcement of Existing Firearms Laws

There are many local, state, and federal laws relevant to the theft, sale, transfer, carrying, and use of both handguns and long guns. These include laws pertaining to the licensing of weapons owners, the licensing of weapons dealers, the carrying of concealed weapons, firearm registration, and the use of firearms against others. Enforcement of these laws should be made a priority, much as is currently being done with drug laws. Such laws should be evaluated for their effect on firearm injury rates.[34-37]

Recommendation: Priority should be given to strictly enforcing current firearm laws at the federal, state, and local levels. All such efforts should be monitored and their outcomes carefully evaluated.

Restrictions on Handguns

Handguns are involved in a large percentage of both intentional and unintentional firearm injuries. Forty-four percent of all homicides are committed with handguns.[7] Among black males 15–24 years of age (the group most at risk for homicide), over 71% of homicide victims die from gunshot wounds and over 76% of the guns used are handguns.[9] Research also shows that a substantial percentage of unintentional firearm deaths and injuries (including those to children) as well as suicides involve handguns.[19,20,38]

Concealability is a major factor in a criminal's choice of gun.[8] The very ease with which handguns can be concealed, carried, and transported, even in bulk, complicates efforts to regulate their possession and limits the effectiveness of jurisdiction-specific gun control measures.[8] Most of the illegal handguns confiscated in New York City, for instance, originate in states with less restrictive gun control statutes.[6] Handguns, and especially illicitly possessed handguns, often cross state lines.[4]

A promising approach to the reduction of firearm injuries may be the selective targeting of several of the previously discussed interventions specifically to handguns. Although rifles and shotguns may be substituted for handguns in some cases, it can reasonably be speculated that the absence of an available lethal weapon may prevent or reduce the severity of the injuries produced in the types of incidents now characteristically associated with handguns.

Recommendation: Local, state, and federal initiatives to restrict the manufacture, sale, possession, and carrying of handguns should be supported and evaluated for their impact on firearms injuries.

Firearm Modifications

A number of product modifications might affect the rate of unintentional injuries. These include designing firearms so that it can easily be determined whether or not they are loaded and designing safety catches "so that they are automatically and always engaged unless held in a disengaged position by the users."[20] Other possible modifications would include manufacturing guns (and especially handguns) that required sufficient trigger pressure to make their firing by a child impossible or difficult.

Recommendation: Research should be conducted on possible product modifications to reduce the potential of unintentional firearm injuries.

Educational Interventions

Firearm safety courses seek to teach individuals how to safely handle, use, maintain, and store firearms. The effects of these courses on firearm injury rates is unknown. An important research question is whether the safety benefits of such courses are outweighed by their ability to promote an interest in firearms, an interest which increases the number of firearms in circulation and the potential for both intentional and unintentional injuries.

One study found that more than half of the Texas families who kept guns at home kept them loaded at all times. Half of these loaded guns were not kept locked away from children. In fact, 10% of the respondents in a study of home firearm storage in Texas reported that their guns were loaded, not locked up, and within reach of children.[39]

The importance of firearm storage becomes obvious when the circumstances surrounding unintentional firearms injuries are examined. The most common scenario involves children playing with a gun.[20] Sixty-seven percent of the victims in a survey of self-inflicted gunshot wounds admitted that they were just "fooling around" with a gun when it went off.[19] Another study found that "the most common case history was of children playing with a gun that had been stored loaded, unlocked, and out of view; the shooting often occurred in the room where the gun was stored."[20]

Recommendation: Little is known about the effects of firearm safety courses and other educational interventions designed to promote the safest possible handling, use, and storage of firearms. Such programs should be monitored and carefully evaluated.

INTERVIEW SOURCES

Robert Campbell, American Firearms Industry Association, Fort Lauderdale, Florida, April 13, 1988

REFERENCES

1. Buchalter G. Why I bought a gun. Parade 1988 Feb 21:5.

2. Accident facts: 1988 edition. Washington, DC: National Safety Council, 1988:13.

3. Jagger J, Dietz P. Deaths and injury by firearms: who cares? JAMA 1986;255:314.

4. Wright J, Rossi P, Daly K. Under the gun: weapons, crime, and violence in America. New York: Aldine, 1983.

5. The use of weapons in committing crimes: special report. Washington, DC: Bureau of Justice Statistics, US Department of Justice, 1986.

6. Fields S. Handgun prohibition and social necessity. In: Katsh ME, ed. Taking sides: clashing views on a controversial legal issue. Guilford, Connecticut: Dushkin, 1983.

7. Crime in the United States: uniform crime reports,

1986. Washington, DC: Federal Bureau of Investigation, US Department of Justice, 1987.

8. Wright J, Rossi P. Weapons, crime, and violence in America: executive summary. Washington, DC: National Institute of Justice, US Department of Justice, 1981.

9. Centers for Disease Control. Homicide among young black males: United States, 1970–1982. Morbid Mortal Weekly Rep 1985;34(41):629–33.

10. Davis H. Workplace homicides of Texas males. Am J Public Health 1987;77(10):1290–3.

11. Kraus JF. Homicide while at work: persons, industries, and occupations at high risk. Am J Public Health 1987;77(10):1285–9.

12. Davis H, Honchar P, Suarez L. Fatal occupational injuries of women: Texas, 1975–84. Am J Public Health 1987;77(12):1524–7.

13. Lester D. Gun control: issues and answers. Springfield, Illinois: Charles C. Thomas, 1984:66.

14. Boyd JH, Moscicki EK. Firearms and youth suicide. Am J Public Health 1986;76(10):1240–2.

15. Brent D, et al. Risk factors for adolescent suicide: a comparison of adolescent suicide victims with suicidal inpatients. Arch Gen Psychiatry 1988;45:581–8.

16. Committee on Trauma Research, Commission on Life Sciences, National Research Council, Institute of Medicine. Injury in America. San Francisco: National Academy Press, 1985:23.

17. Cole J, Patetta M. Hunting firearm injuries—North Carolina. Am J Public Health 1978;78(12):1585–6.

18. Baker S, O'Neill B, Karpf R. The injury fact book. Lexington, Massachusetts: Lexington, 1984.

19. Morrow PL, Hudson P. Accidental firearm fatalities in North Carolina, 1976–80. Am J Public Health 1986; 76(9):1120–3.

20. Wintemute G, et al. When children shoot children: 88 unintentional deaths in California. JAMA 1987;257(22): 3107–9.

21. Christoffel T. The misuse of law to hinder injury prevention: lessons for advocates. Paper presented at the 1988 annual meeting of the American Public Health Association, Boston, Massachusetts, November 1988.

22. Newton G, Zimring F. Firearms and violence in American life: a staff report submitted to the National Commission on the Causes and Prevention of Violence. Washington, DC: National Commission on the Causes and Prevention of Violence, 1969:87.

23. Wright J, Rossi P, Daly K. Under the gun: weapons, crime, and violence in America. New York: Aldine, 1983.

24. Promoting health, preventing disease: objectives for the nation. Washington, DC: US Department of Health and Human Services, 1980:85.

25. Surgeon general's workshop on violence and public health: report. Washington, DC: US Department of Health and Human Services, 1986:53.

26. Baker S. Without guns, do people kill people? Am J Public Health 1985;75(6):587–8.

27. Committee on Accident and Poison Prevention. Injury control for children and youth. Elk Grove Village, Illinois: American Academy of Pediatrics, 1987:136.

28. Mercy J, Houk V. Firearm injuries: a call for science. N Engl J Med 1988;319(19):1283–4.

29. Sloan J, et al. Handgun regulations, crime, assaults, and homicide: a tale of two cities. N Engl J Med 1988; 319(19):1256–62.

30. Jones E. The District of Columbia's "Firearms Control Regulation Act of 1975": the toughest handgun control law in the United States—or is it? Ann AAPSS 1981;455: 138–49.

31. Zimring F, Hawkins G. The citizen's guide to gun control. New York: Macmillan, 1987.

32. Kleck G, Bordura D. The assumptions of gun control (chapter 2). In: Kates D, ed. Firearms and violence: issues of public policy. Cambridge, Massachusetts: Ballinger, 1984.

33. Cook P, Bloise J. State programs for screening handgun buyers. Ann AAPSS 1981;455:80–91.

34. Moore M. The bird in hand: a feasible strategy for gun control. J Pol Anal Mgmt 1983;2(2):190–1.

35. Moore M. Controlling criminogenic commodities: drugs, guns, and alcohol. In: Wilson JQ, ed. Crime and public policy. San Francisco: Institute for Contemporary Studies, 1982.

36. Moore M. The police and weapons offenses. Ann AAPSS 1980;542:22–32.

37. Moore M. Keeping handguns from criminal offenders. Ann AAPSS 1981;455:92–109.

38. Gerberich S, Hays M, Mandel J, Gibson R, Van der Heide C. Analysis of suicides in adolescents and young adults: implications for prevention. In: Laaser U, Senault R, Viefhues H, eds. Primary health care in the making. Berlin: Springer-Verlag, 1985.

39. Patterson P, Smith C. Firearms in the home and child safety. Am J Dis Child 1987;141:221–3.

Part 3: Tertiary Prevention

Chapter 18: Trauma Care Systems

You think people die from heart attacks or accidents, but they really don't. Not directly. Those things produce shock, which is sluggish or nonexistent circulation, and that's what kills you . . . if you stay in shock for very long, you're dead. Maybe you'll die in ten minutes or maybe you'll die next week, but you're dead. So, if you're in shock we have to work fast. You've got, at most, 60 minutes. If I can get to you, and stop your bleeding, and restore your blood pressure, within an hour of your accident . . . then I can probably save you. I call that the golden hour.

—Dr. R. Adams Cowley[1]

The recognition that time is a critical factor in determining whether an individual can survive major trauma is an old one. Attempts to decrease the time from wound to treatment on the battlefield have marked military medicine since Baron Larrey's lightweight, 18th-century *ambulances volantes* (flying ambulances) evacuated Napoleon's wounded soldiers to the rear for care.[2]

During World War I, it took from 12 to 18 hours before an injured soldier received surgical care. In World War II, the time was cut to 6–12 hours; in Korea, 2–4 hours lapsed before surgery. In Vietnam, the concentration and organization of medical personnel and the abundance and flexibility of helicopters for evacuation meant that "a combat casualty characteristically [underwent] definitive surgical care within one and one-half to two hours following injury."[3] As the delay decreased, death rates dropped. Eight percent of soldiers evacuated to aid facilities during World War I died. During World War II, the percentage dropped to 4.5%. It was 2.5% in Korea and less than 2% in Vietnam.[4]

Yet the application to civilian medicine of these principles of systematic trauma care—the rapid evacuation of the severely injured to well-staffed and well-equipped facilities—lagged. "Wounded in the remote jungle or rice paddy of Vietnam," wrote one specialist in 1967, "an American citizen has a better chance for quick, definitive surgical care by board certified specialists than were he hit on the highway near his home town in the continental United States."[3] More than a decade later, the federal government's chief emergency medical care specialist could still lament, "It is somewhat bewildering and the cause for much concern that, over the past decade, the lessons learned so dearly in the

military conflicts over the last half century have not been more effectively incorporated in the peacetime civilian environment."[5]

Great progress has been made in improving the provision of emergency medical services through the efforts of the federal government, especially the National Highway Traffic Safety Administration (NHTSA) and the states.[6] Nevertheless, this chapter is about the still-unfinished business of developing trauma care systems to apply fully the lessons of military medicine to saving the lives of the severely injured. It also examines the relationship of trauma care to the systems approach to injury prevention.

DEFINITIONS

Until now, we have been concerned almost exclusively with primary and secondary prevention. Generally speaking, primary prevention seeks to forestall events (e.g., auto crashes) that might result in injuries; secondary prevention is directed at modifying the consequences of such events to either prevent or reduce the severity of an injury (e.g., the air bag that inflates upon impact). Acute medical care and rehabilitation—also called tertiary prevention—are directed at "the return of a functioning patient to society" after an injury.[7] The Haddon Matrix on page 8 includes other examples of primary (precrash phase), secondary (crash phase), and tertiary (postcrash phase) prevention measures.

Although the terms "emergency medical services systems" (EMS) and "trauma care systems" are somtimes confused, they define related but different entities. The Emergency Medical Services Systems Act of 1973, the basic federal legislation in the field, defines EMS as "a system which provides for the arrangement of personnel, facilities, and equipment for the effective and coordinated delivery in an appropriate geographical area of health care services under emergency conditions (occurring either as a result of the patient's condition or of natural disasters or similar situations)."

A trauma care system is defined by NHTSA as "a system of health care delivery [that combines] prehospital EMS resources and hospital resources to optimize the care, and, therefore, the outcome of traumatically injured patients."[8] This includes the medical and psychosocial rehabilitation care necessary to return patients to the highest possible functional level.[9,10]

Trauma care systems are, therefore, one of several elements included in EMS systems. The existence of a well-organized and effective EMS system is critical to the development of systematic care for the severely injured. The word "system" is the key element in both definitions. Emergency care and trauma care have existed as medical specialties for hundreds of years, yet whether a victim of severe injury reached an appropriately equipped and staffed hospital was all too often randomly determined. In 1966, in *Accidental Death and Disability: The Neglected Disease of Modern Society*,[11] the National Academy of Sciences reported that half of the country's ambulances were operated by morticians and staffed by poorly trained attendants. There was no direct communication between the ambulance and hospitals or fire and police departments. In addition, ambulance service providers had no knowledge of the level of emergency care available at hospitals in their areas. In short, the components of an emergency care system were woefully inadequate, and there was little or no coordination among them. It is such conditions and the preventable deaths that resulted that EMS and trauma care systems are designed to address.[12] Before considering the nature and development of these systems and their role in reducing injury-related death and disability, a word is in order about their place in the broad, systematic approach to injury prevention outlined in this book.

INJURY PRACTITIONERS AND TERTIARY PREVENTION

There are several important reasons that injury practitioners, including those whose sole focus is primary prevention, should understand the role of and work with specialists in trauma care and rehabilitation. It has been emphasized throughout this book that to be effective injury prevention requires a systems approach. As emphasized in Chapter 1, to identify a community's injury problem and the resources with which it can be addressed, it is necessary to work with all the individuals and institutions that have knowledge of the problem. This clearly includes health professionals, especially those involved in emergency care and rehabilitation.

The importance of expertise in trauma care and rehabilitation goes beyond the specific roles that these specialities play in an injury prevention plan. Tertiary prevention systems provide ongoing surveillance of an injury problem and may be able to

help determine the effectiveness of injury countermeasures as well. Trauma care and rehabilitation are concerned, after all, with the failures of primary and secondary prevention. The very success of trauma care systems can shift the burden of injury, as more severely injured persons survive to require prolonged and expensive acute care and rehabilitation.[10] That, too, must be factored into the development of a comprehensive approach to injury.

Because the role of trauma and rehabilitation specialists goes beyond "picking up the pieces," there are many opportunities for collaboration across the spectrum of injury prevention. Injury prevention is one of the specific components around which trauma care systems are organized,[13] and this chapter examines some examples in which emergency medicine specialists have played key roles in primary prevention efforts. Closer collaboration is possible, however, and there are several possibilities for action.

Finally, the continuing growth of trauma registries holds promise for injury-prevention researchers. Exploring ways to link trauma registries, which were developed primarily as tools to evaluate the quality of medical care, to other injury data bases is an important task. So, too, are continuing efforts to use information about injury severity and outcome to advance the understanding of the effects of injury and the opportunities for prevention.

TRAUMA CARE AND EMERGENCY MEDICAL SERVICES

Trauma deaths have a trimodal distribution. Approximately half of all deaths ("immediate" deaths) occur within minutes of the injury. These individuals, with lacerated brains, disrupted spinal cords, or ruptured hearts, could not be saved by even the most effective trauma care system. "Early" deaths occur within 2 hours of injury and account for another 30% of trauma deaths. "Late" deaths, the remaining 20%, occur two to three weeks after injury and usually result from sepsis or multiple organ failure syndrome.[14,15]

Since the late 1950s, nearly 20 studies have tried to determine how many such deaths were preventable and under what circumstances. Most have demonstrated that approximately 20% of trauma deaths are preventable[16] and that "the most common causes of preventable deaths are a delay to definitive surgical intervention and the lack of performance of indicated surgery."[15] It is these causes of preventable death that trauma care systems were designed to address.

A systematic approach to trauma care delivery begins at the scene when an emergency medical technician or paramedic determines the nature and severity of the victim's injuries. . . . [The] patient's vital signs and symptoms . . . determine the level of hospital care needed to treat the injuries. . . . Trauma is a surgical disease, with [many] cases requiring immediate surgery. Therefore, severe trauma victims require fast transport with treatment provided en route to a predetermined hospital emergency unit specializing in trauma care, even if other hospitals with lower levels of care are bypassed.[5]

As noted above, trauma care systems are one of several elements of EMS systems. Within the overall caseload of EMS systems, injuries account for approximately 40% of the cases. Major trauma (i.e., immediately life-threatening injury) is involved in 10% (R. H. Cales, personal communication). The vast majority of EMS cases involve cardiac, behavioral, perinatal, or pediatric emergencies (Figure 1).[5,12]

Trauma care systems involve four primary, clinical components: access to care, prehospital care, hospital care, and rehabilitation.[7] According to the most recent standards developed by the Committee on Trauma of the American College of Surgeons (ACS) and those of the American College of Emergency Physicians (ACEP), the four components can be described as follows.

Access to Care

This includes both technological elements, such as a 911 emergency telephone number, radio transmitters, central dispatch, the identification of special resources, and a knowledge among potential users of how to gain access to the system. Studies support the value of "a single coordinated system

that accesses all ambulance service providers in the area through the commonly known 911 emergency telephone number." Nonetheless, the General Accounting Office reported, "many areas find this difficult to accomplish, due to fragmentation among both service providers and local governments . . . as well as the high initial cost of installing central telephone and dispatch equipment."[6]

Prehospital Care

Ambulance, helicopters and fixed-wing aircraft, and the personnel trained to provide initial resuscitation and triage are the elements of prehospital care. "Those who advocate regionalized trauma care," wrote one specialist, "also emphasize the importance of prehospital care. . . . Early intervention by trained individuals can extend life and reduce mortality."[17]

The quality of trauma care, including prehospital care, is the ethical and legal responsibility of the medical profession. However, the evolution of prehospital care systems has required new mechanisms through which accountability can be maintained. "The medical profession has responded to the need for accountability for the field care provided by emergency medical technicians (EMTs) and nurses by developing prehospital training programs, trauma treatment protocols, radio communication networks, and system monitoring procedures."[18]

Under this medical direction, EMTs at several levels of training, including specialized trauma training, provide prehospital care for the vast majority of seriously injured Americans. Under the federal Highway Safety Act of 1966, the Department of Transportation is responsible for developing standards for emergency prehospital care for the acutely ill and injured. NHTSA has published curricula to train EMTs at the EMT-Ambulance (EMT-A), EMT-Intermediate (EMT-I), and the most advanced EMT-Paramedic levels.[19]

Triage. "Although 3.7 million persons are hospitalized annually for injuries, only 1 individual per 1,000 population, or a total of 250,000 (7 percent) of hospitalized injury patients require treatment by a trauma care system."[20] Triage, the classification of patients according to medical need and the matching of those patients with available care resources, is thus an essential part of prehospital care.[7] Triage can "result in life or death for the trauma victim."[20] However, triage poses enormous challenges for even the best-trained clinician. At the scene of a severe injury—a late-night traffic collision, for example—both time and information are short.

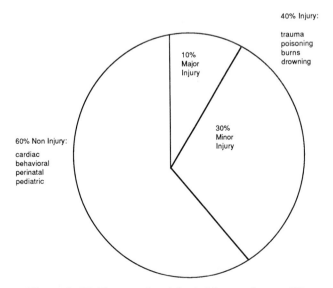

Figure 1. EMS case mix. Adapted from reference 12.

40% Injury:

trauma
poisoning
burns
drowning

10%
Major
Injury

30%
Minor
Injury

60% Non Injury:
cardiac
behavioral
perinatal
pediatric

Injury scoring systems. Generally, information is available at the scene about the patient's physiological status, the nature and probable severity of any obvious (i.e., not internal) anatomic injury, and the mechanism of injury. Several attempts have been made over the years to adapt physiologic scoring systems to facilitate triage decisions in the field.[21] (Anatomic scoring systems, such as the Injury Severity Score, which combines ratings of a patient's various injuries into a single score of severity, are used retrospectively and, in combination with physiologic measures, are an important injury research tool.[22]) For example, the Glasgow Coma Scale is a highly regarded method of combining several variables (e.g., eye opening, verbal response, and motor response) that relate to central nervous system functioning.[23] The Trauma Score, developed in 1981, combines the Glasgow Coma Score with cardiovascular and respiratory variables,[24] "thereby combining information on the three organ systems vital to survival."[20]

Although these and other scoring systems have achieved some success in correlating injury severity with outcome, none has successfully combined the practicality and reliability required for use in the field.[20,21] The American College of Surgeons (ACS) recommends, therefore, that the field determination of injury severity be based not on a single, unified scoring system but on a combination of easily obtained physiologic measurements (including the Glasgow Coma Score), the obvious anatomic characteristics of the wound(s), information about the mechanism of injury, and factors such as the patient's age and any known history of cardiac or respiratory disease.[7] The steps embodied in the ACS Triage Decision Scheme are shown in Figure 2.

Hospital Care

The essence of a trauma care system is the understanding that seriously injured patients survive in greater numbers if they are taken rapidly to specialized facilities rather than to the nearest available hospital. The designated trauma center is distinguished by the immediate availability of specialized surgeons, physician specialists, anesthesiologists, nurses, and resuscitation and life support equipment on a 24-hour basis. Three levels of trauma centers are defined on the basis of capability. A level III center can receive, resuscitate, and stabilize a patient while arranging for transfer to a hospital that can provide definitive surgical care. Level III centers are usually found in smaller community hospitals and in rural areas. A level II center can resuscitate the patient, perform emergency surgery, treat the patient in an intensive care unit, and pro-

vide rehabilitation. A level I center performs the same functions as a level II facility but has surgeons and anesthesiologists on-site 24 hours a day. In addition, level I centers usually include specialized facilities such as burn units and spinal cord injury units and conduct trauma education and research activities.[15] And because most level I facilities are in teaching institutions, they are able to use medical residents to meet trauma staffing requirements.[8]

In developing its trauma care system, planners in the state of Oregon were faced with the need to provide services not only in urban and rural areas, but in sparsely populated, "frontier" regions as well. There, a 3- to 4-hour trip from an emergency room to a tertiary care facility is not uncommon. To address this problem, the planners designated level IV centers, which provide resuscitation and stabilization of severly injured patients until they can be transferred to another trauma system hospital.[8] "There are many states with similar problems," commented an Oregon health official. "In these states a trauma care system can only be called comprehensive if it addresses urban, rural, and frontier needs" (K. Gebbie, personal communication).

Each center must maintain its expertise by treating a minimum number of patients (600–1,000 for a level I hospital or 350–600 for a level II hospital) each year. And each surgeon should treat at least 50 trauma patients annually to remain proficient.[7,25]

Rehabilitation

These services seek to return the trauma victim to the fullest physical, psychological, social, vocational, avocational, and educational level of functioning of which he or she is capable, consistent with physiologic or anatomic impairments and environmental limitations.[9] Rehabilitation services have assumed even greater importance within trauma care systems because of "the increasing survival rate for seriously injured patients [and] because of the patients' ages, which average in the mid-20s."[10]

ADMINISTRATION OF TRAUMA CARE SYSTEMS

In addition to these clinical components, trauma care systems are divided into a number of administrative components, including medical direction, communications, training, public participation, medical evaluation, and others.[25] In general, the adminstrative components of a trauma care system are the same as those into which emergency care systems are divided. This is helpful in developing trauma care systems because the necessary compo-

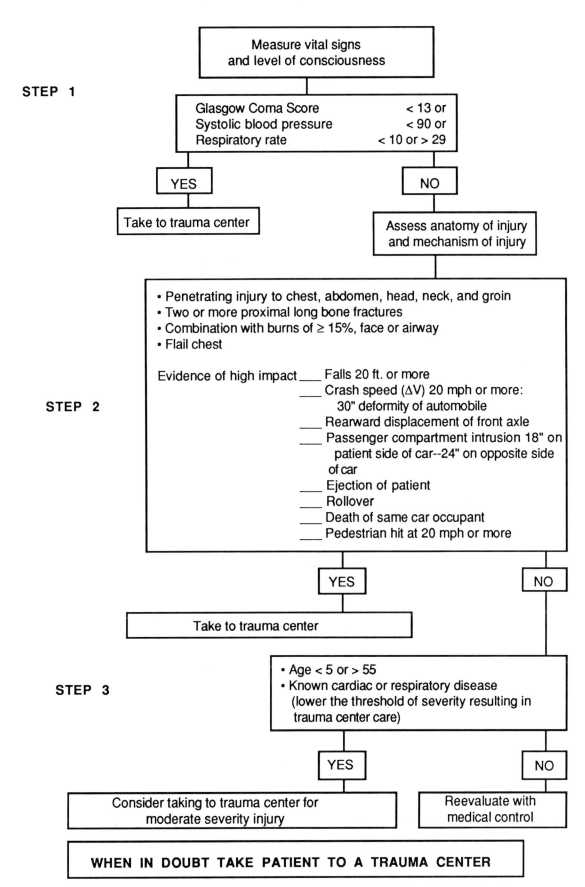

Figure 2. Triage decision scheme. Reprinted from Trauma Care Systems: A Guide to Planning, Implementation, Operation and Evaluation by R. H. Cales and R. W. Heilig, Jr., p. 102, by permission of Aspen Publishers, Inc., © 1988.

nents already exist to some degree in most EMS systems. One important difference, however, is that prevention is treated as an explicit component of trauma systems, although this is not true of EMS systems generally.[13]

The efficacy of trauma care systems in reducing injury-related mortality and morbidity has been demonstrated in a number of studies.[26] The first attempt to compare trauma care in two communities was made in 1979 by West et al.[27] In Orange County, California, where trauma victims were routinely taken to the nearest available hospital, 39 of 90 deaths could have been prevented had a specialized trauma facility been available. In San Francisco, by contrast, the use of a regional trauma center reduced preventable deaths to one in 92. Based on these findings, the authors became the first to suggest that trauma care systems could save lives. Following the implementation of a trauma care system in Orange County, preventable deaths decreased by 50%.[28]

In Washington, D.C., the implementation of a trauma care system has been credited with reducing trauma deaths by 50% over 5 years.[29] And very favorable results also have been reported in studies of the San Diego, California,[30] and Dade County, Florida,[16] trauma systems.

Unfortunately, however, far too few communities are served by comprehensive trauma care systems.[31] To understand why, it is necessary to look briefly at the history of EMS systems in the United States and the factors that inhibit trauma system development.

THE GROWTH OF EMERGENCY MEDICAL SERVICES

The increasing carnage on America's highways in the 1960s and the growing constituency for traffic safety described in Chapter 6 fueled the movement for improved emergency medical services. In his 1966 state of the union address, President Lyndon Johnson announced his intention to submit a Traffic Safety Act to Congress. Nine months later, the bill creating the National Highway Safety Bureau became law. "In the history of medical and social science there are few, if any, instances of a complex and novel idea that has made its way so quickly from the mimeograph papers of academic meetings and the journals of learned societies to the statute books of the nation. . . . Rarely have Americans witnessed a more dramatic instance of responsive and responsible government in action," commented one expert.[5]

The act required states to develop highway safety programs in accordance with specific standards,

one of which, Standard 11, covered emergency medical services. "Complementing state standard-setting and regulating efforts, states were also authorized to use highway safety grant funds to improve EMS equipment and personnel training at the local service provider level."[6] Since 1968, states have used more than $137 million in highway grant funds for EMS programs. In addition, NHTSA has been a major source of technical assistance for state EMS systems.[5]

The highway safety EMS effort vastly improved ambulance and other prehospital services. At the same time, a parallel effort was under way, through the Department of Health, Education, and Welfare (DHEW, later Health and Human Services) to improve available medical services and, most important, to organize the provision of emergency medical care into comprehensive, integrated systems. DHEW funded demonstration projects in several regions. A major initiative from the Robert Wood Johnson Foundation provided competitive funding for demonstration projects in 32 states and Puerto Rico. These efforts to encourage a regional, comprehensive approach to EMS were enshrined as national policy with the passage of the Emergency Medical Services Systems (EMSS) Act of 1973.

The EMSS Act provided, for the first time, a mechanism and funding with which communities could develop regional emergency medical care. The law emphasized, indeed mandated, the systems approach, including the creation of the Division of Emergency Medical Services within DHEW as the federal lead agency. The law also specified the components that any EMS system must include.

Some $30 million in federal grants was made available annually from 1974 to 1981 through the EMSS Act. The country was divided into 303 regions, each of which was eligible to receive funding for up to 5 years, after which the programs were to become self-funding.[6]

The Omnibus Budget Reconciliation Act of 1981 dramatically altered the way in which EMS and many other government programs were funded. Twenty-five categorical health and human services programs were folded into seven block grants. EMS became part of the Preventive Health and Health Services Block Grant. Within the established ceiling for each grant, states are free to set their own priorities. The result, according to the General Accounting Office, has been a significant decrease in funding for EMS (more than a third in some states). And while some states have reversed the downward trend recently, "Total funding has not returned to 1981 levels."[6]

Despite the changes in funding, EMS specialists are hopeful about the continued growth and improvement of EMS systems. According to EMS pioneer Dr. David R. Boyd, "Over the past decade, emergency medical services systems have evolved from a few isolated experiments to a national program of high visibility and acceptance. . . . A systems approach to their delivery is now well appreciated by those who provide the services and by the consumer public as well."[5]

THE SLOWER GROWTH OF TRAUMA CARE SYSTEMS

It must be reemphasized that trauma care systems are only one component of EMS. Unfortunately, the growth in EMS systems described above has not extended to trauma care. A 1986 General Accounting Office (GAO) investigation of EMS and trauma care in six states and 18 local areas within those states pointed to a key distinction.

> An effective EMS system routes the critically ill or injured to specialized medical facilities skilled in treating such cases as major trauma and acute cardiac problems. Appropriate care for cardiac victims is widely available in most hospitals. However, the availability of trauma care is limited to specialized facilities, known as trauma centers, and many local systems have not developed procedures to assure that trauma victims are taken there. As a result, these areas may not be providing trauma patients with the best care available.[6]

The GAO carried out its investigation in 1985 and early 1986. A more recent survey concluded that only two states, Maryland and Virginia, had developed all of the essential elements of a regional trauma system. Twenty-two states had no system at all; 19 had incomplete systems. Twenty-nine states had not yet even begun the process of designating trauma centers. "It is clear that in the 22 years since the [National Academy's 1966 report] progress in implementing regional trauma systems has been slow," commented the authors. It was not, they concluded, due to a lack of direction, but rather because "what is missing is a practical strategy for the implementation and management of such a system."[31] And, as both they and the GAO report and other commentators make clear, the impediments to developing that strategy are both economic and political. Before discussing those impediments, however, it is necessary to understand the process of trauma care system development.

DEVELOPING A TRAUMA CARE SYSTEM

"We were [fortunate] to have a number of military surgeons in San Diego who were aware of what the standard of care should be for trauma victims. The media got interested and that really started the ball rolling. It is difficult to educate politicians unless they begin to see it on television."[30] In 1982, growing public concern in San Diego about the quality of trauma care led the County Hospital Council to commission an outside study to assess the need for a trauma care system.[32] The study revealed that nearly 50% of patients were receiving less than optimal care and that 21.2% of the deaths were preventable.[33] In August 1984 trauma care was regionalized in San Diego County, and an evaluation program was put in place. A medical audit committee was created to monitor and evaluate medical care; it included surgeons, physicians, and nurses from the system's six trauma centers, anesthesiologists and other medical specialists, the president of the county medical society, the chief pathologist and deputy coroner, and key EMS staff members. A system advisory committee added representatives from the fire chiefs' and emergency nurses' associations, among others.[34] "Regionalization of trauma care," reported an evaluation of the new system, "significantly reduced delays, inadequate care, and preventable deaths due to trauma."[32]

"Although the ultimate goal of trauma system development is to improve patient care, it must be accomplished politically," cautioned two experts. "If development is to proceed smoothly, those responsible must be adept at balancing medical need and political necessity."[35]

The development of a trauma care system proceeds through five steps: justification (needs assessment), planning, implementation, operation, and evaluation.[35] San Diego provides an example of justification through the combined activities of trauma surgeons, the media, the public, and the county hospital council. The planning stage, during which authority is established, responsibility is identified, and criteria are developed,[35] can be illustrated by activities in Oregon.

In 1985 the Oregon legislature authorized the state health division to develop a statewide trauma care system. Approximately 1,300 people die in Oregon each year from serious injury; the sparse population and rural character of the state increase the challenge of trauma care system development. A state trauma advisory board was created, with both medical and public representatives, "to help develop standards and policies for the trauma system and to serve as a liaison between state and local planners."[36]

The state was divided into ten trauma areas for planning and service delivery. An advisory board was appointed in each area (again with public representation) to develop a plan to coordinate response, care, and the transportation of patients. In February 1987 the state published the *Oregon Trauma System Administrative Rules,* which established minimum standards to be embodied in the area plans and procedures for designating trauma hospitals.[37]

The development of Oregon's system is still under way; the state has set a deadline of 1990 for improved care for all patients. However, the process has already received praise. A 1988 NHTSA report concluded that, "Oregon appears to have been able to take the lessons learned in other areas into consideration in its development of an approach to trauma systems. This state's approach has included strong legislation, involvement of representatives of all interested parties at the state level, and the promotion of a local decision-making process at the community level."[8]

It should be pointed out that Oregon is unique, not only in its attempt to develop a trauma care system for sparsely populated rural areas, but also because the state health division is serving as the lead agency in system development. In most instances, this function is undertaken by whatever agency is responsible for EMS in the area or region.[8,35]

Implementation is often the most difficult and controversial step in system development. During this step proposals are prepared and reviewed, and hospitals are designated as trauma care facilities. After this step is completed, the system components discussed earlier (e.g., medical direction, communications, training) become operational, as does the ongoing evaluation of trauma care.[35]

As it did with EMS systems generally, NHTSA has played a prominent role in stimulating the development of trauma care systems. A new curriculum is being developed for government officials at the federal, state, and local levels who are responsible for EMS, health systems, and highway safety. The guide provides the conceptual framework around which trauma care systems can be developed.[38] NHTSA is also sponsoring a series of national conferences in 1989 to encourage the growth of trauma care systems (B. T. Scheib, personal communication).

BARRIERS TO THE GROWTH OF TRAUMA CARE SYSTEMS

The essence of a trauma care system is the rapid identification and transportation of a patient to a highly specialized facility that will be ready to provide essential life-saving treatment and services 24 hours a day. It is the opposite of the nonsystematic approach, in which all patients are taken to the nearest hospital without regard for the critical, special needs of trauma victims.

As the ACS and ACEP standards cited earlier emphasize, the ability of a facility to render effective care to trauma patients depends on its resources (e.g., 24-hour availability of surgeons and anesthesiologists) and on its continuing expertise, reflected in the care of a minimum number of trauma patients each year. Thus, in any region, the designation of trauma centers is a particularly critical element in building the system. It can also be a particularly controversial element in the process.

The designation of one hospital as a trauma center is often threatening to other local hospitals and physicians.[6] Systematizing trauma care as a public health measure may raise suspicion and animosity among private medical practitioners.[39] Administrators of community hospitals may seek to block the designation process, fearing they will lose patients to the trauma center. An overassessment of injury severity at the scene, they argue, will result in patients being inappropriately directed. And trauma centers, it is further argued, will use their enhanced status to attract more than their share of nonemergency cases. A recent study, however, indicates that these fears are unfounded.[40]

Economic fears can fuel political positions. In Florida in the 1970s, a bill was introduced to establish a statewide trauma system. It was defeated because of opposition from the Florida Hospital Association. As a result, Florida now approves hospitals as trauma centers without requiring minimum caseloads or the direct transport of victims to a center. As a result, too many hospitals claimed trauma center status (seven in 1987). Then too few hospitals claimed such status (one level I center remains), as most dropped out of the system.[8] In other areas, in an an attempt to compromise, too many hospitals have been designated. The recent review by West et al. calls overdesignation a common problem, citing Missouri and Washington, D.C., as examples.[31]

Finally, there is concern that the federal government's health financing programs may retard the growth of trauma care. Under the Diagnostic Related Groups (DRG) system, the federal government reimburses hospitals based on the average cost to treat patients in each group. Because trauma centers carry a greater load of the most severely injured and frequently provide the most resource-intensive care, it is feared that DRG payments will severely underestimate the actual costs of providing trauma care. There are not as yet sufficient economic data to demonstrate whether this is the case. A recent study of trauma centers in five areas cites economic benefits as well as disincentives.[8] Nonetheless, as in Dade County, a number of designated hospitals around the country have removed themselves from the trauma care system, citing financial distress (B. T. Scheib, personal communication). There are also impediments to growth in the rehabilitation component of trauma care systems.

REHABILITATION

Rehabilitation is the final component of a systematic approach to injury prevention and control. Its importance for individual patients and for society can be illustrated by the story of Peter S. Peter was a transportation planner and part-time radio disk jockey in Boston. Bicycling home from work, he was hit by a car and severely head-injured. For a month, Peter lay unconscious in an acute care hospital. When he awoke and his physical condition was stabilized, he was transferred to a rehabilitation hospital. "By the time someone reaches us, it's clear that he or she is going to live; it's not clear what that life is going to be like. Head injury is a uniquely complex rehabilitation problem because it affects everything," said head injury specialist Dr. Elaine Woo.[41]

Today, after intensive help from physician specialists, physical and occupational therapists, speech pathologists, and his family, Peter is once again planning transportation systems and spinning classic rock and roll recordings. His recovery has been both dramatic and rather unusual. Patients with severe head injuries do not often reclaim so much of their former lives. More often, despite their gains, they remain seriously disabled.

But although Peter's story is not typical, it does highlight the role and importance of rehabilitation. Rehabilitation treats people with disabling injuries or illnesses in special hospitals or units of hospitals with the goal of returning them to their communities, families, and jobs[10] (B. Gans, personal communication). Treatment is carried out by a team that draws on a broad range of disciplines.

The Rehabilitation Team

"Since rehabilitation is a holistic and comprehensive approach to medical care, the combined expertise of an interdisciplinary team is necessary."[9] Rehabilitation is a relatively young discipline, but one measure of its influence is that the concept of team care, developed in the late 1930s and the 1940s, has diffused into other areas of medical care.

Rehabilitation programs and services are provided through freestanding rehabilitation hospitals, inpatient rehabilitation services of general hospitals, comprehensive outpatient programs, and school-based and home-care services. Each type of service is the subject of one or more accreditation procedures. (For example, the Commission on Accreditation of Rehabilitation Facilities and the Joint Commission on Accreditation of Health Care Organizations promulgate standards and review the performance of freestanding and inpatient facilities.) One measure of a community's access to appropriate rehabilitation programs is whether it has access to accredited prgrams.

In the larger facilities and inpatient services, the rehabilitation team includes physicians, nurses, and a wide variety of therapists. Physiatrists (physicians specializing in physical medicine and rehabilitation) serve as team leaders. There is, however, a shortage of physiatrists in the United States, and teams are often led by other physician specialists. Orthopedists, neurologists, neuropsychologists, and psychiatrists are among the specialists represented on the team. Rehabilitation nurses play a major role both in providing and coordinating care. The day-to-day work of rehabilitation is carried out by physical and occupational therapists, speech–language pathologists, recreational therapists, vocational counselors, social workers, and other specialists. It is also axiomatic in rehabilitation that the patient's family be an essential part of the process.

Impediments to the Growth of Rehabilitation

There have long been two major problems with the provision of rehabilitation services to patients. One is the issue of whether a patient receives services; the other is when such services are provided. The problem of whether to provide services has been conditioned by the all-too-common view that rehabilitation is a discretionary service. What happens in the acute care hospital is related to saving the patient's life; anything after that is often treated as less critical and therefore discretionary. Many health maintenance organizations, for example, do not cover comprehensive rehabilitation programs and services.

The question of when rehabilitation services become available to a patient is not dissimilar to the discussion of whether the services are necessary. As in Peter's case, the norm has been for the acute care hospital to complete its work before transferring the patient for rehabilitation. That pattern is changing, albeit slowly, with the recognition that an early start to rehabilitation can be beneficial.

Physiatrist Dr. Bruce Gans explains:

Rehabilitation is being integrated into the principles of the acute trauma care system. Trauma surgeons and health care systems recognize—and this is a distinct change from the relatively recent past—that the process of rehabilitation doesn't start after everything else is done. It is the role of rehabilitation physicians and specialists to participate in some aspects of acute care decision making and thereby prevent secondary complications (e.g., bed sores, muscle contractures) and accelerate the rehabilitation process. These are much better dealt with by prevention than by treatment and management after they occur (personal communication).

Indeed, there is some anecdotal evidence—although no hard data as yet—that earlier rehabilitation may beneficially affect a patient's length of stay. And as one recent review article predicted, the integration of rehabilitation and acute care will increase dramatically in the next decade.[42]

OPPORTUNITIES FOR COLLABORATION

Trauma care systems (including rehabilitation services) are created at the state and local levels and can play an important role in the overall approach to injury prevention and control. Injury prevention practitioners and agencies also can play important roles in helping to establish effective trauma care systems in their communities by providing data on injuries as well as advocacy and political support.

Emergency physicians and other EMS practitioners "are often placed in a critical role in the lives of individuals and play an integral part in public education and prevention in such areas as occupational and sports injuries, trauma associated with handguns, all-terrain vehicles, high-performance motorcycles, motor vehicle injuries, burns, poisonings, and interpersonal violence including spouse abuse."[43]

In New Mexico in 1985, for example, the state chapter of ACEP and the EMT and emergency nurses associations organized both legislative testimony and public rallies in support of a mandatory safety belt law. In preparation for their advocacy efforts, ACEP and the University of New Mexico initiated research comparing the higher emergency department and hospital costs for nonusers versus users of safety belts. As a result of these and other efforts, the legislation was passed and took effect in January 1986.[43]

NHTSA has developed programs to involve emergency medical personnel in injury prevention. In 1986 it funded a series of projects to provide technical assistance on trauma prevention to EMS providers. In Montana, for example, training programs were conducted through the state EMS association "to familiarize the participants with the importance of preventing trauma and to generate enough enthusiasm so that participants would

present training programs in their local communities."[44] NHTSA's final report on the Montana project concludes that one of the important reasons for its success "was the fact that in many rural communities EMS providers are looked upon as community leaders [and] EMS personnel are already active in various forms of prevention programs," including heart disease prevention and poison control.[44]

Injury-prevention practitioners can build on the success of such programs by developing or increasing contacts with EMS providers and emergency medical personnel in their communities. They can make presentations to and organize other training events with the associations to which EMS personnel belong. In Oregon alone, there are a state EMT association, a volunteer ambulance association, a paramedic association, an emergency nurses association, an ambulance association, and a chapter of ACEP. Similar groups can be found in every state. In addition to training courses, injury-prevention and emergency services personnel can explore joint funding for projects and participate in the development of trauma care systems and trauma registries (H. Weiss, personal communication). And they can explore ways to link trauma registry data with injury data bases to "provide the capability for evaluating trauma care systems, including injury control and epidemiology, patient care and quality assurance, resource utilization and cost, and medical research and education."[45-48]

Like their EMS colleagues, rehabilitation specialists sometimes have been involved in primary prevention efforts. In Seattle in the mid-1970s, for example, doctors at the University of Washington Spinal Cord Injury Center noted an increase in adolescents with broken necks. Tracing the injuries to the use of trampolines in Seattle schools, the rehabilitation specialists, led by Dr. Walter Stolov, mounted a successful effort to have the equipment removed. Several other spinal cord centers have also been active in injury-prevention education (B. Gans, personal communication).

The need for the injury-prevention community to support the development of trauma care services extends to rehabilitation as well. No systematic approach to injury prevention is complete unless it includes the reintegration into the community and the family of those whom injury prevention has failed. And, as was discussed in the introduction, substantial components of the economic toll of injury are the continuing needs of disabled trauma patients. Reducing those disabilities to whatever extent possible through the application of rehabilitation medicine can only be beneficial.

When injury practitioners support the continued development of rehabilitation services, they should pay particular attention to two factors. First, rehabilitation services should be considered essential components of the health care system and should be funded accordingly. Second, rehabilitation services should be carried out in concert with acute care treatment to the maximum extent feasible.

Recommendations:

• Support for the development of comprehensive trauma care systems by all levels of government and by the health care and public health professions must be a priority. Trauma care systems have been proven effective in reducing injury-related mortality and morbidity. They are an essential component of a systematic approach to injury prevention from primary prevention through rehabilitation.

• Injury-prevention practitioners and EMS and trauma care specialists must collaborate in community education projects to increase awareness of and support for the development of regional trauma care systems.

• A single coordinated telephone access system (911) has been demonstrated to increase rapid access to emergency medical and trauma care services. However, too few communities are covered by such systems. A federal loan program to finance initial 911 start-up costs is recommended to promote broader coverage, particularly in rural areas.

• A rational system for designating trauma care centers is of critical importance to the growth of trauma care systems. Using model legislation available through NHTSA, medical professionals and their organizations, EMS providers, state legislators, local governments, and others must design ordinances to guide the designation and medical audit programs for trauma centers.

• Better research is needed on the financial disincentives and benefits of trauma center designation. The NHTSA, the Health Care Financing Administration, ACS, and other agencies should cooperate in funding, carrying out, and disseminating the results of such research.

• Trauma registries are important in quality assurance, but their existence is scattered and fragmentary. Their role in general injury prevention and surveillance is not well understood. Injury-prevention experts, trauma specialists, and trauma-registry designers must cooperate in programs for system design and use.

• Injury prevention practitioners and emergency medical specialists should advocate the designation of rehabilitation services as essential components of the health care system, to be funded accordingly. Rehabilitation services should be carried out in concert with acute care treatment to the maximum extent feasible.

INTERVIEW SOURCES

Richard H. Cales, MD, Chief, Emergency Services, San Francisco General Hospital, California, November 30, 1988

Bruce Gans, MD, President and CEO, Rehabilitation Institute, Detroit Medical Center, Michigan, October 24, 1988

Kristine Gebbie, MSN, Assistant Director for Health, Oregon State Health Division, Portland, February 8, 1989

Comdr. B. Thomas Scheib, Chief Emergency Medical Services Division, NHTSA, Washington, DC, January 5, 1989

Hank Weiss, MS, MPH, Director, Injury Control Unit, Wisconsin Division of Health, Madison, November 30, 1988

REFERENCES

1. Franklin J, Doelp A. Shock-trauma. New York: St. Martin's, 1980.

2. Cleveland HC. Transportation. In: Cales RH, Heilig RW Jr, eds. Trauma care systems: a guide to planning, implementation, operation, and evaluation. Rockville, Maryland: Aspen, 1986;129–42.

3. Eiseman B. Combat casualty mangement in Vietnam. J Trauma 1967;7:53–63.

4. Heaton LD. Army medical service activities in Vietnam. Milit Med 1966;131:646.

5. Boyd DR. The history of emergency medical services (EMS) in the United States of America. In: Boyd DR, Edlich RF, Micik SH, eds. Systems approach to emergency medical care. Norwalk, Connecticut: Appleton-Century-Crofts, 1983:1–82.

6. Health care: states assume leadership role in providing emergency medical services. Washington, DC: US General Accounting Office, 1986(Sep).

7. Hospital and prehospital resources for optimal care of the injured patient and appendices A through J. Chicago: American College of Surgeons, 1987(Feb).

8. Woolf P, Barth L. Trauma system development. Washington DC: National Highway Traffic Safety Administration, 1988(Mar).

9. DeLisa JA, Martin GM, Currie DM. Rehabilitation medicine: past, present, and future. In: DeLisa JA, ed. Rehabilitation medicine: principles and practice. Philadelphia: J.B. Lippincott, 1988:3–24.

10. Epperson-SeBour MM, Rifkin EW. Psychosocial rehabilitation. In: Cales RH, Heilig RW Jr, eds. Trauma care systems: a guide to planning, implementation, operation, and evaluation. Rockville, Maryland: Aspen, 1986:163–80.

11. National Research Council. Accidental death and dis-

ability: the neglected disease of modern society. Washington DC: National Academy of Sciences, 1966.

12. Cales RH. Concepts. In: Cales RH, Heilig RW Jr, eds. Trauma care systems: a guide to planning, implementation, operation, and evaluation. Rockville, Maryland: Aspen, 1986:3–18.

13. Wintemute GJ. Prevention. In: Cales RH, Heilig RW Jr, eds. Trauma care systems: a guide to planning, implementation, operation, and evaluation. Rockville, Maryland: Aspen, 1986:37–64.

14. Trunkey DD. Trauma. Sci Am 1983;249:28–35.

15. Kreis DJ Jr. Trauma systems: an interdisciplinary approach to prevention, treatment, and education. Bull NY Acad Med 1988;64:835–7.

16. Kreis DJ Jr, Plasencia G, Augenstein D, et al. Preventable trauma deaths: Dade County, Florida. J Trauma 1986;26:649–54.

17. Stewart RD. Prehospital care. In: Cales RH, Heilig RW Jr, eds. Trauma care systems: a guide to planning, implementation, operation, and evaluation. Rockville, Maryland: Aspen, 1986:111–28.

18. Cales RH. Medical direction. In: Cales RH, Heilig RW Jr, eds. Trauma care systems: a guide to planning, implementation, operation, and evaluation. Rockville, Maryland: Aspen, 1986:21–35.

19. Selfridge J, Sigafoos J, Trunkey DD. Training. In: Cales RH, Heilig RW Jr, eds. Trauma care systems; a guide to planning, implementation, operation, and evaluation. Rockville, Maryland: Aspen, 1986:65–78.

20. Champion HR. Triage. In: Cales RH, Heilig RW Jr, eds. Trauma care systems: a guide to planning, implementation, operation, and evaluation. Rockville, Maryland: Aspen, 1986:79–110.

21. Cales RH. Injury severity determinations: requirements, approaches, and applications. Ann Emerg Med 1986;15:1427–33.

22. Baker SP, O'Neill B, Haddon W, et al. The injury severity score: a method for describing patients with multiple injuries and evaluating emergency care. J Trauma 1974;14:187–96.

23. Teasdale G, Jennett B. Assessment of coma and impaired consciousness: a practical scale. Lancet 1974;1: 81–4.

24. Champion HR, Saco WJ, Carnazzo AJ, et al. Trauma score. Crit Care Med 1980;8:201–8.

25. American College of Emergency Physicians. Guidelines for trauma care systems. Ann Emerg Med 1987;16: 459–63.

26. Cales RH, Trunkey DD. Preventable trauma deaths: a review of trauma care systems development. JAMA 1985; 254:1059–63.

27. West JG, Trunkey DD, Lim RC. Systems of trauma care: a study of two counties. Arch Surg 1979;114:455–60.

28. Cales RH. Trauma mortality in Orange County: the effect of implementation of a regional trauma system. Ann Emerg Med 1984;13:1–10.

29. National Highway Traffic Safety Administration. The NHTSA emergency medical services program and its re-lationship to highway safety. Washington, DC: Department of Transportation, 1985(Aug).

30. Shackford SR, Mackersie RC, Hoyt DB, et al. Impact of a trauma system on outcome of severely injured patients. Arch Surg 1987;122:523–7.

31. West JG, Williams MJ, Trunkey DD, Wolferth CC. Trauma systems: current status–future challenges. JAMA 1988;259:3597–600.

32. Shackford SR, Hollingworth-Fridlund P, Cooper GF, Eastman AB. The effect of regionalization upon the quality of trauma care as assessed by concurrent audit before and after institution of a trauma system: a preliminary report. J Trauma 1986;26:812–20.

33. Amherst Associates. Trauma needs assessment study. San Diego County. Los Angeles: Hospital Council of San Diego and Imperial Counties, 1982(Dec).

34. County of San Diego. Trauma system annual report. San Diego, California: County of San Diego, 1987(Jan).

35. Heilig RW Jr, Cales RH. Development. In: Cales RH, Heilig RW Jr, eds. Trauma care systems: a guide to planning, implementation, operation, and evaluation. Rockville, Maryland: Aspen, 1986:283–92.

36. Oregon trauma system summary. Portland, Oregon: Oregon State Health Division, 1989(Jan).

37. Oregon trauma system administrative rules. Portland, Oregon: Oregon State Health Division, 1987(Jun).

38. Cooper G. Development of trauma systems: a state and community guide. Washington DC: National Highway Traffic Safety Administration (in press).

39. Institute of Medicine. The future of public health. Washington, DC: National Academy Press, 1988.

40. Cales RH, Anderson PG, Heilig RW Jr. Utilization of medical care in Orange County: the effect of implementation of a regional trauma system. Ann Emerg Med 1985; 14:853–8.

41. Cohen S. Will to survive. In: Talley RG, ed. Inside the MGH. Boston Globe 1986 May 25:1–24.

42. Harvey RF. The future of rehabilitation: delivery of rehabilitation services in the 1990s. In: Maloney FP, ed. Physical medicine and rehabilitation: state of the art reviews. Philadelphia: Hanley and Belfus, 1987:321–30.

43. Bernstein E, Roth PB, Yeh C, Lefkowits DJ. The emergency physician's role in injury prevention. Pediatr Emerg Care 1988;4:207–11.

44. Technical assistance for EMS providers in Montana on trauma prevention: final report. Washington, DC: National Highway Traffic Safety Administration, 1988.

45. Cales RH, Kearns ST. Trauma registers. Trauma Q (in press).

46. Centers for Disease Control. Report of the trauma registry workshop. J Trauma (in press).

47. Cales RH, Kearns ST, Jordan LS. Trauma registers: a national survey of purpose, content, and resources. J Trauma (in press).

48. Cales RH, Bietz DS, Heilig RW Jr. The trauma registry: a method for providing regional system audit using the microcomputer. J Trauma 1985;25:181–7.

Appendix A: Data Sources

FEDERAL

National Center for Health Statistics Databases
Scientific and Technical Information Branch
National Center for Health Statistics
Center Building, Room 1-57
3700 East-West Highway
Hyattsville, MD 20782
(301) 436-8500

The National Center for Health Statistics (NCHS) produces a wealth of data relevant to injury control, much of which is available for purchase on tape. The NCHS Data Tape Catalog presents a detailed description of its data bases, available tapes, and costs.

National Vital Statistics Mortality Data: Data from death certificates are reported by month and year and by state. Number of deaths, incidence by type, and leading causes of fatalities are included. Deaths due to injury are categorized by "motor vehicle" and "other." Contains E codes. Published as series 20 and 21 of *Vital Statistics and Health, Monthly Vital Statistics Report,* annual *Vital Statistics of the United States.*

National Health Interview Survey: Annual survey of approximately 42,000 household interviews conducted by NCHS. Includes information on the incidence of illness, accidental injuries, and prevalence of chronic diseases. The focus changes yearly. Some breakdown of injuries by motor vehicle, non-motor–vehicle, and place. Not state-specific. Unclear definition of "accident." Published as Series 10 of *Vital Statistics and Health* and as special reports. Available on tape.

National Hospital Discharge Survey: Continuous nationwide survey of inpatient short-stay hospitals. Data abstracted from a sample of 431 hospitals. Reliable data not available on specific injuries. Includes E codes when available, but this information is grossly underreported. Published in Series 13 of *Vital Statistics and Health.* Available on tape.

National Ambulatory Medical Care Survey: National sample of visits to physicians with information on nature, but not causes, of injuries. Published in Series 13 of *Vital Statistics and Health.*

National Highway Traffic Safety Administration Databases
National Center for Statistics and Analysis NRD-30
National Highway Traffic Safety Administration
400 Seventh Street, SW
Washington, DC 20590
(202) 366-5820

Fatal Accident Reporting System (FARS): Contains information on all fatal motor vehicle injuries reported to police. National and state data. Data include occupant injury, time and location of incident, vehicle make and model, driver record data, weather conditions, restraint use, and blood alcohol reports. Information compiled from police records, driver licensing files, death certificates, and hospital records. Quality control and data vary by state. Annual report. Available on tape.

National Accident Sampling System (NASS): (1) The NASS General Estimates System contains information taken from a sample of all police-reported "accidents." Data include previous driving records, vehicle damage, occupant injuries, and consequences. Information is compiled from visits to crash sites, interviews with victims, and reviews of police and hospital records. Not state-specific. Inconsistent definitions. High rates of missing data on some variables. Annual report. Available on tape. (2) The NASS Crashworthiness System contains information on the crashworthiness of light passenger vehicles involved in traffic "accidents." Designed to emphasize incidents resulting in serious injuries and involving late model vehicles. Data focus on vehicle performance and occupant injuries. Annual report. Available on tape.

Others: (1) NHTSA has "State Accident Data Files," transcribed from police accident reports, available for the following states: Alabama, Arizona, California, Florida, Georgia, Illinois, Indiana, Kansas, Maryland, Michigan, Missouri, New Jersey, Ohio, Pennsylvania, Tennessee, Texas, Virginia, and Washington. Years covered and variables included vary by state. Reporting of lower-severity accidents may change. (2) NHTSA also produces special studies, such as the National Crash Severity Study (NCSS), the Pedestrian Injury Causation Study (PICS), and the Boston Small Car Study.

National Injury Information Clearinghouse
Directorate for Epidemiology
U.S. Consumer Product Safety Commission
5401 Westbard Avenue, Room 625
Washington, DC 20207
(301) 492-6424

National Electronic Injury Surveillance System (NEISS): Injuries involving consumer products taken from a national sample of emergency rooms. Not state-specific. Data include type of product involved, date of treament, age and sex of injured person, and location of incident. Also produces special studies based on more in-depth investigations of specific product-related injury problems. Excludes alcohol, tobacco, firearms, automobiles, other transportation, public equipment, and homemade products. No brand names. *NEISS Data Highlights* published annually. Will answer data requests.

Accident Investigations: Reports based on interviews with victims and witnesses to injuries associated with selected consumer products. Includes description of environment, victim, and product. Only incidents that occurred after mid-1972. Computer printouts of data extracted from original reports and microform or paper copies of original reports available on request. Hazard analyses and special reports also available.

Occupational Injury and Illness Statistics Program
Office of Safety, Health Statistics, and Working Conditions
Bureau of Labor Statistics
Patrick Henry Building, Room 4014
601 D Street, NW
Washington, DC 20212
(202) 272-3459

Annual Survey of Occupational Injuries and Illnesses: National survey sample of employers to document workplace injuries, including fatalities. Stratified by state, industry, and employment size prior to selection. Illness and injury reported by employers only on a summary basis. Includes only those injuries that result in loss of consciousness, restriction of work or motion, transfer to authorities, or treatment beyond first aid. Data include industry, company size, and lost workdays. No information on occupation, age, sex, or race of the injured or ill worker or characteristics of the injury or illness. No information on causes of incidents resulting in injury or illness of the worker. Excludes public employees, self-employed persons, and companies with less than 11 employees. Annual survey estimates for 1976–85 available on tape.

Supplementary Data System (SDS): Occupational injury and illness data from states based on employers' first reports of injury submitted to state workers' compensation agencies. Over 30 states participate in the program. States process information from workers' compensation records under uniform coding procedures on the cost and characteristics of occupational injuries and illnesses and the characteristics of the injured or ill worker. Data are supplemental to annual survey data. Information on detailed characteristics of the injury/illness event and the worker involved. Data include worker characteristics, source of injury, part of body affected, "accident type," and associated object or substance. Interstate comparison of data limited to the distribution of patterns of causes by given characteristics. Lack of exposure information (hours worked or length of employment) precludes the development of rates for comparison purposes. National estimates solely based on SDS cannot be directly developed because of differences in state workers' compensation laws and participation of fewer than all states. Biannual reports. Microdata files of current and closed cases and summary tabulations for these groups of data are available on microfiche from the National Technical Information Service (NTIS) of the U.S. Department of Commerce. Procedures for obtaining these data may be obtained from the BLS Office of Occupational Safety and Health Statistics. Individual state tabulations available.

Work Injury Reports (WIR) Survey Program: Includes special surveys of occupational injuries to obtain causal information tailored to specific problem areas, particularly with regard to OSHA standards. May not be representative of workplaces outside the scope of the survey. For program reports, contact BLS.

National Traumatic Occupational Fatality Database (NTOF)
National Institute for Occupational Safety and Health
Division of Safety Research

944 Chestnut Ridge Road
Morgantown, WV 26505

State- and industry-specific fatality rates based on death certificates since 1980. Does not include motor vehicle–related occupational fatalities. Includes information on industry type, incidence rate of lives lost, occupation, time of day, and causes. E codes collected but not available to the public. Problem with underreporting. Does not include figures on persons under 16 years of age. Annual report. The current issue covers 1980–84 data.

Indian Health Service
Office of Planning, Evaluation and Legislation
Division of Program Statistics
Parklawn Building, Room 6A-30
5600 Fishers Lane
Rockville, MD 20857
(301) 443-1180

Computer-based system covering all inpatient and outpatient encounters at IHS facilities as well as all treatments of IHS clients on a contract basis. National data on both mortality and morbidity. Annual report. Study on injuries published in 1988. Both available from the U.S. Government Printing Office.

Behavioral Risk Factor Surveillance
Centers for Disease Control
1600 Clifton Road, NE
Atlanta, GA 30333
(404) 639-3075

Random sample of adults 18 years of age or over is surveyed each month by telephone. CDC has a core set of questions. Some states add additional questions. Data include age, sex, race, employment status, education, income, seat belt use, drinking and driving. No data on children. No incidence data. Quarterly reports in *Morbidity and Mortality Weekly Report*.

National Fire Incident Reporting System (NFIRS)
National Fire Data Center
National Emergency Training Center
1682 South Seaton
Emmitsburg, MD 21727
(301) 447-6771

Aggregation of information voluntarily submitted by fire departments to state fire marshals in 40 states. Data include date of fire, location, alarm time and arrival time, construction type, estimated dollar loss, cause and nature of injuries, and presence or absence of smoke detector. Provides useful sentinel data on new fire hazards in a community. Annual report. Available on tape.

Uniform Crime Reports
Federal Bureau of Investigation
Gallery Row Building
Corner of 7th and D Streets, NW
Washington, DC 20535
(202) 324-5015

Collects information concerning homicide, sexual assault, and other crimes. System based on monthly reports submitted by more than 15,000 city, county, and state law enforcement agencies in the United States. In 1984 more than 96% of the U.S. population fell within the jurisdiction of the participating agencies. In each state the coordination of data is handled by local or state sheriffs' associations and associations of chiefs of police. (To locate the coordinating body in any given state contact either the National Sheriffs' Association, 1450 Duke Street, Alexandria, VA 22314, or the International Association of Chiefs of Police, 13 Firstfield Road, Gaithersburg, MD 20878.) Information includes age, sex, and race of the victims, weapons used, circumstances, relationship between victims and assailants, and demographic data on the assailants. Also provides arrest information for murder, rape, robbery, aggravated assault, arson, larceny theft, and motor vehicle theft. No injury-specific data. No information on victims and assailants as regards ethnicity or family structure. Annual report. Public use data tapes available through National Criminal Justice Data Archive, Inter-university Consortium for Political and Social Research, University of Michigan, P.O. Box 1248, Ann Arbor, MI 48106, (313) 763-5010.

National Criminal Justice Reference Service
Bureau of Justice Statistics
Box 6000
Rockville, MD 20850
(800) 732-3277

National Crime Surveys: Survey of a national sample of victims of six selected crimes: rape, robbery, assault, burglary, larceny, and auto theft. Unit of analysis is the household. Combination of samples from police records and random-digit dialing. Data include type of crime, medical treatment, victim and offender characteristics, and offender-to-victim relationship. No homicide data. Children under 12 years of age not interviewed. Annual reports. Periodic bulletins. Available on tape.

National Clearinghouse for Criminal Justice
Information Systems
925 Secret River Drive, Suite H
Sacramento, CA 95831
(916) 392-2550

Operates automated index of over 1,000 criminal justice information systems maintained by state and local governments throughout the nation; issues technical publications; operates a computerized Criminal Justice Information Bulletin Board; operates the National Criminal Justice Computer Laboratory and Training Center. Annual report.

Criminal Justice Statistics Association
444 North Capitol Street, NW, Suite 606
Washington, DC 20001
(202) 347-4608

Catalog and library of statistical reports produced by state criminal justice analysis centers.

Boating Accident Reporting System (BARS)
State Boating Law Administration
United States Coast Guard
G-NAB-2
2100 Second Street, SW
Washington, DC 20593
(202) 267-0955

Reports of boating incidents resulting in loss of life, personal injury requiring medical treatment beyond first aid, damage greater than $200 to vessel, or complete loss of vessel. Data include time, date, use of alcohol, boat type, cause of accident, and age. Voluntary reporting with less than 10% of nonfatal incidents reported. Annual report.

Drug Abuse Warning Network (DAWN)
National Institute on Drug Abuse
Division of Epidemiology and Statistical Analysis
Parklawn Building, Room 11A-55
5600 Fishers Lane
Rockville, MD 20857
(301) 443-6637

Drug abuse–related surveillance data collected by the National Institute on Drug Abuse for 27 metropolitan areas across the country. Sample of emergency room drug abuse episodes. Data include sex, race, age, motive, reason for emergency department contact, drug source, drug route, and disposition. Methods and procedures to identify cases may vary by facility. No elimination of repeat visits. Cannot calculate population-based rates. Annual report.

STATE AND LOCAL

Vital Statistics Records
State Offices of Vital Statistics

Data collected from death certificates. Information includes all causes of injury and identifies the condition that triggered the event leading to death. Includes E codes. Town-specific. Some limitations include: no location data, no alcohol data, small numbers, limited range of etiologies, and no cost information. The nature of the injury that appears on the death certificate is also not included. Annual reports available.

Uniform Hospital Discharge Data Sets (UHDDS)
State Rate Setting Commission Offices, State
Departments of Public Health, or State Hospital
Associations

Data collected annually for hospital discharges. Every state acute care hospital collects this information for Medicare reimbursement, but these data are not centralized in all states. Data are hospital-specific and can be used for cost analysis. Some limitations include incomplete E coding, underrepresentation of certain injury causes and outcomes, poor race data, and possible duplication of cases. Abstracts and computerization of hospital discharge data are frequently delayed 2–4 years.

Medical Examiners'/Coroners' Reports
County Medical Examiners' or Coroners' Offices

Includes all deaths due to unintentional injury, homicide, or suicide. Supporting material including any toxicology or autopsy report is included with the coroner's report. Toxicologic findings include blood tests for the presence of alcohol and drugs; and oral, anal, and vaginal swabs tested for the presence of sperm and prostatic fluid. Reports frequently contain details on the circumstances surrounding death: place of injury, time of injury, and an abbreviated account of the circumstances as reported by the medical examiner. Reports are usually timely—available 48 hours after the death. Degree of computerization of the data varies by county. Reports and procedures may not be consistent from medical examiner to medical examiner and frequently do not contain vital epidemiologic data about the external cause of injury.

PRIVATE

National Burn Registry
National Burn Information Exchange (NBIE)
National Institute for Burn Medicine
9600 East Ann Street
Ann Arbor, MI 48104
(313) 769-9000

Voluntary reporting of information from specialized burn care facilities. Data include age, sex, race, date, location and cause of injury, and medical treatment.

National Institute of Handicapped Research, Pediatric Trauma Registry
Coordinating Center
Tufts–New England Medical Center
171 Harrison Avenue, Box 75 K-R
Boston, MA 02111
Attn.: Carla DiScala

A multicenter pediatric trauma registry with data collected from trauma centers located throughout the United States and Canada. Pediatric trauma is described in terms of its etiology, methods of treatment, and functional outcome. Criteria for inclusion include children under 20 years of age admitted to the trauma center with a diagnosis of trauma, excluding a sole diagnosis of burn or poisoning; all children dying from trauma after admission; and those dead on arrival. Annual report.

National Study on Child Neglect and Abuse Reporting
National Resource Center for Child Abuse and Neglect
American Association for Protecting Children
P.O. Box 1266
Denver, CO 80201
(800) 227-5242

Annual study using reports from state child protective services. Data include birth date, sex, ethnicity, placement, perpetrator(s), relationship to child, language, reported condition, and report source. Variability among data reported by states is great. Annual report. Will answer data questions by phone at the 800 number.

Big Ten Injury Surveillance Survey
Steve Troester or John P. Albright, MD
1189 Carver Pavilion
University of Iowa Hospitals
Iowa City, IA 52242
(319) 338-0581, Ext. 425

Tracks injuries in selected men's and women's sports. Athletic trainers from Big Ten Conference schools complete injury forms on players who miss at least part of a practice because of injury. Injury severity defined by time lost from practice or competition. Annual report published and presented at the annual meeting of Big Ten Team Physicians.

National Head and Neck Injury Registry
Joseph S. Torg, MD, or Joseph Vegso, MD
% University of Pennsylvania Sports Medicine Center
Weightman Hall E-7
235 South 33rd Street
Philadelphia, PA 19104
(215) 662-6943

Data on football-related cervical spine and head injuries that cause players to be hospitalized for at least 72 hours or result in death or paralysis. Includes information dating back to 1971 collected for all levels of competition using a newspaper clipping service, a survey of National Athletic Trainers' Association members, the National Association of Secondary School Principals, and football helmet manufacturers. Annual reports published in a variety of journals. Will respond to data requests from physicians, coaches, and researchers.

National High School Athletic Injury Registry
National Athletic Trainers' Association, Inc.
1001 E. Fourth Street
Greenville, NC 27858
(919) 752-1725

Collects data on high school football and girls' basketball injuries from 126 schools. Data are gathered by school athletic trainer and include information on age, height, weight, experience in the sport, level of education, anatomic site of the injury, and characteristics of environment, such as condition of floor or ground surface. An injury is defined as any incident that precludes a person from completing a session or causes a person to miss a session the following day.

Regional Spinal Cord Injury Systems
Samuel L. Stover, MD
National Spinal Cord Injury Statistical Center
University of Alabama at Birmingham
University Station
Birmingham, AL 35294
(205) 934-3330

Made up of 17 regional spinal cord injury centers around the United States. Data are available dating back to 1973. A book on the network providing information on more than 100,000 spinal cord injuries is available for $49.95.

American Association of Poison Control Centers National Data Collection System (AAPCC)
National Capital Poison Center
Georgetown University Hospital
3800 Reservoir Road, NW
Washington, DC 20007
(202) 625-3333

Data aggregated from 57 participating poison centers. Data include age, sex, location, time of incident, substance, and medication taken. Extrapolations from the number of reported poisonings to the number of actual poisonings occurring annually in the United States cannot be made from these data alone, as considerable regional variations exist. Data published annually in the September issue of *American Journal of Emergency Medicine*. Current list of certified regional poison centers published in most issues of *Veterinary and Human Toxicology*.

Appendix B: Selected Journals That Frequently Publish Articles on Injuries and Injury Prevention

Accident Analysis and Prevention
Bimonthly
Pergamon Press, Inc.
Journals Division
Maxwell House
Fairview Park
Elmsford, NY 10523

American Journal of Diseases of Children
Monthly
American Medical Association
535 North Dearborn Street
Chicago, IL 60610

American Journal of Public Health
Monthly
American Public Health Association
1015 15th Street, NW
Washington, DC 20005

Annals of Emergency Medicine
Monthly
American College of Emergency Physicians
Box 619911
Dallas, TX 75261-9911

Child Abuse and Neglect
Quarterly
Pergamon Press, Inc.
Journals Division
Maxwell House
Fairview Park
Elmsford, NY 10523

Journal of the American Medical Association
4 per month
American Medical Association
535 North Dearborn Street
Chicago, IL 60610

Journal of Interpersonal Violence
Quarterly
Sage Publications, Inc.
2111 West Hillcrest Drive
Newbury Park, CA 91320

Journal of Public Health Policy
Quarterly
Meriden-Stinehour Press
% *Journal of Public Health Policy*, Inc.
Lunenburg, VT 05906

Journal of Trauma
Monthly
Williams & Wilkins
428 East Preston Street
Baltimore, MD 21202

Morbidity and Mortality Weekly Report
Weekly
Superintendent of Documents
U.S. Government Printing Office
Washington, DC 20402

Pediatrics
Monthly
American Academy of Pediatrics
141 Northwest Point Road
Box 927
Elk Grove Village, IL 60007

The Physician and Sportsmedicine
Monthly
American College of Sportsmedicine
401 West Michigan Street
Indianapolis, IN 46202

Public Health Reports
Bimonthly
U.S. Public Health Service
Department of Health and Human Services
Hubert Humphrey Building
Room 721-H
200 Independence Avenue, SW
Washington, DC 20201

Suicide and Life-threatening Behavior
Quarterly
Guilford Publications, Inc.
72 Spring Street, 4th Floor
New York, NY 10012

Appendix C: Injury Newsletters

Bulletin on Alcohol Policy
Building One, Room 400
San Francisco General Hospital
San Francisco, CA 94110

Campaign Update
SAFE KIDS
c/o 111 Michigan Avenue, NW
Washington, DC 20010

CHIP Information and Education
Vermont Child Injury Prevention Program
Vermont Department of Health
60 Main Street, Box 70
Burlington, VT 05402

Clinical Toxicology Review
Massachusetts Poison Control System
300 Longwood Avenue
Boston, MA 02115

Community Safety and Health
National Safety Council
444 North Michigan Avenue
Chicago, IL 60611

Consumer Reports
Consumers Union
2001 S Street, NW
Suite 520
Washington, DC 20009

CSPN Newsletter
Consumer Federation of America
1424 16th Street, NW
Washington, DC 20036

Dear Colleague
Program Development and Implementation Branch
Division of Injury Epidemiology and Control
Centers for Disease Control
Atlanta, GA 30333

Family Safety and Health
National Safety Council
444 North Michigan Avenue
Chicago, IL 60611

Fatal Accident Reporting System
U.S. Department of Transportation
National Highway Traffic Safety Board
Washington, DC 20590

Focus
Injury Prevention and Research Center
Dartmouth Medical School, MCH
Hanover, NH 03756

For Kids' Sake
University of Virginia Medical Center
Box 484, Medical Center
Charlottesville, VA 22908

Injury Prevention Network Newsletter
Trauma Foundation
Building One, Room 400
San Francisco General Hospital
San Francisco, CA 94110

Injury Prevention Research and Practice
Injury Prevention Center
Department of Maternal and Child Health
Harvard School of Public Health
677 Huntington Avenue
Boston, MA 02115

IPRC News
University of North Carolina
Injury Prevention Research Center
CB #3430, CTP
Chapel Hill, NC 27599

Massachusetts Passenger Safety Update
Massachusetts Passenger Safety Program
Division of Family Health Services
Massachusetts Department of Public Health
150 Tremont Street, 3rd Floor
Boston, MA 02111

MECAP NEWS
U.S. Consumer Product Safety Commission
Washington, DC 20207

Nation's Health
American Public Health Association
1015 15th Street, NW
Washington, DC 20005

National Safety Council Recreational Newsletter
National Safety Council
444 North Michigan Avenue
Chicago, IL 60611

National Coalition to Reduce Car Crash Injuries
National Coalition to Reduce Car Crash Injuries
1667 K Street, NW, Suite 500
Washington, DC 20006

National Safety Council Product Safety Update
National Safety Council
444 North Michigan Avenue
Chicago, IL 60611

NDPN News
The National Drowning Prevention Network
P.O. Box 16075
Newport Beach, CA 92659-6075

News
U.S. Department of Transportation
Office of the Assistant
Secretary for Public Affairs
Washington, DC 20590

Occupational Health & Safety Letter
Environews, Inc.
1331 Pennsylvania Avenue, NW
Suite 509
Washington, DC 20004

Ohio Monitor
The Industrial Commission of Ohio
Division of Safety and Hygiene
246 North High Street
Columbus, OH 43215

Public Health Currents
Ross Laboratories
625 Cleveland Avenue
Columbus, OH 43215

Report
National Child Passenger Safety Association
1050 17th Street, NW
Suite 770
Washington, DC 20036

Safe Ride News
American Academy of Pediatrics
P.O. Box 927
Elk Grove Village, IL 60007

School Bus Fleet
Bobit Publishing
Artesia Boulevard
Redondo Beach, CA 90278

Small Craft Advisory
Empire Publishing, Inc.

P.O. Box C-19000
Seattle, WA 98109

SPIG Newsletter
Injury Control and Emergency Health Services Forum
American Public Health Association
1015 15th Street, NW
Washington, DC 20005

Status Report
Insurance Institute for Highway Safety
Watergate 600
Washington, DC 20037

Totline
Highway Safety Research Center
University of North Carolina
Chapel Hill, NC 27514

Traffic Safety Update
Traffic Safety Now
300 New Center Building
Detroit, MI 48202

Traffic Safety Newsletter
National Highway Traffic Safety Administration
400 Seventh Street, SW
Washington, DC 20590

WCCIPP News
Wisconsin Comprehensive Injury Prevention Project
1 West Wilson
P.O. Box 309
Madison, WI 53701-0309

Appendix D: Spinal Cord Injury

INJURY SURVEILLANCE RESOLUTION, ACUTE TRAUMATIC SPINAL CORD INJURY, COUNCIL OF STATE AND TERRITORIAL EPIDEMIOLOGISTS

This resolution was passed at the June 1985 meeting of the Council of State and Territorial Epidemiologists (CSTE).

Reporting of Traumatic Spinal Cord Injuries

There are between 7,000 and 8,000 traumatic spinal cord injuries in the United States each year. The public health impact of spinal cord injury is underscored by (a) the high cost of acute care (between $35,000 and $75,000); (b) the age of victims (over half of the injuries occur in the 15 to 24 age group); (c) the preventable etiology of these injuries (46 percent vehicular, 13 percent gunshot, 13 percent sports/recreation related); and (d) the permanence of disability (less than 7 percent of paraplegics and quadriplegics completely recover from their initial injury). Epidemiologic surveillance of spinal cord injuries allows the determination of the chain of events which result in spinal injuries, the determination of causes of spinal injuries, and the identification of preventable intervention for spinal cord injuries.

Position to be Adopted

CSTE recommends that public health agencies make the occurrence of traumatic spinal cord injury a reportable health condition. Basic epidemiologic and demographic information collected for each spinal cord injury should include time, place, person, type of injury, causation, circumstances, and medical care. Sources of data should include medical examiners' case records, death certificates, and patient records from local and regional spinal cord rehabilitation facilities. Reports of spinal cord injury should be forwarded to the Centers for Disease Control (CDC) on a regular basis through an adaption of the Electronic Surveillance Project. CSTE requests that CDC develop a case definition for spinal cord injury and provide financial assistance for the development of injury surveillance systems. Other injuries which should receive high priority for future surveillance activities include head injuries, burns, falls, immersion injuries, and gunshot wounds.

Background and Justification

Injuries are responsible for more morbidity and mortality in the 1 to 44 age group than all communicable and other conditions combined; however, in most states, there is no system for the collection and analysis of nonfatal injuries other than motor-vehicle related injuries. The reporting of spinal injuries is an important first step in the establishment of an injury surveillance system. Spinal injuries are important, costly, and amenable to prevention and intervention. These injuries have a particularly severe impact on the public's health because of the costs associated with treatment, the age of most victims, and the permanent disability they inflict. Additionally, many spinal injuries —such as those caused by motor vehicles and gunshot wounds—may be preventable. The establishment of an epidemiologic surveillance system for spinal cord injury will lead to a better understanding of the nature of these injuries and the preventive interventions that may reduce their impact. Such a system will be the foundation for surveillance of other important injuries such as head injuries, burns, falls, immersion injuries, and gunshot wounds.

CASE DEFINITION, ACUTE TRAUMATIC SPINAL CORD INJURY, CENTERS FOR DISEASE CONTROL AND COUNCIL OF STATE AND TERRITORIAL EPIDEMIOLOGISTS

A case of spinal cord injury (SCI) is defined as a person who suffers an acute, traumatic lesion of neural elements in the spinal canal, resulting in any degree of sensory deficit, motor deficit, or bladder/bowel dysfunction. The deficit or dysfunction can be temporary or permanent. (Intervertebral disc disease [ICD9-CM 722] should not be included.)

The following ICD9-CM rubrics can be used to identify potential cases:

342 Hemiplegia
344 Other paralytic syndromes
805 Fracture of vertebral column without mention of spinal cord lesion
806 Fracture of vertebral column with spinal cord lesion
839 Other, multiple and ill-defined dislocations
952 Spinal cord lesion without evidence of spinal bone injury
953 Injury to nerve roots and spinal plexus

Reference for case definition: Kraus JF, Franti DE, Riggins RS, Richard D, Borhani NO. Incidence of traumatic spinal cord lesions. J Chron Dis 1975;28:471–92.

Appendix E: E Code Categories for Injuries

Adapted from Childhood Injury Prevention Resource Center, Harvard School of Public Health

Motor Vehicle–Occupant
E810.0, E810.1, E811.0, E811.1,
E812.0, E812.1, E813.0, E813.1,
E814.0, E814.1, E815.0, E815.1,
E816.0, E816.1, E817.0, E817.1,
E818.0, E818.1, E819.0, E819.l,
E822.0, E822.1, E823.0, E823.1,
E824.0, E824.1, E825.0, E825.1
Operator or passenger in motor vehicle
  Includes: Collision and noncollision traffic injuries; occupant or passenger of car, bus, truck, van, or construction machinery in transport under its own power; off-road motor vehicle operating on a public highway
  Excludes: Occupant or passenger of bicycle, motorcycle, off-road vehicle not on highway, motorized bicycle; pedestrians

Motor Vehicle–Pedal Cycle Rider
E810.6, E811.6, E812.6, E813.6,
E814.6, E815.6, E816.6, E817.6,
E818.6, E819.6, E822.6, E823.6,
E824.6, E825.6
Any person riding on a pedal cycle involved in a collision or noncollision injury associated with a motor vehicle
  Includes: Public highways, parking lots, driveways, and other locations

Motor Vehicle–Pedestrian
E810.7, E811.7, E812.7, E813.7,
E814.7, E815.7, E816.7, E817.7,
E818.7, E819.7, E822.7, E823.7,
E824.7, E825.7
Pedestrian injured in any collision or noncollision traffic incident involving a motor vehicle
  Includes: Public highways, parking lots, driveways, and other locations

Motorcyclist
E810.2, E810.3, E811.2, E811.3,
E812.2, E812.3, E813.2, E813.3,
E814.2, E814.3, E815.2, E815.3,
E816.2, E816.3, E817.2, E817.3,
E818.2, E818.3, E819.2, E819.3,
E822.2, E822.3, E823.2, E823.3,
E824.2, E824.3, E825.2, E825.3
Operator or passenger on motorcycle involved in collision or noncollision traffic accident
  Includes: Motorized bicycle (moped, dirt bike, scooter, tricycle); occupant of sidecar

Non-MV Pedal Cycle
E826.0–826.9*
Any person injured in a non-motor–vehicle incident involving a pedal cycle
  Includes: Falls from bicycle; bicycle colliding with pedestrian, other bicycle, animal, object, or other road vehicle (e.g. streetcar)
  Excludes: Injuries involving motor vehicles, railroad trains, and aircraft

Off-Road Motor Vehicle
E820.0–E821.9
Nontraffic injuries involving off-road motor vehicles
  Includes: Snowmobiles and all-terrain vehicles (ATVs)

Other or Unspecified Transport or Person
E800.0–E807.9, E810.4, E810.5,
E810.8, E810.9, E811.4, E811.5,
E811.8, E811.9, E812.4, E812.5,
E812.8, E812.9, E813.4, E813.5,
E813.8, E813.9, E814.4, E814.5,
E814.8, E814.9, E815.4, E815.5,
E815.8, E815.9, E816.4, E816.5,
E816.8, E816.9, E817.4, E817.5,
E817.8, E817.9, E818.4, E818.5,
E818.8, E818.9, E819.4, E819.5,
E819.8, E819.9, E822.4, E822.5,
E822.8, E822.9, E823.4, E823.5,
E823.8, E823.9, E824.4, E824.5,
E824.8, E824.9, E825.4, E825.5,
E825.8, E825.9, E827.0–E829.9,
E831.0–E831.9, E833.0–E838.9,
E840.0–E845.9, E846–E848
All other transport injuries
  Includes: Any motor vehicle nontraffic injury involving any person; occupants of streetcars and animal riders; all railway injuries except those with motor vehicles; any person involved in any air or water transport vehicle injury; animal rider or rider of animal-drawn vehicle; motor vehicle nontraffic injury while boarding or alighting off-road vehicles such as snowmobiles and ATVs

Misadventure
E870.0–E876.9
Misadventures to patients during surgical or medical care
  Excludes: Unintentional overdose of drugs

Postoperative Complications
E878–E879
Surgical and medical procedures as the cause of abnormal reaction without mention of misadventure at the time of the procedure
  Includes: Organ rejection, malfunction of devices, postoperative renal failure, intestinal obstruction
  Excludes: Adverse reactions to anesthesia without mention of misadventure; infusion and transfusion reactions without mention of misadventure

Falls
E880.0–E886.9, E887,† E888
Includes: Falls on same level (slipping, tripping); falls on

different levels (stairs, bed ladder); diving into swimming pool; tackles in sports

Excludes: Falls from horses in sports or transport; falls from burning buildings; falls from moving vehicles and bicycles; falls while boarding or leaving vehicles; falls onto sharp objects; falls into water

Environmental
E900.0–E909

Injuries due to natural and environmental factors

Includes: Excessive heat, cold, air pressure changes, travel and motion, hunger, thirst, exposure, neglect, venomous animals and plants

Excludes: Ingestion of poisonous animals or plants; plant puncture wounds, animal bites; road vehicle accidents involving animals

Drowning and Submersion
E830.0–E830.9, E832.0–E832.9,
E910.0–E910.9

Includes: Unintentional drowning and submersion, swimmer's cramp, water sports, any falls or injuries involving watercraft

Excludes: Transport injuries other than watercraft; injuries due to diving into swimming pool and striking side or bottom; drowning as a result of a motor vehicle incident

Fire, Flames and Hot Substances (Burns)
E890.0–E899, E924.0–E924.9

Injuries caused by fire and flames; hot appliances, objects or liquids; steam; acid burns

Includes: Unintentional burning by fire, secondary fires due to explosion; asphyxia due to fire; smoke and fumes due to fire; fall or jump from burning structure; ignition of clothing or bedclothes

Excludes: Fire in machinery in operation; transport vehicle except stationary; fire in other vehicle (e.g., ski lift); watercraft fires; intentional fires; electric current; radiation burns; internal burns

Poisonings (Accidental)
E850.0–E869.9

Unintentional poisonings by drugs, medicinals, and biological substances

Includes: Solids, liquids, gases, vapors

Excludes: Administration with suicidal or homicidal intent or with intent not determined as unintentional or intentional; adverse reactions to correct drugs properly administered; food poisoning (bacterial); poisoning and toxic reactions to plants; carbon monoxide poisoning from motor vehicle, watercraft, or aircraft in transit; carbon monoxide poisoning from smoke and fumes in fire

Suffocations (Accidental)
E911–E913.9

Inhalation and ingestion of food or objects causing obstruction of respiratory passage or suffocation; unintentional mechanical suffocation

Includes: Smothering, choking

Excludes: Intentional suffocations and those not determined as unintentional or intentional; any ingestion of a foreign body without mention of asphyxia or obstruction of respiratory passage

Foreign Body
E914–E915

Foreign body in the eye, adnexa, or other orifice

Includes: Injury to respiratory tract without obstruction or asphyxia

Excludes: Corrosive liquids in eye

Struck by Object
E916–E918

Struck by falling object, striking against or struck by persons or objects, caught unintentionally between objects

Excludes: Striking against object or persons with fall; injury involving machinery or motor vehicle in operation; assault; cutting or piercing instrument; incident resulting in drowning or submersion

Machinery in Operation
E919.0–E919.9

Injury caused by machinery in operation

Cutting/Piercing
E920.0–E920.9

Injury caused by cutting or piercing instrument or object

Includes: Fall on sharp object; injury caused by powered or nonpowered hand tools or appliances

Firearms
E922.0–E922.9§

Unintentional firearm injuries

Excludes: Air rifles, BB guns

Explosives
E921.0–E921.9,
E923.0–E923.9

Injury caused by explosion of pressure vessel or explosive material; fireworks

Excludes: Explosions due to or causing fire, machine explosions, explosions in transport vehicles or machinery in operation

Electricity
E925.0–E925.9

Injury caused by electrical currents

Overexertion
E927

Overexertion and strenuous movements from excessive physical exercise, recreation, lifting, pulling, pushing

Other
E926.0–E926.9,
E928.0–E928.9

Unspecified environmental and accidental causes, exposure to radiation

Includes: Infrared and ultraviolet lamps

Excludes: Sunburns

Late Effects
E929.0–E929.9, E959, E969,
E977, E989, E999

Late effects of injuries, assaults, injury undetermined as unintentional or intentional, legal intervention, suicide, war operations

Adverse Effects
E930.0–E949

Drugs, medicinal and biological substances causing adverse effects in therapeutic use

Includes: Correct drug properly administered in therapeutic or prophylactic dosage as the cause of any adverse effect

Suicide and Self-Inflicted
E950.0–E958.9

Suicide and self-inflicted injuries

Includes: Suicide, attempted suicide, intentional self-inflicted injuries

Homicide and Injury Purposely Inflicted by Other
E960.0–E968.9
Injuries inflicted by another person with intent to injure or kill, by any means
Includes: Child battering, rape

Legal Interventions
E970–E976, E978
Injuries inflicted by the police or other law enforcement agents on duty; includes legal execution by any method

Undetermined Intent
E980.0–E988.9
Injury undetermined whether unintentionally or intentionally inflicted
Includes: Self-inflicted injuries but not poisoning by corrosive and caustic substances

Wars
E990–E999
Injuries resulting from operations of war
Includes: Injuries to military personnel and civilians caused by and occurring during war and civil insurrections
Excludes: Training injuries

OTHER DIAGNOSTIC GROUPINGS (N CODES)

Infectious/Parasitic 001–139

Cancer (Neoplasms) 140–239

Metabolic/Immunity 240–279

Blood/Blood Organs 280–289

Mental Disorders 290–319

Nervous System/Sense Organs 320–389

Circulatory System 390–459

Respiratory System 460–519

Digestive System 520–579

Genitourinary System 580–629

Skin/Subcutaneous Tissue 680–709

Musculoskeletal/Connective Tissue 710–739

Congenital Anomalies 740–759

Perinatal Conditions 760–779

Symptoms, Signs and Ill-Defined Conditions 780–799

Injuries 800–999

NOTES

† Fracture, cause unspecified (E887), has been included as a fall, although it could have a different cause.

§ Firearms appear in several categories. If examining firearms as a separate category, the grouping would be as follows:

Firearms
E922.0–E922.9 Unintentional
E955.0–E955.4 Self-inflicted
E965.0–E965.4 Assaultive
E970 Legal intervention
E985.0–E985.4 Undetermined if unintentional or intentional
E991.2 War

Excludes: air rifles, BB guns (E917.9)

* E code entries in the 810–819 and 822—825 ranges that end in .8 and .9 refer to "other" and "unspecified" injury victims, respectively. Often these are injuries to motor vehicle occupants, and some researchers classify them as such. We have elected not to do so because of the possible introduction of bias into the data. This practice overestimates the importance of occupant injury and underestimates the problem of injuries among pedestrians, bicycles, and motorcycles.

Appendix F: Sources of Technical Assistance

CDC-Funded Injury Prevention Research Centers

Injury Prevention Research Center
Stephen Teret, JD, MPH, Director
School of Hygiene and Public Health
The Johns Hopkins University
615 North Wolfe Street
Baltimore, MD 21204
(301) 955-3995

Bioengineering Center
Albert I. King, PhD, Director
Wayne State University
818 West Hancock
Detroit, MI 48202
(313) 577-1347

The Injury Prevention Center at Harvard University
Bernard Guyer, MD, Director
Harvard School of Public Health
677 Huntington Avenue
Boston, MA 02115
(617) 732-1080

Injury Prevention Research Center
Carol Runyan, PhD, Director
University of North Carolina
CB #3430 UNC-Chapel Hill
Chapel Hill, NC 27599-3430
(919) 962-2202

Harborview Injury Prevention and Research Center
Frederick P. Rivara, MD, MPH, Director
Harborview Medical Center
325 Ninth Avenue
Seattle, WA 98104
(206) 223-3408

CDC Collaboratives

Center for Prevention of Injury and Violence
D. Blair Justice, PhD, Director
University of Texas School of Public Health
P.O. Box 20186
Houston, TX 77225
(713) 792-4364

Center for Injury Research, Control, and Education
Julian A. Waller, MD, Co-director
Mansfield House
University of Vermont
25 Colchester Avenue
Burlington, VT 05405
(802) 656-2528

Injury Prevention Research Center
Philip R. Fine, MD, Director
University of Alabama at Birmingham
UAB Station/THT 433
Birmingham, AL 35294
(205) 934-7845

Regional Injury Prevention Center
Susan Goodwin Gerberich, PhD, Director
School of Public Health
University of Minnesota
420 Delaware Street, SE
Box 197, Mayo Building
Minneapolis, MN 55455
(612) 626-0900

National Highway Traffic Safety Administration
Michael M. Finkelstein, Associate Administrator
Research and Development
400 Seventh Street, SW, Room 6206
Washington, DC 20590
(202) 366-1537

Injury Prevention Research Center
James M. Hassett, MD, Professor and Chairman
Department of Surgery
State University of New York at Buffalo Clinical Center
Erie County Medical Center
462 Grider Street
Buffalo, NY 14215
(716) 845-2781

Injury Prevention Research Center
Lewis Flint, MD, Co-Principal Investigator
Department of Surgery
State University of New York at Buffalo
516 Capen Hall
Amherst, NY 14260
(716) 898-3962

Injury Prevention Research Center
Alden H. Harken, MD, Director
Department of Surgery
University of Colorado Health Sciences Center
4200 East Ninth Avenue
Denver, CO 80262
(303) 270-8055

School of Medicine
Frank R. Lewis, MD, Professor of Surgery
University of California, San Francisco
San Francisco, CA 94143
(415) 821-8818

Center for Injury Prevention
Andrew McGuire, Executive Director
Building One, Room 306
San Francisco General Hospital
San Francisco, CA 94110
(415) 821-8209

UCLA Injury Prevention Research Center
Jess F. Kraus, PhD, Director
School of Public Health
University of California, Los Angeles
405 Hilgard Avenue
Los Angeles, CA 90024
(213) 825-7066

Federal Agencies

Bureau of Maternal and Child Health and Resources Development
Arthur Funke, PhD
United States Public Health Service
Parklawn Building, Room 9-21
5600 Fishers Lane
Rockville, MD 20857
(301) 443-4026

Division of Injury Epidemiology and Control
Jerry Hershovitz, BA, Program Services Section
Center for Environmental Health
Centers for Disease Control
Atlanta, GA 30333
(404) 488-4662

National Highway Traffic Safety Administration
Joan Quinlan, MA, Prevention Program Coordinator
NTS-21
400 7th Street, SW
Washington, DC 20590
(202) 366-2721

Glossary

Abuse A pattern of violence occurring in the course of a domestic (e.g., parent–child, husband–wife) or caregiver–client relationship.

Active countermeasure A preventive measure requiring action on the part of the individual being protected.

Administrative per se license suspension Legislation requiring the immediate surrender of a license by a driver who refuses to submit to a chemical test or whose test records a blood alcohol concentration (BAC) higher than the state's legal limit.

Age-adjusted injury rate An injury rate calculated to reflect a standard age distribution.

Age-specific injury rate An injury rate calculated for a group of a defined age range.

Agent The form of energy that damages body tissues in an injury.

Aggravated assault An unlawful attack by one person on another for the purpose of inflicting serious or aggravated bodily injury (FBI).

Aggregation A process in which data collected from a number of geographic areas are combined to provide a more comprehensive picture.

Air bag An inflatable crash protection device concealed in the steering wheel or dashboard of a car until it is activated by a crash. In a serious frontal crash, the bag fills to create a protective cushion between the person and the steering wheel, dashboard, and windshield (NHTSA).

Assault An act of violence resulting in injury.

Attenuation of effect The decreasing impact of an education/behavior change intervention as it is carried out in a population.

Baseline data Data collected for a period before the implementation of an intervention that are then used for comparison with data collected during or after implementation.

Biostatistics The discipline concerned with the organization and use of biological and medical data.

Case An individual incident or person of the type about which data are collected.

Case definition A description used to specify incidents or persons about which data are collected.

Child abuse A general term encompassing physical abuse, psychological or emotional abuse, sexual abuse, or sexual exploitation and neglect.

Child neglect The systematic disregard for the physical, psychological, or emotional needs of a child by a caregiver.

Child sexual abuse Any sexual contact between a child and an adult. In certain circumstances, sexual contact between children.

Coalition An organization of individuals representing a variety of interest groups who come together to share resources and plan and work together.

Community diagnosis A synthesis of injury morbidity and mortality data and information about the community that is used when designing a program.

Confidence interval The degree of certainty that can be claimed for the accuracy of a statistical calculation.

Conspicuity-enhancement devices Reflective materials and objects (e.g., flashlights, vests, headbands, anklebands, etc.) worn by nighttime pedestrians, bicyclists, motorcycle riders, and joggers to make them more visible to drivers.

Control (or comparison) group A group of individuals as similar as possible to an experimental group who are not exposed to a given intervention.

Convincer, The A carlike device that runs down an incline and simulates the effect of a low-impact-speed motor vehicle crash.

Data abstraction The process of translating information presented in narrative form into variables.

Data linkage The process of matching data on the same cases from more than one source.

Deterrence An attempt to prevent crime through the threat of punishment.

Disaggregate The process of separating data for a particular geographic area or population variable from a more comprehensive collection of data.

Domestic violence Spouse abuse and woman battering.

Dram shop law Legislation that makes civilly liable those who serve alcoholic beverages to a minor or to an individual already intoxicated.

E codes Numerical designations of the external cause of injury developed by the World Health Organization for its International Classification of Diseases system.

Early death A death occurring within 2–3 hours of an injury.

Education/behavior change interventions Preventive measures involving the education of the population at large, targeted groups, or individuals and efforts to alter specific injury-related behaviors.

Elder abuse Physical, psychological, emotional, or sexual abuse or financial exploitation of an elderly person by a caregiver.

Elder neglect Systematic disregard for the physical, psychological, or emotional needs of an elderly person by a caregiver.

Emergency medical services system A system that provides for the arrangement of personnel, facilities, and equipment for the effective and coordinated delivery of health care services in an appropriate geographical area under emergency conditions (occurring as a result of the patient's condition or of natural disaster or similar situations) (EMSS Act of 1973).

Engineering/technology interventions Preventive measures involving changes in the design of products or in the physical environment.

Environment The physical and psychosocial setting in which injuries occur.

Epidemiology The study of the occurrence and distribution of diseases and injuries.

Ergonomics The study of the interaction between worker and machine.

Evaluation The collection and analysis of data to determine the effectiveness of a given program. (See also **Outcome evaluation** and **Process evaluation**.)

Experimental design A type of evaluation design in which individuals are randomly assigned to control and experimental groups.

Experimental group A group of individuals exposed to a given intervention.

Exposure rate A measure of the amount of time an individual or group is involved in an activity or subjected to a hazard or environment that is associated with an injury risk.

Goal A statement of a program's intention to bring about long-term improvement in an injury problem.

Home injury Injury to occupants or guests that occurs in the home (including apartments and boardinghouses) and immediate surroundings (e.g., garages and yards). Does not include suicides, homicides, drownings, playground injuries, or assaults.

Homicide The killing of one person by another.

Host The injured individual.

Ignition interlock A device for preventing a driver from starting a vehicle unless he or she passes an alcohol-detecting breath test.

Immediate death A death occurring within minutes after injury.

Impaired driving Drunk and/or drugged driving.

Incapacitation Imprisonment intended to prevent crime by removing the criminal from society.

Injury Unintentional or intentional damage to the body resulting from acute exposure to thermal, mechanical, electrical, or chemical energy or from the absence of such essentials as heat or oxygen. The terms "injury" and "trauma" are used interchangeably in this book.

Injury rate A statistical measure describing the number of injuries expected to occur in a defined number of people (usually 100,000) within a defined time period (usually 1 year). An expression of the relative risk of different injuries or groups.

Institutionalization The process by which a program achieves ongoing financial support and commitment from the agency and community in which it is based.

Instrument A questionnaire, survey, test, or other data collection form to gather information about injury incidence and/or knowledge, attitudes, or behavior related to injuries.

Intervention A specific prevention measure or activity designed to meet a program objective. The three categories of intervention are legislation/enforcement, education/behavior change, and engineering/technology.

Labor loading chart A table estimating the number of days each staff person will devote to each process objective during a given time period, usually 1 year.

Late death A death occurring within 2–3 weeks after injury.

Lead agency An organization that serves as the focal point for injury prevention expertise on the local or state level. The lead agency offers technical assistance and resources to other groups and serves as a broker of information among groups.

Legislation/enforcement interventions Preventive measures involving the enactment or enforcement of laws or regulation.

Monitor See **Process evaluation**.

N codes Numerical designations of the nature of injury developed by the World Health Organization for its International Classification of Diseases system.

Near-drowning Immersion injury resulting in brain damage from oxygen deprivation.

Nonpowder firearms Guns that use a spring mechanism, a gas cartridge, or air pressure to fire small steel balls (BBs) or lead pellets.

Objective A statement of changes sought in an injury problem in terms that are measurable, time limited, and specific to a given target population. (See also **Outcome objective** and **Process objective**.)

Outcome evaluation A process that seeks to measure a program's progress toward improving injury morbidity and/or mortality, knowledge, attitudes, behavior, physical environments, or public policy and practice.

Outcome objective A statement of the desired impact of an intervention on injury morbidity and/or mortality, knowledge, attitudes, behavior, physical environments, or public policy and practice.

Passive (or automatic) countermeasure A preventive measure requiring little or no action on the part of the individual being protected.

Permissive licensing (of firearms) System in which firearm licenses are issued to all except those legally prohibited from owning a gun on the basis of a previous felony conviction, history of mental illness, etc.

PRECEDE A diagnostic health promotion model focusing on predisposing, enabling, and reinforcing factors that influence health behavior.

Primary enforcement (of safety belt use laws) A stipulation of a safety belt use law that allows law enforcement officials to stop a driver solely on the basis of a safety belt law violation. (See also **Secondary enforcement**.)

Primary prevention Efforts to forestall or prevent events that might result in injuries.

Problem identification The process of determining the nature of an injury problem, the characteristics of the population, the community's perception of the problem, the resources available to address it, and the political environment.

Process evaluation A method of documenting the achievement of proposed program activities, whether and how interventions were conducted, what portion of the target population was reached, total cost of the program, etc.

Process objective A statement of the desired level of achievement of program activities.

Program A coordinated effort organized by a lead agency to reduce an injury problem among a target population.

Program description A written summary describing the magnitude and characteristics of the injury problem(s) to be addressed; program goals, process and outcome objectives, interventions, program strategy, and evaluation measures; and the rationale for selecting a given approach to address the problem(s).

Program design A process in which program goals and outcome and process objectives are established, interventions are selected, and a program strategy is identified.

Program monitoring See **Process evaluation**.

Program targeting The selection of feasible program goals, objectives, and interventions and an appropriate, narrowly defined injury type and target population.

Protocol The outline or plan of a data collection procedure.

Proxy measure An alternative or substitute outcome that has been proven by research or is generally accepted to be associated with reduced injury morbidity or mortality.

Quasi-experimental design A type of evaluation design in which individuals are assigned in a nonrandom manner to control and experimental groups.

Rape All forms of sexual victimization, including forcible rape, attempted rape, and other acts of unwanted sexual aggression.

Regression to the mean Statistical tendency for variation to average over time, that is, the tendency of statistical extremes to even out over a long period.

Rehabilitation Services that seek to return a trauma victim to the fullest physical, psychological, social, vocational, avocational, and educational level of functioning of which he or she is capable, consistent with physiological or anatomical impairments and environmental limitations.

Residential injuries Includes home injuries as well as those occurring to residents of nursing homes and children in day care centers.

Restrictive licensing (of firearms) System in which a person wishing to purchase a firearm must prove that he or she is not legally ineligible to do so (e.g., by virtue of a previous felony conviction) as well as both an established need for and competency to use such a weapon.

Risk factor A characteristic that has been statistically demonstrated to be associated with (although not necessarily the direct cause of) a particular injury. Risk factors can be used for targeting preventive efforts at groups who may be particularly in danger of injury.

Secondary enforcement (of safety belt use laws) A stipulation of a safety belt use law that allows law enforcement officials to address a safety belt law violation only after a driver has been stopped for some other purpose.

Secondary prevention Efforts to modify the consequences of potentially injury-producing events to prevent the injury or reduce the severity of injury.

Sensitivity The ability of a data collection system to include all cases of a particular injury or event.

Server training A method of teaching waiters, waitresses, and bartenders to employ techniques for not serving already intoxicated customers and for preventing customers from becoming intoxicated.

Specificity The ability of a data collection system to exclude all injuries or events that do not fit the case definition.

Spouse abuse Violence within an intimate relationship directed by one partner at the other.

Strategy An overall plan for meeting a program's goals and objectives that combines a set of interventions with the program's resources, a plan for the evaluation of its process and outcome, and a method of securing the necessary community and financial support to stay in operation.

Suicide cluster A group of suicides or suicide attempts, or both, that occur closer in time and space than would normally be expected in a given community (CDC).

Surveillance The ongoing and systematic collection, analysis, and interpretation of health data in the process of describing and monitoring a health event (CDC).

Systems approach A comprehensive, systematic method to address injury problems through the combined, coordinated expertise of individuals and agencies knowledgeable about the magnitude of the problem, the nature of the community, and the resources available for prevention. Also, more generally, a process that incorporates primary, secondary, and tertiary prevention.

Target population The group of persons (usually those at high risk) that program interventions are designed to reach.

Tertiary prevention Acute medical care and rehabilitation directed at the return of a functioning patient to society (ACS).

Trauma See **Injury.**

Trauma care system A system of health care provision that integrates and coordinates prehospital emergency medical service resources and hospital resources to optimize the care and therefore the outcome of traumatically injured patients (NHTSA).

Trauma center A specialized hospital facility distinguished by the immediate availability of specialized surgeons, physician specialists, anesthesiologists, nurses, and resuscitation and life support equipment on a 24-hour basis (ACS).

Trauma registry A collection of data on patients who receive hospital care for certain types of injuries (e.g., blunt or penetrating trauma or burns). Such collections are primarily designed to ensure quality trauma care process and outcomes in individual institutions and trauma systems but have the secondary purpose of providing useful data for the surveillance of injury morbidity and mortality (CDC).

Triage The classification of patients according to medical need and the matching of those patients with available care resources.

Uniform Hospital Discharge Data Set (UHDDS) Data abstracted from hospital discharge records.

Variable An individual aspect of an entity or phenomenon under investigation that can differ among cases (e.g., the variable "sex" can be "male" or "female").

Vector The mechanism by which potentially injurious energy is transmitted to the host (e.g., a motor vehicle, a gun).

Violence The use of physical force with the intent to inflict injury or death upon oneself or another or the use of, or threat of, physical force to control another.

Woman battering A syndrome characterizing a relationship in which a woman is regularly subjected to violent and controlling behavior by her partner.

Years of Potential Life Lost A statistical measure calculated by subtracting an individual's age at death from a predetermined life expectancy. (The CDC generally uses the age of 65 for this purpose.)

ACRONYMS

AAP American Academy of Pediatrics

ACEP American College of Emergency Physicians

ACS American College of Surgeons

ANSI American National Standards Institute

APHA American Public Health Association

ASTM American Society for Testing Materials

ATV All-terrain vehicle

BMCH Bureau of Maternal and Child Health and Resources Development; formerly the Division of Maternal and Child Health (U.S. Department of Health and Human Services)

BAC Blood alcohol concentration

BJS Bureau of Justice Statistics (U.S. Department of Justice)

BLS Bureau of Labor Statistics (U.S. Department of Labor)

CDC Centers for Disease Control (U.S. Department of Health and Human Services)

CPR Cardiopulmonary resuscitation

CSTE Council of State and Territorial Epidemiologists

DHEW U.S. Department of Health, Education, and Welfare; later divided into the Department of Education and the Department of Health and Human Services

DIEC Division of Injury Epidemiology and Control (Centers for Disease Control)

DMCH Division of Maternal and Child Health (United States Department of Health and Human Services); now the Bureau of Maternal and Child Health and Resources Development

DOT U.S. Department of Transportation

EMS Emergency medical services

EMSS Emergency medical services systems

EMT Emergency medical technician

ER Emergency room

FARS Fatal Accident Reporting System (National Highway Traffic Safety Administration)

GAO General Accounting Office (U.S. Congress)

HMO Health maintenance organization

HRA Health risk appraisal

ISS Injury severity score

MMWR Morbidity and Mortality Weekly Report

NASS National Accident Sampling System (National Highway Traffic Safety Administration)

NCHS National Center for Health Statistics (Centers for Disease Control)

NCS National Crime Survey (Bureau of Justice Statistics)

NCPCA National Committee for the Prevention of Child Abuse

NEISS National Electronic Injury Surveillance System (U.S. Consumer Product Safety Commission)

NHTSA National Highway Traffic Safety Administration (U.S. Department of Transportation)

NIMH National Institute of Mental Health (U.S. Department of Health and Human Services)

NIOSH National Institute for Occupational Safety and Health (U.S. Department of Labor)

NSC National Safety Council

OSHA Occupational Safety and Health Administration (U.S. Department of Labor)

PATCH Planned Approach to Community Health

PSA Public service announcement

SCIPP Statewide Comprehensive Injury Prevention Program (Massachusetts); formerly Statewide Childhood Injury Prevention Program

TEAM Techniques for Effective Alcohol Management

UCR Uniform Crime Reports (Federal Bureau of Investigation)

USCPSC U.S. Consumer Products Safety Commission

USDA U.S. Department of Agriculture

YPLL Years of potential life lost

Subject Index

costs, 116
federal government and, 121
interventions, 123
prevention, 120, 122
problem magnitude, 115
risk groups, 117
Trauma care systems, 271
access to care, 273
administration of, 274
barriers to growth, 278
collaboration, 280
definition, 271
development of, 277
growth, 276
hospital care, levels of, 274
injury scoring systems, 274
practitioners, 272
prehospital care, 273
rehabilitation and, 279

services, 272
triage, 273

U

United States Consumer Product Safety
Commission, 163

V

Vehicles, *see also* Traffic injuries
design, 137
occupant injury, 117
occupant protection, 127
Violence, 192
alcohol and, 197
attitudes, 196
criminal justice system and, 199
defined, 192
drugs and, 198

firearms and, 197
health care system and, 199
injury and, 192
mental health system and, 199
prevention, 198
public health and, 194
social service system and, 199
socioeconomic status and, 197
Wisconsin Comprehensive Child Injury
Prevention Program, 107
Worker training and education programs,
183
Workers' compensation agencies, 181
Workplace, child abuse intervention and,
219

Y

Years of potential life lost (YPLL), 5
Youth gangs, 206